FINANCIAL SECTOR DEVELOPMENT IN ASIA
COUNTRY STUDIES

Edited by Shahid N. Zahid

Asian Development Bank

1995

Asian Development Bank
P. O. Box 789
0980 Manila, Philippines

Published by Asian Development Bank

ISBN 971-561-068-4

FOREWORD

The role and function of the financial sector in fostering growth has been a matter of debate and concern for many years. While most analysts and policymakers will agree that a well functioning and efficient financial system can help promote economic development there continue to be disagreements about the ownership pattern and institutional structure of the financial sector in most economies, both developed and developing alike.

To help understand the issues involved and to contribute to the debate on financial sector development the Economics and Development Resource Center of the Asian Development Bank undertook a regional study of financial sector policies in several of the Bank's developing member countries. The study covered a number of the Bank's member countries ranging from those that have had widespread public sector intervention, with and without accompanying economic success, to economies that have had more liberal financial sector regimes. The countries covered were India, Indonesia, Republic of Korea, Pakistan, Philippines, Taipei,China and Thailand. This book contains edited version of the seven country reports. Separate studies were also done to look at the collective experience of the newly industrialized economies, countries in Southeast Asia and those in South Asia. The results of these studies, supplemented with a study of Japan's financial sector experience, were synthesized into an analytical overview of the development of the financial sector in Asia. These have been published as a separate book by the Oxford University Press for the Asian Development Bank. Several renowned scholars and experts have contributed to this study. Policy recommendations based on the outcome of the study are being used by the Asian Development Bank in financial sector operations being financed by the Bank in its member countries. These publications should help disseminate the findings of the study to a much wider audience and is an important part of the Bank's ongoing effort to develop and promote institutions necessary for the development and growth of the region.

The views and opinions expressed in this book are those of the individual authors and do not necessarily represent the views of the Asian Development Bank.

> V.V. Desai
> Director and Chief Economist
> Economics and Development Resource Center

PREFACE

The importance of the financial sector in facilitating the transfer of resources from savers to investors is widely acknowledged. Efficient financial systems facilitate better mobilization and use of resources, which accelerates the process of economic growth. It is a widely held belief that a more liberalized financial system which relies on market forces, free entry, and a wide array of financial instruments is better able to carry out this function than an interventionist system which relies on directed credit and restricted entry, and provides a limited number of financial instruments.

Nevertheless, experience shows that rapid economic growth and capital accumulation can be achieved with financial systems that are interventionist, and that there are a variety of factors which affect the success of financial liberalization. The Republic of Korea and Taipei,China, two of the most rapidly developing economies in the world, have strictly controlled the distribution of credit and highly regulated the financial sector and it is only in recent years that there has been a move towards some liberalization of the financial sector.

While both countries achieved rapid rates of growth, the overall impact of government intervention in the financial sector is a subject of controversy. As the economy grew in sophistication, the financial system did not. The poorly developed financial system, particularly in the Republic of Korea, is now seen as an impediment to industrial restructuring. Reforms in the financial sector are thus an important priority area in economic policy. Governments in South Asian countries have also followed interventionist policies but, unlike the Republic of Korea and Taipei,China, they have not been accompanied by high growth rates or substantial increases in savings and investment, with the possible exception of India which has a relatively high savings rate. Rapid financial liberalization has been accompanied by poor economic performance in several instances. The Philippines underwent financial liberalization during the 1970s and 1980s but it did not achieve significant economic growth. Financial liberalization in Thailand has, however, been successful in promoting development, possibly because it was accompanied by sound monetary and fiscal management which resulted in overall macroeconomic stability and relatively low inflation.

Major differences of opinion continue on the role of financial sector regulation and reforms in the process of development. With a view to assessing the extent and pace of financial sector deregulation and to examining the institutional framework necessary for the development of the financial sector, the Asian Development Bank undertook a study of the financial sector in several of its member countries. The countries chosen for detailed study included those that have had interventionist policies and those that have undergone financial liberalization both with and without accompanying economic success. Two countries from the newly industrializing economies (NIEs), namely the Republic of Korea and Taipei,China; three from Southeast Asia, namely Indonesia, Philippines, and Thailand; and two from South Asia, namely India and Pakistan were selected. Besides the country studies, three regional reports, one each for the NIEs, Southeast Asia, and South Asia were prepared. There was also a special paper on the financial sector experience of post-war Japan. An overall synthesis report analyzing and comparing the financial sector experience of the seven countries completed the output of the study.

Prior to the commencement of work on the country reports, a methodological framework was prepared by Ashok Khanna and Lawrence J. White, and discussed at a workshop held in Manila in March 1993. Along with this framework, detailed outlines of the country reports and the regional reports, and the overall objectives of the study were discussed and finalized. The country reports were prepared by country consultants and were thoroughly reviewed by Bank staff, the regional authors, and the author of the synthesis report. The regional authors and the author of the synthesis report assisted in the supervision of the country studies as well. The regional reports were reviewed by Bank staff and by the author of the overall synthesis report. The overall synthesis report was reviewed by Bank staff and by the authors of the regional reports. All of the reports were presented at a conference entitled "Financial Sector Development in Asia" held at the Asian Development Bank's headquarters in Manila in September 1993.

This book, contains edited versions of the country reports. A companion book includes the overall synthesis report, the three regional reports, and an annex on Japan (published separately by the Oxford University Press for the Asian Development Bank).

The regional report on the NIEs provides a review and assessment of the ways in which the financial system was used in the NIEs (Republic of Korea, Singapore, and Taipei,China) to further industrial policy and mobilize savings. The reform process in this region is analyzed and the differences in approach of the three economies compared and contrasted. Hong Kong was excluded from this report because of its special circumstances, namely a financial sector that has very few controls or regulations, internally or externally, and is perhaps unique to the world. The report on Southeast Asia reviews the different financial sector reform experiences in four countries (Indonesia, Malaysia, Philippines, and Thailand) using information from the country studies and elsewhere. The report on South Asia is based on a review of recent literature on the financial sectors of several countries in the region (Bangladesh, India, Nepal, Pakistan, and Sri Lanka) and uses information from the country reports on India and Pakistan. The report discusses the important medium-term financial sector goals for which the countries should aim, the state of the financial sector in South Asia, and the problems and constraints in its further development and reforms.

Each country report provides a description of the financial system, covering both financial institutions and financial markets. Both the operations of the institutions and markets and some measure of their importance are covered. The linkages between the macroeconomic policy environment and the operation of the financial system are examined in each country report. The constraints that the macroeconomic policy setting imposes on the development of the financial sector and the implications of financial sector reforms for the conduct of macroeconomic policy have been analyzed. In addition, the legal and regulatory environment for each country has been described. For each country, an agenda for further reforms is discussed and priority areas identified.

The major findings of the study are interesting since they display remarkable similarities between the seven countries studied. Financial intermediaries were the dominant source of finance for all seven countries at their early stages of development, and commercial banks were the primary institutions within the intermediary sector. Further, the evidence indicated that the banking systems of these countries have focused on relatively short-term loans.

The individual country studies reveal extensive and elaborate systems of economic regulation applied to their financial sectors by all seven governments during the 1970s and early 1980s. Simultaneously, these governments tended to neglect the prudential regulation of their banks and other intermediaries and to refrain from information regulation efforts to improve the structure of information revelation of their corporate sectors. Since the early 1980s, the seven countries have moved, at varying speeds, to loosen the economic regulatory strictures on their financial sectors while simultaneously tightening their prudential regulation. Though these efforts are far from complete, they appear to be offering improved outcomes. The quality of intermediation is improving, securities markets are flourishing, and more financing choices are available in these economies.

The contribution of a number of Bank staff and outside consultants to the completion of this book is greatly appreciated. In particular, J. Malcolm Dowling provided valuable advice and assistance from the beginning of the project. Naved Hamid and Jean-Pierre Verbiest provided advice during the formative stages of the project. My deep appreciation also goes to Lulu Antonio and Helen Buencamino for their secretarial assistance, to Mina Paz and Emma Bonoan for their technical assistance, and to Lynette R. Mallery for her editorial advice.

Shahid N. Zahid
Asian Development Bank

TABLE OF CONTENTS

LIST OF CONTRIBUTORS — *xxiii*
ABBREVIATIONS AND ACRONYMS — *xxv*
GLOSSARY — *xxvii*

CHAPTER ONE
KOREA'S FINANCIAL MARKETS AND POLICIES
SANG-WOO NAM

Evolution of the Korean Financial Market	1
Policy Environment	3
Selective Credit Allocation	3
Credit Control System	6
Regulation of Interest Rates	7
Slow Pace of Financial Liberalization	10
Information Asymmetry and Legal Framework	12
Overview of the Financial Sector	14
Growth of Financial Markets	15
Flow of Funds	21
Sources of Corporate Financing	26
Banking System	26
Bank of Korea	26
Banking Institutions and Business	28
Deposit and Loan Portfolios of Deposit Money Banks	32
Efficiency of Banks	35
Supervision for Sound Banking Operation	40
Development Institutions	41
Korea Development Bank (KDB)	41
Export-Import Bank of Korea (KEXIM)	43
Korea Long-term Credit Bank (KLCB)	43
Other Financial Institutions	44
Saving Institutions	44
Investment and Finance Companies	47
Merchant Banking Corporations	48
Insurance Companies	48
National Investment Fund and the National Housing Fund	49
Leasing Companies	49
Venture Capital Companies	50
Credit Guarantee Funds	52
Regulation of the Safety and Soundness of NBFIs	52

Informal Credit Market	54
NBFI Ownership	56
Money Market	57
Call Market	58
Commercial Paper Market	59
Bond Market on RPs	59
Market for CDs	60
Short-term Government Bond Market	60
Monetary Stabilization Bond Market	60
Markets for Commercial and Trade Bills	61
Capital Market	61
Development of the Korean Capital Market	61
Stock Market	62
Bond Market	63
Major Capital Market Institutions	64
Agenda for Financial Reform	66
Policies toward NPLs and Bailing out Troubled Firms	66
Restructuring of the Financial Sector	68
Structure of Ownership and Management Control of Banks	71
Upgrading Financial Infrastructure	73
Summary and Conclusion	76
Bibliography	79

CHAPTER TWO

FINANCIAL SECTOR DEVELOPMENT AND POLICIES IN TAIPEI, CHINA
JIA-DONG SHEA

Introduction	81
Overview of the Financial System	82
Types of Financial Institutions	82
Types of Financial Instruments	84
Flow-of-Funds Relationship	87
Consolidated Balance Sheet of Financial Institutions	93
Distribution of Loans	93
Financial Infrastructure, Regulations, and Macroeconomic Policies	95
Accounting System	95
Legal and Supervisory System	98
Financial Regulation and Deregulation	99
Distress Managing Method and Prudential Re-regulation	104
Macro Stabilization Policies	106

  Banking System   110
    Central Bank of China   110
    Full-Service Domestic Banks   110
    Local Branches of Foreign Banks   113
    Medium Business Banks   114
    Deposits and Loans of Banks   115
    Operational Efficiency of Banks   118
  Nonbank Financial Intermediaries   119
    Cooperatives   121
    Investment and Trust Companies   122
    Postal Savings System   123
    Insurance Companies   123
  Nonintermediary Financial Markets   124
    Money Market   124
    Bond Market   125
    Stock Market   126
    Foreign Exchange Market   131
  Unregulated Financial System   134
    Structure of Unregulated Financial System   134
    Types of Financial Transactions   134
    Interest Rate   136
    Size of the Unregulated System   136
  Growth Promotion Financial Policies   138
    Export Financing   139
    Special Loans   140
    Strategic-Industry Financing   141
  Financing of Small- and Medium-Scale Enterprises   142
    Financing for Small- and Medium-Scale Enterprises   143
    Credit Guarantee for Small- and Medium-Scale Enterprise   145
  Evaluation of the Performance of the Financial System   147
    Development of the Financial System   147
    Stimulation and Mobilization of Savings   149
    Allocative Efficiency of Loanable Funds   151
  Prospects and Lessons   154
  Bibliography   157

CHAPTER THREE

FINANCIAL SECTOR POLICIES IN INDONESIA, 1980–1993
ANWAR NASUTION

  Macroeconomic Setting Before and After Deregulation   163
    Macro Economic Conditions   163

Structure of the Financial System	168
Financial Assets	169
Deregulation and Government Interventions	172
Coverage, Aspects, and Sequence of Economic Deregulation	173
Reforms of the Banking Industry, 1983–1993	175
Sequence of Deregulation	175
Coverage and Aspects of the Reforms	178
Impact of the Banking Reforms	184
Growing Market Competition	184
Internal Problems	190
Monetary Policy	197
Economic Overheating, 1989–1991	197
Capital and Money Markets	198
Capital Market	198
Money Market	203
Interbank Money Market	203
SBI and SBPU Markets	204
Controlling Liquidity Through Nonmarket Mechanisms	205
Nondepository Financial Intermediaries	205
Insurance Industry	205
Noninsurance and Nonbank Financial Companies	212
Infrastructure of Financial Sector	215
Internal Capability of Financial Institutions to Supervise Credit	216
Regulatory and Supervisory Agencies	217
Underdeveloped Legal and Accounting Systems	220
Reorganization and Recapitalization of Financial Institutions	222
Corporatization of State-Owned Financial Institutions	222
Bad Debts and Premium Collections	222
Recapitalization of Banks	223
Operating System	225
Conclusion	225
Bank-Oriented Financial System	225
Growing Market Competition	226
Savings Mobilization and Resources Allocation	226
Risks	227
Weak Financial Institutions	227
Weak Human Resources and System of Operations	227
Inadequate Market Infrastructure	228
Ineffective Monetary Policy	228
Acknowledgments	229
Bibliography	229

Chapter Four

A STUDY OF FINANCIAL SECTOR POLICIES: THE PHILIPPINE CASE
Mario B. Lamberte and Gilberto M. Llanto

Overview of the Financial System	235
Microeconomic Efficiency Issues of the Financial System	238
Credit Rationing in a Liberalized Financial Market	239
Financial Intermediaries	240
Banking System	240
Depositor Protection	251
Efficiency of the Banking System in Mobilizing and Allocating Savings	251
Accounting System	257
Nonbank Financial Intermediaries	258
Finance Companies	258
Other NBFIs	260
Insurance Companies	263
Private Pension Funds	269
Nonbank Government-Owned Credit Guarantee Institutions	271
Financial Firms and Markets	271
Money Markets	271
Capital Markets	274
Foreign Exchange Market	279
Infrastructure Support System	281
Accounting and Auditing Systems	282
Legal Framework for Debt Recovery	288
Credit Rating Institution	290
Review and Assessment of Government Policies and Recommendations	292
Bibliography	298

Chapter Five

FINANCIAL SECTOR DEVELOPMENT IN THAILAND
Pakorn Vichyanond

Introduction	303
Overview and Evolution of the Financial Sector	304
Relative Sizes of the Financial Institutions	309
Evolution	312
Financial Infrastructure	314
Commercial Banks	314

Finance Companies	317
Financial Institutions Development Fund	319
Stock Exchange	320
Accounting Standards	323
Value-Added Tax	324
Commercial Banking System	325
Improving the Effectiveness of Bank Supervision	326
Lines of Business	327
Banking Structure and Current Issues	327
Nonbank Savings Institutions	329
Finance Companies	329
Credit Foncier Companies	334
Life Insurance Companies	334
Agricultural Cooperatives	335
Savings Cooperatives	336
Government Finance Institutions	336
Government Savings Bank	337
Bank for Agriculture and Agricultural Cooperatives	339
Government Housing Bank	340
Industrial Finance Corporation of Thailand	340
Nonintermediary Financial Firms	342
Securities Companies	342
Mutual Fund Companies	342
Macroeconomic Issues	344
Monetary Policy	344
Fiscal Policy	347
Exchange Rate Policy	348
Financial Markets	350
Money Markets	350
Foreign Exchange Market	352
Government Bond Market	352
Commercial Papers	353
Stock Market	353
Recent Reform Experiences	356
Deregulation and Relaxation	357
Dismantling of Exchange Controls	358
Interest Rate Liberalization	359
Fewer Constraints on Financial Institutions' Portfolio Management	359
Rural Credit Policy	360
Branch Opening Requirements	360
Modification of the Reserve Requirement	361
Expansion of Financial Institutions' Scope of Operations	361

    Improvement of Supervision and Examination
        of Financial Institutions   362
    Bangkok International Banking Facilities   363
    Other Recent Moves   365
Conclusion   366
    Financialization   366
    Fundamentals   367
    Sequencing   367
    Consequences and Outlook   368
    Bibliography   370

CHAPTER SIX

REFORM OF INDIA'S FINANCIAL SYSTEM
P. JAYENDRA NAYAK

Introduction   371
Microeconomics: Some Theoretical Issues   372
    Incentive Compatibility of Financial Sector Contracts   372
    Responses to Imperfect Information   372
    Diversification of Creditor Risks   374
    Interest Rate Deregulation   375
Macroeconomics: The Conventional Wisdom   377
Informal Finance   378
    Informality as a Banking Style   378
    Reach of the Informal Sector   378
    Sources of Informal Credit   380
    Industrial Structure of Informal Markets   380
    Comparison with Formal Finance   382
    Equity: The Reach of Credit   384
    Efficiency: Impact on the Real Economy   385
    Regulation of Informal Finance   386
Formal Finance   387
    Financial Deepening   387
    Commercial Banks   389
    Banking Policy and Its Outcome   390
    Development Financial Institutions   394
    Fragility of Institutions   396
Reform of Commercial Banking   397
    Macroeconomic Policy Support   397
    Financial Infrastructure Support   399
    Restructuring of Banks   400
    Short-Term Profitability Support   402
    Longer-Term Competitive Strength   404

Restructuring the DFIs	407
Legal and Regulatory Environment	408
Market Development	409
Clarity in Corporate Strategy	410
Nonintermediary Finance	410
Capital Markets	411
Capital Market Improvements	412
Creation of Debt Markets	413
Corporate Debentures and PSU Bonds	415
Money Markets	416
Foreign Exchange Markets	418
Financial Infrastructure	419
Legal System	419
Accounting Conventions	421
Regulation and Supervision	422
Deposit Insurance and Credit Guarantees	423
Issues in Industrial Finance	424
Market Signaling under Weak Contracts Enforcement	424
Risk Diversification for Working Capital	426
Financing Sick Companies: Mortality or Morbidity	427
Financing Small-Scale Industries	428
Concluding Observations	429
Acknowledgments	432
References	432

CHAPTER SEVEN

DEVELOPMENT OF THE FINANCIAL SECTOR IN PAKISTAN
Nadeem Ul Haque and Shahid Kardar

Introduction	437
Financial Sector: Size, Depth, and Sources of Funds	438
Sources and Uses of Funds	438
Financial Institutions: Type and Performance	442
Commercial Banks	444
Development Finance Institutions	452
Financial Markets and Instruments	455
Money Markets	455
Debt Markets	456
Equities Markets	457
Foreign Exchange Markets	458
Insurance	459
Informal Financial Markets	460

Adequacy of the Financial Infrastructure ... 460
    Institutional Capability of the Country's Regulatory Agencies ... 461
    Weak Enforcement of Regulatory Mechanisms ... 462
    Information Disclosure ... 462
    Auditing and Accounting Standards ... 462
    Prudential Regulations ... 463
    Debt Recovery System ... 463
    Weak Payment and Information Systems ... 465
Economic and Financial Policies ... 466
    Constraints of Financing a Large Fiscal Deficit ... 466
    Financial Repression and the Quasi-Fiscal Deficit ... 466
    Recent Financial Sector Reforms ... 472
Developing the Market for Equities ... 474
    Investor Perceptions and Returns to Stockholders ... 474
    Limited Availability of Stock and the Institutional Investor ... 475
    Privatization ... 476
    Cumbersome Procedures for Trading Stocks ... 476
    Limited Use of Current Information Technology ... 477
Constraints on the Development of Secondary Markets
    for Government Debt ... 477
Increased Competition in Instruments ... 479
    Differential Fiscal Treatment of Financial Instruments ... 479
    Limited Availability of Financial Instruments ... 480
    Potential New Instruments ... 481
    Factors Hindering Development of Housing Finance Markets ... 482
Policies for the Development of Financial Markets ... 483
    Fiscal Prudence and Stable Policy Environment ... 483
    Market-Oriented Monetary and Credit Policy ... 484
    Reform of the Financial Sector ... 485
Conclusion ... 487
Bibliography ... 489

AUTHOR INDEX ... 493
SUBJECT INDEX ... 495

Boxes

7.1	State Bank of Pakistan's Prudential Regulation	464
7.2	Quasi-Fiscal Deficit	471

Text Tables

1.1	Composition of Policy Loans	4
1.2	Regulated and Market Interest Rates	9
1.3	Financial Institutions in Korea	16
1.4	Loans and Deposits by Type of Financial Institutions	18
1.5	Trend of Financial Deepening	21
1.6	Composition of Financial Savings	22
1.7	Financial Assets and Liabilities Outstanding at the end of 1992	24
1.8	Loans and Rediscounts of the Bank of Korea	28
1.9	Assets, Deposits, and Loans and Discounts of Deposit Money Banks	31
1.10	Composition of Loans and Discounts of Deposit Money Banks by Fund	34
1.11	Indicators of Bank Management	36
1.12	Comparison of Commercial Bank Efficiency in Korea, Japan, and the US	37
1.13	Comparison of Shares of Bank Loans, Capital Formation, and GDP by Industry	38
1.14	Sources and Uses of National Investment Fund	50
1.15	Sources and Uses of Funds in the Leasing Industry	51
1.16	Business Group Ownership of NBFIs, 1990	56
1.17	Money Market Trends	58
2.1	The Financial System in Taipei,China	83
2.2	Roles of Financial Markets in Providing Funds, Relative to Financial Intermediaries	84
2.3	Number of Units of Financial Institutions	85
2.4	Shares of Financial Institutions at the End of Year	86
2.5	Composition of Major Domestic Financial Instruments Held by Business and Individuals	88
2.6	Financial Surplus or Deficit by Sector as a Percentage of GNP	89
2.7	Sources and Uses of Funds by Spending Sector, 1965–1988	90
2.8	Proportion of Private and Public Enterprises' Investment Funds from Financial System (1965–1988)	92
2.9	Consolidated Balance Sheet of Financial Institutions	94
2.10	Distribution of Financial Institutions' Loans	95

2.11	Relative Shares of Loans from Financial Institutions and Contributions to GDP, By Economic Activity	96
2.12	Comparison of Domestic Borrowings of Public and Private Enterprises	102
2.13	Percentage Distribution of Sources of Domestic Borrowings, By Private Enterprises in 1983, By Scale of Assets	103
2.14	Sources of Borrowings of Private Enterprises: Large Firms vs. Small- and Medium-Scale Enterprises	103
2.15	Annual Interest Rates and Inflation Rates	109
2.16	Composition of Full Service Domestic Banks	112
2.17	Maturity Structure of Deposits with Domestic Banks	115
2.18	Secured Loans and Medium- and Long-Term Loans as a Percentage of Total Loans of Domestic Banks	116
2.19	Distribution of Loans and Discounts at Domestic Banks, By Borrowing Sector	117
2.20	Rates of Nonperfoming Loans and Dishonored Checks	120
2.21	Stock Market in Taipei,China	128
2.22	Composition of Business Sector Financing by the Financial System	137
2.23	Interest Rate Subsidy on Export Loans	140
2.24	Share of Loans to Small- and Medium-Scale Enterprises	144
2.25	Performance Statistics of Small- and Medium-Scale Business Credit Guarantee Fund	146
2.26	Indices of Financial Development in Taipei,China	148
2.27	Savings, Investment, and Foreign Borrowing as a Percentage of GNP	149
3.1	Chronology of the Adjustment Program, 1983–1993	164
3.2	Indicators of Economic Growth and Structural Change	167
3.3	Structure and Growth of the Financial Sector, 1969–1991	170
3.4	The Chronology of Financial Sector Reforms, 1983–1993	176
3.5	Distribution of Banking Assets	186
3.6	Number of Bank Offices and Population Per Bank Office by Province	188
3.7	Market Shares of Banking Institutions by Ownership	191
3.8	Banking Indicators by Group of Banks in Rupiah and Foreign Exchange	192
3.9	Number of Insurance Companies, 1987–1992	207
3.10	Gross Premiums by Type	208
3.11	Number of Finance Companies by Type and Ownership	213
3.12	Finance Companies' Lease Contract by Economic Sector, 1986–1991	214
3.13	Distribution of Lease Contract Value by Province and Ownership of Finance Companies, 1990–1991	215

4.1	Assets of the Domestic Financial System, 1986–1991	236
4.2	Sources of Funds of the Banking System	243
4.3	Indicators of Deposit Mobilization, 1986–1991	253
4.4	Ratio of Past Due Loans	254
4.5	Interest Rates on Loans and Deposits, Bank Spread Consolidation of all Commercial, Government and Foreign Banks, Herfindhal Concentration Index and Share of Banks in Total Deposits	256
4.6	Rates of Return on Assets by Type of Commercial Banks	256
5.1	Salient Features of Financial Institutions in Thailand at the End of 1992	306
5.2	Total Assets of Financial Institutions	307
5.3	M2, TAFI, and GDP	310
5.4	Credit Extension and Capital Funds Tapped from SET	311
5.5	Revocation of Licenses, Mergers, and Acquisition of Finance and Securities Companies	331
5.6	Market Share of Top Five Companies in Finance and Securities Industry in Terms of Total Assets	332
5.7	Capital Increase of Finance and Securities Industry and that of the Listed Finance and Securities Companies	333
5.8	Capital to Risk Assets Ratio of the Industry	333
5.9	Industry's Required Reserves for Doubtful Debts	333
5.10	Bank Branches in 1992	338
6.1	Urban Informal Credit	379
6.2	Sources of Funds	380
6.3	Sources of Informal Finance	381
6.4	Industrial Organization of Informal Markets	381
6.5	Proportion of Borrowing from the Informal Sector	384
6.6	Indicators of Financial Deepening	388
6.7	Flow of Capital Funds	389
6.8	Growth in Indian Banking	390
6.9	Maturity Pattern of Term Deposits in March 1991	393
6.10	Financial Assets of Banks and DFIs	394
6.11	Simplified Balance Sheet	401
6.12	New Capital Issues By Nongovernment Public Limited Companies	411
6.13	Balance of Payments	418
7.1	Financial Assets in Pakistan	439
7.2	Flow of Funds	441
7.3	Financial Sector Overview, 1991	443
7.4	Size of Commercial Banks, 1991	445
7.5	Profitability and Costs of Administration	446
7.6	Comparison of Privatized and State-Owned NCBs	447

7.7	Growth Rate of Gross Revenues and Total Costs of Banks	447
7.8	Performance of Commercial Banks' Scheduled Deposits	448
7.9	Weighted Average Interest Rates on Deposits and Advances	449
7.10	Productivity Ratios of Banks	450
7.11	Classified Advances of Nationalized Commercial Banks	451
7.12	Estimates of the Quasi-Fiscal Deficit	468
7.13	Existing Lending Rates on Concessionary Financing	470
7.14	Interest Rates on Deposits	471

TEXT FIGURES

2.1	Interest Rate Differentials Between Weighted Average Loan Rates and Deposit Rates of Domestically Incorporated Banks	121
7.1	Degree of Monetization	440
7.2	M2/GDP, 1980–1992	440

LIST OF CONTRIBUTORS

NADEEM UL HAQUE
Senior Economist
Developing Country Studies Division, Research Department
International Monetary Fund
United States

SHAHID KARDAR
Director, Systems Limited
Pakistan

MARIO B. LAMBERTE
Vice President and Coordinator of Monetary and Banking Policies Research Program
Philippine Institute for Development Studies
Philippines

GILBERTO M. LLANTO
Research Fellow
Philippine Institute for Development Studies
Philippines

SANG-WOO NAM
Vice President
Korea Development Institute
Korea

ANWAR NASUTION
Professor, Faculty of Economics
University of Indonesia
Indonesia

P. JAYENDRA NAYAK
Joint Secretary
Ministry of Finance
India

JIA-DONG SHEA
Director and Research Fellow, Institute of Economics
Academia Sinica
Taipei, China

PAKORN VICHYANOND
Research Fellow, Macroeconomic Policy Program
Thailand Development Research Institute
Thailand

ABBREVIATIONS AND ACRONYMS

ADB	Asian Development Bank
ARF	Asset Reconstruction Fund
ASEAN	Association of Southeast Asian Nations
BCCI	Bank for Credit and Commerce International
BIFR	Board for Industrial and Financial Reconstruction
BIS	Bank for International Settlements
BKB	Bangladesh Krishi Bank
BSE	Bombay Stock Exchange
CD	certificates of deposit
CIBI	Credit Information Bureau, Inc. of the Philippines
CPA	certified public accountant
CPI	consumer price index
CRR	cash reserve ratio
DFHI	Discount and Finance House of India
DFI	development finance institution
DICGC	Deposit Insurance and Credit Guarantee Corporation
DMB	deposit money bank
ECGC	Export Credit Guarantee Corporation
EPF	Employee's Provident Fund
ETF	Employee's Trust Fund
FEDAI	Foreign Exchange Dealers Association of India
GDP	gross domestic product
GIC	General Insurance Corporation
GNP	gross national product
GSD	Government Securities Dealer
IBRD	International Bank for Reconstruction and Development
ICB	Investment Corporation of Bangladesh
ICICI	Industrial Credit and Investment Corporation of India
IDBI	Industrial Development Bank of India
IFC	International Finance Corporation
IFCI	Industrial Finance Corporation of India
IFCT	Industrial Finance Corporation of Thailand
IMF	International Monetary Fund
IPO	initial public offering
JSE	Jakarta Stock Exchange
KSE	Korea Stock Exchange
KSE	Karachi Stock Exchange
LIC	Life Insurance Corporation
MPBF	maximum permissible bank finance
MSB	monetary stabilization bond
NABARD	National Bank for Agriculture and Rural Development

NBFC	Nonbanking Finance Company
NHB	National Housing Bank
NIDC	Nepal Industrial Development Corporation
NIE	newly industrializing economy
NSE	National Stock Exchange
OTC	over-the-counter
PMS	portfolio management scheme
RBI	Reserve Bank of Indonesia
RBIn	Reserve Bank of India
RCTC	Risk Capital and Technology Finance Corporation
RP	repurchase (repo) agreement
Rp	rupiah
Rs	rupee
SBI	*Sertifikat* Bank Indonesia
SBIn	State Bank of India
SBP	State Bank of Pakistan
SBPU	*Surat Berharga Pasar Uang*
SCC	Securities and Companies Commission
SCICI	Shipping Credit and Investment Company
SEBI	Securities and Exchange Board of India
SEC	Securities and Exchange Commission
SET	Stock Exchange of Thailand
SFC	State Financial Corporation
SIDBI	Small Industries Development Bank of India
SIDC	State Industrial Development Corporation
SITC	securities investment trust companies
SLR	statutory liquidity ratio
TAFI	total assets of financial institutions
TAFIPS	total assets of financial institutions plus securities
TDICI	Technology Development and Information Company of India
TFCI	Tourism Finance Corporation of India
Tk	taka
USAID	United States Agency for International Development
UTI	Unit Trust of India

References to PRC are to the People's Republic of China.
References to Korea are to the Republic of Korea.
References to US are to the United States.

Unless noted otherwise, in this publication $ refers to US$.

The term country does not imply any judgement by the Asian Development Bank as to the legal or other status of any territorial entity.

GLOSSARY

badla — an informal and unrecorded trading arrangement where a handful of brokers provide the funds.

fui market — rotating mutual credit in Korea.

Janakiraman Committee — committee appointed by the Reserve Bank of India to investigate bank irregularities.

kye market — a rotating credit club in Korea.

Lifeboat scheme — established on 4 April 1984 by the Ministry of Finance and the Bank of Thailand, this scheme rehabilitated 25 finance and securities companies that experienced difficulties. Also known as the April 4 Lifeboat Scheme.

modarabas — limited partnerships in Pakistan in which a managing firm agrees to invest the funds of a fairly large number of passive investors. Must be sanctioned by a religious board as being Islamic.

nidhis — indigenous, single branch, mutual financial institutions found in Tamil Nadu, India dealing with the shareholder members. Similar to the early thrift societies in the west.

CHAPTER ONE

Korea's Financial Markets and Policies

Sang-Woo Nam

EVOLUTION OF THE KOREAN FINANCIAL MARKET

With negligible domestic savings, the intermediary role of commercial banks was of little significance in Korea until the mid-1960s. Until then, the primary purpose of financial institutions was to channel aid funds to rehabilitation projects and to farmers. Two special banks, the Korea Development Bank and the Korea Agriculture Bank, accounted for over 70 per cent of total bank lending.

The role of financial institutions as mobilizers of savings was only recognized after the adoption of a high interest rate schedule in September 1965. The effect of this interest rate reform on mobilizing domestic resources was remarkable. As bank deposits increased rapidly, the M2/GNP ratio jumped from 19 per cent during 1965–1970 to over 30 per cent during 1971–1973. Also, the commercial banks' share of total bank loans rose from 27 per cent in 1964 to 55 per cent within five years. However, the interest rate reform of 1965 was short-lived, as bank interest rates began to be lowered in the late 1960s, reaching levels well below the pre-reform rates by 1972.

The 1960s also saw the establishment of new financial institutions engaged in specialized activities desired by the Government. The Medium Industry Bank (currently the Industrial Bank of Korea) and the Citizens National Bank were both created shortly after the military Government came into power in 1961; the Korea Exchange Bank and the Korea Housing Bank were established in 1967.

Efforts in the early 1970s to reduce the informal credit market by establishing new nonbank financial institutions and to promote the capital market, contributed significantly to the diversification of Korea's financial market. In connection with the Presidential Emergency Decree of August 1972, which froze the curb market, investment and finance companies and mutual savings and finance companies were established and credit unions were modernized. Although the growth of these institutions has been constrained by various operational restrictions, including interest rates, they have succeeded in attracting funds which otherwise might have been supplied to the curb market.

Throughout the 1970s, however, Korea's banking sector was under increasing repression. With interest rate controls under accelerating inflation due mainly to two oil price shocks, the growth of the banking sector stagnated relative to the real sector. The Government made sure that adequate funds were available to such favored sectors as exports, other foreign exchange earning activities, and heavy and chemical industries.

These preferential loans, together with other directed credit, critically limited the access to funds by the less-preferred sectors; this resulted in inefficiency in the allocation of bank credit. Government interference in bank asset management was also responsible for the neglect of serious credit evaluation by the banks and the accumulation of nonperforming loans in later years. In particular, the shipping and overseas construction industries, which also received extensive financial support in the 1970s, became a major source of nonperforming loans after the second oil shock and had to go through a massive industry rationalization in the mid-1980s.

The Korean banking system had, to a large extent, been under government ownership until the early 1980s, as commercial banks were nationalized immediately after the military coup in 1961. The Government generally held more than 20 per cent of the shares, while legislation limited the voting power of any private shareholder to a maximum of 10 per cent. Since the privatization of these banks in the early 1980s, the maximum ownership of any single shareholder has been limited to 8 per cent of the total shares. The ownership of local banks and nonbank financial institutions is much more concentrated, with many large business groups holding controlling shares.

The burden of providing medium-term and long-term investment financing has mainly been borne by specialized development institutions, such as the Korea Development Bank, Korea Long-term Credit Bank, and the Export-Import Bank of Korea. These institutions have supplied funds mostly to social infrastructure projects, key manufacturing industries, and export industries. Until the early 1980s, foreign borrowings, government funds, and the National Investment Fund were the major sources of funds for these institutions; the issue of debentures has become more important in recent years.

The Korean capital market was little more than a secondary market for government bonds issued during and after the Korean War until the early 1970s. Only a limited number of equity shares were traded in the market, and no corporate bonds were issued in the market before 1972. Since then, however, the capital market has grown very rapidly thanks to strong promotional measures of the Government. Together with strong incentives to hold equity shares, the Government gave favorable corporate tax treatment to publicly-held firms, and even ordered selected companies to go public.

The money market also gathered a momentum of growth in the early 1970s, with the promulgation of the Short-Term Financing Business Act and the establishment of investment and finance companies. The Government provided a favorable environment for the growth of the money market in an effort to absorb curb market funds. Though its growth has been constrained by the slow pace of interest rate deregulation, the money market has realized much faster growth than the banking sector. Two factors critical for the growth of the market since the mid-1980s include the reintroduction of negotiable certificates of deposit by banks and the large-scale issuance of the central bank's monetary stabilization bonds for the purpose of liquidity control.

The informal credit market used to be important to the financial system in Korea; due to the skewed allocation of credit toward large businesses and the underdevelopment of the financial system, it has continued as an important source of short-term financing for many business firms and individuals. Interest rates in this market are sensitive to the risk differentials among borrowers and to changing market conditions, and have usually been much higher than in the formal credit market due to the high risk premium. The size of the informal credit market, which defies estimation due to its fragmented and illegal nature, has dwindled considerably in recent years with the rapid growth of nonbank financial institutions.

POLICY ENVIRONMENT

Selective Credit Allocation

In the early 1970s, the worsening world trade environment—such as growing protectionist barriers and the emergence of other developing countries as competitors in the export market—as well as changing US foreign policy prompted the Korean Government to promote heavy and chemical industries. The Government wanted to develop indigenous defense industries and to restructure its export composition in favor of more sophisticated and higher value-added industrial goods. The overriding objective of tax, credit, interest rate, and trade policy in the 1970s was to promote the heavy and chemical industries including iron and steel, non-ferrous metals, shipbuilding, general machinery, chemicals, and electronics.

The Government used credit allocation through the banking system as its most powerful means of supporting favored industries. To finance large-scale investment projects in these industries, the National Investment Fund was set up in 1974 by mobilizing public employee pension funds and a substantial share of banking funds. However, as these funds proved insufficient, the banks, in all practicality owned by the Government, were directed to make loans to "strategic" industries on a preferential basis. During the latter half of the 1970s, the share of policy loans in domestic credit for deposit money banks rose steadily from 40 per cent to 50 per cent[1] (Table 1.1).

Table 1.1 Composition of Policy Loans
(per cent)

	1975	1980	1985	1990	1991
Policy loans by banking institutions					
Loans from government fund	6.2	4.3	5.5	5.9	6.3
Loans to National Investment Fund	2.4	3.4	3.6	1.7	1.4
Credit to KDB and KEXIM	3.2	2.1	1.5	12.5	12.0
Loans in foreign currency	15.4	21.9	12.6	13.8	15.2
Credit for foreign trade	15.2	14.5	11.6	3.2	3.3
Credit for small and medium industry	2.9	6.4	4.1	4.8	5.2
Credit for agriculture and fishery	5.1	3.5	4.1	5.4	5.8
Credit for housing	4.9	7.9	9.4	11.2	12.3
Others	17.6	5.4	10.0	16.2	11.1
Subtotal	72.8	69.5	62.0	74.8	72.7
(A)[a]	1,625.1	8,238.7	16,736.3	44.850.4	49,546.4
Policy loans by Other Financial Institutions	27.2	30.5	38.0	25.2	27.3
(B)[a]	607.7	3616.7	10,263.8	15,105.0	18,621.1
Total	100.0	100.0	100.0	100.0	100.0
(A + B)[a]	2,232.8	11,855.4	27,000.1	59,955.4	68,167.5
(A)/Domestic credit of Banking Institutions	40.9	49.1	39.3	46.3	41.9
(B)/Domestic credit of Other Financial Institutions	52.4	43.9	30.2	12.7	12.0
(A+B)/Total domestic credit	43.5	47.4	35.3	27.8	24.9

[a] Amounts in billion won.
Source: National Statistics Office (1992).

---

[1] Policy loans include those lent to earmarked sectors at preferential or nonpreferential rates and non-earmarked loans extended at preferential rates because of policy considerations.

Because of this strong and concerted support in tax, trade, and credit policies, almost 80 per cent of all fixed investment in the manufacturing sector during the late 1970s is estimated to have been directed toward the heavy and chemical industries. Such disproportionate incentives, together with over-optimistic assumptions regarding world trade prospects, led to excessive and duplicative investment in some areas. To correct the situation, the Government intervened in 1980 and coordinated negotiations among participating firms for relinquishing some projects or reducing their capacity. Despite this, the banking sector was loaded with large nonperforming loans by the mid-1980s. The Government had to work out industry-wide rationalization programs to alleviate the problem of over-capacity and growing operating losses, together with other bailout packages for individual firms. Although some firms remained in trouble, shipping and overseas construction industries were responsible for the majority of the nonperforming loans and were the major target of rationalization.

Besides investment inefficiency, the expansion of preferential policy loans caused serious sectoral imbalances and complications in macroeconomic management. First, as the government-favored heavy and chemical industry projects and other preferred sectors received a disproportionate amount of the limited financial resources, credit to other industries was unduly squeezed. Moreover, because of the huge capital requirement needed for heavy and chemical industry development and the weak business position of small- and medium-scale firms, the new heavy and chemical industry projects were "granted" to large business groups, further concentrating economic power.

Furthermore, due to credit expansion to the heavy and chemical industries as well as the boom in Middle Eastern construction activity, the latter half of the 1970s saw a rapid growth in the money supply. As a result, inflation accelerated and speculation in real assets was rampant. Accelerating inflation made financial savings unattractive since real interest rates remained low, very often negative. With the subsequent stagnation of the financial sector, small- and medium-sized firms became more pinched for funds and the informal credit market grew rapidly.

Such problems caused by the Government's overzealous promotion of heavy and chemical industries in the 1970s prompted the policymakers to embark on trade and financial reforms as well as the realignment of other industrial incentives. In the early 1980s, promotion of "strategic" industries with preferential credit and tax treatment gave way to more indirect and functional support of industries in general. The tax reform of 1981 reduced substantially the scope of special tax treatment for key industries. In addition, by June 1982, most policy loans were no longer extended at preferential interest rates, making it easier to scale down those loans. The share of policy loans, however, has not decreased since the mid-1980s, as bank credit has expanded substantially to such sectors as housing, small- and medium-scale industries and agriculture.

Credit Control System

The credit control system and the accompanying "principal transactions bank system" were introduced in 1974 as a part of the Government's efforts to improve the capital structure of large corporations. The principal transactions bank of a business firm was made responsible for urging the firm to improve its capital structure and limiting the bank credit it received. Following the introduction in the form of an agreement of the Council of Banking Institutions, the credit control system was strengthened by a series of ensuing agreements and government measures. During the course of its evolution, the credit control system incorporated other policy objectives such as controlling real estate speculation and excessive business expansion of large business groups, easing the concentration of economic power, and, in recent years, strengthening the competitiveness of the Korean manufacturing industry.

Currently, the system is applied to the 30 largest business groups selected on the basis of the size of their bank borrowings. To make more bank loans available to smaller firms, the collective share of the 30 largest business groups in total bank loans is not to exceed a prescribed level set by the Office of Bank Supervision and Examination each year. However, loans to "major corporations" of each business group are exempted from this credit control. Each business group can select as many as three major corporations among its member companies in consideration of their comparative advantage and future growth potential. This measure was taken in response to the concern that overly tight credit control was hampering the competitiveness of Korean companies by constraining their investment activity.

The principal transactions bank of a business group which is selected by the conference among the banks which have business relations with the group, has various authorities and responsibilities in operating the credit control system. It supervises the overall financial activities of its client corporations and urges them to meet the guided capital ratio by taking appropriate actions. Business groups that are subject to credit control have to get the approval of their principal transactions bank before they purchase real estate of a certain size or make investments in other companies.

The evaluation of the credit control system so far has not been favorable. Though the credit control system was probably successful in preventing the monopoly of bank credit by large business groups—the share of the 30 largest business groups in total bank loans declined from 29 per cent in 1986 to 19 per cent in 1991—such decline was accompanied by an increase in their share in the loans of nonbank financial institutions. The credit control system was certainly not successful in improving the capital structure of large business groups. While the equity ratio of all large manufacturing firms increased from 23 per cent to 26 per cent between 1987 and 1991, that for corporations subject to credit control actually declined slightly, remaining at 19 per cent in 1991. In addition, 20 out of the 30 largest business groups increased the number of their member companies between 1986 and 1991 despite control on corporate acquisitions.

A more serious problem with the credit control system is that it has seriously hampered the development of an autonomous bank–business relationship. The tasks imposed on principal transactions banks are complicated and burdensome. Despite the wide range of authority given to these banks, in reality they lack effective tools for enforcing credit controls on business groups. They are also subject to penalties from the Office of Bank Supervision and Examination for the violation of credit control provisions. Consequently, banks tend to implement credit control only passively with little attention directed to establishing a long-term client relationship.

Recognizing that the current credit control system unduly restricts normal corporate activities, the new Korean Government envisages gradual relaxation of the control. Credit control procedures will be simplified, and the current system of approval by the principal transactions banks for corporate investment in other businesses and real estate would be phased out. Further, only the ten largest business groups would remain subject to credit control. Ultimately, credit control would focus mainly on maintaining prudence and soundness in corporate financing and banking operations, freeing itself from serving other objectives. Easing of excessive concentration of economic power and specialization of business groups would be induced by the Fair Trade Act and the Industrial Development Law, while speculation in real estate would be dealt with by strengthening related tax laws.

Regulation of Interest Rates

The Korean monetary authorities determined the whole spectrum of interest rates on financial assets and liabilities in the organized financial markets until the late 1980s. The only exceptions to this were bond yields in the secondary market and some money market and trust instruments whose returns depend largely on investment performance on securities. Until June 1982, the interest rate structure was complicated: rates were differentiated depending on the source (government, banking, or foreign), use, and supplier of the funds, and preferential interest rates were given to policy loans.

Although the monetary authorities seem to have changed interest rates in consideration of the need to absorb liquidity, to boost investment activity, or in line with changing inflation rates, the determination has, to a large extent, been arbitrary. In general, interest rates have been maintained well below the market rate and have not been adequately flexible. With the excess demand for funds in the organized financial markets, there has always been room for the curb loan market.

Bank deposit (one-year time deposit) interest rates in the latter half of the 1970s was mostly negative in real terms. Despite high (20–21 per cent on average) nominal interest rates on one-year time deposits and on general bank loans, real interest rates (using the consumer price inflation rate) were on average about 2 per cent negative during the period 1979–1981. As inflation slowed down dramatically thereafter, interest rates became significantly positive. During 1984–1987, real interest rates on one-year time deposits ranged between 7 and 8 per cent, and those on general bank loans between 8 and 9 per cent. With accelerated inflation and rigid official rates during 1988–1991, the average real rates dropped below 3 per cent before modestly rising again in 1992.

Average real corporate bond yields determined in the secondary market, which stood at 7.6 per cent during 1976–1979, rose to 10.7 per cent during 1982–1987 when inflation decelerated significantly, but returned to 8.4 per cent during 1988–1992. Another market interest rate is the curb loan rate which is typically applied to discounting corporate bills by individual money lenders. Compared with the corporate bond yield, the curb loan rate was about 20 per cent points higher during 1975–1978. The gap dropped to 15 per cent points during 1979–1980 and stabilized within a 10–13 per cent range during 1981–1987. Thereafter, the gap narrowed sharply to a 2 per cent level by 1991, although it widened again to 4.6 per cent points in 1992. The real curb rate has also shown a steady drop. The average real rate declined from 26 per cent during 1976–1979 to 21 per cent during 1983–1987, and 11 per cent during 1990–1992 (Table 1.2).

The first attempt at full-scale liberalization of interest rates was made in December 1988, taking advantage of favorable economic conditions at the time, such as low and stable inflation since 1983, high national savings in excess of domestic investments, and the enhanced competitiveness of Korean firms. Most bank and nonbank lending rates and some long-term deposit rates were deregulated. Deregulated interest rates or yields also included those on financial debentures, corporate bonds, asset management accounts, commercial papers, certificates of deposit, and large repurchase agreements (RPs).

The result of deregulation, however, fell far short of expectations. The financial institutions which had been accustomed to government direction for interest rate determination were not ready for competition in a liberalized environment. In addition, the Government continued "window guidance" for fear of a steep rise in interest rates. As a result, bank lending rates and most rates in the primary securities market remained rigid and unresponsive to market conditions.

Table 1.2 Regulated and Market Interest Rates
(year average, per cent)

	Regulated Rates		Market Rates		Average borrowing cost (manufacturing)	Consumer inflation rate
	One-year time deposit	General bank loan	Corporate bond yields	Curb rate		
1970	22.8	24.0	...	49.8	14.7	15.4
1975	15.0	15.5	20.1	41.3	11.3	25.4
1976	15.5	16.5	20.4	40.5	11.9	15.3
1977	15.8	17.3	20.1	38.1	13.1	10.0
1978	16.7	17.7	21.1	41.2	12.4	14.5
1979	18.6	19.0	26.7	42.4	14.4	18.2
1980	22.7	23.4	30.1	44.9	18.7	28.7
1981	19.3	19.8	24.4	35.3	18.4	21.6
1982	10.9	12.5	17.3	30.6	16.0	7.1
1983	8.0	10.0	14.2	25.8	13.6	3.4
1984	9.1	10.6	14.1	24.8	14.4	2.3
1985	10.0	11.5	14.2	24.0	13.4	2.5
1986	10.0	11.5	12.8	23.1	12.5	2.8
1987	10.0	11.5	12.8	23.0	12.5	3.0
1988	10.0	11.5	14.5	22.7	13.0	7.1
1989	10.0	11.5	15.2	19.1	13.6	5.7
1990	10.0	11.5	16.4	18.7	12.7	8.6
1991	10.0	11.5	18.8	21.0	13.0	9.3
1992	10.0	11.5	14.0	18.6	12.3	6.2

... data are not available.
Sources: Bank of Korea, *Monthly Bulletin* (various issues); Bank of Korea, *Financial Statements Analysis* (various issues).

In August 1991, the Government released another timetable for interest rate liberalization, which included previously deregulated interest rates. According to this plan, lending rates would be deregulated faster than deposit rates to maintain stability in banking in the early stage of deregulation. Also, deposits with longer maturities and of greater size would be deregulated first to deter any abrupt transfer of funds among different financial sectors and to encourage long-term deposits.

The first phase of deregulation was implemented in November 1991. It included interest rates on bank overdraft loans, discount/selling rates on commercial bills (not eligible for central bank rediscount), commercial papers, and trade bills, and deposit rates on negotiable CDs, large RPs, and some other long-term deposits. Primary market rates for corporate bonds with maturities of two years or more were also liberalized.

The second phase of liberalization was to be implemented by the end of 1993, which would have included interest rates on all lending other than that by the Government or rediscounted by the Bank of Korea, as well as rates on deposits with maturities of two years or longer. In addition, rates on corporate bonds with maturities of two years or less and financial debentures with maturities of two years or more are scheduled to be deregulated. During the third phase (1994–1996), all remaining lending rates together with time and saving deposit rates with maturities of less than two years will be liberalized. Interest rate deregulation for financial debentures with maturities less than two years and monetary stabilization bonds is also scheduled in this phase.

Slow Pace of Financial Liberalization

Serious financial liberalization efforts were begun in Korea by lifting restrictions on bank management and divesting government equity shares in all five nationwide commercial banks to transfer the ownership to private hands in the early 1980s. Further, entry barriers into the financial market were lowered and more diversified financial services were provided by various intermediaries. Significant progress has also been made in interest rate and credit management as previously described.

Despite privatization and greater managerial autonomy by the commercial banks, the banking system is still subject to heavy government intervention. First, the Government continues to play a role in banking because of the heavy burden of nonperforming loans, many of which are attributable to government intervention in bank lending since the 1970s. Given the deteriorated loan portfolios of Korean banks, drastic financial liberalization could threaten the soundness and the safety of these banks. Complete interest rate deregulation, for example, was illogical because banks would be in a substantially disadvantaged position in competing with nonbank financial intermediaries. A different approach, such as relieving banks of the burden of nonperforming loans and exposing them to vigorous competition, might have been a better solution, providing it were politically feasible.

Second, in the absence of effective tools for indirect monetary control, credit ceilings and other direct controls have been imposed on Korean financial intermediaries. The discount window plays only a limited role because much of the central bank loans are automatic rediscounts of policy loans by the banking sector. Since banks have suffered from chronic reserve shortage, changing the required reserve ratio has generally been difficult and ineffective. Also, open market operations have been constrained by an underdeveloped money market, inadequacy of traded securities, and the absence of a secondary market, all of which are largely attributable to interest rate regulation.[2]

Third, policy loans by banks still account for almost half of their domestic credit. Such loans include discounts of commercial bills for small- and medium-scale firms, credit to development institutions such as Korea Development Bank, credit to the housing and agriculture sectors, loans to small- and medium-scale firms, and foreign currency loans mainly for capital goods imports. Until the commercial banks are freed from the obligation of extending policy loans, the extent of financial liberalization will be limited. Preferential policy loans may be provided with social considerations in mind and to complement market imperfections. This obligation, however, should mainly be fulfilled through the government budget rather than by commercial banks.

Finally, the major thrust for still regulating most of the interest rates in Korea is to keep the lending rates from rising too high and to avoid any financial panic for business firms whose capital structure is generally very weak. Interest rate deregulation is constrained by a low level of financial development, which in turn is the result of prolonged interest rate controls. The desire by the Government to borrow cheaply itself with the issuance of national and public bonds must have also affected its interest rate policy.

Government intervention in the management of banks was most clearly manifested by the *de facto* appointment of presidents and high ranking officials of nationwide commercial banks by the Government, even after these banks were privatized. Since the ownership share in nationwide commercial banks has been limited to a maximum of 8 per cent for a shareholder. Under this circumstance, the Government maintained management control, seriously hampering the bank's autonomy. This practice of government appointment of bank presidents is going through a transition, however, and with the government declaration of a hands-off policy, commercial banks recently began to select their new presidents in an autonomous manner through nomination by an independently chosen "president nomination committee."

Contrary to the heavily regulated banking sector, nonbank financial institutions such as investment and finance companies, mutual savings and finance companies, and credit unions which were introduced in the early

---

[2] The Monetary Stabilization Bond issued by the central bank has recently become the major tool for sterilizing liquidity. However, these bonds have been issued at below market rates and sold mainly to financial institutions by coercion.

1970s have been allowed to operate in a more liberalized environment. They have been allowed to offer relatively attractive interest rates and are rarely asked to provide policy loans. This freer environment was based on the Government's desire to nurture these institutions in order to absorb curb market funds and bring about financial deepening. Thanks to this policy, the nonbank financial institutions have shown rapid growth since the 1970s.

In Korea, capital flows in and out of the country have been controlled to a large extent. This is a result of concern about capital flight, excessive monetary expansion, and inefficient use of foreign borrowing and foreign exchange. Foreign capital has been introduced under government guidelines with a view toward maximizing the efficiency of foreign borrowing in both its cost and use. For nonfinancial private corporations, foreign borrowing has mainly been confined to financing importation of capital goods and raw materials, while that of financial intermediaries has been primarily restricted to funds for relending to firms in foreign currencies. With current account surpluses in the latter half of the 1980s, Korea has gradually relaxed controls on capital outflows.

With the limited role of market-determined interest and exchange rates in balancing the demand and supply of funds and foreign exchange, discretionary controls might be inevitable to prevent excessive capital flows in one direction. Since March 1990, Korea has been maintaining a market average exchange rate system, under which the rate is determined by market forces within a specified range. Beginning in early 1992, foreign investors were allowed to purchase stocks in the domestic market. Currently, combined foreign ownership of a stock is limited to 10 per cent of its total outstanding shares listed in the Korea Stock Exchange, while a single individual or institution may buy only up to 3 per cent of the total.

Information Asymmetry and Legal Framework

Problem of Information Asymmetry

A basic function of banks in the world of incomplete and asymmetric information is to collect and analyze information about potential borrowers and evaluate their creditworthiness. In Korea, however, banks have been neglecting this function to a large extent. First, Korean banks have been obliged to extend a substantial share of their total credit as policy loans or other directed loans. Naturally, in this situation, they do not have much incentive to undertake serious credit analysis or monitoring of the borrowing firms.

Second, under government control of interest rates, Korean financial intermediaries usually ask borrowers to put real estate as collateral. Typically, the appraisal value of collateral is required to be 30–50 per cent higher than the loan amount. Since the appraisal value usually remains at about 70 per cent of the market value, a borrower has to provide collateral worth roughly twice the borrowed amount. When a borrowing firm goes bankrupt, the bank disposes the collateral and collects the principal, interest at a high penalty rate, and incidental costs. What is problematic is that a bank can be better off if a borrowing firm goes bankrupt rather than fares well. As a result, banks have no reason to be serious about credit evaluation (moral hazard), and actually have an incentive to lend to risky firms as long as the loans are secured by collateral (adverse selection), and to let firms go bankrupt as soon as they default on their debts.

This problem seems to be less serious for smaller local banks and nonbank financial intermediaries such as mutual savings and finance companies, community credit cooperatives, and credit unions. Based on local communities, these intermediaries as well as curb lenders generally have accurate information about the healthiness and creditworthiness of borrowers. Thus, relatively new and risky businesses have a better chance of being supported by them.

The problem of asymmetric information also exists between depositors and banks. In Korea, however, despite large nonperforming bank loans, no serious solvency question is raised, because bank depositors perceive that the Government provides implicit deposit insurance. A more serious problem of asymmetric information seems to be between bank equity holders and management. As noted earlier, until recently, presidents of nationwide commercial banks were virtually appointed by the Government, even after the early 1980s when these banks were privately-owned. Naturally, top management has been concerned more about the direction of the Government (officials) rather than improving bank profits, quality of services, or bank management.

Some Legal Aspects

As for the legal environment relevant for recovering financial claims, one thing unique to Korea is the measure against firms that issued dishonored checks. Korean firms with a checking account at a bank frequently issue post-dated checks as a way of getting trade credit. If a check is defaulted, the issuer is considered to be a criminal under the Dishonored Check Control Law, and is investigated and/or taken into custody. Once the business owner is in custody, normalizing the firm's operation is usually impossible and the firm is led to bankruptcy. Without the law, public confidence in post-dated checks would weaken, but the service of credit evaluation may be more in demand to replace the role played by the law.

Another legal system in Korea is that of putting business firms in difficulty under the management of a court. The objective is to minimize the social and economic cost of bankruptcies by rehabilitating firms that are in financial distress but have a good chance of recovery. The problem, however, is the fact that out of 143 firms which were under court management during 1962–1990, only 10 were successfully rehabilitated. If a firm is put under the management of a court, its liabilities are frozen for a period of up to 20 years. Given that many firms apply for court management simply to avoid or delay defaults and bankruptcies, this requirement may cause large losses to creditors and equity holders. On the part of creditor banks which have their loans secured by collateral, letting the firms go bankrupt may be far less costly than putting them under court management for many years.

Finally, an Emergency Presidential Order was promulgated in August 1993 to institute the Real-Name Financial Transaction System. The Presidential Order mandates the use of real names for all transactions with financial institutions. Given that holding financial assets under fictitious and borrowed names has often been linked with tax evasion, bribery, rebates, and other irregularities and corruption, the new system is expected to contribute to a "clean society." If the system is successful, it will greatly enhance the transparency of corporate management, particularly that of accounting and internal fund management. Moreover, this system is a precondition for including interest and dividend income in the global income tax system, which is scheduled to be in force in 1996.

OVERVIEW OF THE FINANCIAL SECTOR

Financial institutions in Korea may be classified into four different types: the central bank, banking institutions, nonbank financial institutions, and the securities market. Banking institutions accounted for the largest share of total financial assets until the early 1970s. They include commercial banks (14 nationwide banks, 10 local banks and 51 foreign bank branches at the end of 1992) and six specialized banks (Table 1.3, Table 1.4).

The five largest nationwide commercial banks have a relatively long history; all were privatized in the early 1980s. The local banks were established during the period 1967–1991 and operate in the provincial markets. Foreign banks were allowed to open branches in Korea as early as 1967. Established mostly during the 1960s, specialized banks provide funds to particular sectors such as small-scale and medium-scale industries, households, housing, agriculture, fisheries, and livestock industry.

Many of nonbank financial institutions were introduced in the 1970s with a view to attracting curb market funds into the organized market. Nonbank financial institutions as a whole have grown much faster than banks, and now account for more than half of the Korean financial market. They include development institutions, savings institutions, investment companies, and other institutions such as insurance, leasing, and venture capital companies.

Three development institutions, including the Korea Development Bank, provide medium-term and long-term loans for development of strategic sectors. Savings institutions include diverse savings outlets such as trust accounts of banks, mutual credit, two new types of institutions introduced in 1972 to absorb small-scale curb market funds (mutual savings and finance companies, and credit unions), community credit cooperatives, and postal savings. Investment companies comprise investment and finance companies—also introduced in 1972 to develop the money market and to attract curb market funds—and merchant banking corporations established as joint ventures with foreign financial institutions to provide a wide range of financial services.

Institutions in the securities market include the Securities Supervisory Board, securities companies, the Korea Securities Finance Corporation, and securities investment trust companies. Sales of beneficiary certificates by securities investment trust companies for investment mainly on stocks and bonds now account for over 10 per cent of total financial deposits in Korea. The Korea Securities Finance Corporation is the sole institution specializing in securities financing.

Growth of Financial Markets

Financial development more or less stagnated during the 1970s due to the negative interest rates which persisted throughout most of the period as a result of two oil shocks. The M2/GNP ratio, which stood at 32 per cent in 1970, remained almost the same ten years later. The ratio rose from 34 per cent in 1980 to 41 per cent in 1992 thanks to the deceleration of inflation and the resulting positive real interest rates on deposits since 1982. However, the growth of M2, the major monetary target, has been constrained by the restrictive monetary policy particularly during the first half of the 1980s and introduction of new financial assets such as commercial paper, RPs, variants of trust deposits and CDs (which are not part of M2).

The M3/GNP ratio thus grew much faster, rising from 43 per cent in 1979 to 128 per cent in 1992.[3] The rapid growth of nonbank financial intermediaries can be explained by lower entry barriers to these markets, the introduction of new financial instruments, and more attractive interest rates offered to depositors (often circumventing interest rate controls by the authorities). This favorable environment for the growth of nonbank financial intermediaries can be attributed to the desire of the Government to absorb curb market funds and thus maximize the mobilization of financial resources through organized markets. Rapid growth of nonbank financial intermediaries has mainly been manifested by investment and finance

---

[3] M3 is defined to include M2, deposits of nonbank financial institutions, bank debentures issued, commercial bills sold, CDs, and RPs.

Table 1.3 Financial Institutions in Korea
(As of the end of 1992)

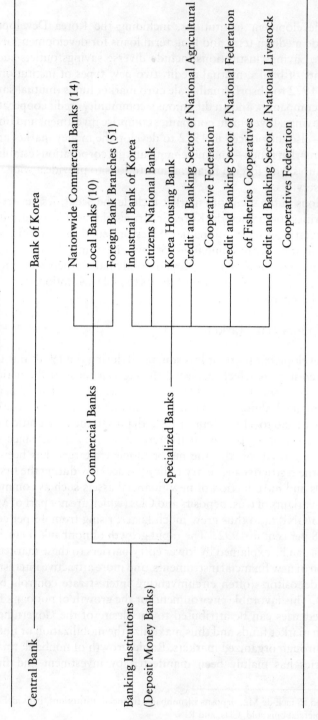

- Central Bank
  - Bank of Korea
- Banking Institutions (Deposit Money Banks)
  - Commercial Banks
    - Nationwide Commercial Banks (14)
    - Local Banks (10)
    - Foreign Bank Branches (51)
  - Specialized Banks
    - Industrial Bank of Korea
    - Citizens National Bank
    - Korea Housing Bank
    - Credit and Banking Sector of National Agricultural Cooperative Federation
    - Credit and Banking Sector of National Federation of Fisheries Cooperatives
    - Credit and Banking Sector of National Livestock Cooperatives Federation

- Nonbank Financial Institutions
  - Development Institutions
    - Korea Development Bank
    - Export-Import Bank of Korea
    - Korea Long-term Credit Bank
  - Savings Institutions
    - Trust Accounts of Banking Institutions (56)
    - Mutual Savings and Finance Companies (237)
    - Credit Unions (1,461)
    - Mutual Credit Facilities (1,704)
    - Community Credit Cooperatives (3,200)
    - Postal Savings (Post Offices)
  - Investment Companies
    - Investment and Finance Companies (24)
    - Merchant Banking Companies (6)
  - Other Institutions
    - Life Insurance Companies (32)
    - Non-Life Insurance Companies (16)
    - The National Investment Fund
    - Leasing Companies (20)
    - Venture Capital Companies (58)
    - The National Housing Fund
    - Credit Guarantee Funds (3)
    - The Korea Non-Bank Deposit Insurance Corporation

- Securities Market
  - Securities Supervisory Board
  - Securities Companies (32)
  - The Korea Securities Finance Corporation
  - Securities Investment Trust Companies (8)
  - Korea Stock Exchange

Note: Figures in parentheses represent the number of institutions.

Table 1.4 Loans and Deposits by Type of Financial Institution

	Loans				Deposits			
	1985		1992		1985		1992	
	(billion won)	Per Cent	(billion won)	Per Cent	(billion won)	Per Cent	(billion won)	Per Cent
Deposit money banks	33,810	56.8	102,797	45.0	31,023	55.9	107,246	35.6
Commercial banks	19,800	33.3	61,532	26.9	18,157	32.7	63,060	20.9
Specialized banks	14,010	23.5	41,265	18.0	12,866	23.2	44,186	14.7
Nonbank financial institutions	24,925	41.9	123,346	53.9	18,979	34.2	156,088	51.8
Development institutions	6,566	11.0	18,467	8.1	220	0.4	1,492	0.5
Korea Development Bank	6,158	10.3	14,275	6.2	191	0.3	920	0.3
Export-Import Bank of Korea	1	0.0	560	0.2	0	0.0	0	0.0
Korea Long-term Credit Bank	407	0.7	3,632	1.6	29	0.1	572	0.2
Savings institutions	9,800	16.5	70,588	30.9	9,376	16.9	108,739	36.1
Trust Accounts at Banks	3,133	5.3	25,194	11.0	3,928	7.1	53,022	17.6
Mutual Savings and Finance Companies	2,895	4.9	15,287	6.7	947	1.7	14,909	5.0
Credit Unions and Mutual Credits	3,772	6.3	30,107	13.2	4,292	7.7	38,022	12.6
Postal Savings	0	0.0	0	0.0	209	0.4	2,786	0.9

Investment companies	4,575	7.7	13,289	5.8	4,109	7.4	8,540	2.8
Investment and finance companies	3,721	6.3	11,057	4.8	3,239	5.8	6,547	2.2
Merchant Banking Corporations	854	1.4	2,232	1.0	870	1.6	1,993	0.7
Life insurance companies	3,984	6.7	21,002	9.2	5,274	9.5	37,317	12.4
Capital market institutions Securities Investment	790	1.3	2,509	1.1	5,454	9.8	37,741	12.5
Trust Companies	540	0.9	433	0.2	5,399	9.7	35,741	11.9
Korea Securities Finance Corporation	250	0.4	2,076	0.9	55	0.1	2,000	0.7
Total	59,525	100.0	228,652	100.0	55,456	100.0	301,075	100.0

Source: Bank of Korea.

companies (corporate bills discounted and resold), securities investment trust companies (sales of beneficial certificates), money in trust offered by banks, life insurance companies (reserves), mutual savings and finance companies, mutual credit, and credit unions.

The role of the securities market in Korea's total financial savings has increased steadily throughout the 1970s and the 1980s. The share of securities issues in the total outstanding financial assets rose from 13 per cent in 1972 to 21 per cent in 1979 and to 31 per cent in 1989. Although the role of the equity market declined upon entering the 1980s, this decline was more than compensated for by the increasing role of public and financial debentures and corporate bonds. The stock market recovered strongly in the latter half of the 1980s, and accounted for more than 13 per cent of total outstanding financial assets during 1988-1990.

The ratio of total domestic financial assets to GNP, the broadest indicator of financial development, also stagnated during the 1970s, but showed rapid growth during the 1980s, rising from 2.16 in 1979 to 4.49 in 1992 (Table 1.5). Significantly positive interest rates, the introduction of new financial assets, an increasingly extensive network of financial services, development of the securities market, and sustained economic growth during the 1980s all contributed to this rapid financial development. Net accumulation of financial assets, however, is not exactly paralleled by the national savings rate. As analyzed by Nam (1990), although positive real interest rates certainly help, the national savings rate is predominantly influenced by changes in income. This is clearly shown by the substantial drop in the national savings rate in the early 1980s and a drastic rebound since 1986.

The GNP share of financial services may serve as another indicator of financial deepening. The ratio of financial sector value added to GNP (in 1985 constant prices) rose steadily from 2.2 per cent during 1971-1980, to 4.3 per cent during 1981-1990, and to 6.6 per cent during 1991-1992. In current prices, the ratio rose from 2.9 per cent during 1971-1980 to 5.6 per cent during 1991-1992.

The composition of financial savings (which excludes currency and demand deposits) shows that time and saving deposits of deposit money banks has declined significantly in importance (Table 1.6). Their share in total financial savings dropped from 46 per cent in 1980 to 21 per cent in 1992. This trend has been matched by a substantial increase in the share of nonbank deposits, from 38 per cent in 1980 to 64 per cent in 1992. Savings in securities, including public debentures, stocks, and corporate bonds accounted for 28 per cent of the total in 1992. Among nonbank financial deposits, money in trust showed the highest growth, increasing its share to over 16 per cent in 1992. Insurance and securities investment trust have also shown substantial gains in their shares, recording 14 per cent and 10 per cent, respectively. The only category of nonbank deposits which showed a relative decline was that of investment companies, whose share remained below 7 per cent in 1992.

Table 1.5 Trend of Financial Deepening

	M2/GDP	M3/GDP	Domestic Financial Assets/GNP	Financial Services/ GNP[a] (per cent)	National Savings Rate (per cent)
1970	0.32	0.37[b]	2.12	1.3	18.0
1975	0.31	0.39	2.17	1.7	18.2
1976	0.30	0.38	2.05	1.9	24.3
1977	0.33	0.42	2.12	2.4	27.6
1978	0.33	0.43	2.14	2.7	29.7
1979	0.32	0.43	2.16	2.8	28.4
1980	0.34	0.49	2.40	3.8	23.1
1981	0.34	0.51	2.57	3.3	22.7
1982	0.38	0.59	2.90	3.3	24.2
1983	0.37	0.61	2.94	3.3	27.6
1984	0.35	0.65	3.09	3.5	29.4
1985	0.37	0.71	3.28	4.1	29.1
1986	0.37	0.78	3.31	4.3	32.8
1987	0.38	0.87	3.51	4.6	36.2
1988	0.39	0.94	3.59	5.1	38.1
1989	0.41	1.06	4.10	5.5	35.3
1990	0.40	1.15	4.22	5.8	36.0
1991	0.41	1.18	4.22	6.2	36.3
1992	0.42	1.28	4.49	6.9	34.9

[a]GNP in 1985 prices.
[b]1971 figure.
Notes: Stock of financial assets is on year-end basis.
Sources: Bank of Korea, *Monthly Bulletin* (various issues).

Flow of Funds

A flow of funds table shows how each sector—financial, government, business and individual—made financial transactions through what kinds of instruments or accounts. The table can be prepared for each year showing net changes in financial assets and liabilities, or can be presented in terms of financial assets and liabilities outstanding at a given point. For the purpose of this chapter, it is more meaningful to look at the table of outstanding balances (Table 1.7).

Table 1.6 Composition of Financial Savings
(per cent)

	1980	1985	1990	1992
Time and savings deposits	45.9	33.9	22.7	21.1
Certificates of deposits	0.0	1.8	2.9	3.6
Nonbank deposits	37.8	52.7	60.3	64.2
Insurance	6.1	11.6	13.0	13.7
Money in trust	5.6	6.6	13.2	16.3
Short-term finance	11.2	11.5	9.5	6.7
Mutual credit	5.2	5.1	6.1	7.2
Mutual savings and finance	2.1	4.6	3.9	4.5
Credit unions	3.4	3.3	4.2	4.9
Investment trust	3.4	9.6	9.6	9.9
Others	0.7	0.5	0.8	1.0
Public debentures	4.8	8.3	6.3	6.8
Stocks	7.2	3.6	13.3	10.6
Corporate bonds	9.8	12.0	10.3	10.5
Total	105.5	112.3	115.8	116.8

Note: The sum of individual items exceeds 100 per cent due to intersectoral transactions.
Source: Ministry of Finance, *Chaejung Kumyung Tongge* (Financial and Monetary Statistics), (various issues).

At the end of 1992, Korean individuals held financial liabilities worth 131 trillion won and financial assets of 283 trillion won (123 per cent of GNP), leading to net financial assets of 152 trillion won. Of the total liabilities of the individual sector, 84 per cent were borrowings from financial institutions with banks accounting for 28 per cent, and 10 per cent were trade credit. Holdings of financial assets by individuals at the end of 1992 were divided into currency and deposits (52 per cent, with 30 per cent for the banking sector), life insurance and pension funds (18 per cent), securities including stocks and other equities (26 per cent) and others (4 per cent).

The business sector is the only deficit sector in Korea at least in terms of accumulated financial assets and liabilities. Total liabilities including equities of business firms amounted to 381 trillion won (166 per cent of GNP) at the end of 1992. With the holdings of financial assets worth 184 trillion won, the net liabilities of the business sector reached 197 trillion won (86 per cent of GNP).

Borrowings from financial institutions accounted for 38 per cent of the total financing of the business sector, almost half of which were bank loans. Issuance of securities including equities financed 36 per cent of the total as of the end of 1992. Debt instruments (19 per cent of the total) were utilized a little more than stocks and other equities (17 per cent of the total). In addition, trade credit and foreign debts accounted for 10 per cent and 6 per cent of the total accumulated financing, respectively. Holdings of financial assets by business firms were mainly deposits at financial institutions and trade credit, accounting for 32 per cent and 27 per cent of the total respectively. Compared with individuals, business firms held relatively more demand deposits, negotiable certificates of deposit, and deposits at investment and finance companies, and almost as much in trust deposits. Holdings of securities including stocks and other equities remained at 16 per cent of the total.

Government debt outstanding at the end of 1992 was 17 trillion won, equivalent to 7 per cent of GNP. This included issuance of securities (21 per cent), borrowings from financial institutions (35 per cent) and foreign debts (23 per cent). Government holdings of financial assets were 53 trillion won. Government loans and equity investment accounted for 30 per cent and 28 per cent of the total, respectively. Holdings of other assets included deposits at the Bank of Korea and deposit money banks, trust, and securities.

Total outstanding fund mobilization by the financial sector at the end of 1992 reached 530 trillion won, equivalent to 231 per cent of GNP. Deposits at deposit money banks, including negotiable certificates of deposit and trust accounted for 28 per cent of the total, while deposits at nonbank financial institutions accounted for 16 per cent. Another 10 per cent or so was mobilized by life insurance and pension funds. Mobilization through securities, mainly financial debentures and beneficiary certificates, amounted to 15 per cent of the total, while stocks and other equities remained at slightly over 3 per cent. Finally, loans from other financial institutions, the Government, and abroad accounted for 11 per cent of the total.

Foreign debt outstanding at the end of 1992 was worth 45 trillion won (57 billion US dollars). Subtracting foreign claims as well as foreign exchange holdings, net foreign debt stood at slightly over 10 trillion won (13 billion US dollars), equivalent to 4.5 per cent of the GNP.

Table 1.7 Financial Assets and Liabilities Outstanding at the end of 1992
(billion won)

	Total		Financial Sector		Government		Business		Individual		R.O.W.	
	Uses	Sources	Uses	Sources	Uses	Sources	Uses	Sources	Uses	Sources	Uses	Sources
Total	1,111,499	1,093,621	543,602	530,296	53,451	16,569	184,272	380,843	283,268	131,044	46,909	34,869
Gold	26	26	26	0	0	0	0	0	0	0	0	26
Currency and transferable deposits	37,405	37,405	12,584	37,405	5,128	0	8,963	0	10,729	0	0	0
Deposits at BOK	8,399	8,399	8,399	8,399	0	0	0	0	0	0	0	0
Currency	8,581	8,581	1,381	8,581	0	0	2,078	0	5,121	0	0	0
Demand deposits	16,006	16,006	2,804	16,006	959	0	6,634	0	5,609	0	0	0
Government deposits at BOK	4,419	4,419	0	4,419	4,168	0	251	0	0	0	0	0
Other deposits	213,510	213,510	17,603	213,510	7,210	0	52,838	0	135,859	0	0	0
Deposits at deposit money banks	71,672	71,672	3,335	71,672	3,035	0	16,875	0	48,428	0	0	0
Negotiable CDs	10,672	10,672	2,970	10,672	560	0	3,984	0	3,157	0	0	0
RP	8,976	8,976	4,173	8,976	174	0	819	0	3,810	0	0	0
Trust	48,705	48,706	2,827	48,706	3,028	0	20,366	0	22,484	0	0	0
Deposits at investment & fin. co.	6,169	6,169	1,071	6,169	259	0	2,743	0	2,096	0	0	0
Others	67,316	67,316	3,226	67,316	156	0	8,050	0	55,883	0	0	0
Life Insurance and pension funds	51,537	51,537	0	51,537	0	0	0	0	51,537	0	0	0
Short-term securities	46,406	46,406	33,438	25,063	335	1,580	7,558	19,763	5,076	0	0	0
Government and public funds	6,631	6,631	6,125	0	0	1,580	352	5,051	154	0	0	0
Financial debentures	25,063	25,063	22,902	25,063	220	0	280	0	1,663	0	0	0
Industrial papers	14,712	14,712	4,410	0	115	0	6,926	14,712	3,260	0	0	0

Long-term securities	111,516	111,516	66,069	56,342	2,985	1,977	9,896	53,197	32,566	0	0	0
Government and public bonds	11,544	11,544	7,510	6,439	246	1,977	1,664	3,128	2,125	0	0	0
Financial debentures	15,797	15,797	8,780	15,797	389	0	3,194	0	3,434	0	0	0
Debentures	50,215	50,215	44,945	146	734	0	3,296	50,069	1,240	0	0	0
Beneficiary certificates	33,960	33,960	4,835	33,960	1,615	0	1,743	0	25,767	0	0	0
Stocks	76,843	58,965	34,810	12,661	3,557	0	10,218	46,304	26,604	1,655	0	0
Loans by financial institutions	287,468	287,468	287,468	27,549	0	5,878	0	144,008	0	110,033	0	0
BOK loans	17,435	17,435	17,435	16,865	0	570	0	0	0	0	0	0
Bank loans	116,520	116,520	116,520	7,981	0	1,393	0	70,442	0	36,704	0	0
Insurance company loans	27,345	27,345	27,345	68	0	1,654	0	13,672	0	11,951	0	0
Investment & fin. company loans	12,025	12,025	12,025	0	0	0	0	10,853	0	1,172	0	0
Other loans	114,144	114,144	114,144	2,635	0	2,262	0	49,041	0	60,206	0	0
Government loans	16,060	16,060	0	10,163	16,060	0	0	3,966	0	1,931	0	0
Equities other than stocks	21,751	21,751	574	4,546	11,531	0	1,487	17,205	8,159	0	0	0
Trade credit	49,881	49,881	0	0	0	0	49,881	36,699	0	13,183	0	0
Foreign exchange holdings	13,498	13,498	13,498	0	0	0	0	0	0	0	0	13,498
Foreign claims and debts	66,598	66,598	12,529	19,093	872	3,881	7,944	22,279	0	0	45,254	21,345
Miscellaneous	119,000	119,000	65,002	72,427	5,774	3,253	35,487	37,422	12,737	5,897	0	0
Difference (uses−sources)	17,878		13,304		36,882		(196,571)			152,224		12,040

Source: Bank of Korea (1993).

Sources of Corporate Financing

The capital structure of Korean manufacturing firms was at its worst in 1980, with the ratio of net worth to total assets reaching 17 per cent. This was mainly attributable to the large-scale provision of subsidized bank loans to heavy and chemical industries and other favored industries in the second half of the 1970s, followed by the worldwide recession in the wake of the second oil price shock. The ratio of net worth improved thereafter, stabilizing within a 22–23 per cent range during 1984–1987, and jumping to over 28 per cent in 1989 thanks to improved corporate liquidity and a boom in the stock market. With the declining stock market and economic slowdown, the ratio slid back to 24 per cent in 1992.

Examination of the capital structure of manufacturing firms by firm size, by export versus light industries, and by heavy and chemical versus light industries, reveals that small enterprises have relatively weaker capital structures. At the end of 1992, the ratio of net worth was 25 per cent for large firms, while it remained at 19 per cent for small firms. Firms in heavy and chemical industries, which tend to be larger than light industries, exhibited a slightly higher net worth ratio. The factor most responsible for the weak capital structure of small firms was their negligible accumulation of capital reserves.

Among liabilities, large firms had more long-term liabilities, while small firms relied heavily on current liabilities. At the end of 1992, current liabilities accounted for 61 per cent of the total liabilities for large firms, and 71 per cent of the total for small firms. Though small firms relied much more heavily on long-term borrowings from financial institutions, their bond issues and use of foreign debt were relatively negligible. Among current liabilities, small firms had a larger portion of notes payable and accounts payable as well as short-term borrowings from financial institutions. One noticeable difference between export and domestic enterprises is that domestic ones rely much more on notes payable.

BANKING SYSTEM

Bank of Korea

The Bank of Korea—the country's central bank—was established in 1950, and consists of the executive body, the Monetary Board, and the Office of Bank Supervision and Examination. As the supreme policy-making organ, the Monetary Board has the ultimate responsibility of formulating monetary and credit policies as well as supervising the activities of banking institutions. The Monetary Board is composed of nine members including two *ex officio* members (the finance minister and the Governor of the Bank of Korea) and seven members recommended by banking institutions and Ministers (of the Economic Planning Board, Agriculture, Forestry and Fisheries, and of Trade and Industry).

However, particularly with the amendment of the Bank of Korea Act in 1962, the Government exercises substantial influence over monetary and credit policies. The Minister of Finance presides over the Board's meeting, and can request reconsideration of resolutions adopted by the Monetary Board. The President of the Republic has the final decision if the request for reconsideration is rejected by the Board.

The Bank of Korea's monetary policy is exercised through both orthodox indirect policy tools as well as more direct control of bank credit. The Bank also sets maximum interest rates on some of the deposits and loans of banking institutions whose rates are not yet deregulated. Since 1982, the monetary authorities have tried to replace direct credit control—which imposes ceilings on credit extension of individual banks—with indirect instruments such as Bank of Korea rediscounts, open market operations and reserve requirement policy. However, the effectiveness of these indirect instruments has been limited, so that the authorities have frequently relied on direct credit control in periods of accelerating inflation or pronounced monetary expansion.

Rediscount Policy

In Korea, rediscount policy has played only a limited role as a monetary control tool. This results from the fact that the Bank of Korea gives automatic rediscounts to many policy loans handled by banking institutions and that provision of other loans is not very flexible because of chronic excess demand for bank credit and the consequent shortage of bank reserves. Changes in rediscount rates have also had little effect on other interest rates since the monetary authorities have usually changed most interest rates directly.

Reserve Requirement Policy

To restrain credit expansion, the required reserve ratios used to be maintained at fairly high levels. These high reserve requirement ratios not only reduced the profitability of banks, but also brought about chronic reliance of banks on Bank of Korea loans to make up their reserve deficiency, weakening the effectiveness of this tool of monetary control. Currently, the required reserve ratios are 11.5 per cent for both demand and time and savings deposits, and 1 per cent for foreign currency deposits (though an 11.5 per cent marginal reserve requirement is in effect temporarily).

Open Market Operations

Monetary stabilization bonds—Bank of Korea's special negotiable securities—have been the major tool of operations. The Bank of Korea has issued and repurchased these bonds in the captive market (mainly financial institutions) on a discount basis at a rate comparable to other regulated interest rates. At the end of 1992, the outstanding amount of these bonds reached 20.6 trillion won, equivalent to 21 per cent of M2.

Table 1.8 shows the principal accounts and the composition of the loans and discounts of the Bank of Korea.

Table 1.8 Loans and Rediscounts of the Bank of Korea
(billion won)

	1985 Amount	1992 Amount	Rate
Rediscounts on commercial bills	1,321	5,873	7.0
Loans for foreign trade	2,039	1,031	7.0
Loans for agriculture, fishery, and livestock	174	489	3–15
General loans	5,463	6,138	3–15
Other loans	644	3,468	–
Total	9,641	16,999	–

– not applicable.
Source: Bank of Korea, *Monthly Bulletin* (various issues).

Banking Institutions and Business

Commercial Banks

With funds acquired through offering liabilities in the form of deposits, securities, and other evidences of debt, commercial banks engage in short-term and long-term financing, sales of commercial bills, securities investment, foreign exchange business, and other financial services.

As of April 1993, the commercial banking sector consisted of 14 nationwide commercial banks, 10 local banks with provincewide branching, and 51 foreign bank branches. Five of the nationwide commercial banks have relatively long histories and a commanding position in Korea's financial market. Since the end of Japanese rule in 1945, these banks were owned by the Government until they were transferred to private hands by 1957.

The military coup of 1961, however, resulted in the Government's possession of the major equity shares of these banks again until they were sold to the public by 1983. With a view to encouraging competition among banking institutions, new entries have been allowed since the early 1980s: two banks during 1982–1983, three in 1989 (besides the Korea Exchange Bank's change of status from a specialized bank to a nationwide commercial bank) and three more in 1992 (two of which were converted from investment and finance companies).

All ten local banks were established during 1967–1971 in an effort to promote regional balance in access to credit and economic development. They have all been privately owned from the outset. Branching has basically been limited to the provinces in which their head offices have been located, except for a few branches allowed in Seoul and neighboring provinces.

The Seoul branch of Chase Manhattan Bank, which opened in 1967, was the first foreign bank in Korea. New entry of foreign banks was most active in the second half of the 1970s. Since 1977, the unit branching policy for foreign banks has been relaxed. Their major role in the earlier years was facilitating foreign capital inducement through their close relationships with international financial markets. With the improvement in Korea's balance of payments in the mid-1980s, however, Government policy toward foreign banks shifted toward placing them on an equal footing with domestic banks. Special treatment such as the Bank of Korea's swap facilities with generous guaranteed yields have been reduced, while restrictions applied to their business activities have been eliminated. They now have access to the Bank of Korea's rediscount and lending facilities and handle trust business and negotiable certificates of deposits (which had been prohibited until the mid-1980s).

Specialized Banks

Established mostly during the 1960s, specialized banks include the Industrial Bank of Korea, the Citizens National Bank, the Korea Housing Bank, the credit and banking sector of the National Agricultural Cooperative Federation, the National Federation of Fisheries Cooperatives and its member cooperatives, and the National Livestock Cooperatives Federation. Even though specialized banks can borrow from government funds, they rely mainly on deposits from the public and issuance of debentures for their loanable funds. They compete directly with commercial banks in receiving deposits.

— The Industrial Bank of Korea specializes in supporting small- and medium-scale enterprises by extending credit, making equity investment or underwriting debentures.

— The Citizens National Bank provides banking services to households and small enterprises mainly in the form of small loans. Mutual remuneration loans extended to those holding mutual installment deposit accounts are its major type of financing.

— The Korea Housing Bank extends loans mainly for the construction and purchase of houses or for the acquisition of housing sites. It also lends to local governments and small and medium-scale firms producing basic housing materials.

— The credit and banking sectors of the National Agricultural Cooperative Federation, the National Federation of Fisheries Cooperatives and member cooperatives, and the National Livestock Cooperatives Federation serve as banking institutions mainly for their members, thereby enhancing the economic and social status of the members and promoting the development of their respective industries.

The Industrial Bank of Korea, the Citizens National Bank, and the Korea Housing Bank were all established with more than half of their paid-in capital subscribed by the Government. They are also allowed to issue their own debentures up to 10 times (20 times for the Korea Housing Bank) of their paid-in capital and reserves, though the Citizens National Bank has not issued debentures so far. Still, these specialized banks rely mainly on deposits from the public for their loanable funds. Loans from the Government (including the National Investment Fund) are relatively important for the credit and banking sectors of the National Agricultural Cooperative Federation, the National Federation of Fisheries Cooperatives, and the National Livestock Cooperatives Federation.

Table 1.9 shows the shares of commercial banks and specialized banks in total assets, deposits, and loans and discounts of deposit money banks.

In 1992, deposits in won accounted for 45 per cent of the total liabilities and net worth of commercial banks. The shares of deposits in foreign currencies and negotiable CDs were about 7 per cent, increasing sharply during the late-1980s. On the other hand, reliance on the Bank of Korea and foreign currency borrowings declined steadily over the same period. The share of paid-in capital and reserves and profits in the total commercial bank fund sources more than doubled to 11 per cent between 1985 and 1992. This was due mainly to substantial recapitalization during the capital market boom in the late 1980s and improved banking profitability resulting from a decrease in the share of nonperforming loans. On the asset side, commercial banks, in recent years, tended to increase the share of securities holdings, while reducing that of loans and discounts.

Table 1.9 Assets, Deposits, and Loans and Discounts of Deposit Money Banks (per cent)

	Year	Commercial Banks	Commercial Banks Nationwide Commercial Banks	Local Banks	Foreign Bank Branches	Specialized Banks
Assets[a]	1980	61.7	45.4	7.2	9.2	38.3
	1985	60.4	43.5	8.2	8.7	39.6
	1990	69.6	52.9	11.5	5.2	30.4
	1992	67.2	50.6	11.1	5.5	32.8
Deposits[b]	1980	62.7	51.9	9.5	1.2	37.3
	1985	58.8	46.8	10.8	1.3	41.2
	1990	61.0	46.6	13.4	1.0	39.0
	1992	58.8	45.1	13.1	0.6	41.2
Loans and discounts[c]	1980	61.3	48.4	7.8	5.0	38.7
	1985	58.6	44.1	8.6	6.0	41.4
	1990	60.6	45.1	11.3	4.2	39.4
	1992	59.9	45.4	10.5	3.9	40.1

[a]Excluding acceptances and guarantees.
[b]Including interbank deposits but excluding deposits in foreign currencies.
[c]Excluding loans in foreign currencies.
Source: Bank of Korea, *Monthly Bulletin* (various issues).

Local banks rely much more on deposits in won than do nationwide commercial banks, as their business in foreign currencies is very small. Other noticeable features in the balance sheet of local banks include less reliance on borrowings from the Bank of Korea and more investment in securities. Foreign banks have portfolios of assets and liabilities which are very different from those of domestic banks. Interoffice borrowings in foreign currencies, which represented over 70 per cent of their total liabilities and net worth in 1985, still accounted for 46 per cent in 1992. The share of paid-in capital and reserves and profits in total liabilities and net worth more than doubled between 1985 and 1992, to 14 per cent. Foreign banks acquire won funds for their operations through the sale of foreign currencies to the Bank of Korea under swap agreements. In 1992, loans and discounts in won represented 35 per cent of their total assets, followed by loans in foreign currencies accounting for 31 per cent.

Specialized banks in Korea now rely more heavily on won deposits for their operations than do commercial banks, due to a sharp reduction in their borrowings in foreign currencies. Borrowings from the Bank of Korea and the Government represented 14 per cent of the total fund sources, and debentures issued also remained at 3 per cent of the total funds in 1992. Paid-in capital and reserves and profits represented only 1.4 per cent. On the asset side, about 60 per cent of the assets of specialized banks were in loans and discounts in 1992, of which only 8.4 per cent were extended with government funds.

Deposit and Loan Portfolios of Deposit Money Banks

Deposit Portfolios

Of the total deposits in domestic currency at deposit money banks, demand deposits accounted for 34 per cent in 1992, and the remainder consisted of time and savings deposits. More than half of the total demand deposits are temporary deposits. They are mainly the proceeds of bank checks or other funds which are temporarily in the custody of banks before they are withdrawn or transferred to other accounts. Bank checks are widely used in Korea for the settlement of transactions involving relatively large sums of money. Interest is not paid on checking and temporary deposits; though interest of 1 per cent per annum is paid on passbook deposits and, in exceptional cases, on temporary deposits.

Among many time and savings deposits, time deposits, preferential savings deposits and installment savings deposits are most popular. Interest rates on time deposits with maturities of more than two years are deregulated. Preferential savings deposits are given much higher interest rates than regular savings deposits when deposits are held longer than three months. Installment savings deposits are usually held by those who want to save a target amount for the repayment of bank borrowings or other purposes.

Other types of time and saving deposits include mutual installment deposits, housing installment deposits, workmen's property formation deposits, and company saving deposits. A mutual installment depositor is entitled to borrow up to the contract amount once installment deposits accumulate to exceed a stipulated amount. Workmen's property formation deposits resemble installment saving deposits except that they are given preferential interest rates and are available only to workers whose monthly wages do not exceed a specified amount. Finally, company saving deposits were introduced in 1988 to attract short-term idle funds of business firms. They are similar to passbook deposits but are paid preferential interest rates.

Loans and Discounts of Deposit Money Banks

Of the total loans and discounts of deposit money banks at the end of 1992, general loans accounted for 42 per cent of the total, and other major loans included discounts on commercial bills and loans for housing (handled mainly by the Korea Housing Bank), which accounted for 13 per cent and 12 per cent, respectively. Overdrafts also reached 7.5 per cent of the total loans and discounts of deposit money banks (Table 1.10).

General loans are made in exchange for promissory notes or other credit instruments issued by borrowers, and are used mainly to provide operational funds. General loans are usually extended for a period of one year, but are frequently rolled over. Thus, business firms often rely on these loans to meet their long-term capital needs. Of the commercial bills purchased by banks, those issued or received by eligible small- and medium-scale firms and bills guaranteed by the Korea Credit Guarantee Fund and the Korea Technology Credit Guarantee Fund are eligible for rediscount by the Bank of Korea.

Loans for foreign trade have decreased, as automatic export financing based on letters of credit has been extended only to small- and medium-scale firms and, to a lesser extent, large firms not affiliated with conglomerates since 1988. However, loans for equipment of export industries have shown substantial expansion. Banks get partial refinancing from the Bank of Korea for loans they extend for foreign trade and equipment of export industries.

Of the loans from government funds, which represented 5.4 per cent of the total deposit money bank loans and discounts, a majority is provided for equipment in the agriculture sector. Also included are loans for housing, fishery, small and medium industry, and improvement of the distribution system. The National Investment Fund, established in 1974 mainly to support heavy and chemical industries, extends part of its credit through deposit money banks. The size of the Fund grew rapidly until 1985, but was decreased thereafter in an effort to reduce policy loans.

By industry, 9 per cent of the total deposit money bank loans and discounts were provided to the primary industry sector at the end of 1992. The manufacturing sector received 44 per cent of the loans, much more than its share in the gross national product. Reflecting structural changes in the Korean manufacturing industry, loan shares for textile, apparel and leather, wood products, and basic metal industry declined, while the portion occupied by loans for fabricated metal products, and machinery and equipment increased significantly. Among the services sectors, the loan share for wholesale and retail trade plus restaurants and hotels has declined, while that for financial, insurance, real estate and business services has risen significantly. Loans to households accounted for 22 per cent of the total deposit money banks' loans in 1992, of which slightly more than half were extended for housing financing.

Table 1.10 Composition of Loans and Discounts of Deposit Money Banks by Fund
(billion won)

	1985				1992			
	Operational Funds[a]	Equipment Funds[b]	Total	Per Cent	Operational Funds	Equipment funds	Total	Per Cent
Loans from Banking Funds	26,161	5,101	31,263	92.8	85,588	10,840	96,428	93.8
General loans	12,163	338	12,501	37.0	40,096	3,479	43,575	42.2
Discounts of commercial bills	2,672	0	2,672	7.9	12,806	0	12,806	12.5
Overdrafts	1,475	0	1,475	4.4	7,724	0	7,724	7.5
Loans for foreign trade	3,130	0	3,130	9.3	2,542	0	2,542	2.5
Loans for equipment of export industries	0	595	595	1.8	0	3,044	3,044	3.0
Loans related to installment savings deposits	1,781	0	1,781	5.3	3,101	0	3,101	3.0
Mutual remuneration loans	2,055	61	2,116	6.3	3,660	124	3,784	3.7
Loans for housing	103	2,226	2,329	6.9	11,726	270	11,996	11.7
Loans for agriculture	404	137	541	1.6	1,190	1,424	2,615	2.5
Loans for fisheries	221	16	237	0.7	484	24	508	0.5
Loans for livestock	71	252	323	1.0	377	499	876	0.9
Other loans	2,086	1,477	3,564	10.6	1,882	1,976	3,857	3.7
Loans from Government Funds	91	1,492	1,583	4.7	742	4,826	5,568	5.4
Loans from National Investment Fund	8	957	966	2.9	7	796	803	0.8
Total	26,261	7551	33,811	100.4	86,337	16,462	102,799	100.0

[a]Operational Funds: funds provided as the working capital.
[b]Equipment Funds: funds provided for equipment and facility investment.
Source: Bank of Korea, Monthly Bulletin (various issues).

Efficiency of Banks

Operative Efficiency

International comparisons of some of the measures of banks' operative efficiency show that Korean banks are fairly inefficient in terms of their loan–deposit interest rate margin and the average amount of financial services handled by a bank employee. The interest rate margin of Korean (domestic) commercial banks stayed above 4 per cent. However, when bank reserve deposit and receipt of certified checks of other banks are excluded from total deposits, and loans repaid with other banks' certified checks are added in total loans, the average interest rate margin is known to have declined to a level slightly over 2 per cent since 1991, which is more comparable to that of other countries (Table 1.11, Table 1.12).

During 1988–1992, the return (net profit) on assets of Korean commercial banks averaged around 0.6 per cent and return on equity was stable, ranging from 6.1 per cent to 6.7 per cent. This bank profitability, which compares rather favorably with that of other countries, together with the relatively large interest rate margin, seems to indicate that Korean commercial banks are run in an environment where competition is weak and management is loose. Actually, total assets, deposits or loans per Korean bank employee in 1985 and 1990 remained between one seventh and one tenth of those of their Japanese counterparts.

In particular, local banks seem to be noticeably less efficient than nationwide commercial banks. In 1992, the average total assets per employee of local banks remained at 63 per cent of that of nationwide commercial banks. Similarly, employees of local banks typically had 31 per cent less deposits and 38 per cent less loans than those of nationwide commercial banks. However, despite this large difference in the productivity of employees, the return on assets of local banks (nearly 0.7 per cent) was substantially higher than that of nationwide commercial banks (a little over 0.5 per cent). This can be partly explained by the much higher loan rates and interest rate margin of local banks (4 per cent) than that of nationwide banks (nearly 1.8 per cent), based on the new method of calculation.

The inefficiency of Korean banks can be largely attributed to heavy intervention of the Government in bank management. As the Government dictated the interest rates and interfered in lending decisions, implicitly guaranteeing the solvency of banks, there was little room for competition among banks and, thus, little incentive for them to become more efficient.

Particularly notable is the share of nonperforming loans and its impact on banks' profitability. The industries which received most of the policy loans in the 1970s and early 1980s, such as heavy and chemical industries, and shipping and overseas construction, faced difficulties as a result of the

Table 1.11 Indicators of Bank Management[a]
(per cent)

	1980	1985	1987	1988	1989	1990	1991	1992	1992 (Nationwide)	1992 (Local)
Growth of total assets	...	...	...	19.2	22.6	29.6	19.2	17.1	19.0	6.6
Growth of total profit[b]	...	...	...	73.0	73.0	10.5	30.2	31.2	38.5	2.9
Growth of net profit[c]	...	...	...	121.0	84.0	36.0	16.9	12.9	20.3	-13.1
Total profit/Total assets	2.32	0.87	0.68	0.96	1.41	1.24	1.30	1.42	1.39	1.64
Net profit/Total assets	0.76	0.19	0.20	0.37	0.58	0.63	0.59	0.56	0.54	0.68
Total profit/Own capital	38.56	20.95	14.36	15.91	16.20	12.43	14.49	16.78	17.39	14.15
Net profit/Own capital	12.68	4.65	4.33	6.13	6.65	6.28	6.58	6.69	6.88	5.87
Average interest rate on										
Won loans	18.96	10.32	9.27	9.79	10.48	10.74	10.28[f]	10.82	10.49	12.22
Won deposits[d]	12.07	5.73	6.54	5.95	5.87	6.21	8.08[f]	0.59	8.7	8.22
Interest rate margin	6.89	4.59	2.73	3.84	4.61	4.53	2.20[f]	2.24	1.79	4.00
Own capital ratio	n.a.	4.9	5.0	6.8	10.5	9.1	8.7	11.2	10.4	16.3
Total assets[e] per employee[g]	305	585	831	994	1,104	1,285	1,468	1,724	1,883	1,184
Deposits per employee[g]	179	336	496	664	754	857	1,006	1,192	1,284	881
Loans per employee[g]	180	345	491	513	569	677	758	928	1,017	625

... data not available.
[a] For commercial banks excluding foreign bank branches.
[b] Total profit=Net profit before tax + reserves for loan loss allowance and employee retirement allowance.
[c] Net profit after tax.
[d] Including certificates of deposit.
[e] Excluding payment guarantees.
[f] Calculation method was changed in 1991.
[g] In million won.
Source: Bank of Korea, *Bank Management Statistics* (various years).

Table 1.12 Comparison of Commercial Bank Efficiency
in Korea, Japan, and the US
(per cent)

	1985			1990		
	Korea	Japan	US	Korea	Japan	US
Return on assets	0.19	0.22	0.69	0.63	0.19	0.49
Return on equity	4.65	9.00	10.70	6.28	6.00	7.60
Average loan rate	10.32	...	10.79	10.74	5.89[a]	9.92
Average deposit rate	5.73	...	8.24	6.21	4.69[a]	7.61
Interest rate margin	4.59	...	2.55	4.53	1.21[a]	2.31
Total assets/employee[b]	585	1,168	...	1,285	1,901	...
Loans/employee[b]	345	670	...	677	1,085	...
Deposits/employee[b]	336	911	...	857	1,438	...

... data not available.
[a] In 1989.
[b] Million won for Korea and million yen for Japan.
Sources: Bank of Korea, *Bank Management Statistics* (various years); OECD (1992); Board of Governors of the Federal Reserve System, *Federal Reserve Bulletin* (various issues).

second oil shock and ensuing worldwide recession. The share of nonperforming loans in the total credit of the seven nationwide commercial banks rose from 2.4 per cent during 1976–1980 to 10.5 per cent during 1984–1986. During the same period, the return (net profit) on assets of these banks plummeted from 0.8 per cent to 0.2 per cent. On the other hand, as the share of nonperforming loans decreased to 5.9 per cent in 1989, the return on assets showed a recovery to nearly 0.7 per cent. Though the share of nonperforming loans was not the only determinant of bank profitability, it was obviously one of the most important factors.

Efficiency of Credit Allocation

It is not an easy task to measure allocative efficiency of bank credit. Industrial allocation of bank loans compared with industrial composition of capital formation or value added may be of some help. The fact that the share of loans to heavy and chemical industries in total bank loans was about twice their share in capital formation or GDP in 1980 (unlike the situation in 1990) seems to indicate that bank credit allocation was skewed toward these industries or the manufacturing industry in general in the 1970s (Table 1.13).

Table 1.13 Comparison of Shares of Bank Loans, Capital Formation, and GDP by Industry
(per cent, current prices)

Sector	Loans and Discounts	Capital Formation	GDP
1970			
Agriculture, forestry, fishing, mining and quarrying	12.6[a]	10.8[b]	28.0
Manufacturing	46.1	24.9	21.3
Light industry	23.5	10.8	12.7
Heavy and chemical industry	22.6	14.1	8.6
Public utilities, construction	12.7	7.8	6.5
Services[c]	28.6	56.5	44.2
Total	100.0	100.0	100.0
1980			
Agriculture, forestry, fishing, mining and quarrying	7.8[a]	8.9	16.2
Manufacturing	53.8	22.9	29.7
Light industry	21.7	6.9	13.2
Heavy and chemical industry	32.1	16.0	16.5
Public utilities, construction	14.6	12.4	10.4
Services[c]	23.8	55.8	43.7
Total	100.0	100.0	100.0
1990			
Agriculture, forestry, fishing, mining and quarrying	6.6[d]	7.7[e]	9.6
Manufacturing	44.0	33.0	29.2
Light industry	13.8	9.6	11.1
Heavy and chemical industry	30.2	23.4	18.1
Public utilities, construction	9.3	6.1	15.4
Services[c]	40.1	53.2	45.8
Total	100.0	100.0	100.0

[a] Loans and discounts of deposit money banks and the Korea Development Bank.
[b] 1975.
[c] Services include government services, private nonprofit services, and import duties less imputed bank service charges.
[d] Loans and discounts of deposit money banks and other financial institutions.
[e] 1988.

It is generally believed that Korea's financial system saw a fair degree of liberalization and integration in the 1980s without much real liberalization of banks. This was possibly due to the relatively rapid growth of the more liberalized nonbank financial sector. Nonbank financial intermediaries (nonbank financial institutions) have been much less constrained than banks in terms of their interest rates and sectoral allocation of credit. Cho and Cole (1991) note that this liberalization/integration of the Korean financial markets in the 1980s was accompanied by a sharp reduction in the spread of borrowing costs across 69 different manufacturing industries. They also observed more equitable access to bank loans by the different sectors of the economy, as well as narrower disparities in rates of return across different sectors. They interpret this as an unmistakable indication of liberalization and higher efficiency of credit allocation, though this evidence may have nothing to do with the efficiency of bank credit allocation.

Amsden and Euh (1990), however, argue that a smaller variance in borrowing costs across different industries is not necessarily good and is not indicative of greater allocative efficiency. The more highly developed the industrial sector, they maintain, the fewer the number of new industries requiring special treatment (preferential loans). Basically, the reduced interindustry variance of borrowing costs is viewed as a result of industrialization, and not necessarily an input into it. They also believe that the policy of promoting the heavy and chemical industries was far from being a failure. The share of heavy and chemical industry products in total exports rose from 24 per cent in 1977 to 54 per cent in a ten-year period, and heavy industry developed as Korea's new leading sector. Meanwhile, the manufacturing sector accounted for only a small share of the total nonperforming loans, with the share for heavy and chemical industries being far lower than their share of total outstanding credit.

Yoo (1989) estimates capital efficiency directly with an indicator which evaluates the efficiency of capital investment in each industry. The indicator measures the value added per unit of capital (nonlabor). During 1971-1978, the capital efficiency of the favored heavy and chemical industries is estimated to be lower than that of other manufacturing industries, while the opposite is true for the 1979-1985 period. The average capital efficiency of the favored heavy and chemical industries (excluding petroleum refinery) remained at 23 per cent throughout the period, while that of other industries (excluding tobacco) declined from 28 per cent to 21 per cent between the two subperiods.

These results strongly suggest that cheap capital in the 1970s led to excessive (inefficient) investment in the heavy and chemical industries. Though, as Amsden and Euh (1990) argue, the Government's efforts to promote heavy and chemical industries can hardly be called a failure in view of the modest share of nonperforming loans for these industries, it seems evident that heavy and chemical industry promotion in the 1970s was pushed too far. As investment in the favored industries slowed and wages rose sharply around the end of the 1970s, it appears that other manufacturing industries made vigorous efforts to substitute capital for labor. However, already high wages in these labor-intensive industries, together with accumulation of capital stock much faster than value-added growth, resulted in relatively low measured capital efficiency during 1979–1985.

Supervision for Sound Banking Operation

Major provisions for the sound operation of banks include those applying to capital and reserve funds as well as to prohibited activities. According to the General Banking Law, banking institutions are to maintain equity capital (capital, reserve funds, and other surpluses) equivalent to at least one twentieth of outstanding liabilities arising from guarantees or assumed obligations. Some liabilities are not eligible for inclusion in this purpose, namely those guaranteed or insured by the Government, other banking institutions, the Korea Credit Guarantee Fund or insurance companies, and those on behalf of the central or local governments. The minimum capital requirement in absolute amount is also prescribed for nationwide and local banks.

Further, when banks dispose of their net profits earned in each fiscal term, at least 10 per cent should be credited to the reserve fund until the fund reaches the size of total capital. The central bank's Superintendent of the Office of Bank Supervision and Examination may require a bank to adjust its book value of assets to set up reserves against bad assets or to write off any assets of little value.

According to the provisions concerning bank operation recently enacted by the Monetary Board, equity capital of a bank should exceed 8 per cent of its risk-weighted assets, which include some off-balance sheet items as well. In addition, banks are to maintain liquid assets of more than 30 per cent of their total deposit liabilities, and loans of less than the combined amount of loanable deposits and loanable equity capital. Most Korean banks are known to have already met the Basle capital ratio. Active capitalization of banks during the stock market boom in the late 1980s led to a sharp rise in the ratio of net worth to total assets for seven nationwide commercial banks from 3.5 per cent to 11 per cent during the 1987–1989 period.

Prohibited activities for a bank include: (i) possessing real properties except those necessary for the conduct of its business (with involved investment not exceeding the equity capital) or held temporarily through the exercise of mortgages; (ii) purchasing or retaining permanent ownership of stocks issued by banking institutions, or stocks in excess of 10 per cent of the shares issued by any other corporation; (iii) granting loans for purposes of speculation in commodities or securities; (iv) granting loans, directly or indirectly, to enable a borrower to buy the bank's own stocks, or on the pledge of its own stocks or stocks in excess of 20 per cent of the issued stocks of any other corporation; (v) granting loans to a borrower in excess of 25 per cent of its equity capital; and/or (vi) guaranteeing or assuming the obligations of a single individual or juridical entity in excess of 50 per cent of its equity capital.

Acceptance of shares of a bank and its investment in bonds or other securities with maturities exceeding three years are not to exceed 25 per cent of its accepted deposits payable on demand. This, however, does not apply to government bonds and the central bank's monetary stabilization bonds. Banking institutions engaged in the trust business are required to keep all associated assets distinct from its banking business and to maintain separate books and records.

Further, no single person is allowed to hold stocks of a nationwide commercial bank in excess of 8 per cent of the total voting stocks of the bank. An officer engaged in the regular business of a banking institution is prohibited from engaging simultaneously in the business of other profit-making corporations.

DEVELOPMENT INSTITUTIONS

The Korean war in 1950 destroyed most of the industrial base that was left in Korea after the Japanese occupation. A large amount of capital was required for the rehabilitation of the economy; the commercial banks could not mobilize enough capital because of the extremely low income and savings level of households. Under such circumstances, the Government needed specialized development institutions to channel government and foreign aid funds to various reconstruction and industrialization projects. Also the chronic shortage of capital which accompanied the rapid growth of the economy during the 1960s and 1970s necessitated other institutions for the supply of long-term capital for continued industrialization.

Korea Development Bank

The Korea Development Bank was established in 1954 to supply funds to major industrial projects for the rehabilitation of the economy. The Bank has supplied its funds through loans, equity investments, and repayment guarantees, mostly to heavy and chemical industries and social infrastructure projects. The share of loan guarantees showed a declining trend to account for 12 per cent of the total assets at the end of 1992, while the shares of loans and equity investments showed substantial increases to 67 per cent and 12 per cent, respectively.

During the 1960s, most Korea Development Bank loans went to social infrastructure projects in electricity, gas and water, and to heavy and chemical industries. The textiles industry and construction industry also received a substantial share of the Bank's loans during that period. The bias toward heavy and chemical industries and infrastructure projects continued throughout the 1970s, reaching over 60 per cent of Korea Development Bank's total loans. Entering the 1980s, the share of the combined industries of transport, storage, and communications increased rapidly to reach about 30 per cent in the mid-1980s before it started to decline from the late 1980s. The heavy and chemical industry loan share, which dropped to a mid-30 per cent level in the mid-1980s rose steadily again to reach 44 per cent in 1992. The loan share of electricity, gas, and water infrastructure projects continued to decline, from 35 per cent in 1975 to a little over 10 per cent in 1992.

As demonstrated by the Government's subscription of all the initial equity capital, the Government was the single most important source of funds for Korea Development Bank until the early 1970s. However from the mid-1970s, foreign loans and the National Investment Fund became more important sources of funds. At the end of 1985, the outstanding amount of loans from this Fund and foreign sources accounted for 35 per cent of total liabilities and shareholders' equity. While their share started to decline from the late 1980s, the issue of Industrial Development Bonds increased rapidly and is now the largest source of funds for the Korea Development Bank. At the end of 1992, the outstanding volume of the bond issue was 39 per cent of total liabilities and shareholders' equity.

When a borrowing corporation becomes insolvent, Korea Development Bank designates it as a bank-managed corporation and participates directly in its management by providing priority financial support, often converting its loans to equity shares, dispatching the Bank's managers, reviewing business strategies and implementation, readjusting production and profit targets, and conducting periodic management diagnosis. The Korea Development Bank also provides start-up funds through equity investment when an investment project is considered very important for the economy but requires too much capital to be carried out by private enterprises. Once the operation of the Bank-invested enterprises is normalized, its equity shares are sold off to the public: Korean Air Lines and Korea Electric Power Corporation are good examples. During 1969–1982, of the 22 corporations in which the Korea Development Bank had equity investment, 14 were sold to the public and six were released from the Bank's management control. During 1974–1982, 22 of the 28 corporations under bank management were normalized.

Entering the 1980s, Korea faced an increasing need for restructuring inefficient industries and upgrading the industrial structure of the economy into a more sophisticated and technology-based orientation. Korea Development Bank played a major role in the industry rationalization programs of the mid-1980s utilizing its close relationship with its client corporations. Currently, Korea Development Bank is concentrating its efforts on supporting industries with high growth potential and aiding technology development. In addition, Korea Development Bank has been increasing its business boundaries through subsidiaries in leasing, merchant banking, and venture capital.

Export-Import Bank of Korea

The main operations of the Export-Import Bank of Korea since its establishment in 1976 has been to supply export financing, especially for deferred payment export of heavy and chemical industry products such as ships, heavy machinery, and industrial plants. In addition, it has supplied funds to overseas investment projects and overseas natural resources development projects. Until recently, it had also operated export insurance business on behalf of the Government. Of the newly disbursed loans in 1992, 38 per cent went to shipbuilding industry and 6 per cent for the export of industrial plants.

After reaching its peak in 1985, the total asset size of the Bank declined sharply due to the decrease in credit to the shipbuilding industry and early repayment of foreign currency loans by borrowers. Though the asset size increased from 1990 on, it is still well below its 1985 level. The Bank's major sources of funds have been borrowings from the National Investment Fund, issuance of Export-Import Finance Debentures, and sales of promissory notes in international markets. Borrowings from the National Investment Fund peaked in 1983 while the issue of Export-Import Finance Debentures have become the most important source of funds in recent years, sharing 24 per cent of total liabilities and shareholders' equity in 1992.

Korea Long-term Credit Bank

The Korea Development Finance Corporation which was established in 1968 as the first private institution to supply plant and equipment financing, was reorganized to become the Korea Long-term Credit Bank in 1980. The Bank supports enterprises by way of making long-term facility loans of three years or longer and related long-term working capital loans, underwriting bonds and share issues, issuing repayment guarantees, and providing management and technical consulting. The Korea Long-term Credit Bank Act also requires it to allocate more than 30 per cent of its loans to small-scale and medium-scale firms. The major source of funds for the Long-term Credit Bank has been the issue of Long-term Credit Bonds and borrowings from domestic and foreign sources.

While the Korea Development Bank's operation has been closely linked to the Government's industrial policy and policy loans, the Korea Long-term Credit Bank has operated on the private level. The average size of loans from the latter is much smaller than from the former and the Korea Long-term Credit Bank makes few policy loans. The client relationship with the Credit Bank is formed on a project-by-project basis and is not as stable as in the case of the Development Bank. The success of the Credit Bank until now seems to have been largely attributable to its strict evaluation of enterprises and projects. As such, Korea Long-term Credit Bank is rarely concerned with supplying emergency funds to save corporate clients in financial difficulty.

OTHER FINANCIAL INSTITUTIONS

The 1980s witnessed a phenomenal proliferation of nonbank financial institutions/intermediaries as manifested in the steep climb of the M3/GNP ratio from 43 per cent in 1979 to 115 per cent in 1990. This drastic expansion can be attributed to lower entry barriers to these markets, the introduction of new financial instruments, and more attractive interest rates offered to depositors by frequent relaxation of interest rate controls. These conditions, largely conducive to nonbank financial institution growth, reflect the Government's intentions to absorb curb market funds by expanding mobilization of financial resources in organized markets.

Manifestations of the rapid expansion of nonbank financial institutions include investment and finance companies (corporate bills discounted and resold), securities investment trust companies (sales of beneficial certificates), money in trust offered by banks, life insurance companies, deposits in mutual savings and finance companies, and mutual credit of agricultural cooperatives.

The remainder of this section discusses various types of nonbank financial institutions in Korea, with particular attention to their roles, growth, funding, fund utilization, and major instruments.

Saving Institutions

Savings institutions include trust accounts of banking institutions, mutual savings and finance companies, credit unions, mutual credit facilities, community credit cooperatives, and postal savings. These institutions offer various loans of small sizes with funds raised through deposits (mainly time deposits).

Trust Business

At present, trust business in Korea is handled by all deposit money banks, 24 foreign bank branches, and all development institutions with the exception of the Export-Import Bank of Korea. This is a result of recent governmental efforts to diversify the business of banks by redistributing the trust business which it had formerly assigned only to the Korea Trust Bank.

Nonspecific money in trust—which accounts for most of the money in trust—follows a scheme very similar to time deposits in that the investment of entrusted funds is not specified by the trusters. For trust contracts whose terms of maturity are under two years, dividend rates are fixed by banks within a ceiling set by the Minister of Finance, but are scheduled to be deregulated by the end of 1996. More recently, additional trust schemes have been developed in which the dividend rates are determined by the performance of fund operations. Examples of such trust contracts include household money in trust, company money in trust, money in trust for old-age benefits, and national corporations' stock trust.

Between 1985 and 1992, the distribution of trust fund sources shifted significantly: the share of money in trust grew substantially to 56 per cent of the outstanding balance of all trust accounts of banks. Meanwhile, the relative weight of securities investment trust decreased to 37 per cent. Asset distribution, on the other hand, remained relatively constant during the same period. Securities holdings accounted for about 65 per cent of the total outstanding balance of bank trust accounts, while loans amounted to 25 per cent.

Mutual Savings and Finance Companies

Mutual savings and finance companies were introduced in 1972 to specialize in financial services to small businesses and households. This constituted part of the Government's strategy to absorb small savers' funds from the informal credit market into the organized financial market.

Although there were 400 of these companies in the beginning, this number dwindled to 199 by the early 1980s due to mergers, government-guided consolidations, and insolvencies. In the early 1980s, the Government lowered entry barriers, allowing some mutual savings and finance companies to open branch offices: at the end of 1992, the number of companies stood at 237 with a total of 97 branch offices.

The main business of mutual savings and finance companies comprises mutual installment savings, entitlement to borrow funds, mutual time deposits, small unsecured loans and discount of bills for mutual installment savers. In 1988, a time deposit scheme similar to that of banks was introduced, and subsequently showed a rapid increase.

Interest rates for the different types of saving schemes are fixed by the companies within a ceiling set by the Minister of Finance after consultation with the Monetary Board. These yields are somewhat higher than bank rates on similar deposits. As in the case of trust accounts of banks, interest rates on time deposits with maturities of one year or more and installments with maturities of two years or more will be deregulated by the end of 1993.

Credit Unions, Mutual Credit Facilities, and Community Credit Cooperatives

Many cooperative associations have been organized within private groups since 1960 to promote mutual economic benefits for their members. Credit unions, organized since 1960, came under Government supervision in 1972. At the end of 1992, there were 1,461 unions with 2.8 million members. Community credit cooperatives, which are organized among people in the same neighborhood, are similar to credit unions in their operations but are under the supervision of the Ministry of Internal Affairs. There were 3,200 of these cooperatives with a membership of 7.6 million people at the end of 1992. Agricultural cooperatives, fisheries cooperatives, and livestock cooperatives operate mutual credit facilities to promote savings and financial services within their respective communities. Currently, there are 1,704 mutual credit facilities in the nation, of which 1,441 were affiliated with agricultural cooperatives, 80 with fisheries cooperatives and 183 with livestock cooperatives.

A feature that distinguishes credit institutions from banking institutions is that only members are eligible for transaction with their respective credit institutions. Interest rates on the various deposits have been fixed within ceilings set by the Minister of Finance, and are somewhat higher than bank interest rates. The assets and liabilities of credit unions, community credit cooperatives, and mutual credit facilities grew more than fivefold between 1985–1992. Almost all the funds are derived from the deposits of their members with very little borrowing, and are used as loans.

Postal Savings

The postal savings system was abolished in 1977 and reintroduced in 1983. It is administered by the Ministry of Communications and offered by all post offices throughout the country. Financial services of post offices include different types of deposits, sale of government and public debentures under repurchase agreements (RPs), postal giro, domestic money orders, and issuance of post office checks.

Interest rates on deposits and investments are determined by the Ministry of Communications after consultation with the Ministry of Finance, and are slightly higher than bank deposit rates. At the end of 1992, post offices had deposit liabilities of 2,786 billion won and sales of bonds under repurchase agreements amounting to 1,456 billion won. Since the current regulation prohibits the postal savings system from making loans, most of its funds are invested in the deposits at financial institutions and government and public bonds.

Investment and Finance Companies

Investment and finance companies were introduced in 1972 with the aim of attracting funds from the curb market and developing the money market. The companies engage principally in short-term business financing, with funds raised by selling papers which they themselves issue or by dealing in papers issued by other firms. Additionally, with the approval of the Minister of Finance, these companies may engage in such securities business as underwriting and selling securities and acting as brokers or agents. Leading investment and finance companies serve as primary dealers for monetary stabilization bonds and treasury bills.

In 1981, these companies were allowed to deal in commercial papers with maturities of three to six months. In 1984, they also began to deal in guaranteed commercial papers, which are issued by small- and medium-scale firms with payment guarantees of banks and other financial intermediaries. Later, the different forms of commercial papers were merged into a standard form. The companies further broadened their business scope in 1984 by engaging in international factoring, which involves purchasing and collecting foreign currency denominated claims from exporters who undertake international transactions on credit. They also began to offer cash management accounts whose yields are dependent on the performance of the funds which are usually invested in corporate papers and monetary stabilization bonds.

Investment and finance companies began to offer brokerage service of commercial papers from 1991, besides their existing business of discounting and selling them. They intermediate commercial papers directly between the issuer and individual investors without holding the papers themselves. The outstanding amount of these intermediated papers grew from 1,238 billion won at the end of 1991 to 3,015 billion won a year later. In 1991, the Government designated eight investment and finance companies located in Seoul as the intermediators in the call market. These eight companies operate both as dealers and brokers in the call market where about 600 financial institutions participate.

The number of investment and finance companies dropped from 32 in 1985 to 24 in 1992 as three were turned into two commercial banks (two merged to become a bank) and another five became securities companies. At the end of 1992, 53 per cent of the total assets of these companies were in the form of loans and discounts of papers. However, papers resold to investors (which do not appear in the balance sheet) during 1992 slightly exceeded total loans and discounts. On the liability side, deposits received through the cash management accounts and short-term borrowings, such as call money, constituted large shares.

Merchant Banking Corporations

Merchant banking corporations, first established in 1976 to provide a wide range of financial services to business firms, were all founded as joint ventures with foreign financial institutions. They concentrated on short-term financing business and inducement of foreign capital in the earlier years, but their activities also include securities investment trust, underwriting, and brokerage of securities sales, and consultancy services including business management counseling.

The main sources of funds for merchant banking corporations include borrowings of foreign capital, sales of beneficiary certificates for securities investment trusts, and the issuance of debentures. The shares of borrowings and debentures issued in total liabilities and net worth amounted to 33 per cent and 21 per cent, respectively, in 1992, while the share of deposits received was 23 per cent. As a consequence of steady diversification efforts in their business since the early 1980s, short-term financing, which had formerly been the major operation, has dwindled in their business share.

Insurance Companies

The insurance industry is presently divided into life and non-life insurance segments; companies are not allowed to undertake business in both segments concurrently. Before 1980, the life insurance segment consisted of six companies and the post office. With a significant reduction in the entry barriers to the insurance market in the late 1980s, Korea now has diverse forms of insurance institutions: 20 local companies, three foreign company branches, six joint ventures, two wholly owned subsidiaries of foreign companies, the National Life Insurance of the National Agricultural Cooperative Federation, and the Postal Life Insurance.

The Korea Insurance Development Institute, which was established with members from both life and non-life insurance companies for the purpose of safeguarding clients' interests and promoting restructuring and rationalizing the insurance industry, is assigned to test the premium rates for insurance companies before the authorization by the Ministry of Finance.

The life insurance industry has recently experienced phenomenal growth: its assets jumped almost sevenfold from 1985 to 1992. Most of the outstanding insurance contracts are held by the six long-established companies. Loans constitute almost half of life insurance companies' assets, although securities have been occupying a steadily increasing share, reaching 26 per cent in 1992.

The non-life insurance industry consists of 13 primary insurance companies (two foreign), two fidelity and surety insurance companies and one reinsurance company. It has also exhibited substantial expansion, with total assets growing more than fivefold, from 1,213 billion won in 1985 to 6,595 billion won in 1992. At the end of 1992, securities holdings accounted for the largest portion of these assets, amounting to 37 per cent of the total, while loans occupied 20 per cent.

National Investment Fund and the National Housing Fund

Instituted in 1974, the National Investment Fund supports investment in heavy and chemical industries, projects aimed at food production and deferred payment exports. Fund sources include compulsory deposit from banking institutions, insurance companies, and various public funds together with transfers from various government budgetary accounts. However, the Fund began repaying existing deposits and other borrowings in 1989, and has been relying mostly on the collection of its previous loans as its current fund source. As a result, loans from the National Investment Fund in 1991 fell below half of the loans in 1985. The Fund's operation is under the direction of the Minister of Finance, but actual management is entrusted to the Governor of the Bank of Korea (Table 1.14).

The National Housing Fund finances national housing projects such as construction of small houses (national housing units) and development of housing sites. Major funding sources include National Housing Preemption Subscription Deposits held by prospective purchasers of national housing units, National Housing Bonds and borrowings from the Government. Being under the direction of the Minister of Construction, its management is entrusted to the Korea Housing Bank.

Leasing Companies

Leasing companies were introduced in 1973, and there are now 20 companies including 15 local companies, of which six are joint ventures with foreign investors. The local leasing companies are under an obligation to provide at least 50 per cent of their leasing services to regional enterprises outside of Seoul. Besides leasing companies, merchant banking corporations and venture capital companies are allowed to undertake leasing business.

Besides lease financing, leasing companies provide payment guarantees for international leasing, and may engage in conditional sales. In 1992, 34 per cent of the total funds came from leasing fees, while the rest was mostly mobilized by borrowing from domestic and foreign financial institutions and selling debentures. Half of the funds were used to purchase equipment and almost 40 per cent were used for debt servicing (Table 1.15).

Table 1.14 Sources and Uses of National Investment Fund
(billion won, per cent)

	1980 Amount	1980 Per Cent	1985 Amount	1985 Per Cent	1991 Amount	1991 Per Cent
Sources						
National savings association	42	8.1	18	3.8	0	0.0
Public funds	91	17.6	(88)	(18.5)	0	0.0
Banking institutions	215	41.4	84	17.6	0	0.0
Insurance companies	24	4.6	(10)	(2.1)	0	0.0
Collection of loans	131	25.3	454	95.7	569	99.2
Carried-over from previous year	16	3.0	17	3.5	5	0.8
Total	519	100.0	475	100.0	574	100.0
Uses						
Heavy and chemical industry	267	51.4	319	67.4	187	32.6
Shipbuilding	–	–	–	–	19	3.3
Technology development	–	–	–	–	44	7.7
Purchase of domestic machinery	–	–	–	–	124	21.6
Electric industry	120	23.1	40	8.4	0	0.0
Projects to increase food production	22	4.2	25	5.3	20	3.5
Deferred payment exports	30	5.8	80	16.9	9	1.5
Repayment of debentures and deposits	0	0.0	0	0.0	328	57.2
Others	0	0.0	0	0.0	28	4.9
Carried-over to following year	81	15.6	10	2.1	2	0.3
Total	519	100.0	474	100.0	574	100.0

Source: Korea Leasing Association, *Lease Sanup Jungbo* (Lease/Industry Information), (various issues).

Venture Capital Companies

The venture capital industry of Korea can be divided into two distinct categories: one that provides financial support for technology advancement, and one that supports the launching of small- and medium-scale enterprises.

Table 1.15 Sources and Uses of Funds in the Leasing Industry
(billion won, per cent)

	1985		1992	
	Amount	Per Cent	Amount	Per Cent
Sources				
Income from leading fee	757	30.8	4,653	36.4
Contract deposits	17	0.7	76	0.6
Borrowings	1,560	61.9	7,759	60.7
In foreign currency	867	35.3	2,086	16.3
In local currency	693	28.2	5,673	44.3
Carried over				
from previous year	120	4.9	304	2.4
Total	2,454	100.0	12,792	94.6
Uses				
Purchase of equipment	1,628	62.5	6,808	50.4
Repayment of borrowings	589	22.6	4,163	30.8
In foreign currency	356	13.6	1,158	8.6
In local currency	233	8.9	3,005	22.2
Interest payment	167	6.4	1,151	8.5
Overhead expenses	21	0.8	90	0.7
Others	95	3.6	813	6.0
Carried over				
to following year	107	4.1	483	3.6
Total	2,606	100.0	13,507	100.0

Notes: Leasing contracts in 1987: 1,986 billion won;
in 1992: 9,385 billion won.
Leasing executions in 1987: 1,547 billion won;
in 1992: 6,696 billion won.
Source: Korea Leasing Association, *Lease Sanup Jungbo* (Lease/Industry Information), (various issues).

As for the first category, there are four "new technology enterprise financial support corporations" promoting technology advancement by financing research and development projects for the development of new products and the improvement of product quality, import of new technology, and the manufacturing and marketing of new products. Financial support is given in the form of equity investments, loans, leasing and factoring.

"Small and medium enterprise start-up support companies," on the other hand, provide support for the establishment and survival of small- and medium-scale enterprises by underwriting securities and offering management consultancy. These companies may also convert to new technology enterprise financial support corporations by satisfying specified conditions, though there have not been any such companies yet. At the end of 1992, there were 54 small and medium enterprise start-up support companies, with an aggregate outstanding investment of 710 billion won.

Credit Guarantee Funds

There are three different credit guarantee funds. The Korea Credit Guarantee Fund, established in 1976, provides credit guarantees, credit information and management, and technical assistance, and invests in small- and medium-scale firms. Contributions from the Government and banking institutions comprise its major fund sources. The Korea Technology Credit Guarantee Fund, founded in 1987, is similar in its operation to the Korea Credit Guarantee Fund, except that it mainly concentrates on guarantees for technology developments. Finally, the Housing Finance Credit Guarantee Fund extends guarantees for loans for the purpose of purchasing, constructing, and renting dwelling houses.

Regulation of the Safety and Soundness of Nonbank Financial Institutions

For the safety and soundness of operation, each type of financial institution is constrained in its assumption of liabilities and use of assets by various laws and regulations.

The Korea Nonbank Deposit Insurance Corporation plays a significant role in this respect. This deposit insurance scheme was introduced in 1983 to protect depositors and to promote sound operation of nonbank financial institutions. It covers investment and finance companies, merchant banking corporations, and mutual savings and finance companies. Its funds are raised through contributions from member institutions based on their insured deposits, and depositors are compensated for losses resulting from insolvency of member institutions. Other services of the corporation include taking deposits from, and extending loans and credit guarantees to member institutions. The following are some of the most important regulations designed to ensure safety and soundness of nonbank financial institutions.

Restrictions on Liabilities: Capital Adequacy

— Investment and finance companies must keep their total indebtedness below fifteen times their own capital, i.e., paid-in capital plus reserves. The amount of own papers issued and short-term borrowings are limited to 100 per cent and 200 per cent of own capital, respectively.

— Total indebtedness of merchant banking corporations and leasing companies cannot exceed 20 times their own capital. Merchant banking corporations cannot issue debentures more than five times their own capital. However, leasing companies are allowed to issue debentures up to 10 times their own capital.

— Mutual savings and finance companies can have total indebtedness of 20 times their own capital and borrow as much as three times their own capital, provided that the borrowing is from financial institutions.

Requirement for Payment Reserves

— Investment and finance companies and merchant banking corporations have to maintain at least 5 per cent of the combined amount of own-paper issues, deposits in cash management accounts, and sales of commercial papers with recourse as payment reserves in the form of cash, savings deposits, certificates of deposit, checking accounts, deposits at the Credit Guarantee Fund, or government, public and financial debentures.

— A newly established insurance company should deposit 30–50 per cent of its paid-in capital at the Insurance Supervisory Board for the protection of its clients. This deposit is returned when the company has accumulated enough reserves and begun to show profits.

— Mutual savings and finance companies must maintain 10 per cent of their installment receipts plus 5 per cent of the combined amount of long-term savings for workers, pass book deposits, and time deposits in excess of their own capital as payment reserves in the form of cash, deposits in financial institutions, and securities such as government bonds and monetary stabilization bonds.

— Community credit cooperatives must maintain at least 10 per cent of their deposits as payment reserves, of which 50 per cent should be deposited at the national federation and the rest at other financial institutions. The required reserve ratio is the same for credit unions, but they can deposit all of the reserves in other financial institutions or invest in the monetary stabilization bonds.

Other Restrictions on the Uses of Assets

— Investment and finance companies and merchant banking corporations cannot make loans, equity investments, or guarantees to a single person or a company in excess of 25 per cent of their own capital. In addition, they cannot hold securities in excess of 35 per cent (30 per cent for merchant banking corporations) of their own capital. They also cannot hold more than 10 per cent of the outstanding shares of a single company. Investment and finance companies and merchant banking corporations are prohibited from extending credit to business areas such as real estate, and some services related to conspicuous consumption including bars, luxurious saunas, and hair salons.

— Insurance companies can purchase stocks only up to 40 per cent of their total assets. The limitation on loans to a single person is set at 3 per cent of total assets. The acquisition of stocks or bonds of a single company or the provision of loans against the collateral of stocks or bonds of a single company cannot exceed 5 per cent of total assets.

— For mutual savings and finance companies, loans to a single person or a company is restricted to 500 million won for companies and 30 million won for others. They can discount bills up to 50 per cent of their total loans or 200 per cent of own capital and purchase stocks as much as 10 per cent of own capital.

— Loans to a single member by community credit cooperatives and credit unions are not allowed to exceed 10 per cent of their capital and surpluses. In addition, total loans extended by credit unions cannot exceed 10 times their capital and surpluses.

— Nonbank financial institutions are prohibited from purchasing real estate except for that directly related to normal business activities. Investment and finance companies cannot purchase business-related real estate in excess of 40 per cent of their own capital. However, the limit is 50 per cent for leasing companies, and 100 per cent for mutual savings and finance companies and community credit cooperatives.

Informal Credit Market

The informal credit market has been an important source of short-term financing for Korean firms and individuals. Unlike organized financial institutions, professional money lenders and relatives and friends usually do not insist upon collateral. The professional money lenders seem to be efficient in evaluating the creditworthiness of borrowers. Interest rates are sensitive to the risk differentials among borrowers, changing demand and supply conditions in the money market, and the rate of inflation. However, this market is very inefficient in other respects: it is an extremely fragmented market and information is much more costly than in the formal credit market. Money lenders do business without license and usually do not pay income tax. Because of the illegal nature of the informal credit market, market information is costly, and risk premium is rather high.

There were several attempts, mainly in the 1970s, to estimate the size of the curb market based on corporate tax return data and household budget and other surveys. The curb market may be defined as the one where business firms borrow, usually by having their promissory notes discounted, from individuals often through professional money lenders. A rather reliable estimate was obtained by the Presidential Emergency Decree in August 1972, which required all curb loans outstanding to be reported to the National Tax Office. The reported volume of credit was equivalent to over a quarter of total domestic credit to the private sector. One particular feature disclosed by the Decree was that almost 30 per cent of the total curb loans were so-called disguised curb loans: those made by major equity holders or executives of the borrowing firms. This phenomenon arose from the repressed nature of the financial regime as well as inadequate information in the curb market.

These disguised money lenders not only had superior information about the firm but also had influence over corporate (financial) decisions: they could have their loans repaid ahead of others when the borrowing firm was in danger of going bankrupt. Despite this privilege of virtually subordinating other loans, the insider fund suppliers, disguising their identity, charged fairly high (market) interest rates. The corporate insiders capitalized on their positions for their own benefit, but there was little evidence that the firms could save the cost of borrowing by relying on insiders' funds.

The relative importance of the curb market, however, is known to have declined thereafter, as new nonbank financial institutions were established which offered interest rates more in line with the curb market rates. Investment and finance companies now accommodate the short-run working capital needs of business firms in a rather speedy and flexible manner. With growing markets for financial assets whose return is "market" determined, the supply of funds in the curb market appears to be dwindling.

Some money lenders have been observed to lend money through financial institutions by making deposits on the condition that the money be lent to a borrower designated by the depositor. The interest rate differential between the organized and unorganized financial markets is directly settled between the lender and the borrower. Unlike in a pure curb transaction, which is not reported to the tax office, the borrower gets a tax deduction on the interest payment to the financial institution.

Nonbank Financial Institution Ownership

The ownership of nonbank financial intermediaries by large business groups in Korea is extensive. For example, as of 1990, 37 per cent of the outstanding shares of all life insurance companies was owned by the five largest business groups. This applies to other types of nonbank financial institutions as well: the 10 largest business groups hold 23 per cent of the outstanding shares of merchant banking corporations, while the 30 largest groups own 30 per cent of all investment and finance companies, 45 per cent of non-life insurance companies, and 63 per cent of securities companies. This ownership pattern is possible because there are no restrictions on ownership concentration for nonbank financial institutions. Naturally, this affiliate structure can be beneficial as well as harmful (Table 1.16).

Table 1.16 Business Group Ownership of NBFIs, 1990
(per cent)

	5 largest business groups	10 largest	30 largest
Investment and Finance Companies	7.2	10.1	29.9
Merchant Banking corporations	12.8	23.3	23.3
Life Insurance Companies	36.5	36.5	38.4
Non-life Insurance Companies	28.0	41.4	44.5
Securities Companies	26.3	36.5	63.1
Mutual Savings and Finance Companies	1.2	1.6	4.7

Source: Jung and Yang (1992).

The benefits of this arrangement come from increasing economies of scope: mixing financial services with nonfinancial services renders input factors more productive than they would be in either sector alone. In this particular mix, increasing economies of scope can result in very high benefits, since the information-gathering process required in financial services can be useful for many other services, such as product marketing.

However, the business ownership of financial intermediaries may also have detrimental effects, especially if entry into the financial sector is limited. Large business ownership of nonbank financial institutions is likely to result in undue concentration of economic power and a perversion of the political process. If entry is restricted, then financial affiliates are likely to amass great profits, which they can subsequently use to further restrict entry and protect their interests. Moreover, the mix of financial and nonfinancial products increases opportunities for exploiting conflicts of interest. Profitable exploitation of conflicting interests is possible when a business group enjoys market power in at least one of the two (financial and nonfinancial product) markets.[4] Many of the nonfinancial markets have an oligopolistic structure, where large business groups command market dominating power. They may try to capitalize on their market power by tying in financial services with the sale of nonfinancial products. In an imperfect financial market, they may try to discriminate against competing nonfinancial firms by limiting or charging higher prices for the supply of credit.

In this connection, it is often argued to be irrational that restrictions on ownership concentration are imposed on banks but not on nonbank financial institutions. To the extent that bank ownership restriction is due to concern about concentration of economic power, similar restrictions may also be extended to nonbank financial institutions, since banks no longer dominate the Korean financial market. Further, with interest rate deregulation and progress toward a universal banking system, the uniqueness of commercial banks will be weakened and banks will no longer be particularly susceptible to temptation to abuse conflicts of interest.

MONEY MARKET

The money market is typically characterized by transactions of marketable short-term instruments at liberalized interest rates by a wide range of market participants. With the introduction of money market instruments such as monetary stabilization bonds and treasury bills in the 1960, as well as government efforts to promote the money market, the market is now fairly diverse and includes markets for call money, commercial paper, bonds on repurchase agreements, negotiable certificates of deposit (CDs), and others.

The money market has grown rapidly in relation to nominal GNP, rising from 15 per cent in 1985 to 30 per cent in 1992. Commercial papers, even with a substantial decline in importance in the money market, still

---

[4] However, a business group providing many products including financial services may be more careful about, and refrain from, abusing conflicts of interest, because such action is likely to damage the reputation and business of the entire group.

accounted for the largest share (35 per cent) of the total. The share of the RP market also declined sharply to 7.4 per cent in 1992. The submarkets that grew most rapidly between 1985 and 1992 include those for monetary stabilization bonds and negotiable CDs, which accounted for 30 per cent and 18 per cent respectively of the total money market in 1992 (Table 1.17).

Table 1.17 Money Market Trends
(billion won, per cent)

	1985		1992	
	Amount	Per Cent	Amount	Per Cent
Call market	432.7	3.6	3,330.4	4.9
RPs	2,562.7	21.3	5,017.8	7.4
Negotiable CDs	1,080.9	9.0	11,943.0	17.6
Monetary Stabilization Bonds	504.1	4.2	20,264.1	29.8
Treasury bills	0.0	0.0	1,579.7	2.3
Commercial paper	7,395.8	61.5	24,072.3	35.4
Commercial bills	46.4	0.4	–	0.0
Trade bills	0.0	0.0	3,285.3	4.8
Total	12,022.6	100.0	69,492.6	102.2
Total/M1 (per cent)	–	159.1	–	276.2
Total/M3 (per cent)	–	22.0	–	23.0
Total/Nominal GNP (per cent)	–	15.4	–	29.5

– not applicable.
Source: Bank of Korea (1993).

Call Market

The call market was established in 1975 to facilitate the adjustment of financial institutions with shortages or surpluses of funds for very short periods. Until 1989, the call market was divided into two submarkets: the exchange market, mainly for banks, and the over-the-counter market, mainly for nonbank financial institutions. Now the market is fully integrated with the participation of all financial institutions including foreign bank branches, the fund management department of the Bank of Korea, the Korea Securities Finance Corporation, and credit guarantee funds.

Eight investment and finance companies are now nominated as brokers and dealers of call transactions, even though call transactions may be conducted either through brokers or directly between participating institutions. Maturities for call transactions are standardized into seven different periods, ranging from overnight to 15 days.

Commercial Paper Market

In 1972, the Government took the first serious measure to promote a modern money market. With the intention of channeling curb market funds into the organized financial market, investment and finance companies were introduced; they began to deal in commercial paper. Different types of papers are traded in this market: (i) "own paper" issued by investment and finance companies and merchant banking corporations, which are in the process of a gradual phase-out; and (ii) resold notes issued by business firms with or without recourse (payment guarantees) by the dealing companies.

Business firms issue commercial paper usually as a substitute for short-term bank loans. Maturity ranges from one to 180 days. Investment and finance companies and merchant banking corporations serve as dealers in commercial paper. Although there is no active secondary market in commercial paper, it can be redeemed by the dealers before maturity at specified loss of interest. Credit ratings for prospective issuers of commercial paper are important, since commercial paper is mostly unsecured. The credit evaluation is undertaken by individual investment and finance companies and three credit rating companies: the Korea Investors' Service Inc., the National Information & Credit Evaluation Inc., and the Korea Management Credit Rating Company.

Bond Market on Repurchase Agreements

A repurchase agreement (RP) involves sales of bonds with a simultaneous commitment to purchase the same bonds on a specified date within one year at a specified price. It was 1977 that RPs were first introduced, as the Korea Securities Finance Corporation started such transactions with securities companies. Securities companies, banking institutions, and post offices were allowed to engage in this business one after another in the early 1980s. While these institutions are the main borrowers in the RP market (RP sellers), the main lenders (RP purchasers) are individual investors and nonprofit corporations.

Government, public, and financial debentures as well as corporate bonds serve as instrument securities. The maturities and minimum denominations differ by institution. More recently, financial institutions tend to use RPs as a means of adjusting their short-term liquidity. The Bank of Korea seeks to control the short-term liquidity of banking institutions through sales of government and public bonds under repurchase agreements in the open market.

Market for Negotiable Certificates of Deposit

A negotiable certificate of deposit (CD) is evidence that a certain amount of money has been deposited for a fixed period of time and will be redeemed with interest at maturity. After being discontinued in 1982, CDs were reintroduced in 1984 with a view to promoting banks' competitiveness against nonbank financial institutions in attracting short-term deposits, by offering higher interest rates than those applied to ordinary time deposits.

The maturity periods range from 30 days to 180 days and the minimum denomination of CDs was lowered from 100 million won to 50 million won in 1987. Though not very active, CDs are traded in the secondary market by such financial institutions as investment and finance companies, merchant banking corporations, and securities companies.

Short-Term Government Bond Market

This market is where government securities with maturities of less than one year are issued and traded. Such securities include (i) treasury bills issued by the Government to cover temporary deficits in the budget or to control money supply; (ii) Foreign Exchange Equalization Fund Bonds issued to stabilize the nation's foreign exchange market as well as to control liquidity supply; and (iii) Grain Management Fund Bonds issued to stabilize grain prices.

Treasury bills, first introduced in 1967, were not issued during 1969–1976 and 1983–1985 because of improvements in the nation's fiscal balance. Since 1986, however, the issuance of the bills was resumed as a means of easing rapid expansion of the money supply resulting from large surpluses in the balance of payments and continued strong demand for domestic credit. It was only in 1987 that the Foreign Exchange Equalization Fund Bonds began to be actively issued. Although the common maturity for Equalization Fund Bonds used to be one year, the maximum maturity was extended from three years to five years with a view to alleviating the pressure of reissuance and burden of redemption after maturity. The maturities of the Grain Management Fund Bonds range between three months and five years.

Monetary Stabilization Bond Market

The issuance of monetary stabilization bonds, which are liabilities of the Bank of Korea, increased drastically since 1986 for the control of monetary growth. They serve as the most important instruments for the open market operation of the Bank of Korea. Monetary stabilization bonds have sometimes been issued by public offerings to individual investors and financial institutions. However, without deregulation of interest rates on such bonds, they have usually been assigned to (underwritten by) financial institutions including banks, securities companies, insurance companies, investment and finance companies, and merchant banking corporations.

The maturities for monetary stabilization bonds are not more than two years, with the most common maturities being six months and one year. Many securities companies, investment and finance companies and banks were given approval to act as primary dealers in such bonds in an effort to promote the secondary market.

Markets for Commercial and Trade Bills

Since 1982, banks have been permitted to sell commercial bills discounted by themselves to individual and corporate investors. The market, however, has been inactive, due to unattractive interest rates, nonstandardized maturities and amount of bills, and unavailability of rediscount at the Bank of Korea except for bills issued and received by small- and medium-scale firms.

Trade bills were introduced in 1989 to help export firms raise funds. Banks, investment and finance companies, and merchant banking corporations engage in the acceptance, discount, and sale of trade bills. Eligible trade bills must be the bills of exchange issued against export letters of credit or export contracts on D/P and D/A, and their maturities should not exceed 180 days.

CAPITAL MARKET

Development of the Korean Capital Market

The capital market was inactive in Korea during most of the 1960s, because there were few good enterprises in which the public could invest. Enterprises which were mostly owned and run by founding families were dependent on bank loans as their funding source and were not interested in going public for fear of losing control of their businesses.

The Government's efforts to promote the capital market started with the enactment of the Act on Fostering the Capital Market in 1968 and the Public Corporation Inducement Act in 1972, which empowered the Ministry of Finance to force a designated corporation to go public. Tax incentives were given to encourage corporations to go public, and securities investment trusts were introduced to boost demand for securities. Owing to such promotional measures, the capital market grew rapidly during the early 1970s.

This growth prompted the need for a proper regulatory body to ensure orderly and fair transactions and, thereby, protect investors. In

1977, the Securities and Exchange Commission and its executive body, the Securities Supervisory Board, were set up to oversee the securities market and institutions.

Various government efforts were also made to improve the efficiency and soundness of the market. In 1983, a new system was introduced requiring stocks to be issued at market prices rather than at face prices. An over-the-counter market was established in 1987 to provide small companies and venture businesses access to the capital market. Another important measure contributing to enlarging the capital market was the partial sales of the shares of two large public enterprises, Pohang Iron and Steel Company and the Korea Electric Power Company, to the public in the late 1980s under the Government's National Citizen's Share Program.

Progress was also made in the internationalization of the capital market during the 1980s. Indirect investment vehicles, such as beneficiary certificates, were offered to foreigners by securities investment trust companies. Corporate type funds were set up in Europe and the US for investment in Korean stocks. Samsung Electronics issued convertible bonds in the Euro market in 1985. By the end of 1992, there were 43 cases of overseas issues of convertible bonds, bonds with warrants, and depository receipts raising a total of 2.2 billion dollars. The most recent measure taken for the internationalization of the Korean capital market was the opening of the stock market to foreign investors in 1992, which allowed foreigners to buy up to 10 per cent of the shares of individual companies traded on the Korea Stock Exchange.

Stock Market

The Korean stock market grew tremendously in the latter half of the 1980s, thanks to the booming economy fueled by a favorable external environment. However, with the slowdown of the economy as well as the flood of stock issues, the market cooled down rapidly since the spring of 1989. The stock price index, which reached 1007 in April 1989, fell as low as 459 in August 1992, before it recovered to around 700 in September 1993.

At the end of 1992, stocks of 693 companies worth 84.7 trillion won of market value were being traded in the Korea Stock Exchange. Major requirements for listing in the Korea Stock Exchange are as follows: (i) The company must be five years or older; (ii) Its paid-in capital and total equity capital must be at least 3 billion won and 5 billion won, respectively; (iii) Its average sales for the previous three years must be at least 15 billion won (20 billion won in the most recent year); (iv) Return on paid-in capital must be at least 1.5 times the interest rate on one-year time deposits for the most recent year, and at least the same for the preceding two years. In 1989, when the stock market was at its peak, new listings worth 4.9 trillion won were made by 124 newly listed companies, which contrasts with only 114 billion won raised by four newly listed companies in 1992. Total corporate financing in the stock market, raised through either public offering or offering to shareholders, reached as much as 14.7 trillion won in 1989, before it dropped to only 1.8 trillion won in 1992.

Trading in the Exchange also showed tremendous growth during the same period. The value of traded stocks grew almost 25 times between 1985 and 1992 to reach 90.6 trillion won. During the same period, the number of shareholders grew from 772,000 people to 1,332,000. Trading usually takes place on a three-day settlement basis. The number of registered companies in the over-the-counter market has also increased steadily, reaching 126 companies in 1992. Trading volume in the market slightly exceeded 100 billion won in 1992, representing a little over 0.1 per cent of the trading volume at the Exchange.

Bond Market

Bonds are broadly classified into three types in Korea: government, public, and corporate bonds. Public bonds are issued by local governments, public enterprises, and financial institutions including the central bank and specialized banks. According to the Commercial Code, corporations can issue bonds up to twice the size of their paid-in capital and reserves. Securities companies usually serve as the principal underwriter of most bond issues, even though banks, investment and finance companies, securities investment trust companies and merchant banking corporations can also underwrite bond issues. Currently, most corporate bond issues are underwritten on a firm commitment basis. Most government and public bonds have maturities of less than five years. Corporate bonds usually have three-year maturities, and are issued with or without repayment guarantees of financial institutions. Coupon rates and other terms of corporate bond issues are more or less deregulated, while those of government and public bonds are still regulated by the Ministry of Finance.

The bond market showed substantial growth during the 1980s, with the value of newly issued bonds growing thirteenfold between 1980 and 1992. In particular, the issue of government and public bonds has grown rapidly since the mid-1980s, due to the massive sales of monetary stabilization bonds by the central bank in an effort to absorb excess liquidity arising from the current account surplus. The monetary stabilization bonds' share in newly issued government and public bonds rose from 39 per cent in 1985 to 56 per cent in 1990. The issue of corporate bonds picked up momentum entering the 1990s, as corporations had no other choices in the face of tight bank credit and a sluggish stock market. The value of outstanding corporate bonds grew almost fivefold between 1985 to 1992 to reach 35.4 trillion won.

The value of total outstanding bonds reached 95.2 trillion won in 1992, representing 41 per cent of GNP and 12 per cent more than the market value of listed stocks.

Trading in bonds has grown even faster than the issue market. The value of traded bonds grew almost 20 times between 1980 and 1992. Contrary to the stock market, bond trading took place almost entirely in the over-the-counter market, partly due to unrealistic commissions for brokerage firms in the Exchange.

Major Capital Market Institutions

Securities Companies

At the end of 1992, there were 32 securities companies: most were controlled by large business groups and banks. Since 1991, foreign securities companies have been allowed to operate in the Korean market as branch offices or as joint ventures with Korean companies. Currently, there is one joint venture securities company. Securities companies may undertake all or part of major securities businesses, such as dealing, brokerage, underwriting, and making arrangements for public offering. Securities companies with paid-in capital over a certain size are also allowed to provide payment guarantees for corporate bond issues, intermediate negotiable CDs and undertake overseas underwriting.

The total assets of securities companies grew fourteenfold between 1985 and 1992, reflecting rapid growth of the capital market since the mid-1980s and the addition of seven new companies during 1991–1992. The average sizes of paid-in capital and assets of the securities companies at the end of 1992 were 109 billion won and 700 billion won, respectively. The main sources of income for securities companies have been the commissions and gains on security trading. Commissions including brokerage commissions and underwriting fees have generated about half of the operating revenue. During the stock market boom in the latter half of the 1980s, income from brokerage commissions exceeded underwriting fees, reflecting the boom in the stock market, while the opposite was the case during 1991 and 1992.

Securities Investment Trust Companies

Securities investment trust companies were first introduced in 1970. With the addition of five companies in 1990, there are now eight investment trust companies. In addition, all six merchant banking corporations are allowed to engage in this business. Securities investment trust companies raise funds from the public through the issuance of beneficiary certificates, put these funds in the custody of banks, invest them mainly in stocks and bonds and distribute the resulting returns to the beneficiaries. A most noticeable change in the balance sheet of securities investment trust companies in recent years has been a sharp increase in the shares of stocks and borrowings. This was the result of a government policy to boost the sagging stock market in December 1989 by letting securities investment trust companies buy up stocks in the stock market with a fairly large amount of funds borrowed from the banks. The heavy burden of interest payment since then has dragged these companies' net worth to below zero.

Korea Securities Finance Corporation

The Korea Securities Finance Corporation was established in 1955 to provide securities financing, but became active only in 1971 with the introduction of margin transactions. It makes loans to underwriters, lends money or securities to securities companies and individual investors, undertakes bond trading under repurchase agreements, and provides a securities custody service. The Corporation's fund sources include deposits from securities companies, issuance of its own paper, and borrowings from other financial institutions. The Korea Securities Finance Corporation grew rapidly recently due mainly to government measures in December 1989 and May 1990 to revive the stagnant stock market which induced large increases in deposits and issuance of its own bills.

Securities and Exchange Commission and Securities Supervisory Board

The Securities and Exchange Commission was formed in 1977 to deliberate and decide on matters related to the issuance and trading of securities and supervision of securities institutions. The Commission is composed of nine members: the Central Bank Governor, the Chairman and Chief Executive Director of the Korea Stock Exchange, the Vice-Minister of Finance, and six members appointed by the President, three of whom serve full time. The Commission must report to the Minister of Finance who has the power to annul the Commission's decisions and to request reconsideration of the Commission if he deems it necessary for public interest or investor protection.

The Securities Supervisory Board is the executive body of the Securities and Exchange Commission. Its major operations include registration of securities issuers, examination of registration statements, supervision of listed companies, inspection of securities institutions, and regulation of the over-the-counter market.

The Korea Stock Exchange had been operated as a government-controlled public corporation since the early 1960s, but was transformed into a membership organization run by member securities companies in 1988.

AGENDA FOR FINANCIAL REFORM

Policies Toward Nonperforming Loans and Bailing out Troubled Firms

The Government's efforts to promote "strategic" industries in the 1970s produced mixed results. After the second oil shock, some industries, such as electronics and passenger cars, grew into leading industries in the 1980s, while others, such as shipbuilding, overseas construction, and shipping became seriously ill.

Nonperforming loans of banks have mainly been rooted in government-directed credit allocation at subsidized interest rates. With assured government backing, lending banks had little incentive for serious credit evaluation or *ex post* monitoring. Availability of cheap credit and tax benefits caused many firms to fail to give enough attention to appraising their investment projects. These overly leveraged firms were vulnerable when the industry experienced recession.

Often, the Government and creditor banks delayed appropriate actions against nonperforming loans, keeping the borrowing firms alive and allowing the loans to prolong and snowball. Liquidation of large firms was often out of the question because of the concerns over massive unemployment and visible losses to domestic financial institutions, that might trigger a chain reaction of defaults and extreme instability in financial markets. There was also concern that such a situation would threaten confidence in domestic firms and financial institutions in the international market, perhaps bringing about requests for early repayment of foreign debts as well as reduced availability of foreign borrowing and a sharp rise in country risk. There were, however, no guidelines concerning the identification of industries eligible for government-initiated restructuring.

As was the case for overseas construction and shipping industries, restructuring (rationalization) programs in Korea typically entailed a reduction of the number of firms, with troubled or insolvent firms being absorbed by healthier firms. Usually, the debts of the absorbed firms were rescheduled and additional credit was infused.

Because of the central role played by the Government in initiating and shaping the restructuring programs, it failed to impose market discipline. The Government continued to be viewed as an implicit risk partner, since creditor banks usually played only a minor role in restructuring decision-making even though they generally had the best information about the prospect of the loans. As bank losses were to be assumed partially by the Government, banks had little alternative but to accommodate whatever packages it came up with. In addition, the banks, which had been under government control for a long time, didn't have enough expertise to work out such restructuring programs completely on their own. The Government's continuing role as risk partner failed to eliminate the "moral hazard" problem. The restructured firms were reluctant to reduce capacity and tended to pursue risky strategies in anticipation of another government rescue in the case of failure.

Certainly, the Government's role as a restructuring promoter, with all the tax and financial incentives, helped impose on the firms restructuring measures that they otherwise would not be willing or able to undertake. As such, agency problems caused by conflicts of interest between creditors and debtors and the associated costs in the process of prolonged negotiations could be reduced. If the Korean Government and banks had reacted more quickly to the problems, the cost of restructuring would have been much less.

However, timeliness was not the only factor. The government-imposed industry restructuring in Korea put major emphasis on mergers rather than industrial exit or conversion, which tended to delay needed adjustments and entail inefficient resource allocation. In addition, the troubled firms were mostly taken over by large business groups, which contributed to the concentration of economic power. Absence of clear criteria related to takeovers as well as inadequate information made available to the public about the performance of those acquired sick firms has weakened credibility toward government policy on industrial restructuring.

With the second oil shock, several Korean industries suffered severely from declining orders, overcapacity, and financial distress. Thus, the Korean Government attempted investment adjustment in 1980 in areas where duplicative and excessive investment was most evident, including heavy power-generating equipment, heavy construction equipment, motor vehicles, vessel diesel engines, electronic exchangers, heavy electrical equipment and copper smelting. During 1984 and 1985, the two major ailing industries, shipping and overseas construction, were rationalized. Another round of industry restructuring occurred between 1986 and the first half of 1988 under the guidelines of the Industrial Development Deliberative Council created by the Industrial Development Law. Financial support played a critical role in the restructuring process which involved a total of 78 corporations.

Between December 1985 and May 1987, the Bank of Korea provided 1.7 trillion won of special credit to banks, at a low interest rate of 3 per cent per annum, to alleviate their financial burden due to corporate restructuring.

Other central bank loans to banks extended in connection with industrial rationalization have usually carried an interest rate of 6 per cent. By the end of 1987, the total central bank credit to commercial banks related to industry rationalization programs reached 3.0 trillion won, which was almost equivalent to the total increase in reserve money during the period 1985–1987.

On the other hand, tax exemptions given to industry restructuring during the period 1986–1988 amounted to 241 billion won. Corporate income taxes and acquisition taxes were not levied in connection with the sales of real estate held by the troubled firms and takeovers of such real estate and equity stock. Withholding taxes on presumptive donated income of the acquiring firms was also exempt.

The current policy toward nonperforming loans is to let the banks, which have accumulated the most information concerning the borrowing firms, dispose of their nonperforming loans on their own account. On the part of the Government, priority is given to inducing the banks to write off bad debts early on and to reinforcing supervision to prevent new nonperforming loans. The scope for writeoffs treated as losses for tax purposes is to be expanded, and a change in the size of the nonperforming loans together with the ratio of write-offs to total nonperforming loans will be considered as important items in bank performance evaluation.

Although the problem of nonperforming loans is not as serious now as it was in the second half of the 1980s in terms of their share in total bank credit, a more serious approach might be required if moral hazards are to be avoided and market discipline is to be imposed. An alternative might be to work out a deal between the Government and an individual bank, by which the Government is completely exempt from further taking care of nonperforming loans in return for a lumpsum payment estimated to be proportional to the size of nonperforming loans attributable to past government intervention. Debt-equity swaps may also be encouraged to induce banks to strengthen their monitoring of the nonperforming loan holders and to restructure the management of these firms, if necessary.

Restructuring of the Financial Sector

Korea has a specialized banking system in which restrictions are placed on the types of financial services that may be offered by different types of financial institutions. The establishment of a specialized as opposed to a universal banking system was largely an outgrowth of the Government's policy of using commercial banks as conduits for the direct allocation of credit to targeted sectors and industries. Since the early 1980s, however, the business boundaries of some intermediaries have been expanded so that they overlap with one another. One major development, for example, has been the opening up of some short-term securities businesses to both commercial banks and securities companies.

In the absence of an overall blueprint for a desired financial market structure, however, the broadening of business boundaries has been largely motivated by a need for "balanced growth" of different types of intermediaries. More specifically, the Government has sought to prevent the growth of banks from falling too far behind that of nonbank intermediaries in the presence of extensive government controls on banks with policy loans, other directed credit and interest rate regulations.

For long-term efficiency in financial intermediation, the structure of the financial system must be reassessed. This will require the resolution of a number of key questions, one of which is whether commercial and investment banking functions should continue to be largely separate and, if not, how the different functions should be combined.

The major factor justifying commercial bank entry into investment banking through a subsidiary or an affiliated firm is that this is generally expected to enhance competition and enable banks to diversify business risk and enjoy economies of scope. For investors and corporate customers, this would mean more convenience, higher investment returns, and a lower corporate financing cost. Since the customers of nationwide commercial banks are also likely to be customers in their investment banking, the potential economies of scope are expected to be substantial. In the long run, Korea should aim at allowing provision of both commercial and investment banking services under one roof.

However, Korean investment banking institutions require urgent restructuring before they can enter into commercial banking. They are generally overly specialized, and the broadening of business boundaries in recent years for some intermediaries has left others relatively squeezed in their business scope. Thus, priority should be given to broadening their business areas within investment banking.

There are other reasons why commercial bank entry into investment banking might be premature in the short run. Given interest rate controls in the primary securities market and the lack of various options and financial futures markets, underwriting corporate securities is a risky business in Korea. Commercial bank asset portfolios may be exposed to serious risk as the banks extend loans to help their distressed securities affiliates in a futile effort to avoid a crisis of confidence in the banks, or as business groups attempt to transfer unsound assets of their securities subsidiaries or affiliates to banks in which their shares of equity investment are much smaller. In addition, as long as bank loan rates remain controlled, banks may be able to tie underwriting services to bank loans, thus leaving little underwriting business for nonbank competitors. Moreover, without solutions to the large number of nonperforming loans resulting from the Government's past abuse of banks, allowing investment banking institutions to enter into commercial banking would be unfair to existing banks.

Finally, and most importantly, combining commercial and investment banking leaves open the possibility of conflict of interest abuses mainly among different bank customers. While such conflicts are difficult to exploit profitably in a competitive market where relevant information is promptly available to all potential market participants at low cost, in the Korean market, institutions are yet to be firmly established to encourage generation of complete and reliable information and to prevent unfair transactions. The system of corporate disclosure and outside audit and the function of credit rating agencies are still primitive and weak. Insider trading, price manipulation and other unfair securities transactions are widespread and are not effectively regulated. Unless substantial progress is made in this area, conflicts of interest may pose a potentially serious problem when commercial and investment banking are combined. This long list of constraints, however, should not be interpreted as implying that transition to universal banking is a remote possibility within a decade or so. As a matter of fact, such a transition would help ease the constraints to a large extent.

Investment banking institutions in Korea include securities companies, investment and finance companies, merchant banking corporations, and securities investment trust companies. Among these four categories of intermediaries, investment and finance companies are most in need of adjustments, as their major service of providing short-term securities is now fairly open to both securities companies and commercial banks. A logical solution would be to allow these companies to convert to, or merge with, other financial institutions. The remaining investment companies may specialize as dealers or brokers in the money market. In 1990, the Law on Mergers and Conversions of Financial Institutions was enacted and investment and finance companies located in Seoul were allowed to merge and convert to banks or securities companies. Of the 16 located in Seoul, five chose to convert to securities companies, one converted to a bank, and two others merged together to convert to another bank.

Securities investment trust companies also need adjustment since new financial services resembling securities investment trusts, such as Trust for Households and Corporate Trust Accounts, cash management accounts, and bond management funds, have been introduced in recent years by other financial intermediaries. Given the experiences of the securities investment trust companies in long-term securities businesses, they may be encouraged to convert to securities companies, and may have investment management companies as their subsidiaries. Since they are owned mainly by financial institutions, the probability of conflict of interest abuses arising from the combination of a securities company and an investment management company as a subsidiary does not seem high.

Separation of short- and long-term securities businesses in Korea can be traced to the establishment of the investment and finance companies in 1972 for the purpose of absorbing curb market funds by the formal market. These two services are complementary for investors and offer a means of achieving economies of scope when supplied together. The provision of both services by one intermediary will best serve customers by offering more convenience, lower transaction costs, higher investment yields, and reduced financing costs. Thus, the case for blurring the business boundaries of investment and finance companies, merchant banks, and securities companies by permitting them to undertake both short- and long-term securities businesses seems to be very strong.

Structure of Ownership and Management Control of Banks

Before nationwide commercial banks were privatized between 1981 and 1983, the Government owned 20–30 per cent of total shares of these banks. Furthermore, the voting power of any one private shareholder was limited to a maximum of 10 per cent regardless of their equity shares. As the government-held shares were sold to the private sector, there was a great concern over the possibility of large business groups controlling these banks. This concern led to a new provision that limits the maximum equity share of a nationwide commercial bank that can be held by a stockholder (including those with a special relationship to the stockholder) to 8 per cent of the total.

Contrary to this concern, no attempt to control these banks by any business groups has been reported. Initially, there were many large business groups purchasing nationwide commercial bank shares, mostly their principal transaction banks, probably in anticipation of some favors. This initial interest, however, seems to have been largely lost. The number of business firms with equity shares of over 1 per cent in any of the five nationwide commercial banks declined from 19 in 1986 to 7 in 1991.

Tight control over credit to large business groups since 1984 might be partly responsible for their declining interest in holding nationwide commercial bank shares. The share of nationwide commercial bank loans extended to the largest 30 business groups dropped from 29 per cent in 1987 to 17 per cent in June 1993, and the share for the largest five groups declined in a similar manner during the same period, to 9 per cent in June 1993. As the Government continues to intervene in the management of nationwide commercial banks in such areas as credit allocation and selection of bank presidents, business groups seem to find little interest in holding nationwide commercial bank shares for the purpose of controlling these banks.

The average equity holdings of the largest shareholder for each of the five nationwide commercial banks dropped from 8.7 per cent in 1986 to 5.4 per cent in 1991. Combined equity shares for those stockholders owning more than 1 per cent of the total (excluding trust property) decreased from 33 per cent to 24 per cent during the same period. Among these stockholders, the combined shares for business firms and individuals dropped from 16 per cent to 5 per cent, while the share for institutional investors (mainly insurance companies) increased.

Dispersed ownership structure of nationwide commercial banks and appointment of their presidents by the Government were frequently blamed for inefficiency of bank management. However, allowing large business groups to control these banks is certainly not a solution. As entry into banking is restricted and bank loans have a subsidy element (while the burden of nonperforming loans is alleviated by the Government), allowing business groups to hold controlling shares of banks would be an unjustifiable favor to the business groups and would lead to concentration of economic power.

Even when financial markets are competitive with free entry, the mix of financial and nonfinancial products within a business group increases the probability of abuses involving conflicts of interest as long as the involved firms have market power in the nonfinancial products. Tie-in sales are a common means of exploiting conflicts of interest. Given the highly oligopolistic Korean market structure where business groups have market dominating power in many industries, the room for abusing conflicts of interest is potentially large.

In this connection, however, there is no strong rationale for applying the 8 per cent ownership restriction for nationwide commercial banks in an indiscriminate way, if concerns about the concentration of economic power and abuses of conflicts of interest are the major reasons for such restriction. Limiting the restriction only to large business groups, which are so extensively diversified as to have high potential for abusing conflicts of interest, may be an alternative. It will be much easier to monitor and regulate any abuses by relatively small business groups doing business in a limited number of industries. Further, the fact that ownership restrictions apply only to commercial banks (the maximum being 15 per cent for local banks) among financial intermediaries is rather unfair, since commercial banks no longer dominate the financial market and their uniqueness will be weakened in the future. The Government is thinking of introducing restrictions on the ownership of nonbank financial institutions. Of course, to the extent that loans of nonbank financial institutions are less subsidized, and that the impact of a nonbank failure is less serious than that of a bank, ownership regulations will not have to be so restrictive.

Even acceding that large business groups should not be allowed to hold controlling shares of nationwide commercial banks, government appointment of their presidents is not desirable from the perspective of managerial efficiency of these banks. As a transitional solution, the Government is suggesting establishing a nomination committee for candidates for bank presidency. It is suggested that the committee be composed of three former bank presidents, four representatives for stockholders (two large stockholders and two small shareholders), and two representatives of corporate and individual clients (one representing each group). These members will be selected by the expanded board of directors and be approved by the Office of Bank Supervision and Examination. Among the candidates nominated by the committee, the general stockholders' meeting and the board of directors are to select a bank president. Actually, some commercial banks have already adopted this procedure for the selection of their presidents. The selected presidents came mostly from within the banks.

Upgrading Financial Infrastructure

Making accurate corporate information readily available to all potential investors and creditors is essential for the efficiency of the financial market and the protection of investors and creditors. Incentives for fraud and abusing conflicts of interest tend to be stronger when information asymmetry is large, the search for accurate information is costly, and unfair transactions are widespread. In Korea, systems for corporate accounting, disclosure, and credit ratings are yet to be firmly established. In a still shallow securities market, insider trading, price manipulations, and other unfair transactions are not regulated effectively.

Corporate Accounting System

In instances where two or more companies are legally independent but economically linked or interdependent, consolidated financial statements provide more useful information regarding the financial status and management performance of the entire business group. Until recently, consolidated financial statements were not mandatory in Korea. Even when they were prepared, their credibility was questionable since no external audits were performed. It was only in 1992 that the preparation and external auditing of consolidated financial statements became mandatory for all listed companies. Consolidated financial statements will contribute significantly to the prevention of manipulative accounting practices such as window dressing as well as such tax evasion practices as the transfer or dispersal of profits among member companies in a business group.

Currently, consolidated financial statements are required only for firms whose equity share in other companies exceeds a stipulated level. However, certain characteristics of Korean businesses render this simple standard somewhat problematic, as firms with a close relationship might be excluded, if the relationship comes from strong family ties in management or horizontal business affiliations, wherein a single shareholder holds major equity shares in a large number of relatively independent firms. Thus, it would be desirable to require all business groups to prepare consolidated financial statements.

Another way of enhancing the credibility of accounting information is to reduce window dressing by limiting the legitimate scope of accounting methods from which firms may choose. Currently, a variety of legitimate accounting methods are available concerning special depreciation, asset revaluation, writeoffs of bad debt, and reserve for employee retirement, which gives firms enough accounting flexibility to manipulate their business performance.

Audit by independent and qualified outsiders—enhancing public confidence in regularly published financial statements of corporations—is essential for the protection of investors and other interest groups. Currently, all stockholding companies in Korea with equity capital or total assets over specified amounts are required to have an external audit. With the number of firms obliged to have an external audit increasing rapidly, and the consequent overload for auditors, the cases of poor and incomplete audits have increased. Further, the competitive bidding system for audit contracts tends to strengthen the position of the firms to be audited, resulting in undue leniency on the part of auditors in their effort to secure continued business with a company. Institutional improvements thus seem to be needed to prevent collusion between them and the unilateral change of outside auditors by the firms without good reason.

Starting in 1991, a Designated Auditor System has been partially implemented, allowing the Government to arbitrarily assign auditors. Auditors are designated for firms (i) that failed to hire an auditor within a specified time limit; (ii) recommended to go public; (iii) designated for industry rationalization; (iv) that failed to comply with the Securities and Exchange Act; (v) with a high debt-to-equity ratio; and (vi) warned against unacceptable external auditing. However, this raises concerns over excessive government intervention and deterioration of auditing standards. A solution may be to introduce an evaluation system wherein auditors proven to have produced reports of a certain quality are rewarded with contracts. Concurrently, the Government may take stronger punitive measures toward auditors of poor performance.

Corporate Disclosure

Investment information is primarily obtained from corporate disclosure. A good corporate disclosure system is vital for an efficient securities market. Even though the Korean corporate disclosure system was modeled after that of advanced countries, the system is still not well established. First, disclosed information is not sufficient: such instruments as preliminary prospectuses as well as quarterly and occasional reports are not usually used, and information on future prospects is insufficient. Consolidated financial statements are not available for affiliated firms under the same management control. Recently, however, the disclosure obligation of firms listed on the stock exchange has been extended substantially to include changes in equity capital, suspension of operation, start of management control by creditor banks, major claims received, large unit orders received, and development of new materials or products.

Second, the nationwide information delivery system is inadequate, and coordination and cooperation are poor among information generating and handling institutions such as the Securities Supervisory Board, the Korea Securities Dealers' Association, and the Korea Stock Exchange.

Finally, Korean corporations tend to be reluctant to disclose corporate policies and strategies, try to evade or delay disclosing unfavorable information, and pay little attention to the accuracy of disclosure. Stepped-up efforts have been made recently to strengthen the enforcement of corporate disclosure. The Korea Stock Exchange is empowered to inquire of relevant firms about rumors and unconfirmed reports which may affect the prices of their stocks. For corporations which violate disclosure obligations, the Securities and Exchange Commission may recommend dismissing the officers responsible, or may restrict the issuance of securities for a specified period, suspend listing, or bring a lawsuit, depending on the severity of the offense.

A reliable credit rating by an independent institution would compensate for the limitations in the corporate disclosure system by reducing information asymmetry between corporate insiders and investors through the analysis of both undisclosed and disclosed information and industry prospects. In Korea, however, corporate credit rating institutions have been operating only since 1985, and are yet to play their full expected role, mainly due to rigid government control of interest rates and repayment guarantees of financial institutions for corporate bonds which largely obviate the need for credit ratings.

Guaranteed bonds account for the majority of total corporate bond issues in Korea. Demand for unguaranteed corporate bonds has been limited due to interest rate regulations in the primary market, an underdeveloped credit information system based on corporate disclosure and credit ratings, and weakness in investor protection in bond covenants. Continued reliance on guaranteed corporate bonds, with the guarantee fee fixed regardless of the creditworthiness of the issuers, results in rationing of repayment guarantees to the best-known large corporations (eliminating promising small- and medium-scale firms), weakens demand for credit rating, and limits the availability of reliable credit information.

Fair Trade

The regulation of insider trading is grossly inadequate in Korea. Corporate officers, employees or major stockholders with *de facto* ownership of 10 per cent or more of a firm's total shares (regardless of the registered owners) are prohibited from earning excessive profits by using confidential information obtained while fulfilling their duties or by their positions. Recently, regulations on insider trading have been strengthened to cover not only corporate insiders but also certified public accountants and others with a special relationship to the firm. The fact that only a few cases of insider trading have ever been disclosed in Korea indicates the difficulty of proving the use of inside information and whether or not an investor is a major stockholder. A recent amendment of rules which has shifted the burden of proof from the securities authorities to the suspects of insider trading will help enforce the law more effectively. Relying entirely on the reports of insiders for the identification of their changing equity ownership is another critical limitation.

The Korean stock market has also suffered from frequent price manipulation through self-dealing, bull cornering, and dumping. Due to the poor stock price monitoring system, most of these transactions have not been exposed. Currently, there is little computer-assisted monitoring of unfair transactions, with common patterns of unfair practices being preprogrammed. In addition, responsibilities for market supervision are overlapping, and sometimes inconsistent and poorly coordinated by the Securities Supervisory Board, the Korea Stock Exchange, and the Ministry of Finance.

SUMMARY AND CONCLUSION

In the 1970s, building heavy and chemical industries was the highest policy priority of the Korean Government. To support this policy, Korean banks and some other financial institutions were urged to channel most of their resources to the preferred sectors. Preferential loans were extended at low interest rates and other guidelines were imposed for the management of financial assets. These policy efforts paid off later by rapidly upgrading the industrial and trade structure of the Korean economy. This achievement, however, was largely possible at the sacrifice of the financial sector.

A rapid expansion of policy loans was mainly responsible for the chronic inflation in the 1970s, which made the financial policies all the more repressive. Furthermore, the burden of the disappointing performances of some of the supported industries was mostly borne by banks. As a result, the development of the Korean financial sector has been seriously retarded, as it suffered from slow growth of financial savings, excessive operational regulations, lack of competition, managerial inefficiencies, and a sizable accumulation of nonperforming loans.

The Korean Government had to help these banking institutions by protecting them from vigorous competition from nonbank financial institutions. Interest rate controls have continued, and entry barriers into the financial sector have hardly been lifted. Because of the continued role of the Government in the bail-out decisions for troubled firms as well as the inertia and incompetence of banks, the Government has maintained its influence on the selection of the top management of nationwide commercial banks. These policies and practices have delayed financial liberalization. The Government's desire to keep private investment activities vigorous for sustained economic growth and to save its own borrowing cost has been another factor behind its cautious approach toward interest rate deregulation. For similar reasons, as well as growing concern over social equity, reduction of policy loans has also been slow.

On the other hand, the Government's efforts to maximize mobilization of domestic financial resources and absorb the informal curb market into the organized financial market led to the introduction of new nonbank financial institutions in the early 1970s. Nonbank financial institutions in general have been treated favorably with respect to interest rate controls, the burden of providing policy loans, and ownership regulation. With this policy, the Government has been successful in absorbing the informal credit market into the organized market and in realizing a fair degree of financial deepening. As these nonbank financial institutions, which operate in a relatively free environment, have grown much faster than banks, the Korean financial market as a whole has become fairly liberalized. This experience shows that promotion of nonbank financial institutions can be a pragmatic way of deepening the financial market, if a country has to maintain intervention in bank management for some reason.

Development institutions have played a major role in the process of Korea's economic development. During the 1960s and 1970s, the Korea Development Bank supplied the majority of the long-term capital needed for economic rehabilitation and industrialization. It efficiently channeled government and public funds such as the National Investment Fund to the Government's priority sectors. It also played an important role as a channel for foreign capital. Even though its relative importance has decreased as other sectors of the financial system grew, it still remains the largest supplier of long-term capital.

The Korea Long-term Credit Bank and the Export-Import Bank of Korea, which were established later, complemented the functions of the Korea Development Bank in a more specialized way. The Export-Import Bank, whose size has decreased significantly since 1986, concentrated mainly in financing external trade and overseas investments. Like the Korea Development Bank, the Long-term Credit Bank also supplied its funds for long-term investment in facilities. However, it was not influenced as much by the Government's industrial policies as the Development Bank in its choice of clients, and was much less dependent on governmental and public funds.

Until the 1980s, the development of the capital market was rather slow. In an environment of high inflation and low savings, there was little room for the formation of long-term capital. The corporate sector relied on bank loans for its capital requirements, and large investment projects were financed by policy loans. As the economy continued to grow rapidly, the Government implemented various measures to foster the growth of the capital market so that it could complement the banking sector. Thanks to the booming economy and governmental efforts, the capital market began to grow rapidly from the mid-1980s. The stock market showed remarkable growth during the latter half of the 1980s, while the bond market also grew rapidly though to a lesser extent than the stock market.

The internationalization of the capital market started in the 1980s. Since 1985, Korean companies have been issuing bonds in the overseas capital market. Currently, foreigners are allowed to hold up to 10 per cent of the outstanding shares of a listed company. In addition, Korea Fund, Korea Europe Fund, and Korea Asia Fund have been set up to specialize in investment in the Korean capital market.

Despite the fast growth in size in the 1980s, further structural improvements in the system are required to ensure the Korean capital market's soundness and efficiency and to prepare for its further internationalization. One such improvement is to enhance the credibility of corporate accounting information through strengthening the external auditing system as well as the implementation of obligatory disclosure of consolidated financial statements. There should be improvements in corporate disclosure and credit rating systems as well. Finally, insider trading should be more strictly regulated.

The Government's efforts to absorb curb market funds into an organized market resulted in the introduction of new money market instruments for nonbank financial institutions, which offered attractive interest rates. However, even though it was less subject to governmental interest rate controls than the banking sector and was thus able to grow fairly rapidly, the slow pace of interest rate deregulation has negatively affected the development of the money market. As demonstrated by the results of interest rate deregulation in 1988, without complete liberalization of interest rates, development of the money market will be inevitably limited.

Bibliography

Amsden, Alice, and Yoon-Dae Euh. 1990. *South Korea's Financial Reform: What are the Lessons*. Paper prepared for the United Nations Conference on Trade and Development (UNCTAD) Secretariat.

Bank of Korea. (various years). *Bank Management Statistics*.

_____. (various years). *Financial Statements Analysis*.

_____. 1990. *Financial System in Korea*. Seoul.

_____. 1993. *Flow of Funds*. March.

_____. (various issues). *Monthly Bulletin*.

_____. 1993. *Call Shijang ui Hyunhwang gwa Kinungjego Bangan*.

Board of Governors of the Federal Reserve System. (various issues). *Federal Reserve Bulletin*. Vol.78, No.7; Vol.72, No. 9.

Cho, Yoon-Je, and David C. Cole. 1991. "The Role of the Financial Sector in Korea's Structural Adjustment." in *Structural Adjustment in a Newly Industrialized Country: Lessons from Korea*. Edited by V. Corbo and S. Suh.

Cole, David C., and Yung Chul Park. 1979. *Financial Development in Korea*. Korea Development Institute, Working Paper No. 7904. Seoul.

Insurance Supervisory Board. (various issues). *Monthly Insurance Review*.

Jung, Byung-Hyu, and Young-Sik Yang. 1992. *Hanguk Chaebol Bumun ui Kyungje Bunsuk (Economic Analysis of Korean Chaebols)*. Seoul: Korea Development Institute.

Kang, Moon-Soo. 1990. *Money Markets and Monetary Policy in Korea*. Seoul: Korea Development Institute, Working Paper No. 9020.

Korea Development Bank. (various issues). *Monthly Economic Review*.

Korea Leasing Association. (various issues). *Lease Sanup Jungbo (Lease/Industry Information)*.

Korea Stock Exchange. (various issues). *Stock.* Seoul.

Ministry of Finance. (various issues). *Chaejung Kumyung Tongge (Financial and Monetary Statistics).*

———. (various issues). *Kigum Kyulsan Bogoseo (Fund Management Report).*

———. 1991. *Ministry of Finance Bulletin.* No. 97. September.

Nam, Sang-Woo. 1988. *Readjustment of the Business Boundaries of Financial Intermediaries in Korea.* Seoul: Korea Development Institute, Working Paper No. 8822.

———. 1991. "Korea's Financial Policy and Its Consequences." A paper presented at the Workshop on Government, Financial System and Economic Development: A Comparative Study of Selected Asian and Latin American Countries. Honolulu: East-West Center.

———. 1992. *Korea's Financial Reform Since the Early 1980s.* Seoul: Korea Development Institute, Working Paper No. 9207.

———. 1993. "The Principal Transactions Bank System in Korea and Its Comparison with the Japanese Main Bank System." A paper presented for the Economic Development Institute (EDI)/World Bank Workshop on the Japanese Main Bank System and Its Relevance for Developing Market and Transforming Socialist Economies. Washington, DC.

National Association of Investment and Finance Companies. (various issues). *Tuja Kumyung (Investment and Finance).*

National Statistics Office. 1992. *Korea Economic Indicator.*

Organisation for Economic Co-operation and Development (OECD). 1992. *Bank Profitability—Financial Statements of Banks, 1981–90.* Paris: OECD.

Securities Supervisory Board. (various issues). *Monthly Review.*

Yoo, Jung-Ho. 1989. "The Korean Experience with an Industrial Targeting Policy." Seoul: Korea Development Institute. Mimeo.

CHAPTER TWO

FINANCIAL DEVELOPMENT AND POLICIES IN TAIPEI,CHINA

JIA-DONG SHEA

INTRODUCTION

Over the past four decades, Taipei,China has enjoyed rapid economic growth, stable prices, and equally distributed income. From 1951 to 1992, the country's real GNP increased at an average annual growth rate of 8.8 per cent, and per capita GNP grew 11.5 times in real terms, reaching $10,215 in 1992. During this period of rapid growth, prices remained relatively stable. From 1953 to 1992, the average annual growth rate of the GNP deflator was 6.3; the wholesale price index grew 4.8 per cent, and the consumer price index, 6.5 per cent during the same period. If 1973–1974 and 1979–1980—periods in which prices were highly affected by the energy crises—are excluded, the average annual growth rates are only 5.0 per cent for the GNP deflator, 2.6 per cent for wholesale price, and 5.0 per cent for consumer prices.

Income distribution became more equal throughout the 1950s, 1960s, and 1970s. Although the trend reversed in the 1980s, the level remains one of the most equal by world standards. In the early 1950s, the income of the richest fifth of Taipei,China's households was estimated to be more than 15 times that of the poorest fifth. This ratio stood at 5.33 in 1964, then fell steadily to 4.17 in 1980, before rising to 4.50 in 1985, 15.18 in 1990, and 4.97 in 1991. By comparison, the number for the US was 8.91 in 1985, for Japan, 4.55 in 1989, for former West Germany, 5.69 in 1984, and for the Republic of Korea, 5.72 in 1988. Measuring in terms of the Gini coefficient, the figure in Taipei,China fell from 0.326 in 1968 to 0.277 in 1980, and then turned upward to 0.290 in 1985, further to 0.321 in 1990, and then fell to 0.308 in 1991. This coefficient for the US in 1985 was 0.354, for Japan in 1989 was 0.288, for former West Germany in 1984 was 0.301, and for the Republic of Korea in 1988 was 0.336.

While there is a vast literature studying the experiences of Taipei,China, few studies are available in English on the development of the financial sector and its role in economic growth in Taipei,China. This chapter will help to reduce this gap.

To serve this purpose, this chapter first provides a brief overview of the financial system in Taipei,China, covering the structure of the system, the types of financial institutions and instruments, the flow-of-funds relationship among sectors, the consolidated balance sheet of financial institutions, and the distribution of loans. The chapter then discusses the country's accounting, legal, and supervisory systems, government regulations, methods for managing distress, and macro financial policies which affect the performance and development of the financial sector. The structure, evolution, and performance of the banking system, nonbank financial intermediaries, and nonintermediary financial markets are then introduced. In addition, the structure, types of financial transactions, interest rate and size of the unregulated informal financial system are discussed. Growth promotion financial policies, and financing for small and medium enterprises are explained. The development of the financial system, its roles in stimulating and mobilizing savings as well as allocating funds, are evaluated, as are the prospects of the country's financial sector and several lessons regarding financial development and financial policies from Taipei,China's experience.

OVERVIEW OF THE FINANCIAL SYSTEM

The financial system in Taipei,China is summarized in Table 2.1. An important feature of the system is financial dualism, which means that besides the regulated formal financial system, the unregulated informal system also plays an important role. The regulated system includes all institutions and markets established according to financial laws or rules, and subject to regulations by the financial authorities. The unregulated system is composed of all markets not set up according to financial laws or rules; it operates lending and borrowing activities without being under the direct regulation or supervision of the financial authorities.

The regulated financial system consists of financial intermediaries, nonintermediary financial markets, and other institutions and markets. Financial intermediaries or financial institutions can be further grouped into two categories: banks and nonbank financial intermediaries. Non-intermediary financial markets include the money market, bond market, stock market, and foreign exchange market.

The relative roles of the financial institutions, money market, bond market, and stock market in providing funds are reported in Table 2.2. This table clearly indicates that financial institutions have always been the dominant source of funds.

Types of Financial Institutions

There are eight types of financial institutions in Taipei,China: domestic banks, foreign bank branches, medium business banks, credit cooperative associations, credit departments of farmers' and fishermen's associations, investment and trust companies, postal savings system, and insurance companies. The number of units of the various financial institutions is shown in

Table 2.1 The Financial System in Taipei, China

Regulated Financial System	Unregulated Financial System

I. Financial Intermediaries
  Banks
    Central Bank of China
    Full Service Domestic Banks
      Commercial Banks
      Specialized Banks
    Local Branches of Foreign Banks
    Medium Business Banks
  Nonbank Financial Intermediaries
    Cooperatives
      Credit Cooperative Associations
      Credit Departments of Farmers' Association
      Credit Departments of Fishermen's Association
    Investment and Trust Companies
    Postal Savings System
    Insurance Companies
      Life Insurance
      Property and Casualty Insurance
      Central Reinsurance Corporation
 	
II. Nonintermediary Financial Markets
  Money Market
    Bills Finance Companies
    (Interbank Money Center)
  Capital (Bond and Stock) Market
    Taiwan Stock Exchange Corporation
    Fuh-Hua Securities Finance Company
    Security Dealers
      Brokers
      Traders
  Foreign Exchange Market
  Foreign Currency Call-Loan Market
    (Offshore Banking Units)
 	
III. Other Institutions
  Small-and-Medium Business Credit Guarantee Fund
  Central Deposit Insurance Corporation
	Organization
	  Installment Credit Companies
	  Leasing Companies
	  Investment Companies
	  Credit Unions
	  Rotating Credit Clubs (*huis*)
	  Unorganized Markets
	    Money Lenders
	    Pawnbrokers
	    Others
	
	Other Types of Transactions
	  Unsecured Credit
	  Secured Credit
	  Loans Against Post-Dated Checks
	  Deposit with Firms
	  Margin Credit for Stock Purchases

Table 2.2 Roles of Financial Markets in Providing Funds, Relative to Financial Intermediaries

	1970	1980	1990	1992
Total loans and discounts of Financial Institutions (NT$ billion)	94	1,023	4,982	7,704
As a percentage of total loans and discounts of Financial Institutions				
Money market instruments outstanding	–	8.47	13.29	12.21
Corporate and Government Bonds outstanding	10.08	4.25	4.82	8.00
Total par value of shares of Listed Companies	8.95	10.62	10.16	9.55

– nil.
Source: Central Bank of China.

Table 2.3. From 1961 to 1992, the number of financial institutions (head offices only) increased by 29 per cent, from 400 to 515, while the number of locations, including head offices and branches, more than tripled from 1,359 to over 4,400. The number of persons served per location thus fell steadily from 8,204 or 12,279 in 1961 (if the postal savings system is included as a financial institution), to 4,705 or 6,511 in 1992. Although this number is still much larger than in developed countries, the network of financial institutions is nonetheless dense, especially when the fact that Taipei,China has one of the highest population densities in the world is taken into account.

The distribution of assets, loans and investments, and deposits among different types of financial institutions is shown in Table 2.4. Based on this table and Table 2.3, it is evident that domestic banks have always been the backbone of the country's financial system, and the scale of domestic banks per unit is much larger than other types of institutions.

Types of Financial Instruments

Under strict control from the authorities, the financial system in Taipei,China has been sluggish to bring in new types of nonbank financial intermediaries and new financial instruments. Even during the period of financial liberalization in the 1980s, there were seldom any new instruments introduced.

Table 2.3 Number of Units of Financial Institutions[a]

End of Year	Total Units	Full-Service Domestic Banks		Foreign Bank Branches		Medium Business Banks		Credit Cooperative Associations	
		Unit	Per Cent	Unit	Per Cent	Unit	Per Cent	Unit	Per Cent
1961	1,359	260	19.1	1	0.1	84	6.2	153	11.3
1970	1,821	393	21.6	7	0.4	118	6.5	222	12.2
1980	2,830	536	18.9	21	0.7	165	5.8	274	9.7
1990	4,012	738	18.4	43	1.1	283	7.1	473	11.8
1992	4,410	937	21.2	50	1.1	315	7.1	513	11.6

End of Year	Credit Departments of Farmers' and Fishermen's Assn.		Investment and Trust Companies		Postal Savings System		Insurance Companies	
	Unit	Per Cent	Unit	Per Cent	Unit	Per Cent	Unit	Per Cent
1961	385	28.3	1	0.1	451	33.2	24	1.8
1970	393	21.6	1	0.1	610	33.5	77	4.2
1980	724	25.6	26	0.9	952	33.6	132	4.7
1990	1,053	26.3	54	1.3	1,202	30.0	166	4.1
1992	1,115	25.3	61	1.4	1,223	27.7	196	4.4

[a]Units are physical locations–head offices and branches. The percentages are of total units. The Central Bank of China, Central Reinsurance Corporation, and postal agencies–which do not do business with the general public–are excluded. Number of foreign bank branches is from Central Bank of China.

Sources: Ministry of Finance (1992b). Number of foreign bank branches is from Central Bank of China.

Table 2.4 Shares of Financial Institutions at End of Year[a]
(per cent)

	Total Assets				Loans and Investments[b]				Total Deposits[c]			
	1961	1970	1980	1990	1961	1970	1980	1990	1961	1970	1980	1990
Domestic Banks	80.22	70.48	64.53	53.29	81.95	76.42	67.52	62.50	75.57	65.75	55.58	46.18
Local Branches of Foreign Banks	0.88	2.95	5.49	2.92	0.40	3.05	7.58	3.38	0.03	1.54	0.32	1.10
Medium Business Banks	3.09	4.37	3.71	7.34	3.76	4.92	4.82	9.16	4.90	5.24	4.55	7.40
Credit Cooperative Associations	7.52	7.97	6.84	9.18	7.06	7.10	7.03	8.46	10.13	10.54	10.65	11.48
Credit Departments of Farmers' and Fishermen's Associations	5.21	5.08	5.02	6.85	4.66	4.35	4.71	5.76	5.91	5.42	6.96	8.25
Investment and Trust Companies	1.33	1.44	4.29	4.34	1.58	1.98	6.32	5.49	...	...	6.20	5.18
Postal Savings System	0.84	5.80	8.33	11.66	0.32	0.47	0.17	0.17	3.38	9.70	13.65	14.88
Insurance Companies	0.91	1.91	1.79	4.41	0.27	1.71	1.85	5.06	0.08	1.81	2.09	5.53
Total	100.00	100.00	100.00	100.00	100.00	100.00	100.00	100.00	100.00	100.00	100.00	100.00
Amount (NT$ billion)	26.10	136.20	1,703.30	9,087.30	17.70	97.80	1,132.20	5,490.90	18.50	93.90	1,012.20	7,019.90
As a percentage of GDP	37.26	60.05	114.23	215.24	25.27	43.12	75.93	130.05	26.41	41.40	67.88	166.27

[a] Central Bank of China is excluded.
[b] Loans and investments include loans, discounts, portfolio investments and the holdings of real estate.
[c] Total deposits include deposits held by enterprises and individuals, government deposits, trust funds, and life insurance reserves.
... data not available.
Source: Central Bank of China.

Now, however, banks accept various kinds of deposits, including checking accounts, passbook deposits, passbook saving deposits, time deposits, time saving deposits, negotiable certificates of deposits, foreign currency deposits, foreign exchange proceeds deposits, foreign exchange trust funds, and foreign currency certificates of deposits. Enterprises and individuals can also hold bank debentures issued by specialized banks, savings bonds, and treasury bills issued by the Central Bank, or entrust funds to the investment and trust companies, and buy life insurance policies from insurance companies. The postal savings system also accepts transfer accounts, passbook saving deposits, time saving deposits and life insurance policies. In the money market, instruments transacted include commercial paper, negotiable certificates of deposits, bankers' acceptances, commercial acceptances, treasury bills and government bonds, bank debentures, and corporate bonds, with a maturity of less than one year. Instruments transacted in the capital market include government bonds, corporate bonds, convertible corporate bonds, financial securities, savings bonds/securities of the Central Bank, negotiable certificates of deposits of the Central Bank, and shares and stocks.

Although there exists a variety of financial instruments, bank deposits are always the dominant asset held by the general public. As Table 2.5 shows, deposits in the financial institutions have maintained a roughly constant share at about 80 per cent of the major domestic financial instruments (excluding stocks) held by businesses and individuals over the past 30 years[1]. In addition, while trust funds and life insurance reserves are gaining their weight, their shares have just reached the level of 5 per cent respectively.

Flow-of-Funds Relationship

The financial surplus or deficit of government, public enterprises, private enterprises, and households as a percentage of GNP is given in Table 2.6. According to the data in this table, the household sector has always been the most important source of surplus, although, until 1988, the Government too had provided a huge financial surplus. The financial surplus has been channeled by the financial system to finance the deficit of public and private enterprises over the past 30 years, and also to finance the Government's budget deficit since 1989. Also, since the mid-1980s, a large share of domestic savings has been used to finance the rest of the world through a huge trade surplus, in exchange for the accumulation of foreign exchange reserves.

---

[1] Since the market value of stocks and equities has varied greatly, stock holding is sometimes excluded in the discussion.

Table 2.5 Composition of Major Domestic Financial Instruments Held by Business and Individuals[a]
(per cent)

Financial Instrument	1965	1970	1980	1990
Currency	14.22	13.19	10.77	4.96
Demand deposits[b]	26.26	21.35	27.94	22.07
Time and savings deposits	53.95	56.56	51.81	57.00
Other deposits[c]	1.90	1.65	1.24	3.14
Trust funds	...	0.31	2.19	5.24
Life insurance reserves	1.57	2.20	3.27	5.73
Bank debentures	...	...	0.27	0.03
Government securities	2.03	4.73	0.12	0.74
Commercial paper	...	...	1.74	0.36
Bankers' acceptances	...	...	0.60	0.44
Corporate bonds	0.06	–	0.04	0.28
Total (NT$ million)	40,004	101,433	1,025,054	7,145,127

[a]Stock holdings are excluded.
[b]Includes checking accounts passbook deposits and passbook savings deposits.
[c]Includes negotiable CDs, foreign currency deposits, foreign exchange proceeds deposits, foreign exchange trust funds and foreign currency CDs.
... data not available.
– nil.
Source: Central Bank of China (1992a).

The sum of annual nominal figures of the sources and uses of funds by sector over the period 1965–1988, based on flow-of-funds statistics, are reported in Table 2.7. This table suggests the following characteristics regarding the aggregate financial relationships among sectors:

(i) A large proportion of household saving and total sources of funds had been held in the form of deposits in financial institutions, followed by the holdings of stocks and shares, and lending to enterprises through curb markets. In sum, 102 per cent of the total saving or 70 per cent of the total funds of the household sector were channeled into the financial system (excluding the stock market) during the period 1965–1988.

(ii) Direct investments in infrastructure and purchasing corporate stocks (especially the stocks of public enterprises) constituted the most important use of funds by the Government. A handsome share of government saving (26 per cent) also went to financial institutions as government deposits.

Table 2.6 Financial Surplus or Deficit by Sector
as a Percentage of GNP[a]

Year	Govern-ment	Public Enterprises	Private Enterprises	House-holds	Rest of the World
1951	1.99	−3.83	−0.32	3.40	−1.23
1960	1.69	−2.78	−3.89	2.55	2.43
1965	0.25	−1.28	−7.81	6.84	2.00
1970	0.53	−2.72	−6.94	9.14	−0.01
1971	1.28	−3.62	−5.49	10.41	−2.58
1972	3.76	−3.45	−4.92	11.10	−6.50
1973	3.74	−4.76	−6.66	12.96	−5.28
1974	4.95	−8.14	−15.20	10.70	7.70
1975	2.62	−9.21	−4.97	7.77	3.79
1976	3.52	−7.20	−3.14	8.39	−1.57
1977	2.34	−4.67	−3.06	9.70	−4.31
1978	4.38	−4.20	−3.32	9.27	−6.13
1979	5.33	−4.74	−8.58	8.49	−0.50
1980	3.53	−7.02	−4.88	6.99	1.56
1981	2.29	−5.27	−4.32	8.62	−1.31
1982	0.66	−4.99	−1.99	9.74	−4.84
1983	1.90	−2.41	−1.98	10.62	−8.70
1984	2.24	−0.45	−2.51	11.39	−11.87
1985	1.69	0.13	−0.91	13.13	−14.82
1986	0.40	0.98	0.97	18.24	−21.35
1987	2.51	−0.19	−3.86	19.24	−18.40
1988	2.22	0.10	−6.80	15.85	−11.69
1989	−4.94	−1.57	−6.20	20.47	−8.51
1990	−1.09	−2.99	−4.31	14.95	−7.29
1991	−5.71	−1.08	−5.11	18.61	−7.22

[a]Rows do not sum to zero because of statistical discrepancies in the underlying data.
Source: Computed by author from data in Central Bank of China (1992a), accounts 4–7 (financial surplus by sector from 1965); (1987), Tables 6 and 7 (financial surplus by sector 1951 and 1960); (1992b), Ch 3 table 16 (overseas sector) and Ch 2 table 1 (GNP).

(iii) The major sources of funds for private enterprises were corporate saving, issuing of stocks or shares, and borrowing from financial institutions, followed by borrowing from households, as Tables 2.7 and 2.8 indicate. However, since most of the stocks and shares were not issued in public, the stock market has never been an important source of funds for private enterprises. In total, 72 per cent of the

Table 2.7 Sources and Uses of Funds by Spending Sector, 1965–1988[a]
(NT$ million)

	Households and Nonprofit Institutions	Private Enterprises	Public Enterprises	Government	Financial Institutions	Total Domestic Economy
Source of funds						
Gross domestic saving	4,557,768	1,859,603	1,199,310	1,805,222	236,493	9,658,396
Borrowing–domestic						
Financial Institutions	1,827,775	1,272,783	172,152	45,581	...	3,318,291
Money Market	...	115,873	13,435	...	...	129,308
Bond Market	...	11,508	40,356	173,941	755,893[b]	981,698
Households	...	557,290	9,806	...	3,984,845[c]	4,551,941
Enterprises	18,249	50,057	1,406	...	888,898[c]	958,610
Government	52,982	15,754	5,510	...	468,213[c]	542,459
Corporate stocks and shares of Noncorporate Enterprises	...	1,724,632	507,757	...	...	2,232,389
Trade credit	183,122	294,806	147,207	240,565	...	825,700
Other domestic liabilities	30,081	...	47,081	...	635,491	712,653
Net foreign liabilities	...	...	27,633	...	...	27,633
Total	6,669,977	5,902,306	2,171,653	2,225,309	6,969,833	23,939,078

Uses of funds						
Gross domestic investment	1,192,286	2,770,117	1,889,728	1,072,808	71,319	6,996,258
Currency	226,294	88,735	474	...	-315,503	...
Deposits in Financial Institutions[c]	3,984,845	727,813	161,085	468,213	...	5,341,956
Money market instruments	40,379	31,842	7,066	...	50,021	129,308
Central Bank Securities and Bank Debentures	50,950	31,204	47,574	...	626,165	755,893
Bonds	26,893	7,939	5,094	...	185,879	225,805
Stocks and shares	1,517,445	152,893	17,093	488,166	56,792	2,232,389
Loans	567,096	56,228	13,484	74,246	3,318,291	4,029,345
Trade credit	259,330	245,570	108,867	211,933	...	825,700
Other domestic financial assets	...	225,905	...	73,660	413,088	712,653
Net foreign assets	...	115,493	...	9,384	2,568,158	2,693,035
Total	7,865,518	4,453,739	2,250,465	2,398,410	6,974,210	23,942,342
Discrepancy: Source-Uses	-1,195,541	1,448,567	-78,812	-173,101	-4,377	-3,264

[a]Data in this table are totals of annual nominal figures over the period 1965 to 1988.
[b]Central Bank securities (treasury bill-B, CDs and savings bonds) and bank debentures.
[c]Denote deposits in financial institutions.
... data not available.
Source: Central Bank of China (1989).

Table 2.8 Proportion of Private and Public Enterprises' Investment Funds from Financial System (1965–1988)[a] (NT$ million; per cent)

	Private Enterprises			Public Enterprises		
	Amount	As a Percentage of Investment	As a Percentage of Total Uses of Funds	Amount	As a Percentage of Investment	As a Percentage of Financial Deficit
Financial Institutions	1,272,783	45.95	28.58	172,152	9.11	24.93
Money Market	115,873	4.18	2.60	13,435	0.71	1.95
Bond Market	11,508	0.42	0.26	40,356	2.14	5.85
Curb Markets	607,347	21.92	13.63	11,212	0.59	1.62
Households	557,290	20.12	12.51	9,806	0.52	1.42
Other Enterprises	50,057	1.80	1.12	1,406	0.07	0.20
Total	2,007,511	72.47	45.07	237,155	12.55	34.35

[a]The stock market is excluded.
Source: Central Bank of China (1989).

investment funds or 45 per cent of the total uses of funds of private enterprises came from the financial system (excluding the stock market).

(iv) Tables 2.7 and 2.8 show that public enterprises depended on the financial system to supply their investment funds or financial deficit (investment minus saving) which is much less than private enterprise did. Only 13 per cent of the investment funds, or 34 per cent of the financial deficit of public enterprises were financed by the financial system (excluding the stock market) over the 1965-1988 period. The major sources of public enterprises' investment funds were their internal savings and stock issuing.

(v) Financial institutions used a major part (over three fifths) of the deposits they received to make loans to households, enterprises, and government. However, accumulation of foreign assets, which was almost entirely held by the Central Bank, also constituted a large proportion (over one third) of the total uses of funds. This simply reflects the fact that Taipei,China has been troubled by excess savings and a huge trade surplus, and has accumulated too large foreign reserves in the 1980s.

Consolidated Balance Sheet of Financial Institutions

The assets and liabilities of all financial institutions, excluding Central Reinsurance Corporation and property and casualty insurance companies, are consolidated and reported in Table 2.9. According to the data, total net assets/liabilities of all financial institutions increased more than 10 times during the period 1970-1980, and grew further at 4.6 times in 1980-1990. Also, in the 1980s, there were major changes in the composition of total assets. Foreign asset holdings (mostly by the Central Bank) had been substituting loans and discounts. Portfolio investments (on securities) had also gained more weight.

On the liabilities side, although clear trends were not evident on some shares, currency issued by the Central Bank had been losing its share in the process of financial development, while trust funds and life insurance reserves had been gaining weight. These trends are consistent with the data presented in Table 2.5 (p. 88). The deterioration of the Government's fiscal condition since 1989 has also caused the share of government deposits to decline.

Distribution of Loans

Table 2.7 indicates that 55 per cent of the incremental loans made by financial institutions in the period 1965-1988 went to households and

Table 2.9 Consolidated Balance Sheet of Financial Institutions[a]
(NT$ billion) (per cent)

Item	1970 Amount	1970 Per Cent	1980 Amount	1980 Per Cent	1990 Amount	1990 Per Cent
Total assets (net)	134.7	100.00	1,530.3	100.00	8,549.3	100.00
Foreign assets	30.2	22.42	358.7	23.44	2,471.1	28.90
Loans and discounts	94.1	69.86	1,022.8	66.84	4,982.5	58.28
Portfolio investments	8.4	6.24	113.0	7.38	907.1	10.61
Real estates	0.9	0.67	16.9	1.10	104.6	1.22
Cash in vaults	1.1	0.82	18.9	1.24	84.0	0.98
Total liabilities (net)	134.7	100.00	1,530.3	100.00	8,549.3	100.00
Foreign Liabilities	7.7	5.72	165.0	10.78	301.6	3.53
Currency issued	14.4	10.69	128.5	8.40	431.7	5.05
Deposits held by enterprises and individuals	80.7	59.91	830.3	54.26	5,874.2	68.71
Government deposits	20.5	15.22	127.8	8.35	618.0	7.23
Trust funds	0.3	0.22	64.7	4.23	373.2	4.37
Life insurance reserves	2.2	1.63	33.5	2.19	409.3	4.79
Bank debentures issued	0.0	0.00	3.9	0.25	59.0	0.69
TB-B, CDs & SB Issued by CBC	0.0	0.00	1.5	0.10	87.8	1.03
Other borrowings	0.9	0.67	5.1	0.33	34.8	0.41
Presettlement requirement for imports	3.7	2.75	22.4	1.46	40.1	0.47
Other items	4.3	3.19	147.6	9.65	319.6	3.74

[a] The Central Bank of China is included, but property and casualty insurance companies are excluded.
Source: Central Bank of China.

nonprofit institutions, 38 per cent went to private enterprises, 5 per cent were granted to public enterprises, and 1 per cent was borrowed by the Government. The distribution of financial institutions' total loans for representative years are reported in Table 2.10. This table shows that households and private enterprises are always the major borrowing sectors. Since 1970, the share of private enterprises has remained roughly constant, while the households sector had gained its share at the sacrifice of the government sector before 1980, and then of the public enterprises sector after 1980. The fall of the public enterprises' share in the 1980s was caused, in part, by the Government's endeavor to reduce the role of public enterprises in the economy due to its low economic efficiency. The rapid rise of the Government's share in the late 1980s is again a reflection of the deterioration of the Government's fiscal condition.

Table 2.10 Distribution of Financial Institutions' Loans
(per cent)

End of Year	Households and Non-Profit Institutions	Private Enterprises	Public Enterprises	Government	Total Amount (NT$ million)
1965	41.16	29.78	15.17	13.89	37,685
1970	39.28	38.93	13.03	8.76	94,076
1975	40.99	40.77	14.94	3.30	391,062
1980	42.55	38.28	17.03	2.14	1,019,600
1985	49.51	35.94	11.83	2.72	1,834,741
1990	51.57	37.42	4.74	6.27	4,988,719

Source: Central Bank of China (1992a).

Information about how loans to enterprises had been allocated among different industries in 1980 and 1990 is provided in Table 2.11. As shown in this table, under the development strategy of export promotion, the manufacturing industries have been favored in loan allocation. They have enjoyed a much larger share of loans from financial institutions than their share of contributions to GDP. By contrast, the service sector was less favored in loan allocation until the late 1980s.

FINANCIAL INFRASTRUCTURE, REGULATIONS, AND MACROECONOMIC POLICIES

Accounting System

The existence of reliable financial statements of firms and banks is a prerequisite for a sound financial system. Accounting's inadequacies paralyze banks when it comes to monitoring borrowers, and handicap the authorities when it comes to monitoring banks. As they are under frequent examination from the financial authorities and the public disclosure requirement on banks which are listed on the stock market, the accounting statements of financial institutions are generally believed to be accurate. However, the reliability of firms' financial statement is quite low in Taipei,China. According to a 1991 survey from the National Taiwan University, respondents were less than satisfied by the financial statements and accounting information provided by firms; and the quality of professional accountants' auditing was generally not believed to be satisfactory. Even the group of accountants responding did not have a high degree of satisfaction with the general quality of their profession's auditing.

Table 2.11 Relative Shares of Loans from Financial Institutions and Contributions to GDP, By Economic Activity[a]
(per cent)

Economic Activity	1980		1990	
	Loans from Financial Institutions	Contributions to GDP	Loans from Financial Institutions	Contributions to GDP
Mining and quarrying	0.09	1.25	0.08	0.55
Manufacturing	72.60	47.45	56.90	44.02
Electricity, gas and water	7.89	3.32	5.72	3.69
Construction	2.11	8.25	3.65	6.24
Wholesale and retail trade, restaurants and hotels	10.28	17.32	22.87	19.70
Transport, storage and communication	5.53	7.87	6.34	7.87
Other services	1.49	14.54	4.44	17.93
Total	100.00	100.00	100.00	100.00
Amount (NT$ million)	551,568	1,131,873	2,101,875	3,295,211

[a] Finance and insurance industries are excluded. Besides, since most of the agriculture sector's borrowings are classified as households' borrowings in the data source, the agriculture sector is also excluded in this table.
Sources: Central Bank of China (1991); DGBAS (1992).

The unreliability of the firms' financial statements and accounting information is quite easy to understand. Given the widespread propensity to evade taxes, especially by the majority of small businesses, it is difficult to obtain sufficient, accurate credit information about virtually any but the largest borrowers and a small number of companies listed on the stock market, which are required to have public disclosure. It is even unclear as to what degree of reliability is possible from large companies' financial statements.

It is said to be a common practice for a firm to have at least three versions of its accounting statements: the one that under reports the profits is for the tax collector; the second that over reports profits is submitted to banks for loan application purposes; and the third more or less accurate account is used for management decisions. Sometimes, additional inaccurate versions are prepared for other shareholders or investors.

The situation is further complicated by the Income Tax Law, which indirectly sanctions the practice of not maintaining adequate accounting records by allowing the tax collecting authority to use industry standards for profitability to estimate the income of small businesses that cannot produce account books. In 1986, for instance, 68 per cent of the enterprises in Taipei had their incomes estimated by this superficial tax review system.

Besides the unwillingness of firms' senior-level managers to maintain adequate accounting records, for whatever reason, the poor reputation of the accounting profession has been a bottleneck in the country's ability to have a sound accounting system. This poor reputation is due partly to the shortage of qualified personnel, and partly to the lack of discipline and low professional standards.

At the end of February 1991, there were only 918 acting professional accountants and about 7,000 professional and administrative assistants to serve about 18 thousand* large enterprises and 150,000 small- and medium-scale enterprises which need financial or taxation endorsements, accounting services, and managerial advisory services from accountant offices.

There are two channels for becoming a CPA in Taipei,China. The first is to pass the high-level national CPA examination; the second is to be sanctioned by specific qualified persons such as accounting teachers at college, and retired military or government personnel with training and experience in accounting for government entities. Before 1983, a majority of licensed CPA came to their position through the second route, but in recent years, the situation has been completely reversed. Since 1988, the passing rate of the national CPA examination has risen to more than 17 per cent, and the number of persons passing the exam has surmounted 200 per year.

Though the supply of accountants is expanding, the quality of accountants' auditing skills still needs to be improved. In Taipei,China, certification by a CPA of a false or misleading financial statement is unlikely to result in legal liability. The discipline committee of the CPA union has never recommended that any of its members be reprimanded for improper conduct, and an attempt to assess accountant offices by the CPA union failed due to strong opposition from the accountant offices. An assessment of the firms by the Securities and Exchange Commission has only recently been completed, although it was begun in 1989.

Cooperation between accountants and managers to evade taxes is still a common phenomenon. A profession of underground accountants, such as retired or on-duty tax officials, and professional writers who write legal documents for clients, has existed for a long time to meet the demand for keeping duplicate accounting books. Until the accounting profession has an adequate supply of qualified CPAs, the standards of the profession are raised, and the managers of firms maintain accurate accounting records, the financial system in Taipei,China will be forced to operate with only inadequate credit information available.

Legal and Supervisory System

The legal framework—including civil law, company law, and the judicial system—for dealing with debt recovery and insolvency cases is well established in Taipei,China. The financial institutions have little difficulty in following the legal procedure to protect their rights of claims.

The formal financial system is supervised and regulated by the Ministry of Finance and the Central Bank of China. The former is in charge of matters relating to administration, and is responsible for taking action against and penalizing wrong-doers in the profession, while the latter is responsible for the banking operations of financial institutions. In addition, Central Deposit Insurance Corporation has the power to examine the operations of the insurant, and the Cooperative Bank of Taipei,China is authorized to examine and supervise the operations of credit cooperatives.

The Banking Law, promulgated in March 1931, and as amended many times, is the primary source of law regulating banks in Taipei,China. The Banking Law governs the formalities of incorporation, application for licensing, modification or dissolution of corporate structure, standards of conduct for bank officers, scope of business, ceiling of credit to related persons or enterprises, and regulatory responses to bank failures. Until an amendment in 1989, the Banking Law also regulated the terms on which credit could be provided. Earlier versions of the Banking Law exercised strict regulations on the type of loans a bank could extend. Although some of these regulations have been relaxed in later revisions, others remain. For example, before 1968, a bank was not allowed to extend credit unsecured for a term longer than six months or secured by pledge or mortgage of real estate for a term longer than one year. In the 1968 revision, the terms were extended to one and three years, respectively. In 1975, this requirement was replaced by one requiring that the volume of medium-term credit not exceed the amount of time deposits a bank holds. This restriction was not removed in the 1989 revision. In addition, until 1975, the Banking Law stated that banks could not extend credit for more than 70 per cent of the appraised value of the collateral. Although this requirement was phased out in the 1975 revision, this 70 per cent ceiling of credit on collateral has been a common practice up to now.

Responding to these regulations, as well as the problems of government ownership of major banks and inaccurate accounting statements which will be discussed later, local bankers focused primarily on collateral rather than the profitability and productivity of the borrowers when they were extending loans. The often-criticized pawnshop mentality of banks in Taipei,China, therefore, can be partly attributed to these regulations. Besides, to observe the restrictions on medium- and long-term lending, banks structured their loans as short run, but with an understanding between banks and borrowers that the loans would be rolled over as long as the collateral was in place.

As will be discussed in more depth in the next section, most of the major large-scale domestic banks in Taipei,China are government owned. The regulations on government banks and the supervision of the Control Yuan—one of the five branches of the Government which is entitled to perform the functions of audit and investigation, and to impeach or censure public officials—have also contributed greatly to the conservative and pawnshop behavior of banks. Loans made by a government-owned bank are interpreted as government assets by the Control Yuan. Getting permission from the auditors of the Control Yuan to write off bad loans is difficult. In addition, as the employees of these banks are considered to be civil servants, the responsible lending officer of a government-owned bank who made any uncollectible loan may be subject to administrative sanctions such as demotion or reprimand, or, in theory, be liable for the repayment of the defaulted loan. The fear of such repercussions has naturally strengthened the emphasis on collateral and heightened the risk averse attitude toward extending loans.

Financial Regulation and Deregulation

Besides the supervisory system described above, a high degree of government intervention was imposed before 1980 on the financial system, especially the banking system. Since 1980, Taipei,China has launched financial deregulation to reorganize its financial system so that it may better serve the economy.

Financial Repression Before 1980[2]

Many economists argue that a competitive financial system without intervention can best achieve optimal allocation of resources and promote economic growth. However, few developing countries seem to believe in this "theory." They usually place strict controls on the financial system and intervene extensively in its functions, i.e., setting interest rates on both deposits and loans, allocating loans by selective credit rationing policies, restricting entry and operations, and keeping some or most of their financial institutions under government ownership. The reasons for these interventions are many, including prudential considerations, market failures, social justice objectives, off-budget subsidy arrangements, and political purposes. The most frequently cited justification is that financial interventions can generate more rapid economic growth by increasing the level of investment, and implementing an export-led development strategy and sector-specific industrial policy. In this regard, Taipei,China is no exception.

To reduce the cost of funds and thus stimulate investment and export, the Central Bank often set the bank interest rate lower than the market equilibrium rate, although the real deposit rate usually remained positive. To resolve the problem of resulting excess demand for loanable funds, criteria other than the productivity of borrowers were applied by the financial institutions to ration credits. The consequences of this include inefficient allocation of loanable funds, compensating balance requirements, commission requests, discrimination against small- and medium-scale enterprises, and a prosperous informal financial system. The financial authorities also often ordered financial institutions to expand or restrict their loans to certain economic activities, industries, or borrower groups for purposes such as to promote economic growth, to stabilize the economy, or to equalize the availability of bank loans.

The Government also owned and managed most of the domestic banks (this issue will be discussed in a later section), an ownership which caused great damage to the economic efficiency of the banking industry. Numerous regulations and restrictions on personnel and on the accounting, budgeting, and auditing functions of the banks have turned government banks into bureaucracies which operate inefficiently and are sluggish in adopting financial innovations. The supervision of the Control Yuan on the government banks also has strengthened the conservatism of these banks.

---

[2] Detailed discussions on the financial interventions in Taipei,China can be found in Shea (1993) and Yang (1993).

Moreover, to prevent instability caused by competition among financial institutions, the authorities had maintained strict regulations on the setting up of new domestic financial institutions and branches until 1991, and nonbank financial institutions such as investment and finance companies, mutual saving and finance companies, and merchant banking corporations were not allowed to become established. Besides, the creation of a wide range of financial instruments which are close substitutes was discouraged, and financial services provided by financial intermediaries were seldom expanded and diversified.

As a consequence of these government interventions, the informal or unregulated financial system has played an important role in Taipei,China. In addition, interest-rate control, government ownership, and market-entry regulation have all contributed to the creation of financial-market segmentation, i.e., public enterprises and large firms have been favored by banks in credit rationing, while small- and medium-scale enterprises have been discriminated against and, therefore, depend heavily on informal-system financing. This phenomenon of financial-market segmentation is illustrated in Tables 2.12, 2.13 and 2.14, which clearly indicate that public enterprises mainly borrowed from the formal financial system, and the smaller the scale of private firms, the larger the proportion of borrowing from the informal system.

Financial Deregulation in the 1980s[3]

In the 1970s and early 1980s, financial deregulation had become a fashion in the international community. Taipei,China also launched the movement of financial deregulation to reorganize its repressed financial system in the 1980s. The two major driving forces for liberalization were pressures from the outside world and changes in the domestic economic and political environments, such as the rapid accumulation of domestic liquidity and wealth, and the relaxation of social and political controls.

The most rapid and significant progress in the country's deregulation process was in interest rate liberalization. In November 1980, the authorities adopted a strategy of "planned gradualism" to liberalize interest rate control step-by-step. Finally, the revision of the Banking Law in July 1989 nominally closed the history of interest rate control in Taipei,China. In principle, interest rates on loans, deposits and other financial instruments could, after 1989, be determined by the market although the influence of the Central Bank and the collusion among banks through bank associations cannot yet be completely eliminated.

---

[3] Shea (1993), Chiu (1992), and Liu (1993) contain detailed information on the financial deregulation policies in Taipei,China.

Table 2.12 Composition of Domestic Borrowings of Public and Private Enterprises
(per cent)

End of Year	Formal Financial System				Informal Financial System[a]	Total
	Financial Institutions	Money Market	Bond Market	Sub-total		
Public enterprises						
1980	86.30	1.34	10.56	98.20	1.80	100
1981	82.02	4.82	10.07	96.91	3.09	100
1982	82.13	3.72	10.10	95.95	4.05	100
1983	82.21	4.59	9.24	96.04	3.96	100
1984	79.03	7.64	10.21	96.87	3.13	100
1985	77.62	10.04	9.50	97.16	2.84	100
1986	76.40	9.09	12.09	97.58	2.42	100
1987	71.72	7.57	17.01	96.30	3.70	100
1988	73.18	5.53	16.66	95.37	4.63	100
1989	74.99	4.60	13.03	92.62	7.38	100
1990	64.25	17.78	11.08	93.11	6.89	100
1991	70.51	10.68	11.28	92.47	7.35	100
Average (1964–1991)	87.35	3.43	6.66	97.44	2.56	100
Private enterprises						
1980	54.66	8.87	0.53	64.05	35.95	100
1981	52.54	11.65	0.50	64.69	35.31	100
1982	53.16	13.31	0.50	66.97	33.03	100
1983	54.52	14.22	0.66	69.40	30.60	100
1984	54.14	14.95	1.05	70.14	29.86	100
1985	51.19	13.13	0.97	65.29	34.71	100
1986	50.16	8.35	0.83	59.33	40.67	100
1987	55.86	6.61	0.65	63.12	36.88	100
1988	63.31	5.73	0.57	69.61	30.39	100
1989	64.74	7.27	0.39	72.41	27.59	100
1990	66.14	9.98	0.38	76.50	23.50	100
1991	67.55	9.27	0.53	77.35	22.65	100
Average (1964–1991)	60.31	4.95	0.66	65.92	34.08	100

[a] Borrowings from informal financial system include borrowings from other enterprises and households.
Source: Central Bank of China (1992a).

Table 2.13 Percentage Distribution of Sources of Domestic Borrowings, By Private Enterprises in 1983, By Scale of Assets

Assets (NT$ million)	Source		
	Financial Institutions	Money and Bond Market	Informal Financial System
Under 1	10.5	0	89.5
1–5	31.0	0	68.9
5–10	44.1	0	55.9
10–40	50.9	0.3	48.8
40–100	59.3	1.0	39.7
100–500	66.4	4.4	29.2
500–1000	65.8	15.9	18.3
Over 1000	70.1	19.6	10.3

Source: Derived from Liu (1988), Table 4, which is compiled from data from a 1983 Central Bank of China survey of the financial conditions of private enterprises in Taipei,China.

Table 2.14 Sources of Borrowings of Private Enterprises: Large Firms vs Small- and Medium-scale Enterprises[a]
(per cent)

		Financial Institutions	Money sand Bond Markets	Households and Other Enterprises	Government Agencies and Abroad	Total
1985	Large	62.70	16.29	16.11	4.90	100
	SMEs	41.49	0.27	58.24	0	100
1986	Large	65.41	12.24	18.28	4.07	100
	SMEs	41.63	0.53	57.84	0	100
1987	Large	70.37	8.97	17.81	2.85	100
	SMEs	45.84	0.70	53.46	0	100
1988	Large	73.58	8.30	14.97	3.15	100
	SMEs	55.87	0.63	43.50	0	100
1989 (June)	Large	76.10	7.49	13.67	2.74	100
	SMEs	55.61	0.66	43.73	0	100

[a]For this table, in 1985–1986 (1987–1989), enterprises with annual sales (total assets) that satisfy the following minimum standards are classified as large firms; otherwise, as SMEs (NT$ million): mining 30(25), manufacturing 50(70), electricity, gas and water 50(10), construction 30(60), commerce 30(25), transportation and communications 30(55), other enterprises 50(30).

Source: Calculated from Table 4.13 in Chiu (1992), which is based on *Data for Flow of Funds in Taiwan District, ROC,* Economic Research Department, Central Bank of China.

To promote market competition, the authorities gradually relaxed control on the setting up of branches of both domestic and foreign banks. Since August 1992, the number of branches each domestic bank is allowed to establish went from three to a maximum of five new branches each year. Most of the operational restrictions on foreign banks were also lifted in the late 1980s. Further, the 1989 revised Banking Law liberalized the establishment of new private-owned banks. By April 1993, 16 new banks had been granted charters and begun operation, and China Trust was converted from an investment and trust company to a commercial bank.

Privatization of major banks has so far made little progress. By definition, if the Government controls more than 50 per cent of the capital of a bank, the bank is government-owned and is then subject to a variety of regulations governing public enterprises. To improve the efficiency of these government-owned banks, so they will survive in the foreseeable fierce competition with the newly-established private banks, the authorities originally decided to reduce the shares of government-owned equity in the three most influential provincial government-owned commercial banks (First, Hua-Nan, and Chung-Hwa) to below 50 per cent by selling a portion of the stocks in the market. However, the idea faced strong opposition from the members of the Taiwan Provincial Assembly and some bank employees. The Assemblymen "captured" the banks; and the employees, who enjoy an easier life and more job security in the government banks than in the private banks, are unwilling to abandon their vested interests. The Ministry of Finance has planned to privatize some central government-owned specialized banks.

Distress Managing Method and Prudential Re-regulation

Though safety and stability of the financial system have always been given high priority in the decision-making process of the financial authorities, there have still been about 20 cases of financial distress since 1963. Most of these distress events occurred in the credit departments of farmers' association or credit cooperative associations. Among the seven investment and trust companies licensed in the 1970s, three also experienced major liquidity crises and have been rescued through official intervention.

These financial distresses originated mainly from internal factors, such as poor management, false investment, and concentration of credit to industrial groups. For example, overinvestments on real estate left Asia Trust and Cathay Trust in severe problems in the first half of the 1980s when the real estate market went into recession. The failures of Asia Trust, Cathay Trust, and the Tenth Credit Cooperative Association in Taipei were also, to a great extent, the result of overextending of credit to their affiliated companies.

Before the 1982 Asia Trust incident, all the financial distresses occurred in small-scale cooperatives, and were easily resolved either by simply letting the cooperative go bankrupt or by consolidating it with another cooperative. The lack of experience in dealing with major financial crises, therefore, resulted in a variety of distress managing methods being applied to the following five major events: (i) In 1982, the management of Asia Trust was turned over to the International Commercial Bank of China. After several years, it was returned to its original owner, the Asiaworld group of companies. (ii) Overseas Chinese Bank, which ran into insolvency trouble in 1984, was partly acquired by the Government and run by a government-appointed managing group. (iii) The Tenth Credit Cooperative Association in Taipei was handed over to the Cooperative Bank of Taipei,China for management in 1985, and eventually merged with the Bank. (iv) In 1985, Cathay Trust experienced a run and was taken over by a group of banks; several years later it was sold to new investors. (v) In late 1985, the Overseas Trust was taken over by the United World Chinese Commercial Bank after a run.

These inconsistent distress managing methods resulted in protests from the original investors of the failed institutions, and also in strong criticism from the general public, especially for the cases such as Asia Trust, the Tenth Credit Cooperative Association, and Cathay Trust, in which the authorities spent large sums for the rescue, or designated government-owned banks to resume the rescue responsibility. To institutionalize the distress managing method, the authorities set up the Central Deposit Insurance Corporation in 1985, enriched the contents of its distress managing-related articles, and legalized the managing procedure in the revised Banking Law in 1989. Further, to reduce financial distress, the authorities are designing an early warning system, and strengthening the monitoring and auditing system. The idea of the early warning system is to use an econometric model to analyze the financial status of financial institutions and thus screen out shaky institutions at early stage.

Under the deposit insurance system, the financial intermediaries are free to choose to be insured or not, and a uniform insurance premium is imposed on the insured deposits. In June 1992, 36 per cent of the privately owned financial institutions (164 out of 455) and four of the 13 government-owned banks were insured; the amount of insured deposits reached 39.4 per cent of the total eligible deposits of all financial institutions. The insured ratios of deposits for the privately owned and government-owned financial institutions were 68.2 per cent and 12.6 per cent, respectively. The reason for a much lower insured rate for government-owned banks is that the Taiwan Provincial Assembly is strongly against the idea of allowing the provincial government-owned banks to "waste" money on deposit insurance. The Assemblymen believe that the Government has implicitly assumed the full responsibility of the eternal operation of government-owned banks. The Central Deposit Insurance Corporation thus faces a number of major challenges: how to expand the insurance base; how to solve the traditional problems of adverse selection and moral hazard caused by insurance; and how to strengthen the insurance funds.

Following financial deregulation, the authorities have also adopted several measures to address the effects of entry deregulation on stability and equity. The 1989 revised Banking Law: (i) limits individual ownership stakes to 5 per cent and group shareholdings to 15 per cent of any newly established private bank, so as to prevent any family or industrial group from controlling a bank; (ii) sets stringent limits on related-party lending to prevent financial scandal; (iii) sets a requirement for a risk-based capital ratio of at least 8 per cent, as recommended worldwide by the Bank for International Settlements; and (iv) enforces stricter supervision over banks and imposes heavier penalties against offenders. The Guidelines for the Establishment of Commercial Banks also sets a minimum capital requirement as high as NT$10 billion, equivalent to roughly $370 million, and requires that at least 20 per cent of the capital be issued to the general public.

Despite this prudential re-regulation, effective supervision of the new private banks will be a major challenge for the authorities. There is clearly a shortage of qualified regulatory and monitoring personnel in the Ministry of Finance and the Central Bank of China. The lack of clear and proper definitions on "industrial group" and "related-party," and the common practice of "dummies" or "name borrowing," (borrowing another person's name for the purpose of making transactions or registration), also hinder the effective implementation of ownership and lending-limit controls. As a result of name-borrowing, most of the new private banks are in fact set up by groups affiliated with major industrial or commercial conglomerates in Taipei,China.

Macro Stabilization Policies

Stable prices are a benefit to the development of financial systems. During the decades of rapid economic growth since the mid-1950s, Taipei,China has enjoyed relatively stable prices. This price stability certainly contributed to the development of the financial system.

In the early years after World War II until around 1952, Taipei,China suffered high inflation. The end of this inflationary period was attributed to currency reform, the introduction of gold saving deposits and preferential rate deposits, and the arrival of aid from the US (Tsiang 1982; Lee and Chen 1984). Stabilization of prices in Taipei,China has been a high-priority national goal since 1952. As a result, except for 1955 and 1960, and during two oil shocks, the prices in Taipei,China have been quite stable.

Several policies have contributed to this price stability: hyperinflation in the People's Republic of China during the period 1945–1949, which was caused by printing money to finance the budget deficit, taught the Government of Taipei,China to maintain its budget in balance or in surplus. The government sector was a major supplier of financial surplus until 1989. In addition, the following major financial policies have been adopted to stabilize domestic prices: money supply policy, interest rate policy, and exchange rate policy.

Money Supply Policy

During the period 1962–1992, the money supply (M2 which is defined as the sum of currency, checking account deposits, passbook deposits, and passbook saving deposits) in Taipei,China increased at an average annual rate of 20 per cent, which was much higher than the average growth rate of real GNP (9.1 per cent). However, due to a high income elasticity of money demand, estimated to be 1.3 to 1.5, prices in Taipei,China remained quite stable.

By decomposing the increase in money supply from 1962 to 1992 into four sources—i.e., the financing of government budget deficits, excess expansion of credit to households and enterprises[4], accumulation of net foreign assets by the banking system, and others—the following three outcomes occurred: (i) the financing of government budget deficits has in fact been a contractionary factor to the money supply over the period; (ii) excess credit expansion to the general public was the most important source for the increase in money supply in the 1960s and 1970s; and (iii) the accumulation of net foreign assets by the banking system contributed greatly to money creation in the 1960s and 1970s, and was the primary cause of the rapid increase in money supply in the 1980s.

Before the money market became effective and of substantial scale, the Central Bank of China depended on a required reserve ratio, control of the rediscount rate, redeposits from the postal savings system and the Cooperation Bank, accommodations to domestic banks, and moral influence to control the money supply. Since 1979, open market operations through the money market have become an additional powerful policy tool. For example, when the huge trade surplus caused the supply of high-powered money to expand very quickly, the Central Bank issued treasury bills, CDs and savings bonds in 1986 and 1987, equal to well over half of the money supply of the previous year, to sterilize the impact of the trade surplus on the money supply.

---

[4] Excess credit expansion to households and enterprises is defined as the difference of monetary institutions' claims on households and enterprises over quasi-money (mainly composed of time deposits and time savings deposits) and pre-settlement requirements for imports.

Interest Rate Policy

Interest rate policy is an effective instrument for fighting inflation. Both McKinnon (1973) and Tsiang (1979) have advocated the policy of raising the deposit interest rate to end inflation. Taipei,China is probably the first developing country to set an example by doing this successfully. In the early 1950s, the preferential savings deposits rate, which offered an extraordinary nominal interest rate, was introduced under the suggestion of Tsiang to stabilize domestic prices. The impact was indeed prompt and successful. The details of that policy and its effects can be found in Tsiang (1979, 1982).

This early success, however, did not lead the authorities to accept the principle of letting interest rates equate the demand and supply of loanable funds. Bank interest rates were strictly under the control of the Central Bank before 1980, and were usually set at levels lower than the market clearing rates to stimulate investment, as explained above. Fortunately, the Central Bank exercised its authority by, for the most part, setting interest rates high enough above the inflation rate (Table 2.15), which presumably contributed to the stimulation and mobilization of savings. The Central Bank also raised the bank interest rates as part of the package of stabilization policies when the economy faced the threat of inflation.

Exchange Rate Policy

In its early years of economic development, Taipei,China was seriously troubled by a foreign exchange shortage. The authorities therefore imposed strict foreign exchange control and adopted a dual or multiple exchange rate system at different points of time. In September 1963, a dual exchange rate system was replaced by a fixed, uniform exchange rate system. Subsequently, the New Taiwan dollar (NT$) appreciated relative to the US dollar in February 1973, from NT$40 per US$1 to NT$38. In July 1978, the NT$ again appreciated against the US dollar from NT$38 to NT$36, and a floating exchange rate system was adopted.

The purpose of introducing a floating exchange rate was to make the economy less vulnerable to external disturbances. In 1972, 1973, 1977, and 1978, a huge trade surplus resulted in 33 per cent, 45 per cent, 34 per cent, and 35 per cent increases in the money supply, which in turn intensified the severe inflation in 1974, 1979, and 1980 (the GNP deflator rose by 32 per cent, 11 per cent, and 16 per cent, respectively). Having learned from these experiences, the authorities adopted a floating exchange rate system in the hope that it would ease the impact of foreign disturbances on the domestic money supply and on the stability of the country's economy.

Table 2.15 Annual Interest Rates and Inflation Rates
(per cent)

	Nominal Interest Rate			Real Interest Rate		
	One-Year Savings Deposit	Curb-Market[a]	Percentage Change in Wholesale Price Index	One-Year Savings Deposit	Curb-Market	Interest Rate Difference
	(1)	(2)	(3)	(4)=(1)–(3)	(5)=(2)–(3)	(6)=(2)–(1)
1980	12.50	30.63	21.54	–9.04	9.09	18.13
1981	13.59	30.96	7.62	5.97	23.34	17.37
1982	11.18	28.28	–0.18	11.36	28.46	17.10
1983	8.80	26.69	–1.19	9.99	27.88	17.89
1984	8.31	25.97	0.48	7.83	25.49	17.66
1985	7.23	24.98	–2.60	9.83	27.58	17.75
1986	5.52	22.51	–3.34	8.86	25.85	16.99
1987	5.00	22.90	–3.25	8.25	26.15	17.90
1988	5.12	20.87	–1.56	6.68	22.43	15.75
1989	8.09	21.74	–0.38	8.47	22.12	13.65
1990	9.50	23.44	–0.61	10.11	24.05	13.94
1991	9.15	23.83	0.17	8.98	23.66	14.68
1992	7.91	23.93	–3.05	10.96	26.98	16.02
Average (1961–1992)	10.06	25.54	3.74	6.32	21.81	15.48

[a] Average interest rates on unsecured loans and loans against post-dated checks of Taipei,China, Kaohsiung Municipalities and Taichung City weighted by the amounts of bank clearings of each city.
Source: Central Bank of China.

However, after floating rates were introduced in 1978, the accumulation of net foreign assets by the banking system turned out to be the primary source of the rapid growth of money supply in the 1980s, the reason being that the authorities still took the exchange rate as an important policy instrument to promote exports. In the mid-1980s, as the domestic investment ratio was dropping dramatically and the trade surplus was growing each year, the Central Bank intervened in the foreign exchange market to slow the appreciation rate of the NT$ and thus maintain the profitability of domestic products in the export market. As a result, a huge trade surplus and speculative capital inflow caused the money supply to expand very quickly.

BANKING SYSTEM

Taipei,China's Banking Law defines banks to cover commercial banks (including foreign bank branches), savings banks, specialized banks, and investment and trust companies. There are no savings banks. Most domestic commercial banks have established a savings department, and some have trust departments. Thus, "universal" banking is a common practice in Taipei,China. Even most of the specialized banks are also engaged in commercial banking and savings banking, as well as specialized banking, Since investment and trust companies have a different business focus than the others, they are classified as nonbank financial intermediaries in this chapter, and will be discussed in the next section.

The banks in Taipei,China, excluding the Central Bank, can be classified into three groups: full service domestic banks (including commercial banks and specialized banks), foreign bank branches, and medium business banks. The number of units, and shares of total assets, loans and investments, and total deposits are shown in Tables 2.3 and 2.4 (pp. 85 and 86).

Central Bank of China

After the Government retreated from the People's Republic of China in 1949, the Central Bank of China did not resume operations in Taipei,China until 1961. The Bank of Taiwan, which is a commercial bank controlled by the Provincial Government, performed most of the functions of a central bank. In 1961, the Central Bank resumed operation to improve the country's financial condition and promote economic development. Before 1979, the Central Bank was under the supervision of the President, and enjoyed a high degree of independence in its operations, usually dominating the Ministry of Finance. However, in 1979, the Central Bank was placed under the supervision of the Executive Yuan (the Republic of China Cabinet) to better coordinate with the Cabinet in economic policies. A general impression of the country's economists is that the policies of the Central Bank have been increasingly influenced by the cabinet and legislators since the late 1980s.

Full-Service Domestic Banks

There were 16 domestic banks in Taipei,China by the end of 1990, with 722 branches located around the island. Between June 1991 and April 1993, 16 new private banks, each with five branches, were allowed to enter the market. In July 1922, China Trust, the largest investment and trust company, also received permission to convert into a commercial bank. Thus, at the end of April 1993, there was 33 full service domestic banks with 915 branches.

Aside from commercial banking operations, most of the domestic banks operate savings and foreign exchange departments (except for new private banks), and some have trust departments. Among them, four are specialized banks. They are:

(i) the Bank of Communications: resumed operations in Taipei,China in 1960, and specialized in venture-capital investments and medium- and long-term loans to industries, mining, transportation, and other public enterprises. From 1992, it has been called Chiao Tung Bank in English as well as Chinese;

(ii) the Farmers Bank of China: resumed operations in Taipei,China in 1967. It specializes in agricultural financing;

(iii) the Land Bank of Taipei,China: reorganized and established from an inherited Japanese bank in 1946, it specializes in real-estate and agricultural loans; and

(iv) the Export-Import Bank of China: established in 1979, and responsible for export and import financing. Its services include extending medium- and long-term loans, endorsements, and insurance on exports and imports. It does not accept deposits from the public, thus funds come mainly from capital, undistributed earnings, borrowing from the Central Bank and other financial institutions, support from the Development Fund of the Executive Yuan, and the issuance of bank debentures.

These specialized banks were established mainly to better conduct the country's growth promotion polices, which will be discussed in a later section of this chapter. Except for the Export-Import Bank, the other three specialized banks are, in fact, engaged in a mixture of functions including commercial banking, savings banking, and specialized banking. In addition, the Cooperative Bank of Taipei,China, reorganized from a colonial-period bank, is designated by the Ministry of Finance to provide agricultural finance and is authorized to examine and supervise the operations of credit cooperatives. The Central Trust, founded on the mainland in 1935 and removed to Taipei,China in 1949, is responsible for cooperating with the Government in its purchasing, trade, banking, trust, insurance, storage, freight, and other related needs. It also operates as a commercial bank and as an investment and trust company. These two banks are also sometimes classified as specialized banks.

Among the 33 full-service domestic banks, 12 are government-owned banks (defined as banks where the Government owns more than 50 per cent of the capital), four are old private banks, and 17 are new private banks (including China Trust Bank). The above specialized banks, except for Land Bank, and together with the Central Trust, are under the control of the Central Government, five commercial banks and the Land Bank are controlled by the Provincial Government, and two are affiliated with the Taipei and Kaohsiung Municipal Governments, respectively. The numbers of operating units and the market shares of various groups of banks as of the end of April 1993 are reported in Table 2.16.

Table 2.16 Composition of Full Service Domestic Banks (End of April 1993)

Group of Banks	Operation Units		Market Share (per cent)		
	Head Office	Branches	Total Deposits	Total Loans	Total Assets
Government-owned Banks	12	721	82.43	82.21	79.93
Commercial banks	8	576	67.16	61.68	62.10
Central	1	9	0.95	1.11	2.05
Provincial	5	514	60.39	55.03	55.26
Municipal	2	53	5.82	5.54	4.79
Specialized banks	4	145	15.27	20.53	17.83
Private banks	21	194	17.57	17.79	20.07
Old banks	4	100	8.29	8.10	10.61
New banks	17	94	9.28	9.69	9.46
		(17)[a]	(3.89)[a]	(3.14)[a]	(2.94)[a]
Total	33	915	100.00	100.00	100.00
Amount (NT$ billion)			5,486	5,153	12,003

[a]Figures for China Trust Commercial Bank
Source: China Trust, Ministry of Finance (1993).

This table clearly indicates that government-owned banks far outweigh the private banks in importance. Although there are only 12 government-owned banks out of 33, they have 721 branches, which is 3.7 times as many as the 21 private banks with their 194 branches. The market shares in deposits, loans, and total assets of the government-owned banks are also nearly four times that of the private banks. This table also shows that the full-service domestic banks are dominated by the five banks owned by the Provincial Government. In 1993, these five large banks own more than half of the total number of branches, as well as the market shares in deposits, loans, and total assets. As for the 16 new private banks (excluding China

Trust), all were established within two years; their market share, although still small, is certain to grow in the next few years.

Several factors have contributed to the dominance of government ownership of the domestic banks: (i) six banks were inherited from the Japanese colonial government after World War II, and were put under the control of the Provincial Government; (ii) government-owned specialized banks were set up or resumed operation in Taipei,China to implement the growth-promotion financial policies; and (iii) government ownership of banks has long been considered an effective strategy for preventing the formation of conglomerates and the concentration of economic power. Government-owned banks are also more easily controlled by the authorities and more willing to comply with the prudential rules.

Although domestic banks constituted only about 20 per cent of the total number of units of financial institutions, they have always been the backbone of the country's financial system. As shown in Table 2.4 (p. 86), although the role of domestic banks decreased steadily from 1961 to 1990, they still accounted for the majority of total assets, total loans, and investments, and almost half of the total deposits of all financial institutions in 1990. The main reason their role declined is that the domestic banks have long suffered from severe restrictions on the operations and the setting up of new banks and branches. However, after the entry of new private banks, this role can be expected to increase.

Local Branches of Foreign Banks

Before 1965, the monetary authorities in Taipei,China hesitated to open the domestic market to foreign banks. As US economic aid was about to end in 1965, and as the country's official diplomatic relationships suffered contractions in the late 1970s, the authorities revised their stance and allowed foreign banks to establish branches in Taipei. This was intended to encourage foreign investment, strengthen economic ties with the outside world, and improve domestic banking operations through competition and transfer of advanced banking management technology and knowledge. In addition, because of the country's rapid economic growth, the interest of foreign banks to enter this market intensified. The number of local branches of foreign banks therefore increased dramatically from one (the Nippon Dai-ichi Kangyo Bank) in 1964 to 43 at the end of 1989. Although several banks retrenched in 1990 due to operation losses and mergers of overseas parents, the numbers continued to grow, reaching 51 at the end of April 1993. These 51 branches comprise 37 foreign banks from 14 countries and Hong Kong. Among these, 15 are from North America, 11 from European countries, and the rest from Southeast Asia and other countries.

Before 1986, foreign banks were each allowed to set up only one branch office in Taipei. Then in 1986, foreign banks with a branch office in Taipei were allowed to establish a second branch in Kaohsiung. Later, the number of branches allowed was increased further, so that at the end of April 1993, there were 11 foreign banks with two branch offices each, and Citibank with four.

Following the relaxation of entry barriers, some of the operational restrictions on foreign banks were also gradually lifted in the 1980s. Foreign banks are authorized to handle foreign exchange transactions, extend loans to individuals as well as firms, engage in trust business, and apply for licenses as securities brokers, dealers, and underwriters. They can accept local currency deposits—checking accounts, demand deposits, saving deposits and time deposits, with the total being subject to a ceiling of 15 times their paid-in capital. They are also eligible for the interbank call-loan market. Hence, foreign banks now enjoy the same range of activities as domestic banks.

However, despite the expansion of the operating units and the business scope, the role of the foreign banks in Taipei,China is still limited (Table 2.4, p. 86). In 1990, they constituted merely 1.1 per cent of all financial institutions in terms of operating units and total deposits, and just 3 per cent in total assets and loans. The shares of foreign banks in total assets and total loans and investments in 1990 were even substantially lower than those in 1980. The main reason is that foreign banks suffered great losses from defaulted loans in the early 1980s which led to their adopting a more conservative lending policy.

Medium Business Banks

Medium business banks were developed in 1978 and 1979 from the mutual loans and savings companies, which ran the business of rotating credit. By the end of 1979, the mutual savings still accounted for almost 20 per cent of the total deposits; the mutual loans represented over half of the total loans of medium business banks. However, the roles of mutual loans and savings has declined rapidly ever since, with their shares of total deposits and total loans declining to only 2.2 per cent and 1.3 per cent, respectively, by the end of 1992.

Under the Banking Law, medium business banks are identified as specialized banks which extend medium- and long-term credit to small- and medium-scale enterprises. Loans to these enterprises must represent at least 70 per cent of the total loans. At the end of 1992, each medium business bank extended about 70 per cent of its total loans to small- and medium-scale enterprises, just meeting the minimum requirement. With the exception of the government-owned Medium Business Bank of Taiwan, which operates with islandwide branching, the other seven privately owned medium business banks are restricted in their operations to nonoverlapping regions, and thus do not compete with each other.

Since the early 1960s, the number of units of medium business banks (or mutual loans and savings companies) had maintained a stable share of about 6–7 per cent of all financial institutions. However, their share of total assets, and of loans and investments has more than doubled, while their share of total deposits has increased by about half.

Deposits and Loans of Banks

Banks mainly deal with short- and medium-term credit. They compete with bills finance companies to provide working capital to businesses, and compete with investment and trust companies in markets for long-term savings and credit.

On the deposits side, the household sector has always been the dominant source. At the end of 1992, government agencies, public enterprises, and private enterprises contributed only 12 per cent, 3 per cent, and 16 per cent, respectively, to the total deposits (NT$4,461 billion) of domestic banks, with the remaining 69 per cent contributed by individuals and others. Benefiting partly from the stability of the general price level, the maturity structure of deposits has been improving. As Table 2.17 shows, the share of time deposits of one year or over with domestic banks has been increasing.

Table 2.17 Maturity Structure of Deposits with Domestic Banks[a]

	1961–1970	1971–1980	1981–1992
Checking and passbook deposits	38.6	37.9	27.9
Time deposits of less than 1 year	16.5	13.6	16.3
Time deposits of 1 year or over	44.9	48.5	55.8

[a] Data are averages of annual shares within the period. Time deposits include time savings deposits.
Source: Central Bank of China.

Due to such factors as Banking Law restrictions, government ownership of major banks, and inaccurate accounting records of enterprises (especially the small- and medium-scale enterprises), collateral has been emphasized by banks in extending credit, medium- and long-term loans are in short supply, and private enterprises have been discriminated against in applying for loans. Table 2.18 indicates that secured loans accounted for nearly 59 per cent of the loans made by domestic banks in 1969–1992, and, although the second column of this table shows that the share of medium- and long-term loans (defined as the loans of maturity over one year) in total loans of all domestic banks had increased steadily over time, it always remained at a level of less than 40 per cent before 1980. In the 1980s, the banks—troubled by excess liquidity—began to expand their consumer loans. A major part of the consumer loans had been used to speculate in the real estate and stock market booms, and has been secured by the households' real estate. These secured consumer loans were mostly long term. Therefore, the shares of secured loans, medium- and long-term loans, and loans of more than five years increased sharply in the 1980s. The Central Bank's survey of financial conditions on the private enterprises also shows that medium- and long-term borrowing constituted only 21.5 per cent of total borrowing from financial institutions by private enterprises during the period 1981–1991.

Table 2.18 Secured Loans and Medium- and Long-Term Loans as a Percentage of Total Loans of Domestic Banks[a]
(per cent)

Period	Secured Loans	Medium- and Long-Term Loans	Loans of More Than 5 Years
1961–1965	61.1	30.1	...
1966–1970	63.9	31.4	...
1971–1975	55.4	31.8	13.6[b]
1976–1980	54.0	35.2	18.5
1981–1985	55.8	50.1	28.9
1986–1992	61.7	63.4	47.6
1961–1992	58.8	41.8	31.4[c]

[a]Discounts and advances on imports are not included in total loans.
[b]Average of 1974–1975.
[c]Average of 1974–1992.
... data not available.
Source: Central Bank of China.

Table 2.19 reports the distribution of loans and discounts at domestic banks by borrowing sector.[5] To promote economic growth, most of the loans were used to finance the business sector, while individuals' financing

---

[5] Note that figures in Table 2.19 (see p. 121) differ greatly from those in Table 2.10 (see p. 95). Since loans made by cooperatives, which are restricted to their individuals members, are included in Table 2.10, the share of households in Table 2.10 is much higher than that in Table 2.19.

needs were neglected purposely before 1980. In the 1980s, the decline of the domestic investment ratio to GNP reduced the loan demand of the business sector, and banks expanded loans to households. The share of loans to individuals, therefore, increased rapidly in the 1980s. In addition, the deterioration of the budget starting from late 1980s caused the share of government borrowing to rise.

Table 2.19 Distribution of Loans and Discounts at Domestic Banks, By Borrowing Sector[a]
(per cent)

Period	Government Agencies	Public Enterprises	Private Enterprises	Individuals and Others
1961–1965	6.06	30.96	48.62	14.36
1966–1970	5.24	18.76	63.77	12.23
1971–1975	2.79	17.86	67.79	11.56
1976–1980	2.30	21.05	58.87	17.78
1981–1985	3.38	22.59	46.72	27.31
1986–1992	7.00	9.06	41.71	42.23
1961–1992	4.62	19.36	53.78	22.24

[a]Data are averages of annual shares within the period.
Source: Central Bank of China.

The uneven distribution of lending to public and private enterprises, with public enterprises having been favored by the banks, has been a source of ongoing criticism. According to Table 2.12 (p. 102), which will be discussed in more depth in a later section of this chapter, financial institutions (including banks and nonbank financial intermediaries) funded about 87 per cent of public enterprises' domestic borrowing over the 1980–1991 period (average 1964–1991), but provided only 60 per cent of private enterprise domestic borrowing. According to calculations by Shea (1993), public enterprises had also always been granted more loans by financial institutions per unit of production value or GDP than had private enterprises until the mid-1980s. Over the period of 1965–1988, the ratio of loans from financial institutions relative to gross value added averaged 47 per cent for public enterprises but only 29 per cent for the private.

At the end of 1992, loans to small- and medium-scale enterprises amounted to 36.6 per cent, 74.4 per cent, and 4.1 per cent of the total loans outstanding for full-service domestic banks, medium business banks and foreign bank branches respectively, and averaged 39.9 per cent of the total loans of all banks. To compare the availability of loans from financial institutions by company size, Table 2.13 (p. 103) shows the composition of domestic borrowing of private enterprises classed by asset scale in 1983. This table shows very clearly that the larger the firm, the greater the ability to obtain financing from both financial institutions and the money and bond markets; and the smaller the asset scale, the greater the dependency on high-cost informal-system financing. This will be explained in a later section of this chapter. Table 2.14 (p. 103) also shows that the ratio of borrowing from financial institutions was higher for large firms than for small- and medium-scale enterprises each year from 1985 to 1989.

Operational Efficiency of Banks

The ratio of nonperforming loans is often used as one of the measures of the operational efficiency of financial institutions.[6] Nonperforming loans in Taipei,China can be classified into three groups:

(i) overdue loans: loans having not been repaid on the due date;

(ii) called accounts: overdue loans which meet one of the following two criteria: (a) a loan which is six months overdue; (b) a loan that is less than six months overdue but for which the loan's collateral is claimed by other creditors; and

(iii) bad loans: the amount of unpaid two-year overdue loans, deducted by the estimated repayable parts, such as from selling the collateral.

It is a common practice for banks to renegotiate with about-to-over-due or already-overdue borrowers to pay back part of the loans and roll over the rest, rather than to list them as overdue or bad loans. The restriction on government banks, which says that the bad loan ratio for secured loans and unsecured loans should not be over 1 per cent and 3 per cent respectively, strengthened by other regulations and supervision on government banks and their employees, further intensifies the reluctance of government banks to list any loan as an overdue or bad loan. The conservative behavior caused by the pawnshop mentality of government banks also helps to reduce their ratio of nonperforming loans or bad loans. Therefore, judging the operational efficiency of any type of bank by comparing the ratios of nonperforming loans or bad loans among different types of banks may not be meaningful.

This background aids in the discussion of Table 2.20, which shows the nonperforming loan ratio and bad loan ratio of government-owned commercial banks, government-owned specialized banks, privately owned

---

[6] The discussion on nonperforming loans in this subsection is based mainly on Yang (1993).

banks, and foreign bank branches over the period 1977–1991. The two ratios have exhibited a declining trend since the late 1980s.[7] In recent years, the ratios of nonperforming loans and bad loans of all banks averaged about 2 per cent and 1 per cent, respectively. Also, as expected, since the early 1980s, government banks usually had lower ratios of nonperforming loans and bad loans than privately owned banks and foreign bank branches. Moreover, a greater willingness to take risks and write off bad loans, and their unfamiliarity with the local custom of inaccurate accounting records caused foreign bank branches to have even higher ratios of nonperforming loans and bad loans than privately owned domestic banks.

Table 2.20 also reports the ratio of dishonored checks, which can be used as a measure of the general condition of creditability. This ratio of dishonored checks in 1991 remained at about the same level as that in 1978. However, the ratios of nonperforming loans and bad loans had fallen sharply during the same period. This contrast may indicate that the banks had been improving their operational efficiency.

There are no data available on the spread between deposit and loan rates of banks for the period before 1980. However, as shown in Figure 2.1, which is reproduced from Chiu (1992), the differential between the weighted average loan rates and the weighted average deposit rates of locally incorporated banks continued to decline from 5 per cent in January 1982 to less than 2.5 per cent in March 1989, which is evidence of the success of the efforts of interest rate liberalization in the 1980s.

NONBANK FINANCIAL INTERMEDIARIES

To maintain financial stability, the financial authorities have been conservative in introducing new types of nonbank financial intermediaries. So far, no investment and finance companies, mutual saving and finance companies, or merchant banking corporations have been allowed.[8] Thus, there are only four kinds of nonbank financial intermediaries in Taipei,China: cooperatives, investment and trust companies, postal savings system, and insurance companies. Though cooperatives are allowed to do so, the other three groups cannot accept demand deposits and checking deposits, and

---

[7] The economic recession in 1985 caused the nonperforming loan ratio of that year to rise. The Ministry of Finance allowed firms with financial problems to extend the due date of bank loans in that year to relieve the impact of recession. Without that policy, the nonperforming loan ratio in 1985 would have been even higher.

[8] In the late 1980s, unregulated financial investment companies had been prospering. Though some scholars advocated the legalization of those financial investment companies, the suggestion was rejected by the authorities on the basis that legalizing illegal activities is immoral and unjust. The proper timing for introducing this new type of nonbank financial intermediaries was thus lost.

Table 2.20 Ratios of Nonperforming Loans and Dishonored Checks
(per cent)

End of Year	Nonperforming Loan Ratio[a]					Bad Loan Ratio[a]					Ratio of Dishonored Checks	
	All Banks	GCB[b]	GSB[b]	PB[b]	FBB[b]	All Banks	GCB	GSB	PB	FBB	Ratio in Number	Ratio in Amount
1977	6.00	5.38	6.55	4.48	7.31	2.28	2.04	2.31	2.98	2.51	0.58	0.43
1978	4.71	4.52	5.48	3.10	5.17	2.22	2.03	2.56	2.14	2.30	0.33	0.25
1979	5.13	4.81	6.42	3.79	4.47	2.55	2.43	2.86	2.55	2.16	0.45	0.33
1980	4.68	4.39	5.96	3.47	3.85	2.52	2.33	2.87	2.65	2.32	0.34	0.25
1981	4.09	3.91	5.12	3.83	2.54	2.27	2.05	2.62	2.71	1.90	0.46	0.31
1982	4.26	4.42	4.03	4.07	4.70	2.25	2.09	2.09	2.84	2.97	0.59	0.36
1983	4.55	4.60	3.86	5.00	7.29	2.11	2.06	1.77	2.70	3.42	0.65	0.44
1984	5.51	4.42	3.53	6.50	14.52	1.92	1.89	1.49	2.98	3.15	0.67	0.44
1985	5.64	5.02	3.85	6.93	23.33	1.94	1.77	1.48	2.66	5.87	0.75	0.69
1986	5.08	4.45	4.32	5.59	14.08	1.89	1.79	1.64	0.99	5.81	0.39	0.33
1987	4.43	3.99	3.67	5.31	11.25	2.05	1.90	1.77	2.16	4.92	0.39	0.22
1988	3.15	2.95	2.86	3.20	7.13	1.83	1.75	1.71	1.60	4.00	0.36	0.18
1989	2.52	2.33	2.13	2.57	7.24	1.56	1.58	1.39	1.34	3.40	0.35	0.17
1990	2.27	2.18	2.16	2.52	3.34	1.21	1.12	1.16	1.33	2.52	0.42	0.23
1991	1.78	1.77	1.57	2.32	2.65	0.95	0.94	0.87	0.95	1.91	0.39	0.22

[a]The ratio is calculated by dividing nonperforming (bad) loans by total outstanding loans at the end of the year.
[b]GCB: Government-owned commercial banks; GSB: Government-owned specialized banks; PB: Privately owned banks; FBB: Foreign bank branches.
Source: Ministry of Finance.

therefore cannot create deposit money. The number of operating units and market shares of each group in the total of all financial intermediaries is reported in Tables 2.3 and 2.4 (pp. 85–86).

Figure 2.1 Interest Rate Differentials Between Weighted Average Loan Rates and Deposit Rates of Domestically Incorporated Banks

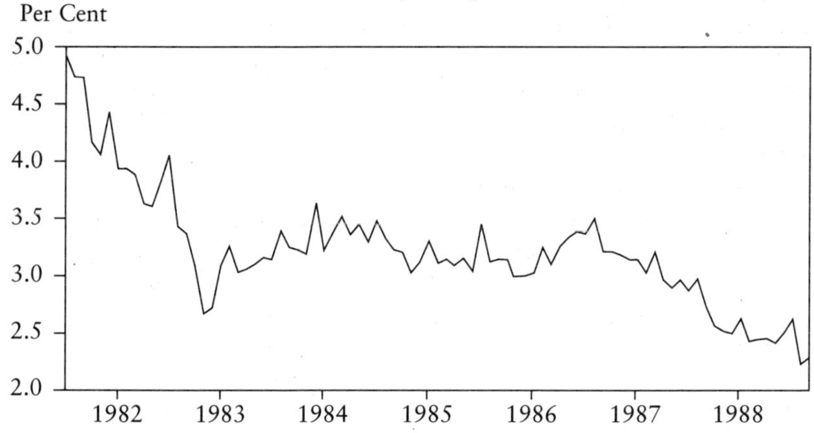

Source: Chiu (1992, Figure 4.8b).

Cooperatives

There are three types of cooperatives in Taipei,China: credit cooperative associations, credit departments of farmers' associations, and credit departments of fishermen's associations. Any individual can become a member of a credit cooperative association, but only farmers or fishermen can become members of their respective professional associations. Cooperatives operate within a certain district and may accept deposits, including checking and demand deposits, from their members and may grant loans to their members only. Their scope of business is the same as commercial banks, but the Banking Law does not include them in the category as banks. Compared to other financial institutions, cooperatives enjoy the advantage of a 5 per cent business tax exemption on their net income.

Under the membership constraint, cooperatives cannot take deposits from or extend credit directly to enterprises, although small- and medium-scale enterprises can borrow from cooperatives indirectly through the proprietors. Thus, cooperatives are usually small in scale. As a consequence, although they represented about 35 per cent in the number of units to all financial institutions, their shares of total assets and loans and investments were always less than 17 per cent, and their deposits accounted for less than 20 per cent.

A special feature of Taipei,China's cooperative financing system is that the operations of all cooperatives are examined and supervised by the Cooperative Bank of Taipei,China, which is a government-owned commercial bank but authorized by the Ministry of Finance and the Central Bank to act in a manner like the central bank to all cooperatives. The Cooperative Bank has the obligation to report any wrong-doings of the cooperatives to the monetary authorities. The authorities then decide how to deal with the problems. In addition, the Cooperative Bank is obliged to take the surplus of funds of cooperatives as redeposits. By redepositing the passbook savings deposits they received with the Cooperative Bank in the form of time deposits, the cooperatives could enjoy a spread of about 3 or 4 per cent per annum. This explains why the loans-to-deposits ratio for the cooperatives decreased in the late 1980s. For instance, at the end of 1990, the total loans of cooperatives constituted approximately 57 per cent of their total assets, while their claims on financial institutions (mostly on the Cooperative Bank) other than the Central Bank amounted to about 34 per cent. The remaining 9 per cent of their total assets were composed of claims on the Central Bank, government securities, and cash in vaults.

Investment and Trust Companies

Before 1971, there was only one investment and trust company, and it was government owned. In 1971 and 1972, seven private investment and trust companies were established. Since then, new licenses have not been granted. The operations of investment and trust companies include trust-deposit management, direct and syndicated medium- and long-term loans and investments, and securities underwriting. They have seldom engaged in trust-funds management on their clients' behalf; instead they offer trust certificates with a minimum maturity of one month, guaranteeing a yield at a premium above the bank deposit rates of similar maturity. At the end of 1992, 60.6 per cent, 28.3 per cent, and 2.5 per cent of their total assets were held in the form of loans, portfolio investments and real estate, respectively. Although they compete with the trust departments of banks for trust funds, their market share of trust funds were maintained at the level of 84.7 per cent at the end of 1992.

In the 1970s, the market share of investment and trust companies in all financial institutions increased rapidly. However, in the first half of the 1980s, three investment and trust companies (Asia Trust, Cathay Trust, and Overseas Trust) ran into troubles due to either overinvestment in the real estate market, which was in recession in that period, or overlending to ill-managed firms that belonged to the same conglomerate groups as the investment and trust companies. The operations of investment and trust companies became less expansionary afterward. In 1992, China Trust, the largest investment and trust company, was allowed to convert into a commercial bank, after the requirements set by the authorities were met. Other large trust companies have the same intention to become commercial banks, since commercial banking is assumed to be more profitable.

Postal Savings System

The Postal Savings System is a heritage of the Japanese colonial period. Its functions are performed by the Directorate General Remittance and Savings Banks. Besides the remittance services conducted inside and outside the country, the Directorate General Remittance and Savings Banks accept savings deposits and conduct life insurance business through an extensive network of post offices, which have separate windows for financial operations. The term "postal savings system" is used to distinguish the financial operations from other business and services.

Since 1961, the operational units of the postal savings system have accounted for about a third of all financial institutions. It accepts the same kinds of savings deposits and pays the same interest rates as other financial institutions. However, since it enjoys the advantages of longer business hours and income tax-free passbook savings up to a certain amount, the system has been very effective in mobilizing savings through its widespread post offices. The shares of total assets and deposits grew rapidly from 0.8 per cent and 3.4 per cent in 1961 to 11.7 per cent and 14.9 per cent in 1990, respectively.

Because the postal savings system is not allowed to extend loans, most of the deposits received were redeposited with the Central Bank before March 1982. The Central Bank paid them according to the one-year, two-year, or three-year deposit rate plus a 0.18 per cent service charge. The Central Bank then established a medium- and long-term loan fund to extend credit to certain investment projects through full-service domestic banks at a preferential interest rate. This redeposit policy was, therefore, a powerful tool for the Central Bank to control the money supply and conduct selective credit rationing policies. Beginning in March 1982, this redeposit policy was modified. Part of the increment of the saving deposits accepted by the postal saving system was redeposited with four specialized banks (Bank of Communication, the Farmers Bank of China, Land Bank of Taiwan, and Medium Business Bank of Taiwan) to supplement their loanable funds. At the end of 1992, redeposits with the Central Bank constituted nearly 75 per cent of the total assets of postal savings system, and specialized banks constituted nearly 22 per cent.

Insurance Companies

For many years and until 1988, there had been eight life insurance companies and 14 property and casualty insurance companies in Taipei,China. In

1989, due to pressure from the US, three foreign life insurance companies and five foreign property and casualty insurance companies, all from the US, were allowed to set up branches in Taipei,China. Since then, more foreign insurance companies have entered the market. At the end of 1992, there were 22 life insurance companies and 22 property and casualty insurance companies with a total of 71 and 81 branches, respectively. In 1993, the market of property and casualty insurance was further opened to domestic investors to set up new insurance companies. The first new license of a life insurance company was also just granted by the Ministry of Finance in July 1993.

The reserve funds of insurance companies are used primarily for lending and real-estate and securities investments. For instance, as of the end of 1990, the total assets of insurance companies, which amounted to nearly 9.5 per cent of GDP of that year, were composed of loans (33.3 per cent), real-estate investments (22.3 per cent), securities investments (19.8 percent), deposits with banks (nearly 20 per cent), cash in vaults (0.1 per cent), and fixed assets (7.3 per cent). As shown in Table 2.4 (p. 86), the roles of insurance companies in Taipei,China have been quite limited. The government-provided labor insurance and government employees' insurance programs have reduced greatly the market for private insurance companies. The tradition of strong family ties, which provides emergency help to family members, and the superstition that insurance brings bad luck also limit the expansion of insurance companies. However, the breaking-down of the traditional family structure and increasing education are benefiting the insurance companies. Their role in Taipei,China can be expected to rise in the future, especially after the relaxation of entry barriers in 1993.

NONINTERMEDIARY FINANCIAL MARKETS

Financial markets channel funds from lenders to borrowers by expediting the creation and trading of financial instruments. They include all the mechanisms for the issuance, trading, and redeeming of claims. According to the maturity of financial instruments, financial markets in Taipei,China are classified into the money market and the capital market. The money market deals with instruments due in one year or less, while the capital market, which is composed of the bond market and the stock market, deals with longer maturities. For convenience, the foreign exchange market is included and discussed in this section.

Money Market

To provide an additional conduit of funds to enterprises, as well as a market-determined reference interest rate for the Central Bank to adjust its monetary policy or the controlled bank interest rate, the authorities established the money market in May 1976 by allowing a bills finance company to be established. Two more bills finance companies entered the market in 1977 and 1978. In the beginning, the market instruments were limited to treasury bills, commercial paper, bankers' acceptances, and negotiable CDs. In the later years, government bonds, trade acceptances, bank debentures, and corporate bonds with one year or less remaining to maturity were added to the list of transaction instruments. The bills finance companies advise firms on the use of money market instruments, and act as the dealers, underwriters, and brokers as well as guarantors or endorser of many kinds of money market instruments. Besides creating a primary market, they also participate in the secondary market to promote the negotiability of bills and notes.

At the end of 1992, the total outstanding volume of money market bills reached 18 per cent of GDP of that year, with the following compositions: treasury bills, 9 per cent; commercial paper, 36 per cent; bankers' acceptances, 6 per cent; and negotiable CDs, 49 per cent. In the same year, total selling of bills finance companies was equivalent to 167 per cent of GDP, and the relative shares of instruments were: treasury bills, 2.6 per cent; commercial paper, 61 per cent; bankers' acceptances, 3.2 per cent; negotiable CDs, 32.4 per cent; government bonds, 0.6 per cent; and the rest, 0.2 per cent. In terms of customers, 53.8 per cent of total selling goes to banks, 25.7 per cent to private enterprises, 8.8 per cent to individuals, and 11.7 per cent to other buyers.

Though no new bills finance company has been allowed to be established since 1979, by the end of 1992, the number of branches had increased to 21, and the size of the market had grown rapidly. As Table 2.2 (p. 84) shows, outstanding money market instruments as a percentage of total loans and discounts of financial institutions increased from 8.5 per cent in 1980 to 12.2 per cent in 1992.

A rapidly growing market which is controlled by only three bills finance companies has brought about huge profits for these three almost every year. To dismantle the oligopolistic market structure and promote further development of the money market, the Ministry of Finance opened the market in May 1992 by allowing banks to enter the secondary market to act as the dealers and brokers, and agreed to allow banks to participate in the primary market as the underwriters and endorsers. Hence, there should be much fiercer competition in the money market in the future.

Bond Market

The bond market and the stock market are both regulated by the Securities and Exchange Commission (SEC), which was set up in September 1960 under the jurisdiction of the Ministry of Economic Affairs; it was subsequently transferred to the supervision of the Ministry of Finance in 1981. Because bond-issuing costs (including interest and administrative costs) were usually higher than bank-financing costs, and because the issuing procedure was complicated and time-consuming, firms preferred bank financing. Further, the restrictive provisions of the Company Law have hindered the issuance of corporate bonds. For example, according to Article 247 of the Company Law, total bonds outstanding can never be greater than the net worth of the company. Article 250 also prohibits a company from issuing bonds if its average after-tax profit rate for the last three years is lower than the interest rate for the proposed bond issues. So far, the majority of corporate bonds were issued by public enterprises and bought by financial institutions. At the end of 1992, the outstanding volume of corporate bonds reached only NT$65 billion, less than 1 per cent of the total loans and discounts of financial institutions.

The Government is committed to maintaining a balanced budget, but this prudent fiscal performance has worked against the development of a government bond market. The outstanding volume of government bonds thus amounted to approximately 3.8 per cent of the total loans and discounts of financial institutions at the end of 1990, or about 4.4 per cent of GNP of that year. However, the deterioration of the budget in recent years (Table 2.6, p. 89), caused for the most part by the execution of the Six-Year National Development Plan starting from 1991—which involves a large amount of government spending—has resulted in a rapid accumulation of government bonds. The outstanding volume of government bonds increased from NT$189 billion in 1990 to NT$451 billion in 1992, and is expected to grow further in the future. Table 2.2 (p. 84) shows that the sum of corporate and government bonds outstanding as a percentage of total loans and discounts of financial institutions increased rapidly from 4.8 per cent in 1990 to 8 per cent in 1992.

To sterilize the impact of the large balance-of-payment surplus on domestic money supply, the Central Bank issued huge amounts of certificates of deposit and savings bonds through the bond market in the late 1980s, which expanded the size of bond market to a great extent. By the end of 1987, the sum of CDs and savings bonds issued by the Central Bank reached a peak of NT$1,123 billion, more than six times the level of outstanding corporate and government bonds. By the end of 1992, this total was still NT$101 billion, about 1.6 times the outstanding corporate bonds. In addition, deposit money banks started issuing bank debentures in the bond market in 1980; by the end of 1992 the amount outstanding was NT$90 billion.

Stock Market

There was no formal stock market in Taipei,China until the Taiwan Stock Exchange Corporation (TSE) began operation in 1962. Although there had been over-the-counter markets operated by securities brokers and traders before then, the number of listed companies and transactions of listed stocks were rather small. After the TSE began to operate, and until the end of 1989, over-the-counter transactions were prohibited. In 1980, the Fuh-Hwa Security Finance Company was established to increase overall market liquidity and the marketability of individual securities, by providing margin loans and stock loans for securities transactions and by serving as a securities custodian.

To promote listings, the authorities provided incentives such as offering listed companies a 15 per cent reduction on corporate income taxes for three years, and allowing shareholders to receive up to NT$360,000 ($270,000 since 1990) of dividends plus interest which is free from income tax. However, the listing criteria set by the authorities were also high. As described in "Criteria for Screening Stocks Listing," listed companies are classified into three categories—A, B and C—according to their capital size, profitability, financial structure, and distribution of shares. The application for listing must pass the examination of TSE and SEC. In addition, the traditions of a low proportion of equity funding, a reliance on bank borrowing, and the desire for close management control and little public disclosure have impeded the growth of the stock market. Even companies which were listed often allowed only a small proportion of shares outstanding to trade on the market. By the end of 1986, when the number of so-called "big companies" with assets in operation over NT$100 million (equivalent to $2.8 million) reached 5,062, there were only 130 listed companies, and the ratio of total par value of listed stocks to total loans and discounts of financial institutions was under 12 per cent.[9] The ratio of market value and trading value to GNP were usually under 20 per cent for most of the years before 1986, as shown in Table 2.21.

Since the mid-1980s, Taipei,China's economy has been troubled by excess liquidity due to the huge trade surplus and speculative capital inflow. Consequently, the stock price index rose from 746 in 1985 to 2,135 in 1987, and total trading value increased from NT$195 billion to NT$2,669 billion. To intensify market competition and to respond to the pressure for market entrance, the Ministry of Finance allowed new brokerage houses to be set up in May 1988. The number of brokers, dealers, and broker branches increased dramatically, from 67 before market opening to 307 by the end of 1989 and to 452 in 1990. This market opening further strengthened the momentum in the market, so that the stock price index rose almost steadily from 2,135 in 1987 to 8,616 in 1989, and further to the peak of 12,495 on 10 February

---

[9] Most of the time, and until the late 1980s, stocks of newly listed companies were issued at par value in the stock market in Taipei,China. Consequently, total capital of all listed companies amounted to NT$761,091 million, only 3.5 per cent higher than the total par value of NT$735,640 million at the end of 1992.

Table 2.21 Stock Market in Taipei, China
(NT$ billion, except as noted)

	Broker Locations[a]	Accounts Number[b]	Per Cent[c]	Listings[d]	Total Par Value[e] NT$	Per Cent[f]	Total Market Value NT$	Per Cent of GNP	Per Cent[h]	Price Index[g]	Turn-over[b]	Trading Value[i]
1962	...	...	...	18	5.9	23.1	6.8	8.9	26.6	...	...	0.6
1970	...	...	...	42	8.4	9.0	20.0	8.8	21.4	134	160	4.8
1980	68	362	1.9	102	108.7	10.6	219.1	14.7	21.4	547	108	10.9
1986	67	474	2.4	130	240.9	21.0	548	19	47.8	945	162	23
1987	67	634	3.2	141	287.3	11.9	1,386	42	57.4	2,135	267	81
1988	149	1,606	8.1	163	343.6	10.2	3,383	94	100.4	5,202	295	219
1989	307	4,209	20.9	181	421.3	9.6	6,174	156	140.7	8,616	524	641
1990	452	5,033	24.7	199	506.4	10.2	2,682	62	54.0	6,775	459	438
1991	470	5,163	25.1	221	616.7	10.1	3,184	66	52.1	4,929	285	201
1992	418	5,078	24.5	256	735.6	9.5	2,545	48	32.9	4,272	146	111

[a]Number of brokers, dealers, and broker branches.
[b]Number of accounts in thousands.
[c]Account as a percentage of the population.
[d]Number of listed companies.
[e]Number of shares times their par value. The par value of all shares is NT$10.
[f]As a percentage of total loans and discounts of financial institutions.
[g]1966 = 100.
[h]Percentage of shares outstanding traded during the year.
[i]As a percentage of GNP.
... data not available.
Source: Ministry of Finance.

1990. In 1989, the average stock price to its after-tax net earning ratio (the P/E ratio) rose to as high as 55.9. The total market value and trading value of listed stocks also grew dramatically. As indicated in Table 2.21, the total market value of listed stocks increased from 19 per cent of GNP in 1986 to 156 per cent of GNP in 1989, or from 47.8 per cent of total loans and discounts of financial institutions in 1986 to 140.7 per cent in 1989; and the ratio of trading value to GNP grew from 23 per cent to 641 per cent during the same period. The number of client accounts as a percentage of population also rose from 3.2 per cent in 1987 to 20.9 per cent in 1989 and further to 24.7 per cent in 1990.[10] However, due to the sluggishness in listing examination and the factors mentioned above, from 1986 to 1989, the number of listed companies had increased only modestly from 130 to 181, which was much less than the number of brokerage houses. After the Central Bank tightened the money supply in 1989 and 1990, the stock market bubble burst in 1990. Although the number of broker locations declined in 1992, there were still 418 branches, compared to just 256 listed companies.

Understandably, most of the brokerage houses are located in areas with high population densities. In 1992, brokerage houses in Taipei City constituted a market share of 52.4 per cent, followed by Kaohsiung City (8.3 per cent), Taipei Hsien (8 per cent) and Taichung City (6.7 per cent).

Another important feature of the stock market in Taipei,China is that it is full of small, uninformed individual investors or speculators, who can be easily manipulated by a small group of large investors and insider traders. For instance, at the end of 1989, the number of individual investors who had accounts at brokerage houses reached 2,809,000, equivalent to nearly 14 per cent of the total population. Domestic individuals also constituted nearly 56 per cent of the sources of capital of listed companies in 1992 (compared to 46.7 per cent in 1989). Among 7.06 million individual shareholders of all listed companies in 1992, 6.19 million (88 per cent) shareholders have an investment amount each of less than NT$50,000 (less than $2,000).

To prevent insider trading and protect small investors, the revised Securities and Exchange Law in 1988 requires listed companies to spread the new shares out to more shareholders when they are making a public offering. A new disclosure regulation has been enacted which requires that all listed companies provide more information to shareholders, including operating results and the timely disclosure of major events. The revised

---

[10] There were many clients who had multiple accounts, so the actual percentage of population involved in the stock transactions was less. For example, at the end of 1989, the number of client accounts was 4.1 million, and the number of investors was 2.8 million, equivalent to nearly 14 per cent of the total population. However, there were only 0.44 million investors in the market in July 1988 before new brokerage houses were allowed to open. In one-and-a-half years, the number of investors increased by 542 per cent.

Securities and Exchange Law also strengthens the prohibition and punishment of insider trading. However, for the following two reasons, the SEC does not have the capacity to prevent insider trading or market manipulation. First, regulations on such activities are not clear, particularly as to definitions. The judiciary department often has different opinions and reaches different conclusions than the SEC. Unless the judiciary department confirms a SEC verdict, any finding of manipulation or insider trading is not sustained. Second, the widespread use of borrowed-name accounts hinders effective implementation. Such accounts are opened in the names of real people, but not the person who actually owns and controls the account.

Speculation on the stock market was further increased by illegal marginal lending provided by brokers with access to the informal financial system. Until 1990, only Fuh-Hwa Securities Finance Company was officially allowed to provide marginal loans against stock, and the standards for marginal credit from government-controlled sources were quite stringent. To attract customers and enlarge transaction volume, many brokerage houses thus provided illegal marginal credit. The revised Securities and Exchange Law in 1988 permits the establishment of qualified integrated securities firms, which are allowed to trade for customers as brokers, to trade for their own accounts as dealers, to engage in the business of underwriting for issuing companies as underwriters, and to offer marginal financing directly to their customers. After drafting detailed regulations governing marginal financing, qualified integrated securities companies and the trust departments of banks began providing marginal credit in October 1990. Consequently, the ratio of marginal transactions to total trading value increased from about 1 per cent in 1989 to nearly 6 per cent in 1990 and further to about 14 per cent and 18 per cent in 1991 and 1992 respectively. This marginal transaction ratio is relatively low compared with developed countries. It is also consistent with the general impression that the purchase of stocks in Taipei,China is mainly financed by investors' own savings, not by borrowing. This practice might be the most important factor which prevented a crisis in the payment system and social stability, when the stock price index fell more than three quarters from its peak of 12,495 on 10 February 1990 to an average of 2,912 in October 1990.

Fortunately, since the turmoil in 1990, factors such as the Central Bank's effective control on the money supply, the implementation of disclosure and insider-trading regulations, and the more rational behavior of small individual investors who had gained experience, have contributed to the stabilization of the stock market. The stock price index has fluctuated from between 3,500 and 5,000 since July 1991.

To introduce foreign input into the stock market, the authorities allowed foreign investors to make indirect investments through the purchase of the beneficiary certificates of four overseas funds in 1983. Since 1986, the beneficiary certificates of securities investment trust funds have been approved to be issued in the domestic market. At the end of 1992, four securities investment trust enterprises have offered publicly and issued 28 funds with a net asset value of NT$73.5 billion. These funds are used to invest in either foreign or domestic securities. The trade in beneficiary certificates is also heavy. In 1992, the shares of trading values of four kinds of securities in the market were as follows: stocks, 94 per cent; beneficiary certificates, nearly 6 per cent; government bonds, 0 per cent; and corporate bonds, 0.06 per cent.

To internationalize the local stock market further, the 1988 Securities and Exchange Law permits foreign investors to participate in the securities business through investment in and management of local securities firms. Further, there is no restriction on foreign investment in securities-investment consulting companies. In June 1989, the SEC took a further step by accepting the applications of foreign securities firms to set up branches in Taipei,China to engage in local brokerage business and overseas securities transactions. In 1991, foreign investment institutions were allowed, with SEC permission, to invest in local securities.

Foreign Exchange Market

The evolution of the exchange rate policy in Taipei,China was discussed earlier. In Taipei,China's early years of economic development, the authorities imposed strict foreign exchange control on nontrade-related outward remittances by local residences. Both outward and inward remittances of direct capital investment were also subject to the approval of the Investment Commission of the Ministry of Economic Affairs. Strict foreign exchange control created an active black market, and invited under-invoicing of exports and over-invoicing of imports. Foreign currency (especially US dollars) could be bought and sold at jeweler's shops. Huge discrepancies between Taipei,China-US trade statistics were also a common phenomenon.

Capital Flow Deregulation

In July 1978, the authorities introduced a floating system to make the economy less vulnerable to external disturbances. However, the Central Bank continued to intervene on the foreign exchange market during the mid-1980s, and thus, nullified the function of a floating exchange rate system. Also, the trade surplus, short-term speculative capital inflow, and balance-of-payment surplus had been expanding continuously, which allowed foreign exchange reserves and the money supply to increase too rapidly.

To reduce the pressure of the trade surplus on the domestic money supply and the value of the currency, the authorities took steps toward foreign exchange liberalization. "Foreign exchange trust funds" and "investment fund beneficiary certificates" were created in late 1986 to enable local residents to invest in foreign securities. Further, most of the foreign exchange controls were phased out in July 1987. Under the new system, current-account transactions were completely liberalized. As for capital-account transactions, an adult or a company was allowed to purchase and remit outward up to an annual limit of $5 million; the ceiling on inward remittances for each person was set at $50,000 per year.

The purpose of the lower ceiling on inward remittances was to fend off "hot money" that had been entering Taipei,China in 1986–1987; and the aim of the large-scale relaxation on the ceiling on outward remittances was to encourage investment abroad and capital outflow, so the balance-of-payments surplus and the growth of the money supply could be effectively reduced. However, strong appreciation expectations regarding Taipei,China's currency, caused by the Central Bank's controlled crawling appreciation strategy in 1986–1987, not only handicapped capital outflow but attracted speculative capital, which came into Taipei,China through both legal and illegal channels.

In 1987, the Central Bank allowed the NT dollar to appreciate more rapidly; this finally stopped the inflow of speculative capital in 1988. The relaxation of control on capital outflow then resumed its function. In 1988–1992, there was a substantial amount of net capital outflow, especially for direct investment in the Southeast Asian countries and in the PRC. In August 1989, the Central Bank raised the ceiling on foreign liabilities of both local and foreign banks by 30 per cent of the original ceiling set in October 1987, which had been intended to block the inflow of hot money. The ceiling on inward remittances was also raised incrementally. Both inward and outward remittance limits were set at $3 million in March 1991, and further raised to $5 million in October 1992.

After the relaxation of foreign exchange controls, individuals and firms were free to hold and make use of foreign exchange. In addition, with such high ceilings on outward and inward remittances, limitations on the capital flow of domestic nationals have been phased out almost completely. The future impact of this deregulation on monetary control by the Central Bank on portfolio managements of commercial banks and the general public, and on direct investment by domestic firms in foreign countries will be formidable.

Offshore Banking Center

Establishing an offshore banking center was a major step toward financial internationalization in Taipei,China. To strengthen international financial activities, the Central Bank promulgated various regulations concerning offshore banking services in late 1983 which allowed banks located within the national boundaries to establish offshore banking units. The Central Bank has granted several preferential tax treatments to these offshore banking units to encourage their development. For example, their income is exempt from tax, and their transactions are free from stamp duty.

In June 1984, the International Commercial Bank of China set up the first offshore banking unit; this was followed by four other domestic commercial banks and two foreign banks in the same year. By May 1993, there were 36 offshore banking units established by 18 domestic banks and 18 foreign banks. Their total assets had expanded from $4.3 billion at the end of 1984 to $17 billion at the end of 1989, and to $26.2 billion at the end of May 1993.

Regarding the market participants, financial institutions have been the main providers and main users of funds. In May 1993, 89 per cent of the liabilities were due to financial institutions, and the share of claims on financial institutions on the assets side also reached 74 per cent. Deposits of and loans to nonfinancial institutions have been relatively limited. According to Liu (1993), the proportion of offshore banking unit's assets accounted for by foreign banks were 40 per cent by the end of 1992; and foreign banks brought in funds from 45 different countries. On a geographical basis, Asia has been by far the largest source as well as the largest user of funds, accounting for 82 per cent and 90 per cent respectively of the total funds at the end of 1991.

Taipei Foreign Currency Call-Loan Market

The Taipei foreign currency call-loan market was set up on 7 August 1989, to better use the foreign exchange reserves held by the Central Bank, to improve the efficiency of the management of foreign exchange funds held by authorized foreign exchange banks, and to promote Taipei as an international financial center. The Central Bank designated part of its foreign exchange reserves ($3 billion and 500 million Deutsche Mark initially, plus a subsequent $4 billion) to finance authorized foreign exchange banks through the call-loan market. Many local branches of foreign banks participate, and nonresident banks have been allowed to take part since August 1990. The number of tradable currencies has expanded from just US dollars to include the 13 others listed on the foreign exchange market. Maturities range from overnight to 12 months. Total transactions in May 1993 were equivalent to $24.6 billion, of which 88 per cent were overnight transactions. The Taipei foreign currency call-loan market has also been linked with the markets in Singapore and Hong Kong since 1991, and with Tokyo since March 1992, all through well-known money brokerage houses in those markets.

UNREGULATED FINANCIAL SYSTEM

Historically, the unregulated informal financial system has played an important role in providing the financing needs of enterprises and households in Taipei,China.

Structure of Unregulated Financial System

The unregulated informal financial system is composed of all the markets or mechanisms where borrowing and lending activities between and within businesses and households occur without being subjected to the direct supervision and regulation of the financial authorities according to the banking laws. In this system, except for financial installment credit companies, financial leasing companies, financial investment companies, and credit unions, most of the markets—such as moneylenders, pawn-brokers, and rotating credit clubs—have no formal organization and are, therefore, fragmented. Financial installment credit companies, leasing companies and investment companies are licensed by the Ministry of Economic Affairs, but they extend credit to or take deposits from the general public without being regulated by the financial authorities. Credit unions were introduced into Taipei,China by the Roman Catholic church in 1964. They were allowed by the Ministry of Finance to operate under certain limits on an experimental basis in 1968, but so far have not been legalized by the banking laws, and have not yet been formally regulated by the financial authorities.

Financial investment companies were little more than Ponzi or pyramid schemes. They prospered during the middle and late 1980s but finally ceased operation in 1990. They collected funds from the general public by paying monthly returns of as high as 9 per cent, although 4 per cent was the norm. They used the funds to speculate in the stock, foreign exchange, and real estate markets. By accepting funds from the public and investing them in real and financial assets, the companies played a role similar to mutual funds.

In 1989, the revised Banking Law prohibited these companies from collecting additional funds from the public. The cessation of new cash infusions and the decline of the stock and real estate markets in 1990 eventually brought about the collapse of financial investment companies.

Types of Financial Transactions

The types of financial transactions in the unregulated informal financial system are many, including, among others, financial installment credit, financial leasing, mutual loans and savings deposits with firms, secured and unsecured borrowing and lending, loans against post-dated checks, rotation mutual credit, and informal marginal credit for stock purchases.

The rotating mutual credit or rotating credit club (commonly called *hui* or *biao-hui* in Taipei,China) has been the most popular practice in the informal market. A large-scale survey by the Ministry of Justice in 1985 indicated that 85 per cent of the population had participated in *huis*. A *hui* involves a group of members who have mutual trust or trust the founder of the club. The size and practices of each *hui* vary widely. Typically, group numbers meet at regular intervals (usually monthly), contribute a fixed amount each, and lend the collected pool of their savings by competitive bidding. Those who currently have outstanding loans generally cannot borrow more, so that each member of the group has a turn at receiving the collected pool at some point. Thus, the member willing to pay a higher interest rate receives the pool earlier than other members, and then repays the loan in installments.

Many firms take deposits from their employees at favorable interest rates to give them an incentive to support the company and also to avoid having to seek outside financing. Some companies extend this deposit-taking to relatives and friends of employees, and even the general public. Until 1990, income tax laws gave an incentive to employees to participate because the interest earned up to a certain amount was excluded from taxable income. Although the 1989 Banking Law tightened the prohibition on deposit-taking by nonbanks, in practice it is difficult to prevent firms from continuing to take deposits from the public. The practice remains legal as regards employees and their families.

Loans against post-dated checks are also widespread because they are usually eligible for discounting at banks. Thus, discounting post-dated checks is a common form of lending for underground money lenders, who can either hold the check until the stated date or present it to a bank for immediate (albeit discounted) payment. Before 1986, loans against post-dated checks were protected by criminal law: a person who bounced more than three checks was subject to criminal penalties. Removal of criminal sanctions in 1986 increased the risks involved, but the loans remain common. The payee (lender) now depends on civil law, collateral, the credit record of the payor (borrower), or even professional money collectors (including gangsters) to protect the claim. The Taipei clearinghouse maintains records which are open to the public of who has bounced checks, and clearinghouse regulations require banks to close checking accounts of customers who have bounced a specified number of checks within a certain period.

Pawnbrokers and moneylenders operate in a fairly competitive environment, often advertising their services in newspaper classified sections. One can discount post-dated personal checks, borrow against cars, and even buy and sell real estate through pawnbrokers and moneylenders.

Illegal marginal lending provided by brokers or professional money lenders for stock purchases is also a common phenomenon. As discussed earlier in this chapter, until 1990, there was only one Securities Finance Company, Fuh-Hwa, officially allowed to make margin loans against stock, and the standards for regulated marginal credit from government-controlled sources were quite stringent: the service of informal illegal marginal lending thus emerged. This illegal marginal lending fueled the speculative bubble in the stock market in the late 1980s. The collapse of the stock market in 1989 and the allowance of qualified integrated securities firms to offer marginal financing directly to their customers in 1990 reduced the role of illegal marginal lending. Nonetheless, it remains popular.

Interest Rate

Although some of the informal system's activities are illegal, the authorities usually refrain from suppressing them. The Central Bank monitors the conditions of the informal financial system, and collects its interest rate data through commercial banks by asking banks' customers the rates they are actually paying for informal-system borrowing. The compiled monthly interest rates for loans against post-dated checks, unsecured loans, and deposits with firms in three major cities (Taipei, Kaohsiung, and Taichung) are then published by the Central Bank in its publication, *Financial Statistics Monthly*.

The annual informal system interest rate or the curb-market interest rate for the period 1980–1992 (average of 1961–1992) is reported in Table 2.15 (p. 109). This table shows that the nominal and real curb-market interest rates averaged as high as 25.5 per cent and 21.7 per cent in the period, and there was a persistent gap between the curb-market rate and the saving deposit rate (between 11 and 18 per cent). A major part of this gap may represent the higher default risk and transaction costs in curb-market lending.

Size of the Unregulated System

Though the unregulated informal financial system has played an important role in the unregulated financial system, there is no detailed information about its size. The only government-published statistics available are on the business sector's borrowing from the unregulated system. Each year, the Central Bank uses a questionnaire to acquire data on enterprises' assets and liabilities, from which it constructs its flow-of-funds tables. Tables 2.12 (p. 102) and 2.22 are based on that data and by defining the borrowing from other enterprises and households as the borrowing from the informal financial system. The tables reveal the composition of the business sector's financing sources.

Table 2.22 Composition of Business Sector Financing by the Financial System[a]
(per cent)

	1964–1990	1964–1970	1971–1975	1976–1980	1981–1985	1986–1990
Financial Institutions	54.5	57.1	67.4	57.2	52.2	54.1
Money Market[b]	7.3	...	0.0	4.5	10.6	6.9
Capital Market[c]	14.4	15.7	9.8	13.9	13.9	15.1
Informal System	23.8	27.2	22.8	24.4	23.3	23.9
Total (NT$ billion)	968.9	52.4	205.5	650.8	1,514.6	2,787.6

[a] The business sector includes both private and public enterprises. The percentages are computed from totals over the entire period. The NT$ amounts in the last line are averages of annual data.
[b] The money market was established in 1975.
[c] The amount of financing from the capital market is defined as the sum of the accumulated net amount of stock issued by listed companies and the outstanding volume of corporate bonds.
... data not available.
Sources: The Central Bank of China (1991); Ministry of Finance (1990).

According to Table 2.22, during the period 1964–1990, financial institutions provided the business sector with 54.5 per cent of its domestic financing. The informal system, capital market, and money market accounted for 23.8 per cent, 14.4 per cent, and 7.3 per cent, respectively. By further disaggregating the business sector into public enterprises and private enterprises (including incorporated and unincorporated businesses), Table 2.12 shows that public enterprises depended primarily on financial institutions as the source of borrowing, while private enterprises borrowed only 60.3 per cent from financial institutions, and borrowed about 34 per cent from curb markets, 5 per cent from the money market, and 1 per cent from the bond market in 1964–1991.[11]

A large proportion of the borrowing from households of private enterprises in fact represents shareholders' financing to their own enterprises. According to Chiu (1992, 184), a survey conducted by the Bank of Taiwan in 1988 indicated that about 82 per cent of the households' financing to small- and medium-scale enterprises of the manufacturing industry consisted of loans by shareholders to their own enterprises.

---

[11] The distinction between public and private enterprises depends on whether the government controls more than 50 per cent of the shares of the enterprises or not.

Table 2.12 (p. 102) further shows that the share of informal system financing of private enterprises had a clear declining trend for the periods 1980–1984 and 1986–1991, and an increasing trend during 1984–1986. Following financial liberalization, including decontrol of bank interest rates starting from 1980, the share of informal system financing declined steadily in 1980–1984 and in 1986–1991. The reasons for the rise of informal system financing in 1984–1986 are believed to be closely related to the economic recession in 1985 and the remaining interventionist policies, such as market-entry restrictions and government ownership of major banks. The slowdown of economic growth in 1985 raised the riskiness of loans, as evidenced by the rise of the ratio of nonperforming loans and the ratio of dishonored checks (Table 2.20, p. 121). As a consequence, financial institutions became even more conservative in extending loans, which left ample room for the informal system to function.

While there are so far no government-published data available on the role of the informal financial system in providing loans to the households sector, the general impression is that—since consumer financing has been purposely repressed except for the period of the late 1980s in which banks were troubled by excess liquidity, and it is difficult for households to get loans from financial institutions for any purposes other than for secured loans using real estate as collateral—the share of households' borrowing from the informal system must be high. Some surveys by academic institutions show that the share of informal system financing might be about the same or higher than the share of financial institution financing. For example, Shea et al. (1985) estimated that the households sector borrowed between 45 and 52 per cent from the informal financial system in 1982.[12]

The informal system has contributed to the stimulation and mobilization of savings as well as to the allocative efficiency of loanable funds. In addition, the informal system is the last resort of households and small- and medium-scale enterprises which seek funds. Therefore, although some of its financial activities—such as accepting deposits and charging a usury fee—are illegal in the legislative sense, the authorities usually have no intention to suppress it, which explains, in part, why the informal system has played such an important role in the country. It is expected, however, that following financial liberalization, the unregulated informal system will be gradually replaced by the formal system.

GROWTH PROMOTION FINANCIAL POLICIES

Financial policies are usually regarded as effective and powerful tools for promoting economic growth. Before interest rate liberalization in the 1980s, the Central Bank often repressed the bank interest rate to keep it

---

[12] This conclusion was based on a survey by Chung-Hwa Institution for Economic Research.

below the market equilibrium level. The purpose was to reduce the interest cost, and thus stimulate investment, exports, and economic growth. In addition, the Central Bank has tended to undervalue the NT$ to increase export competitiveness.

Besides these price intervention policies, the Government adopted several nonprice intervention financial policies to accommodate the financial needs of some specific industries and economic activities and/or to guide the direction of the flow of loanable funds. These policies include export financing, industrial financing, machine-import financing, strategic-industry financing, setting up specialized banks and venture capital investment enterprises, and small- and medium-scale enterprise financing. Among these, the policies aimed at exports, strategic-industries, and small- and medium-scale enterprises deserve some further clarification; the others can be summarized as special loans policies. The financing policies on small- and medium-scale enterprises will be discussed in the next section.

Export Financing

Taipei,China has a small, open economy with a high degree of trade dependency. Since export promotion was proposed as the economic development strategy in the 1960s, export financing has been important, as have other export incentives such as rebates of custom duties and commodity tax on imported raw materials, tax exemptions, and retention of foreign exchange earnings for the import of raw materials and machinery. Export firms were allowed to apply for short-term loans as pre-shipment working capital based on letters of credit. After the products were shipped, the firms were eligible for loans based on letters of credit, documents against acceptance or payment, and shipping documents. Local branches of foreign banks have been allowed to extend pre-shipment export loans in foreign exchange since 1978.

In addition, the Central Bank has provided a special low rediscount rate to designated domestic banks for export loans. Designated domestic banks were allowed to charge borrowers an annual rate, usually one point higher than the rediscount rate. The difference between the export loan rate and the minimum interest rate for secured loans is illustrated in Table 2.23.

As a continuous trade surplus led to the increase in foreign exchange reserves in the late 1970s, the Central Bank gradually reduced the interest rate difference. Moreover, because the rediscount rate for export financing was only one point lower than the export loan rate charged by designated domestic banks, domestic banks often had no particular desire to apply for rediscounts. The major part of export loans by domestic banks was therefore financed by their own loanable funds rather than accommodated by the Central Bank. Thus, the actual benefit of export loans to the export firms came mainly from the availability of funds rather than from the interest-rate subsidy. This was true even in the 1960s and 1970s. Before they were permitted to accept savings deposits and time deposits of over six months maturity in the late 1980s, foreign bank branches' capability to make local currency loans was limited. They thus endeavored to extend export loans in a foreign currency. As a consequence, the share of foreign bank branches in total outstanding export loans surpassed that of domestic banks in 1980, 1981, 1986, and 1987. At the end of 1992, foreign bank branches still contributed nearly 29 per cent to the total outstanding export loans.

Table 2.23 Interest Rate Subsidy on Export Loans
(per cent)

Year	Export Loan Rate (1)	Minimum Rate for Secured Loans[a] (2)	Interest Rate Difference (3)=(2)-(1)
1978	6.50	10.50	4.00
1979	8.75	11.98	3.23
1980	10.53	13.50	2.97
1981	11.55	14.19	2.64
1982	9.79	11.18	1.39
1983	8.05	8.60	0.55
1984	7.86	8.31	0.45
1985	7.41	8.08	0.67
1986	5.79	7.20	1.41
1987	5.50	6.75	1.25
1988	5.50	6.87	1.37
1989	7.54	9.22	1.68
1990	8.75	10.23	1.48
1991	8.31	9.56	1.25
1992	6.88	8.34	1.46

[a]The prime rate of three Provincial Government-owned commercial banks is reported here for the period after the prime rate system was set up in March 1985. Monthly interest rate data are averaged to get yearly data.
Source: Central Bank of China.

Special Loans

The Central Bank has often adopted a special loans policy to accommodate financing needs and to direct the allocation of funds. Through both government-owned commercial and specialized banks, several special loans have been implemented in different periods for various purposes. After extending special loans, banks could rediscount them at preferential rates at the Central Bank.

The loans were to finance the special needs of the agriculture and industrial sectors, and to promote the growth of exporting activities and strategic industries. As regards the promotion of machine imports, the Government promulgated some acts in the late 1980s which allowed firms to borrow indirectly from the Central Bank through the designated banks while buying machines from abroad.

In addition, the so-called "medium-and-long-term credit special fund" of the Central Bank was set up in 1973 to finance basic construction and long-term investments. The funds mostly came from the redeposits of the postal savings system and the Cooperative Bank of Taiwan. Finally, seven government banks (Communications, Farmers Bank, Land Bank, Export Import Bank, Cooperative Bank, Central Trust, and Medium Business Bank of Taiwan) were set up or designated to specialize in industrial, agricultural, real-estate, small- and medium-scale enterprises, and export-import financing; and special loan programs for agriculture, small- and medium-scale enterprises, and venture capital have been executed. These seven specialized banks are discussed later in this chapter. Redeposits from the postal savings system have been one of the main sources of funds for the specialized banks since 1982.[13]

Further, to promote the development of technological enterprises, the authorities encouraged the establishment of venture capital investment enterprises by providing tax exemption incentives. In September 1985, the Development Fund of the Executive Yuan and the Bank of Communications together provided NT$0.8 billion to invest in technological enterprises through the first venture capital investment corporation. As of the end of April 1992, there were 24 venture capital investment corporations with a total capital of NT$15 billion, and the number of invested technological enterprise was at 250.

In addition, when the economy was in recession, the authorities sometimes provided emergency financing to firms to relieve their financial difficulties. For example, in 1974 and 1979, when the energy crises caused GNP and export growth to slow down, the Central Bank provided various special loans and rediscounts to relieve the financial difficulties of export firms and small- and medium-scale enterprises. When the domestic economy again suffered from stagnation in 1982 and 1985, special loans and expansionary monetary policies were adopted by the Central Bank. The Ministry of Finance even allowed firms with financial problems to extend the due date of bank loans and granted some qualified firms an extension of the redemption period for their dishonored checks.

Strategic-Industry Financing

In the 1980s, improvement of the industrial structure was taken as an important policy goal. Thus, measures to promote the growth of strategic

---

[13] A more extensive discussion of the special loans policy can be found in Lee and Chen (1984).

industries were promulgated in 1982. Strategic industries were selected based on six criteria: large linkage effect, high market potential, high technology intensity, large value added, low energy coefficient, and low pollution. Selected at that time were the electrical and nonelectrical machinery, information, and electronics industries; the list was later extended to include biotechnology and *materiel* industries.

The Government provided firms belonging to the strategic industries with special medium- and long-term low interest loans. Seventy-five to 80 per cent of the funds were extended by the Bank of Communications and Medium Business Bank of Taiwan, with another 20–25 per cent provided by the Development Fund of the Executive Yuan. The interest rate difference between the strategic loans and the prime rate was between 175 and 275 basis points. Besides financing strategic industries, the preferential loans were expanded later to finance the purchase of automation equipment and domestically produced machines, and for investment projects such as environmental protection.

The preferential loan policy has been the subject of severe criticism from its beginning. It is regarded as a distortional policy, and aspects of the selection criteria for strategic industries are criticized as being improper or inconsistent.

In sum, the Government tried to effect the allocation of available investment funds and to create additional funds for specific industries and activities through nonprice interventional financial policies. The effectiveness of these policies will be reviewed and evaluated in a later section of this chapter.

FINANCING OF SMALL- AND MEDIUM-SCALE ENTERPRISES

The definition of small- and medium-scale enterprises in Taipei,China has changed five times since first defined in 1967. Generally, they are defined by registered or paid-in capital, total amount of assets or sales, and total number of employees, although sometimes different criteria were applied for different industries. Between July 1982 and December 1992, small- and medium-scale enterprises were defined to cover those enterprises which had a paid-in capital of less than NT$40 million and total assets of not more than NT$120 million for the manufacturing sector; annual sales of not more than NT$40 million for the service sector; and paid-in capital of less than NT$40 million for the mining industry. In December 1992, the definition was relaxed to cover manufacturing enterprises with a paid-in capital of less than NT$60 million or regular employees of no more than 200 persons, and service enterprises with annual sales of less than NT$80 million or regular employees of less than 50 persons.

The small- and medium-scale enterprises have dominated Taipei,China's economy. Such enterprises have most often accounted for more than 97 per cent of the total firms in terms of number, and contributed 57.3 per cent of total exports in 1990. When manufacturing is examined on its own, in 1990, small- and medium-scale enterprises accounted for 98.3 per cent of the number of firms, 60.5 per cent of the value of manufacturing exports, 39.4 per cent of total sales, and 71.0 per cent of the total manufacturing employment.

Despite its importance, however, small- and medium-scale enterprises have always had difficulties in obtaining credit from the regulated formal financial system, due to the lack of a sound accounting system, their inability to provide sufficient collateral, and a lack of expertise in capital planning. As discussed earlier, and shown in Tables 2.13 and 2.14 (p. 103), as these enterprises had difficulties in obtaining financing from the formal financial system, they have depended heavily on informal-system financing.

To assist and promote the growth of small- and medium-scale enterprises, the authorities adopted several measures to increase the availability of loanable funds. The most important of these were the setting up of medium business banks to extend medium- and long-term credit and the establishment of the Small- and-Medium Business Credit Guarantee Fund to share the default risk of small- and medium-scale enterprises with financial institutions; this latter assistance raised the creditability of small- and medium-scale enterprises. In addition, there have been various special loans for specific purposes for which small- and medium-scale enterprises can apply, including Sino-American-Fund loans, special loans for automation, computerization and pollution-prevention, strategic-industry loans, and youth business start-up loans.

Financing for Small- and Medium-Scale Enterprises

Among the eight medium business banks in the country, one is owned by the Provincial Government and operates islandwide, and the other seven are privately owned and operate in restricted nonoverlapping regions.

Medium business banks are required to extend at least 70 per cent of their total loans to small- and medium-scale enterprises. As shown in Table 2.24, the ratio of loans to such enterprises by medium business banks has been maintained at a level never less than 73 per cent since 1986, much larger than the same ratio of full-service domestic banks and foreign bank branches. However, since full-service domestic banks are, in general, much larger than medium business banks in terms of scale and number of branches, they have accounted for about 75 per cent of the total loans to small- and medium-scale enterprises, while the share of medium business banks is only around 22–24 per cent. The data also show that there are four influential large-scale Provincial Government-owned domestic banks—First, Chung-Hwa,

Table 2.24 Share of Loans to Small- and Medium-Scale Enterprises
(per cent)

		Ratio of Loans to Small- and Medium-Scale Enterprises				Share in Total Loans to Small- and Medium-Scale Enterprises by all Banks		
	Total Loans of all Banks (NT$ billion)	All Banks	Full Service Banks	Medium Business Banks	Full Service Bank Branches	Full Service Domestic Banks	Medium Business Banks	Foreign Bank Branches
1986	580	35.63	32.09	82.72	3.76	75.05	24.30	0.65
1987	751	38.82	35.42	79.16	9.55	75.15	23.33	1.52
1988	1,096	41.14	37.82	76.95	13.94	76.48	21.77	2.75
1989	1,376	39.94	36.60	74.37	15.93	76.02	21.85	2.13
1990	1,540	40.43	36.50	75.46	18.84	74.48	23.16	2.36
1991	1,924	40.72	37.38	73.35	7.93	75.42	23.68	0.90
1992	2,391	39.92	36.57	74.41	4.08	75.19	24.38	0.43

Source: Ministry of Finance (1992).

Hua-Nan, and the Cooperative Bank of Taiwan—which have extended approximately 60 per cent of their loans to small- and medium-scale enterprises since 1986. Compared with the high small- and medium-scale enterprise-loan ratio of these four banks, medium business banks can hardly be called specialized banks to finance such enterprises. Also, like most of the other specialized banks in Taipei,China, medium business banks are engaged in a mixture of functions of commercial banking, savings banking, and specialized banking. It is understandable that they are not eager to offer specialized financing to small and medium-scale enterprises, if not required by law.

Credit Guarantee for Small- and Medium-Scale Enterprises

The Small-and-Medium Business Credit Guarantee Fund was set up by the authorities in May 1974 to provide credit guarantees for small- and medium-scale enterprises, so that these enterprises could receive funds from financial institutions. At the end of 1991, the Fund had a net worth of NT$11.8 billion, of which NT$6.7 billion and NT$2.2 billion were contributed by the Government and financial institutions respectively, and the remaining NT$2.9 billion was the accumulated retained earnings.

The objects of credit guarantees include the following: general purpose loans, commercial paper guarantees, guarantee for deferred payment of import duties and taxes, small-scale business loans, export loans, operational input procurement loans, contract performance guarantees, policy-oriented loans, own-brand loans, youth business start-up loans, and other special guarantees. Among these, guarantees for general purpose loans, operational input procurement loans, and export loans are major items, which accounted for 62 per cent, 21 per cent, and 10 per cent, respectively of the total amount of guarantees in 1992.

The application of credit guarantees is usually proposed by the financial institutions after they have received loan applications from small- and medium-scale enterprises. The Fund reviews the application and charges a uniform fare ratio to the guaranteed amounts, usually around 80 per cent of the loans made by financial institutions. The utilization rate of the Fund—defined as the ratio of total outstanding loans to small- and medium-scale enterprises of domestic banks (including full service banks and medium business banks) which has been guaranteed by the Fund—over the period 1974–1991 is reported in Table 2.25. This rate increased steadily from 0.38 per cent in 1975 to 10.41 per cent in 1986, and then fell to 5.82 per cent in 1991. According to Yang (1992), this utilization rate in the late 1980s was about the same as that in Japan and Korea.

Table 2.25 also shows the overdue rate of the guaranteed amount having not been repaid in time. This overdue rate has stabilized at 2–3 per cent in recent years. The Fund's statistics show that a large proportion of the overdue guaranteed amount can usually be recalled by the Fund after several years.

Table 2.25 Performance Statistics of Small- and Medium-Scale Business Credit Guarantee Fund
(per cent)

Year	Utilization Rate[a]	Overdue Rate[b]
1974	–	
1975	0.38	14.79
1976	0.79	4.21
1977	2.05	2.68
1978	2.28	1.64
1979	4.31	2.00
1980	4.88	1.06
1981	5.15	1.34
1982	5.38	2.98
1983	6.58	3.46
1984	9.48	3.05
1985	10.35	6.06
1986	10.41	3.64
1987	9.32	2.77
1877	6.85	2.01
1989	5.68	2.32
1990	5.27	3.30
1991	5.82	2.46

[a]Utilization rate is defined as the total outstanding loans to Small- and Medium-Scale enterprises of domestic banks which have been guaranteed by the Fund at the end of the year.
[b]Overdue rate is defined as the guaranteed amount which has not been repaid by Small- and Medium-scale enterprises in time.
Source: *Small and Medium Business Credit Guarantee Fund* (various issues); Ministry of Finance.

Although the Fund has performed its function well so far, some suggestions concerning its operations can be made, such as to raise the guarantee ratio of loans, to impose a fare rate varying with the degree of riskiness of guaranteed small- and medium-scale enterprises, to lower the fare rate or extend its coverage to more riskier small- and medium-scale enterprises, and to accept more guarantee applications directly from such enterprises instead of from the financial institutions.

EVALUATION OF THE PERFORMANCE OF THE FINANCIAL SYSTEM

The role of the financial system in mobilizing and transferring resources from net savers to net users endows it the power to influence economic growth in many ways. Through a sound and well-functioning financial system within which interest rates are determined by market demand and supply, resources can be mobilized and transferred from lower-return uses to higher-return uses. Further, better utilization of resources will yield a higher return to savers, and thus induce a higher savings rate. The growth rate of the economy will be increased through both better utilization of resources and a rise in the savings rate.

This relationship between the financial system and economic growth can be further explained by the famous Harrod-Domar model. According to this model, the growth rate of an economy equals the investment rate divided by the marginal capital-output ratio. The development of the financial system could contribute to economic growth through the channels of stimulating and mobilizing savings, as well as by allocating efficiently the loanable funds. Stimulating and mobilizing savings could provide more funds for investment purposes and could, therefore, raise the investment rate, while more efficient allocation of loanable funds could lower the marginal capital-output ratio. Both are beneficial in promoting economic growth.

The development of Taipei,China's financial system, and its performance in stimulating and mobilizing savings as well as allocating funds are reviewed in this section.

Development of the Financial System

Two indices are used here to measure the development of the financial system in Taipei,China. One is the ratio of M2 to GDP, which is a commonly used indicator of financial deepening. The other is the contribution ratio of financial sector to GDP, which is defined as the ratio of GDP produced by finance and insurance industries. These two indices are shown in Table 2.26.

The two indices show a clear growing trend over the period 1961–1992 (91). The M2/GDP ratio was only 26.2 per cent in 1961–1965, almost tripled in 1981–1985, and further jumped to 131 per cent in 1986–1992. The contribution ratio of the financial sector to GDP also increased steadily from 1.78 per cent in 1961–1965 to 8.26 per cent in 1986–1991. The increase of this ratio indicates that the financial sector has developed faster than the overall economy. The figure also shows that these two indices grew at a much faster pace in the second half of the 1980s than prior to that year. The increased pace of growth of the contribution ratio of the financial sector to

GDP was mainly attributed to the policy of market-opening to new brokerage houses and prosperity in the stock market. The turmoil of the stock market in 1990 contrarily caused this contribution ratio to stagnate in that year and then to fall in 1991.

Table 2.26 Indices of Financial Development in Taipei,China[a]
(per cent)

Period	M2/GDP	Percentage of GDP Produced by Financial Sector
1961–1965	26.2	1.78
1966–1970	34.5	2.60
1971–1975	43.9	3.16
1976–1980	58.9	4.68
1981–1985	75.3	5.97
1986–1992	131.0	8.26[b]

[a]Data are averages of annual figures.
[b]Average of 1986–1991.
Source: Central Bank of China.

The rise of the M2/GDP ratio implies that the income elasticity of money demand in Taipei,China must be larger than one. Empirical studies in Taipei,China have consistently confirmed this phenomenon of high income elasticity of money demand. There are several explanations for this phenomenon. Tsiang (1977) argues that import and export transactions create extra demand for money which is not taken care of by the national income variable. Shea (1983b) shows that the process of monetization, financial development, and the specialization of the production process have all contributed to the high income elasticity of money demand in Taipei,China. As for the sharp increase of M2/GDP ratio in the second half of the 1980s, it can be explained by the abrupt increase of the transactions in the stock and real estate markets, which have positive effects on the money demand.[14]

---

[14] Empirical evidence to support this can be found in Wu and Shea (1993).

Stimulation and Mobilization of Savings

As Table 2.27 shows, the share of gross domestic savings in GNP has risen steadily from 15 per cent in 1951–1960 to 33 per cent in 1981–1992. In the late 1980s, Taipei,China had one of the highest saving ratios in the world. Because of this, the country's rapid growth had been accompanied by much less foreign borrowing than other developing countries. According to the data in Table 2.27, although the savings rate in 1951–1970 was slightly less than the investment rate—so a small proportion of the domestic investment was financed by foreign borrowing or US aid—investments in the later period were entirely financed out of domestic savings. In 1981–1992, an average of 10.13 per cent of GNP had even been loaned abroad.

Table 2.27 Savings, Investments, and Net Foreign Borrowing as a Percentage of GNP[a]

Period	Savings	Investment	Borrowing[b]
1951–1960	14.91	16.08	1.17
1961–1970	21.07	21.89	0.82
1971–1980	31.85	30.48	−1.37
1981–1992	32.52	22.39	−10.13
1951–1992	25.44	22.70	−2.74

[a]Data in this table are simple averages of annual data.
[b]Net foreign borrowing is gross domestic investment (column 2) minus gross savings (column 1).
Source: DGBAS (1992).

The high savings rate in Taipei,China could be attributed to the high growth rate in real GNP, the stable price level, the large size of government savings (the average share of government savings was 21 per cent of national savings and 20 per cent of government's total current revenue in 1951–1990), the underdeveloped state of the social security system, the low accessibility of consumer financing, the tradition of being frugal in Chinese culture, and/or a strong desire to establish an independent business for the family through saving.[15] In addition, the policy of high real rates of

---

[15] Most of these factors contributing to the high savings ratio have been discussed by Sun and Liang (1982) and others.

interest on deposits, vigorous savings promotion campaigns by financial intermediaries in the early years, and easy access to financial institutions certainly helped.

The high savings rate may also be ascribed to the incentives provided by the tax and banking policies. Between 1960 and 1980, interest income from two-year or longer term deposits was tax exempt. Since 1981, combined interest and dividend income up to a limit of NT$360,000 (NT$270,000 since 1990) was also exempt from income tax. The interest income from postal passbook accounts under the ceiling of NT$700,000 (NT$1 million after March 1990) of deposits has also been tax exempt since 1965. Further, until May 1990, depositors of banks were able to lock their funds into high interest rate long-term savings deposits and were subject to no loss of interest if they canceled the accounts before maturity.

Finally, a widespread and convenient informal financial system which provides a high rate of earning for savings, and the industrial structure which is full of small- and medium-scale enterprises are contributing factors to a high savings rate as well. The majority of the small- and medium-scale enterprises are family owned, thus owners have to save funds to meet most of the working and investment needs of their enterprises.

The accessibility and geographical distribution of financial institutions were discussed earlier in this chapter. The bank deposit rate, curb-market interest rate and inflation rate are reported in Table 2.15 (p. 109). This table indicates that the real interest rate on one-year savings deposits has been positive since 1961 except for the oil-shock periods, with the average at about 6.3 per cent over the period 1961–1992. The real curb-market interest rate averaged as high as 21.7 per cent in the same period.

The flow-of-funds relationship among sectors was also discussed in an earlier section and shown in Tables 2.5 through 2.7 (pp. 85–90). Though private and public enterprises had generated through internal saving almost two thirds of their investment funds in 1951–1988, they were still the major sectors with financial deficits. The household sector spent 26 per cent of its saving on investments in housing, transport, and equipment; however, it has always been the most important source of financial surplus. The Government also provided a large financial surplus, although it had invested a large proportion (some 60 per cent) of its savings in infrastructure.

In sum, during the period 1965–1988, 102 per cent of the total savings or 70 per cent of the total funds of the household sector had been channeled into the financial system (excluding the stock market). If the unregulated informal system is excluded, the ratios of household savings and total funds in the regulation formal financial system were still maintained at 90 per cent and 62 per cent respectively. In addition, 26 per cent of government savings went to financial institutions as government deposits; and 72 per cent (51 per cent) of the investment funds or 45 per cent

(31 per cent) of the total uses of funds of private enterprises came from the financial system (regulated formal financial system). Public enterprises depended mainly on internal savings and stock issuing to finance their investments. Financial institutions used over three fifths of the deposits to make domestic loans. However, accumulation of foreign assets also constituted over one third of the total uses of their funds. Based on the above facts, the conclusion can be drawn that the main function of Taipei,China's financial system in mobilizing savings has been to channel household savings to finance the investment of private enterprises, and this function had been performed reasonable well until the mid-1980s, when a large share of domestic savings began to be used to finance foreign investments, thus allowing a large proportion of deposits to be turned into foreign assets.

The success of the financial system in mobilizing savings is also evidenced by the steady increase of the ratio of M2 to GDP as shown in Table 2.26. Other evidence of the success can be found in Table 2.15 (p. 109), which shows that the share of deposits with longer maturities has been increasing.

As for the factors that have contributed to the successful mobilization of savings, a positive real bank deposit rate, the ease of access to financial institutions and their services, and an active informal financial system which the Government never tried to suppress are all beneficial factors. As noted in Tables 2.7 (p. 90) and 2.13 (p. 103), the informal financial system (curb markets) played important roles in attracting household savings and financing the investment needs of private enterprises.

Allocative Efficiency of Loanable Funds

As far as economic growth is concerned, efficient allocation of loanable funds is as important as the availability of such funds. While the financial system in Taipei,China has a good record in stimulating and mobilizing domestic savings, the performance of financial institutions in allocating these funds has often been criticized. Such criticisms include the following: (i) the emphasis was on collateral rather than the profitability or productivity of the borrowers; (ii) medium- and long-term loans were in short supply relative to short-term loans; (iii) public enterprises and large firms were favored, while private enterprises and small- and medium-scale enterprises were discriminated against; (iv) export industries were privileged in obtaining bank loans and preferential loans compared to import-competing industries and the nontradable sector; and (v) consumption finance was neglected purposely.

This chapter has dealt with each of these criticisms. In addition, Shea (1993) provides empirical evidence through regression analysis which shows that collateral was emphasized more than productivity of the borrowers, that export activities were privileged in applying for bank loans, and that the industries which had been disproportionately favored in the credit rationing process were not consistent with those which had shown higher production growth rates.

Regarding the impact of the above allocating behavior of financial institutions on the allocative efficiency of loanable funds, innovative investment plans that were profitable but which could not meet collateral requirements might have been rejected for loans by the financial institutions. The shortage of medium- and long-term loans must have forced enterprises to cut down on long-term investments, or at best to roll over short-term loans for long-term investment purposes, which is a common phenomenon in Taipei,China. This inevitably increases the liquidity risk of firms. The privilege of export industries in obtaining bank loans and preferential loans has been generally regarded as one of the successful policy measures in stimulating export and economic growth. However, it also distorted fund allocations among export, import-competing and nontradable sectors, and contributed, together with the policy of discriminating against consumption financing, to the problem of the huge trade surplus in the 1980s.

As for the preference of financial institutions in lending to public enterprises and large businesses, this must have also distorted the allocation of investment funds. Public enterprises in Taipei,China have often been criticized for their low operation and investment efficiency. The prosperity and the important role of small- and medium-scale enterprises also indicate that large businesses might not be superior to small- and medium-scale enterprises in operational or investment efficiency.

The above analysis indicates that the performance of financial institutions in allocating funds has been far from satisfactory. The causes of these allocative inefficiencies include the Banking Law restrictions, regulations on government banks, supervision of the Control Yuan, interest rate control, market-entry regulation, inaccurate accounting records of small- and medium-scale enterprises, and the financial policies to promote exports and economic growth. Each has been discussed at some length in this chapter.

As Stiglitz (1993) points out, there are many reasons for governments to intervene in the financial sector. In the developing countries, problems such as a severe shortage of savings and loanable funds, the imperfection of legal and accounting system, and the need to accelerate economic growth, further invite the government to intervene in fund allocation. Similar to other developing countries, the authorities in Taipei,China intervened extensively in the financial sector, and thereby created the above inefficiencies of fund allocation. Although these allocative inefficiencies have been harmful to economic growth in Taipei,China, the negative impacts have not been severe, mainly due to the following two factors.

First, bureaucratic elites had no vested interest in the economic arena, and were therefore willing to adopt an export-promotion policy and financial reforms in the 1950s. They also had the ability to identify the industries with comparative advantages, to which loanable funds were rationed more favorably.

Second, the authorities had no intention of suppressing the informal financial system by seriously implementing the usury law and other related prohibitive rules. As a result, the informal system could remain active to supplement financial institutions in stimulating and mobilizing savings as well as improving the allocative efficiency of loanable funds. The much higher interest rates offered by the informal system, as reported in Table 2.15 (p. 109), should have been one of the contributing factors to the high saving rate in Taipei,China, and have benefited the mobilization of savings for investment purposes. In addition, the informal system, again through higher interest rates, competed away part of the loanable funds that otherwise would have been used by the financial institutions to finance less efficient investment projects. Also, the more efficient investment projects, which had been rejected for loans by financial institutions simply because they could not provide enough assets as collateral or could not meet the Government's criteria of selective credit rationing, could be financed by the informal system. Hence, the informal financial system improved the allocative efficiency of loanable funds.[16]

To conclude, in the 1950s and early 1960s, the economy was short of investment funds, the private sector lacked experienced entrepreneurs and reliable accounting information, and industries with comparative advantages could be easily identified. Under such circumstances, the strategy to depend on the interventional financial policies rather than the price (interest rate) mechanism to allocate investment funds to public enterprises, export activities, and industries supposed to have comparative advantages, might not cause much harm. However, by the 1970s and 1980s, when Taipei,China's economy had more investment funds and private entrepreneurs, and the industries with comparative advantage were no longer easily identified, the Government's continuous dependency on interventional financial policies to allocate investment funds naturally invited distortions and criticisms.

Fortunately, continuous complaints and criticisms on the misallocation of loanable funds from the general public and scholars, aided by pressures from the outside world and changes in the domestic economic and political environments, forced the authorities to launch financial deregulation in

---

[16] Shea and Kuo (1984) provide some empirical results which measure the allocative inefficiency and efficiency loss of financial institutions, and the contribution of the informal system in improving allocative efficiency. Its model and results are briefly summarized in Shea (1993).

the 1980s. Some of the above causes of allocative inefficiency were thus eliminated to a great extent. The allocative efficiency of loanable funds in financial institutions may therefore be improved in the future. However, since interventional financial policies are usually the most cost-saving arrangements to extend off-budget subsidy to borrowers, the authorities' continuing dependence on it as a major component of policies to promote growth or to serve political or social justice purposes is certainly inevitable.

PROSPECTS AND LESSONS

Together with the rapid growth of the overall economy, the financial sector in Taipei,China had shown some progress in development before 1980; however its backwardness had always been criticized as the bottleneck of further economic growth. Although the financial system performed the functions of stimulating and mobilizing savings reasonably well, its performance in allocating investment funds was not satisfactory due to factors such as government intervention, extensive regulations, and an inadequate accounting system. Fortunately, the existence of an active, unregulated informal financial system complemented the regulated formal financial system in improving the allocative efficiency of loanable funds as well as stimulating and mobilizing savings. Thus, the conclusion can be derived that, as Patrick (1993) puts it, "the financial system [in Taipei,China] was rather more the accommodator of this real economic performance than its instigator."

In the 1980s, under domestic as well as foreign pressures, the authorities adopted several measures to liberalize the financial system. The purpose was to convert a regulated, segmented, and protected credit-rationing financial system into a deregulated, more market-oriented competitive system linked to global financial markets. Although the liberalization process has been slow, gradual and piecemeal, few developing countries have managed more durable and far-reaching financial deregulation than Taipei,China.

However, the appropriateness of the sequence of financial liberalization in Taipei,China has often been criticized by local economists. The second-best theory implies that sequencing should be an important issue in the process of liberalization. McKinnon (1982) and others suggest that a country's domestic financial market is better liberalized prior to the capital account of the balance of payments. In Taipei,China, when the domestic financial system was not well developed and still under strict control, whether lifting foreign exchange control rapidly was a wise policy remains an open question. Further, the inadequate accounting system, market-entry regulation, government ownership, and interest rate control were major factors causing financial market segmentation, i.e., public enterprises and large firms have been favored by banks in credit rationing, while small- and medium-scale enterprises have been discriminated against and, therefore, must depend heavily on curb-market financing. Without solving the problems of the first three, deregulating the interest rate alone cannot significantly reduce the degree of market segmentation. As Table 2.12 (p. 102) shows, after several years of attempting interest rate decontrol, the share of private enterprise borrowing from curb markets even rose in 1985–1986, although banks were swamped with excess reserves during that period. Consequently, whether relaxing other financial controls under financial market segmentation will benefit the economy also becomes an issue worth further study.[17]

In addition, 16 new private banks were allowed to be set up prior to the privatization or deregulation of the operations of government banks and the establishment of a sound supervision and monitoring system. This sequence of liberalization is generally believed harmful to the safety and stability of the financial system. How to strengthen prudential regulation and supervision to prevent the misuse of the new private banks by their owners is now a major challenge facing the authorities.

Three additional problems also face the authorities: how to overcome the opposition of vested interest groups, especially the provincial assemblymen, so that government banks may be privatized; how to prevent insider trading in the stock market; and how to improve the accounting system and the quality of businesses' accounting information. Until these problems are all resolved, the efficiency and stability of the financial system in Taipei,China cannot be guaranteed.

Several lessons can be derived from the discussion in this chapter. On the positive side, a conservative fiscal policy and price stability are preconditions for a well-functioning financial system. The widespread systems of postal savings and the credit cooperatives, associated with the policies of a balanced government budget, money supply control, and a positive real interest rate on deposits, have been beneficial to the stimulation and mobilization of savings. When there is interest rate control, fund shortages, and an underdeveloped formal financial system, allowing an informal financial system to be active is helpful in mobilizing savings, financing the needs of those discriminated against by the formal system, as well as improving the allocative efficiency of loanable funds. Using the redeposits from postal savings and the Cooperative Bank of Taiwan to provide medium- and long-term loans is also recommended.

On the negative side, strict entry regulations, and government ownership of financial institutions are factors which adversely affect financial development and the allocative efficiency of investment funds. Interest rate control,

---

[17] The welfare effects of economic liberalization under financial market segmentation have been analyzed by Shea (1992).

which had set bank interest rates lower than market-clearing rates, reduced allocative efficiency. Selective credit control policies or nonprice interventional financial policies—although not severely distorting resource allocation during the beginning stages of economic development—are not recommended for the later stages. For a small open economy unable to control capital inflow and outflow effectively, intervening in foreign exchange markets may bring a disaster to the control of domestic money supply, and is, therefore, not a recommended policy either. Finally, an adequate accounting system, prudential regulations and a sound supervision and monitoring system are essential to a well-functioning financial system.

Bibliography

Annual Report of Small and Medium Business Credit Guarantee Fund.

Asian Development Bank. 1985. *Improving Domestic Resource Mobilization Through Financial Development.* Manila. September.

_____. 1992. "Managing Financial Sector Distress and Industrial Adjustment." Energy and Industry Division, World Bank.

Central Bank of China. 1987. *National Income In Taiwan Area, ROC.* Taipei: Central Bank of China. Tables 6 and 7.

_____. 1989. *Financial Statistics Monthly: Flow of Funds in Taiwan District, ROC* [Republic of China]. Taipei: Central Bank of China.

_____. 1991. *Financial Statistics Monthly: Flow of Funds in Taiwan District, ROC.* Taipei: Central Bank of China.

_____. 1991. *Report of Survey Results on the Financial Conditions of Public and Private Enterprises in Taiwan Area, 1981 and 1991.* Taipei: Central Bank of China.

_____. 1992a. *Financial Statistics Monthly: Flow of Funds in Taiwan District, ROC.* Taipei: Central Bank of China.

_____. 1992b. *National Income In Taiwan Area, ROC.* Taipei: Central Bank of China. Chapter 3; Table 16, and Chapter 2, Table 1.

_____. 1993. *Financial Statistics Monthly: Flow of Funds in Taiwan District, ROC.* Taipei: Central Bank of China.

_____. (various issues). *Annual Report on the Operations of Financial Institutions.* Financial Inspection Department. Taipei: Central Bank of China.

Chang, Chi-Cheng. 1989. "Financial Liberalization in the Republic of China." Keynote address at the First Annual Pacific Basin Finance Conference in Taipei, 13 March.

Cheng, Hang-Sheng. 1986. "Financial Policy and Reform in Taiwan, China." In *Financial Policy and Reform in Pacific Basin Countries*, edited by Hang-Sheng Cheng. Lexington, MA: Lexington Books.

Chiu, Paul C.H. 1982. "Performance of Financial Institutions." In *Experiences and Lessons of Economic Development in Taiwan*, edited by Kwoh-Ting Li and Tzong-Shian Yu. Taipei: Institute of Economics, Academia Sinica.

――――. 1992. "Money and Financial Markets: The Domestic Perspective." In *Taiwan: From Developing to Mature Economy*, edited by Gustav Ranis. Boulder: Westview Press.

Chou, Tein-Chen. 1993. "Government, Financial System and Economic Development in Taiwan." Paper presented at 35th International Atlantic Economic Conference in Brussels. April.

Directorate-General of Budget, Accounting, and Statistics (DGBAS). 1992. *National Income in Taiwan Area, ROC, 1992*. DGBAS.

Ghate, P. 1992. *Informal Finance: Some Findings From Asia*. Hong Kong: Oxford University Press for Asian Development Bank.

Kuo, Shirley W.Y. 1989. "Liberalization of the Financial Market in Taiwan in the 1980s." Keynote address at the First Annual Pacific Basin Finance Conference in Taipei, 14 March.

Lee, Yung-San, and Shang-Cheng Chen. 1984. "Financial Development in Taiwan: Review and Prospective." *Proceedings of the Conference on Financial Development in Taiwan*. Taipei: Institute of Economics, Academia Sinica. December. In Chinese.

Lee, Yung-San, and Tzong-Rong Tsai. 1988. "Development of the Financial System and Monetary Policies in Taiwan." *Conference on Economic Development Experiences of Taiwan and Its New Role in An Emerging Asia-Pacific Area*. Vol 1. Taipei: Institute of Economics, Academia Sinica. January.

Liu, Christina Y. 1993. "Financial Policies in a Maturing Taiwan Economy." Paper presented at the Conference on the Evolution of Taiwan Within a New World Economic Order. Taipei: Chung-Hua Institution for Economic Research. May.

Liu, Shou-Hsiang. 1988. "The Development and Prospect of Financial Institutions in Taiwan." In *Proceedings of the Conference on the Modernization of Service Industries in ROC*. Taipei: Chinese Economic Association. In Chinese.

McKinnon, Ronald I. 1973. *Money and Capital in Economic Development.* Washington DC: The Brookings Institution.

_____. 1982. "The Order of Economic Liberalization: Lessons from Chile and Argentina." In *Economic Policies in a World of Change,* edited by Karl Brunner and Alan H. Meltzer. Amsterdam: North-Holland.

Ministry of Finance. 1990. *1990 Securities and Exchange Commission Statistics.* Taipei: Ministry of Finance.

_____. 1992a. *Financial Statistics Abstract.* Bureau of Monetary Affairs. December.

_____. 1992b. *Yearbook of Financial Statistics of the Republic of China.* Taipei: Ministry of Finance.

_____. 1993. *Financial Statistics Abstract.* Bureau of Monetary Affairs. June.

Ministry of Interior. *Statistical Abstract of the Interior.* Taipei: Ministry of Interior.

National Taiwan University. 1991. "A Study to Improve the Quality of Businesses' Accounting Information in Taiwan." Department of Accounting. Project Report. May.

Patrick, Hugh T. 1993. "Comparisons, Contrasts and Implications." In *The Financial Development of Japan, Korea and Taiwan,* edited by Hugh T. Patrick and Yung-Chul Park, Chapter 8. Oxford University Press.

Shea, Jia-Dong. 1983a. "Financial Dualism and the Industrial Development in Taiwan." In *Proceedings of the Conference on Industrial Development in Taiwan.* Taipei: Institute of Economics, Academia Sinica. March. In Chinese.

_____. 1983b. "The Income Elasticity of Money Demand in Taiwan." In *Academia Economic Papers.* Taipei: Institute of Economics, Academia Sinica. September. In Chinese.

_____. 1992. "The Welfare Effects of Economic Liberalization Under Financial Market Segmentation: with Special Reference to Taiwan."

In *Academia Economic Papers.* Vol. 20, No. 2, Part II. Taipei: Institute of Economic, Academia Sinica. September.

──────. 1993. "Financial Development in Taiwan: A Macro Analysis." In *The Financial Development of Japan, Korea and Taiwan*, edited by Hugh T. Patrick and Yung-Chul Park, Chapter 6. Oxford University Press. Forthcoming.

Shea, Jia-Dong, and Ping-Sing Kuo. 1984. "The Allocative Efficiency of Banks' Loanable Funds in Taiwan." In *Proceedings of the Conference on Financial Development in Taiwan.* Taipei: Institute of Economics, Academia Sinica. December. In Chinese.

Shea, Jia-Dong, Ming-Yih Liang, Ya-Hwei Yang, Shou-Hsiang Liu, and Kun-Ming Chen. 1985. "A Study of the Financial System in Taiwan." *Economic Papers.* 65. Taipei: Chung-Hua Institution for Economic Research. June. In Chinese.

Shea, Jia-Dong, and Ya-Hwei Yang. 1989. "Financial System and the Allocation of Investment Funds in Taiwan." Paper presented at the Conference on State Policy and Economic Development in Taiwan, held by Soochow University in Taipei. December.

Shea, Jia-Dong, and Tzung-Ta Yen. 1992. "Comparative Experience of Financial Reform in Taiwan and Korea: Implications for the Mainland of China". In *Economic Reform and Internationalization*, edited by Ross Garnaut and Guoguang Liu. Allen & Unwin.

Small and Medium Business Credit Guarantee Fund. *Annual Report.* various issues.

Stiglitz, Joseph E. 1993. *The Role of the State in Financial Markets.* Chung-Hua Series of Lectures by Invited Eminent Economists, No. 21. Taipei: Institute of Economics, Academia Sinica. April.

Sun, Chen, and Ming-Yih Liang. 1982. "Savings in Taiwan, 1953–1980." In *Experiences and Lessons of Economic Development in Taiwan*, edited by Kwoh-Ting Li and Tzong-Shian Yu. Taipei: Institute of Economics, Academia Sinica.

Tsiang, Shoh-Chieh. 1977. "The Monetary Theoretic Foundation of the Modern Monetary Approach to the Balance of Payments." *Oxford Economic Paper.* November.

———. 1979. "Fashions and Misconceptions in Monetary Theory and Their Influences on Financial and Banking Policies." *Zeitschrift fur die gesamte Staatwissenschaft.* December.

———. 1982. "Monetary Policy of Taiwan." In *Experiences and Lessons of Economic Development in Taiwan,* edited by Kwoh-Ting Li and Tzong-Shian Yu. Taipei: Institute of Economics, Academia Sinica.

Wu, Chung-Shu, and Jia-Dong Shea. 1993. "An Analysis of the Relationship Between Stock, Real Estate and Money Markets in Taiwan in the 1980s." Paper presented at the 1993 Far Eastern Meeting of the Econometric Society, Taipei, Taiwan. June.

Yang, Ya-Hwei. 1992. "A Study of the Financing of Small and Medium Businesses." Taipei: Chung-Hua Institution for Economic Research, January. In Chinese.

———. 1993. "A Micro Analysis of the Financial System in Taiwan." In *The Financial Development of Japan, Korea and Taiwan,* edited by Hugh T. Patrick and Yung-Chul Park, Chapter 7. Oxford University Press. Forthcoming.

CHAPTER THREE

FINANCIAL SECTOR DEVELOPMENT IN INDONESIA

ANWAR NASUTION

This chapter examines Indonesia's transition from a repressed financial system to one that is more market-based, a transition that occurred between 1983 and 1993 and required numerous financial policy reforms. The objectives of those reforms were to promote domestic savings, improve the efficiency of resource allocation, and provide a framework that allowed for the implementation of effective market-based monetary control.

In addition, this chapter reviews Indonesia's financial sector by examining the following:

(i) the status of financial institutions and financial markets before and following the reforms in October 1988;

(ii) the efficiency and effectiveness of the financial system in mobilizing savings, financial intermediation, and allocating investment, both prior to and after the reforms;

(iii) the constraints faced by the financial system and the obstacles preventing the development of the financial sector. In particular, the chapter analyzes the role of macroeconomic environment, sectoral policy mix (such as credit and tax policies), competition policy and market infrastructure, including prudential regulations and supervision in promoting the development of money and capital markets; and

(iv) the problems with sequencing financial sector reforms.

MACROECONOMIC SETTING BEFORE AND AFTER DEREGULATION

Macroeconomic Conditions

Falling terms of trade, rising international interest rates, and the foreclosing of capital inflows which had risen with external debt repayments coupled with stagnant investment and economic growth in the early 1980s forced Indonesia into pursuing drastic adjustment policies. Initially, between 1983–1985, the focus of these policies was on short-run stabilization, in particular on restoring external creditworthiness. When it became apparent that these measures alone would be insufficient to restore long-term growth, the stabilization policy was complemented by structural reforms aimed at increasing domestic saving mobilization and efficiency of resource use.

In addition, because of the trauma of hyperinflation in 1960–1966, the present Government is concerned about maintaining both internal and external economic balance. The principal tools used by the Government to preserve this macroeconomic balance are a conservative balanced budget policy; an external borrowing strategy which mainly maximizes long-term concessionary loans from a consortium of its donors; a relatively open exchange rate system and sound foreign exchange rate policy; and a willingness to postpone investment projects which it cannot afford. Table 3.1 presents a chronology of this adjustment program.

Table 3.1 Chronology of the Adjustment Program, 1983–1993

Policy Instrument

Exchange Rate
1. Rupiah was devalued by 28 per cent against the US dollar on 30 March 1993 from Rp 703 to Rp 906 per US dollar; since then the exchange rate has been made more flexible.
2. Rupiah was devalued by 31 per cent against the US dollar on 12 September 1986 from Rp 1.134 to Rp 1.644 per US dollar.

Fiscal Policy
1. Tight fiscal policy since 1983 marked by:
   a. large capital and import intensive projects (particularly investment in manufacturing, petrochemicals, and mining) rephased in May 1983;
   b. major cutback in public real capital spending;
   c. more resources for social programs; and
   d. restraints on civil service employment and salaries.
2. Tax reform enacted in January 1983, involving simplification of both tax structure and tax administration of all tax sources, excluding taxes on foreign trade.

Monetary and Financial Policy
1. Financial reform initiated on 1 June 1993, involving removal of credit and rate ceilings for state bank's operations, a reduction in the scope of credit programs, and introduction of new market-oriented instruments of monetary control.

2. New deregulation measures introduced in December 1987, October and December 1988, and March 1989 aimed at enhancing financial sector prudential standards and efficiency, and developing the capital market by, among others, removing barriers to duty.
3. Improved monetary management to control inflation and to curb exchange rate speculation.
4. Removal of central bank's direct credits ("liquidity credits") and major reduction of economic sectors covered by subsidized "priority credits" in January 1990 to curb inflationary pressures and credit fungibility.
5. New regulations introduced on 14 March 1991, which are aimed at strengthening the capital base of banks and tightening supervision over financial institutions. The new measures require the banking system to meet the BIS guidelines on capital adequacy ratio of 8 per cent of the bank assets by December 1993.
6. Relaxation of prudential standards introduced on 29 May 1993 and the deadline to meet the capital adequacy ratio of 8 per cent was extended to December 1994.

Trade Policy

1. Across-the-board reductions in nominal tariffs introduced in April 1985, October 1986, and 28 May 1990.
2. Measures to provide internationally priced inputs to exporters announced on 6 May 1986, and 28 May 1990. This scheme permits exporters and suppliers of inputs for exporters to bypass the import licensing system and import tariff or, if they cannot bypass the system, to reclaim import duties, although the cost imposed by the nontariff barriers cannot be rebated. The import bias of the protective system had been lessened but not uniformly.
3. Major deregulation of import licensing system announced on 25 December 1986, 15 January 1987, 28 May 1990, and July 1992.
4. Additional measures to reduce anti-export bias announced in December 1987 by reducing regulatory framework for exporters.
5. Major removal of nontariff barriers, switch from nontariff to tariff barriers, and general reduction of tariff rates on 28 May 1990 and July 1992. Also covering simplification of licensing procedures in trade, manufacturing, health, and agricultural business, the policy package is aimed at reducing the high cost economy.
6. Further removal of nontariff barriers, general reduction of import tariffs, and reopening of several business fields to new domestic and foreign investors was announced on 2 June 1991. Several major features of the reform cover

outright import bans of cold-rolled steel coils, and other steel products, abolition of the export quota system for palm oil and copra, introduction of an import quota system on built-up commercial vehicles, and reopening of car component manufacture to new investors.
7. In June 1993, for the first time in 23 years, import ban on built-up passenger cars was replaced by prohibitive 300 per cent tariff rate, further reduction in tariff for other commodities, and relaxation of regulations in trade and agriculture products.

Other Regulatory
Framework
1. Reorganization of customs, shipping, and port operations announced in April 1985 to reduce handling and transport cost for exports and to simplify the administrative procedures governing interisland and foreign trade. Further deregulation of maritime activities announced on 21 November 1988 to reduce cost and encourage private sector participation, including foreign capital and foreign shipping companies.
2. Measures to reduce the investment and capacity licensing requirements, relax foreign investment regulation, and reduce the local content program.
3. Measures announced on 6 July 1992 to allow joint venture firms to hold land titles (right to use the land) and use them for credit collateral, liberalized import of used machinery, plant equipment and other capital goods, and liberalize expatriate work permits.
4. On 10 June 1993, the number of investment negative list reduced from 51 to 34.

In general, the successful implementation of the stabilization programs and substantial economic reforms have diversified the productive base of the economy, expanded the role of the private sector, and reduced the country's reliance on oil. Because of these developments, over the past twenty years Indonesia has been hailed as one of the shining lights of the international community.

Despite various external shocks, the average annual inflation rate was controlled at 17.5 per cent in the 1970s and 8.6 per cent in the 1980s. In terms of GDP, the economy grew by an average 7.7 per cent per annum in the 1970s and 5.5 per cent in the 1980s (Table 3.2).

The rapid rate of economic growth combined with a gradual decline in the rate of population growth resulted in rising annual average income per capita by 5.5 per cent in the 1970s and 3.3 per cent in the 1980s. This was accompanied by steady improvements in the reduction of poverty, with the percentage of population living below the official poverty line falling from 40 per cent in 1976 to 20 per cent in 1987 and to 15 per cent in 1990. In 1990, the World Bank found that in two decades Indonesia had achieved the highest annual average reduction in the incidence of poverty among all the countries studied (World Bank 1990). With the level of income per capita at $650 in 1992, however, Indonesia remains a poor country.

Table 3.2 Indicators of Economic Growth and Structural Change

	1970–1980	1981–1990
I. Growth indicators Average growth (per cent per annum)		
1. GDP (at constant 1983 market prices)	7.7	5.5
2. GDP per capita	5.5	3.3
3. Non-oil GDP	5.3	6.3
4. Manufacturing	14.1	9.9
5. Population	2.4	2.0
6. Inflation rate	17.5	8.6
7. Money supply (M1)	35.1	16.9

	1970	1980	1985	1991
II. Structural change (as a percentage of GDP at constant 1983 market prices)				
1. Non-oil GDP	97.4	76.4	78.6	81.3
2. Agriculture	45.5	30.7	22.7	18.4
3. Manufacturing	8.4	15.3	15.8	19.9
4. Gross domestic investment	15.8	24.3	28.0	35.0
5. Gross national savings	14.3	33.0	25.8	31.0
6. Current accounts[a]	–3.5	4.2	–2.2	–3.9

[a] As a percentage of GDP at current market prices.
Sources: Growth indicators and structural change are from World Bank, *World Development Report* (various issues); money and GDP data are from IMF, *International Financial Statistics* (various issues); total assets of financial institutions are from Cole and Slade (forthcoming).

Consistent with the general pattern of development elsewhere in the world, the share of agriculture has declined gradually. On the other hand, the modern industrial sector, including manufacturing, grew by 11 per cent per annum in the 1970s and 5.8 per cent in the 1980s. The manufacturing industry accounted for 11 per cent of GDP in 1980 and 21 per cent in 1992. The financial service sector grew comfortably at 12.1 per cent per annum in the 1980s. The contribution of this sector to GDP has increased from 3.5 per cent in 1985 to 4.2 per cent in 1992.

The rapid rate of economic growth has been possible mainly because over the past 25 years Indonesia has annually invested more than a quarter of its GDP. Since 1987, private sector investment, along with non-oil exports and domestic consumption, has been the engine of growth of the Indonesian economy. The bulk of the resources required for investment came from domestic savings. As a percentage of GDP, the national savings ratio has increased from 9.8 per cent in 1970 to 20.9 per cent both in 1980 and 1990. With the rise in national saving, the annual average investment-savings gap has steadily declined, from 5.3 per cent of GDP in 1970 to 3.8 per cent in 1980 and to 3 per cent in 1990. The investment-savings gap for the public sector has always been positive, while the private sector has always been a net saver. With the growing role of the private sector in the economy, the share of the public sector in total investment has declined steadily from 44 per cent in 1983 to 40 per cent in 1990.

The high share of Indonesia's development financing from domestic sources is also reflected in the relatively low level of its current account deficit which, in terms of GDP, was 3.5 per cent in 1970, negative 3.9 per cent during the "oil boom" in 1980, and 3.0 per cent in 1990. The financing of these twin deficits has been dependent on foreign aid and borrowing and foreign direct investment.

Structure of the Financial System

Prior to October 1988, Indonesia adopted a system of strict functional specialization of financial institutions, consisting of commercial banks, development banks, saving banks, finance companies and other nonbank intermediaries, securities firms, and insurance companies. In terms of total assets and number of offices, however, the core of the Indonesian financial sector is the banking system. Other financial institutions, such as savings banks, leasing companies, insurance firms, securities companies, and pension funds are fast-growing segments of the financial system, but these sectors as a group are relatively small.

The banking system consists of Bank Indonesia (the central bank), and the commercial and development banks, of which seven are state-owned banks. Together, this group of banks held over 90 per cent of the gross assets of the financial system in 1991. The Banking Acts of 1968 defined the major areas of concentration for each state-owned bank, although, in practice, this is not strictly implemented. Bank Indonesia's certificates (SBIs) accounted for 72 per cent of the value of instruments traded in money markets.

Financial market information is imperfect because of weaknesses in the market infrastructure, particularly the underdevelopment of the legal and accounting systems. Under such conditions, it is difficult for financial institutions to acquire information about their clients on the asset side of the balance sheet. This has led banks to rely on developing information through building long-term relationships with their clients (Caprio 1993) and demanding high value collateral for their nonprogram credits. Financial institutions owned by private conglomerates prefer to give credit to their own nonfinancial subsidiaries which they know best and over which they maintain control.

The capital markets (stock exchanges in Jakarta and Surabaya) are in the early stages of development. In 1991, the outstanding value of stocks and bonds issued in these markets was only 31 per cent of the banks' outstanding credit which indicates the minor role played by capital markets in Indonesia. A heavy reliance on debt financing has had adverse effects at the microeconomic level as it has led to unbalanced capital structures. High debt-to-equity ratios present few problems as long as the firms continue to grow. However, because of the rise in interest, accompanied by slow growth since 1991, the heavy reliance on debt has strained companies' cash flows and increased their vulnerability, which has led to problems in the economic adjustment process and has lessened the effectiveness of the interest rate policy.

Financial Assets

In a less-developed country with underdeveloped financial sophistication, such as Indonesia, money (both M1 and M2) is the leading financial asset of the general public. The rising ratios of M1 and M2 to GDP indicate that financial deepening has occurred, particularly after the economic reforms. The rapid rate of growth, particularly in the non-oil export activities, has promoted monetization. This, and the more aggressive financial institutions, have promoted savings in forms of financial assets. Following the October 1988 reforms, M1/GDP, increased from 0.10 in 1988 to 0.12 in 1991 and, during the same period, M2/GDP rose from 0.30 to 0.44 (Table 3.3). These two indicators, however, cover only the liabilities of the banking system in rupiah, the national currency. They exclude national holding of foreign currencies which is quite significant in Indonesia, particularly at times of political and economic instability. Indonesian citizens and foreign residents, under the current open foreign exchange system, are free to open accounts either in foreign currencies at authorized banks or in rupiah.

Table 3.3 Structure and Growth of the Financial Sector, 1969–1991

	Number				Share in Assets (per cent)				Asset Growth (per cent)		
	1969	1982	1988	1991	1969	1982	1988	1991	1970–1982	1983–1988	1989–1991
Bank Indonesia	1	1	1	1	57.7	42.4	36.8	23.8	31.1	18.8	9.2
Deposit Money Banks	179	118	111	195	42.3	52.9	56.9	68.5	37.6	22.4	32.6
State commercial banks	5	5	5	5	30.3	37.9	34.5	30.2	36.6	19.7	20.7
Private banks	126	70	63	129	3.7	5.8	13.1	25.2	34.0	41.5	56.9
Private forex banks	3	10	12	28	...	3.6	8.8	19.7	...	36.1	64.9
Private nonforex banks	123	60	51	101	...	2.2	4.3	5.5	...	32.2	37.3
Foreign banks	11	11	11	29	4.3	3.6	2.8	5.2	32.6	16.8	55.4
Development banks	25	29	29	29	4.0	4.1	4.4	6.3	34.6	22.1	42.2
Savings banks	12	3	3	3	0.1	1.4	2.1	1.6	60.0	27.9	14.5
Nonbank financial inst.	...	13	13	13		2.5	2.7	2.1	...	22.3	14.9
Insurance companies	...	83	106	185		1.6	1.6	3.5	...	21.2	62.3
Leasing companies	...	17	83	88		0.4	1.5	1.8	...	45.4	33.6
Other credit institutions	8,568	5,808	5,783	6,243		0.3	0.6	0.4[a]	...	33.4	15.9[b]
Total	8,748	6,040	6,097	6,725	100.0	100.0	100.0	100.0	34.2	21.2	21.3
Total (Rp trillion)					0.7	32.3	115.5	218.5			

Memo items:	1969	1982	1980	1991
M1/GDP	0.07	0.09	0.10	0.12
M2/GDP	0.09	0.14	0.30	0.44
Total assets of Financial Institutions (TAFI)/GDP	0.26	0.52	0.81	0.96
M2/TAFI	0.33	0.34	0.36	0.45

[a]December 1990.
[b]Average 1989–90.
... data not available.
Sources: Bank Indonesia, *Indonesian Financial Statistics* (various issues); *Annual Report* (various issues); Cole and Slade (forthcoming).

The third rough indicator of financial asset is the total asset of all types of financial institutions (TAFI). As was pointed out by White (in Zahid 1995) and others (Cole and Slade 1993), this indicator has limitations, first, because there may be significant double counting and, second, because it is only available for selected years. Nevertheless, the ratio of TAFI to GDP rose from 0.81 in 1988 to 0.96 in 1991. Part of this increase was due to the rapid rise in the Government's deposits at Bank Indonesia, which resulted in part from the increase in the Government's revenue from oil taxes which rose sharply in 1990–1991, along with the rise in oil prices following the Middle East crisis. The Government's deposit, as a percentage of Bank Indonesia's total assets, went from 24.5 per cent in 1988 to 27.4 per cent in 1991.

The rise in domestic savings mobilization and the country's financial deepening can be attributed to a combination of factors. First, both are influenced by the availability of foreign aid and loans as well as by fluctuations in oil revenues resulting from changes in oil prices. These external resources have been injected into the economy through the public budget and financial system. Second, both are responding positively to income growth and structural changes, including monetization of the economy. Third, both are influenced by increases in interest rates, from highly negative to positive rates as well as the improvements in financial services. A World Bank study, however, has indicated that once positive real interest rates prevail, financial savings do not appear to be positively related to interest rates. In addition, an increase in interest rates may result in increased household savings and lower corporate profits. If corporations have a higher marginal savings rate than households, then aggregate savings may decline as well. This indicates that savings mobilization is also attributable to "non-price" factors, such as expansion of the menu of financial instruments offered by the financial institutions and improvements in the quality of their services.

Deregulation and Government Intervention

Economic deregulation or reform in Indonesia has been moving in two opposite directions. On the one hand, deregulation has significantly strengthened competition and market forces in commodity, asset, and labor markets. In the financial service industry, deregulation promotes competition; as newcomers are allowed to enter into the industry, cross-border restrictions are reduced and demarcation lines between the various financial sectors become increasingly blurred. Deregulation promotes market forces as government controls on prices and quantities, including interest rates, insurance premiums, volume and allocation of banks' credit, and prices of equities traded at domestic stock exchanges, are significantly reduced. At the same time, the authorities removed the discriminatory tax policies, such as discriminatory tax treatments between financial instruments.

On the other hand, deregulation has been accompanied by two types of re-regulations or government interventions. The first type of re-regulation is in the form of strict prudential regulations and supervision by the Government. This type of government intervention is required to ensure a well-functioning market, with low transaction costs, and to ensure equity. The other types of government interventions are growth-retarding; they cause economic distortions. This will be discussed further in the following section.

To achieve market efficiency and equity objectives, improvements in market infrastructure to correct information asymmetry are required, including improvements in the legal and regulatory framework, accounting system, supervisory practices, and human resources. The latter were eroded, in part, because of the long period of financial repression and the development strategy adopted prior to 1988. Re-regulation in the financial sector is needed to ensure that financial institutions are operated in a safe, sound, and law-abiding manner. To achieve these objectives, re-regulation has raised minimum capital requirements for financial institutions. Following the financial sector reforms, for example, banks have been subject to capital adequacy standards and the insurance sector has been subject to solvency margin regulation. These two rulings have linked capital requirements of these financial institutions to risks they assumed. Because of this, deregulation has not eroded government interventions in the economy, including in financial industry.

The rapid rates of growth of the number of financial industries and their assets since 1988 has not been accompanied by the same rate of progress in financial infrastructure strengthening. This is indicated by the failures and insolvencies of a number of banks and an insurance company, misuses of credit and funds owned by state-owned financial companies, and anomalies in securities markets between 1988 and 1993. Financial infrastructure strengthening should cover improvements in the following areas: the ability of lenders to monitor borrowers; the ability of owners to monitor managers; the ability of regulators, savers, and investors to monitor banks and other financial institutions and listed companies; and the ability of a country's citizens to monitor office holders. This final aspect also covers the need for improved monitoring of subsidized loans by state-owned financial institutions to borrowers who cannot or will not repay their loans.

Coverage, Aspects, and Sequence of Economic Deregulation

With the objective of getting the prices right, and with the support of the international community, Indonesia embarked on a broad economic restructuring program to move towards an export-oriented-industrial strategy in the early 1980s.[1] The impetus for the reforms is a combination of domestic needs and external developments. The import-substitution-industrial policy, carried out in the 1960s and 1970s, had reached its limits, as shown by sluggish rates of growth in the early 1980s. The successive external terms of trade and interest rate shocks, in the early 1980s, and currency realignments, in the mid-1980s, had widened Indonesia's internal and external imbalances. Reviving the growth rate and improving external creditworthiness require a change in development strategy to promote domestic savings and non-oil exports.

Economic deregulation or liberalization cannot be achieved overnight. It takes time to improve the market infrastructure. In addition, the speed of adjustment or response to deregulation measures varies in different parts of the economy. This raises questions concerning the efficiency and equity of the adjustment program, and highlights the importance of sequencing.

Indonesia has adopted a gradual and "stop-go" approach in reforming its economy. Deregulation has relaxed many of the barriers to market entry, significantly reduced the Government's control on the economy, and eliminated many distortions in the commodity, asset, and labor markets. In contrast to the "proper sequence" as suggested by some of the economic literature (see e.g., Edwards 1986; McKinnon 1991), Indonesia liberalized its capital account in the early 1970s, long before deregulation in the current account, which was begun in the mid-1980s. In addition, the pace of deregulation in the financial sector has been much faster than in the real sector of the economy (Table 3.3).

Deregulation in the capital account was introduced, along with the unification of the exchange rate, as Indonesia liberalized its foreign exchange rate system in April 1970. Under this system, there were no surrender requirements for export proceeds and taxes or subsidies on the purchase or sale of foreign exchange. Foreign nationals and Indonesian citizens are free to open accounts in either rupiah or foreign currencies with all "bank devisa," which are authorized to deal in foreign exchange transactions. These banks are free to extend credits in foreign currencies with a 15 per cent withholding tax on interest rates.

---

[1] Prior to this, Indonesia had a long period of financial repression with an import-substitution industrial strategy. In the early 1970s, the authorities held new licenses to establish new banks. The requirements for obtaining a license to deal in foreign exchange transactions and to open new branches were extremely restrictive, making such transactions difficult. Operations of the 10 foreign banks were confined to Jakarta. A uniform reserve requirement was set at 15 per cent on 30 December 1977, but the base for calculating the required reserve was differentiated by the type of deposit and ownership of the bank. Since 1967, the public sector (including state-owned enterprises) has been required to hold deposits only with state-owned banks. Because of the dominant role of the public sector in the economy, there is a large captive market for these banks. A more complex credit ceiling-cum-selective credit policy, with subsidized interest rates, was introduced on 9 April 1974 (Nasution 1983).

In other sectors of the economy, deregulation measures included major reforms of the fiscal system, trade policy, investment licensing, transport regulations, and administrative reforms to reduce transaction costs. The customs service, for example, was privatized in April 1985 and placed in the hands of Societe Generale de Surveillance, a Swiss-based private company. This had the effect of reducing red tape, corruption, and inefficiency in external trade. Trade liberalization consisted of a phased reduction in nontariff barriers replaced with tariffs and export taxes, and a rationalization of the tariff structure by lowering the tariff levels and their variance. In the field of investment, wider activities have been opened to domestic and foreign private investors.

Deregulation in Indonesia has not eliminated all economic distortions, however, and, despite deregulation, the Government retains some forms of industrial policy, and controls prices of state-vended products. Further, in some sectors of the economy, deregulation has not been synonymous with strengthening market competition, and in others, including the nearly 200 state-owned enterprises, deregulation has been largely unfelt. In other parts of the economy, deregulation merely transferred monopoly rights from the state to the well-connected private enterprises. Automotive and "strategic industries" are shielded from internal and external competition by layers of protective measures, such as high import tariff and sales taxes, license system, local content programs, and a negative investment list (Nasution 1993b). Some objects of insurance remain exclusively reserved for state-owned and quasi-publicly owned firms.[2] The persistence of such distortions has delayed the efforts to strengthen market competition and improve the efficiency of the economy.

REFORMS OF THE BANKING INDUSTRY

Sequence of Deregulation

Indonesia has also adopted a gradual and "stop-go" approach in reforming its financial system, including its banking sector. As shown in Table 3.4, the first stage of the reform in banking sector was introduced on 1 June 1983 and involved the removal of direct controls on interest rates on most credits and deposits, elimination of most credit programs, and curtailment of the central bank's funding of state-owned banks. In the second stage, the central bank's facilities were further restructured and money markets began to develop.

---

[2] Bank Indonesia leans more towards the interventionist camp (Cole, in Zahid 1995). Intervention theorists, such as Stiglitz (1993), criticize pro-market financial reform and recommend an active role of government in correcting what they perceive as the pervasive market failures in less-developed countries. According to this view, the main role of government is to remove economic distortions and not to create new ones.

Major reforms, known as *Pakto*, were introduced in October 1988; other measures soon followed. The coverage and aspects of financial sector reforms, introduced between June 1983 and May 1993 are summarized in Table 3.4.

Table 3.4 The Chronology of Financial Sector Reforms, 1983–1993

June 1993	Elimination of credit ceilings and the end of control on interest rates for time deposits and credits; and reduction in the types of priority/program credits which are eligible for Bank Indonesia liquidity credits.
February 1984	Reintroduction of Bank Indonesia Certificates (SBI) to support open market operations and introduction of rediscount windows.
January 1985	Introduction of money market instruments (SBPU) consisting of promissory notes of bank customers, banks, nonbank financial institutions, and trade bills. Rediscount facilities to banks were also set up by Bank Indonesia to help both the day-to-day reserve management as well as maturity arrangement.
May 1987	With the objective of improving fund management, Bank Indonesia raised the interest of SBPUs, SBIs, discount facilities, and of the swap premium.
July 1987	Introduction of managed auction system for the trade of SBIs and SBPUs.
October 1988	Introduction of measures to foster competition, i.e., (a) lessening restrictions on entry and of branching for foreign/domestic banks; (b) allowing public enterprises to place up to 50 per cent of their deposits outside state banks; (c) allowing nonbank financial institutions to issue CDs; (d) permitting banks and nonbank financial institutions to raise capital in the stock market; (e) easing entry to leasing, insurance, venture capital, consumer finance, and securities activities; (f) reducing reserve requirements from 15 per cent to 2 per cent; (g) improving other aspects of protective and preventive measures.
March 1989	Introduction of decrees clarifying matters with regard to non-bank financial institutions, legal lending limits, joint venture capital ownership and bank mergers, the definition of bank capital, reserve requirements, bank investment in stocks, and bank exposure to foreign exchange fluctuations.
December 1989	Jakarta Stock Exchange (JSE) was decontrolled and privatized, Bapepam became a regulatory and supervisory agency.

	The special right of PT Danareksa to buy new issues was revoked.
January 1990	Reduction in the scope of credit programs which are eligible for Bank Indonesia liquidity credit. Introduction of legal lending limits requirements. A regulation was stipulated to require all domestic banks to lend 20 per cent of their portfolio to small firms and cooperatives, and foreign and joint-venture banks to extend at least 50 per cent of their credit for export-oriented activities.
December 1990	Bapepam and JSE were reorganized and working mechanism of stock exchange was improved. Capital requirements for stock exchange and securities firms were introduced.
February 1991	Various measures were stipulated to improve banking management and supervision by, among other measures, requiring banks to step up capital base according to BIS regulations.
November 1991	Reimposition of ceilings on offshore borrowing for the public sector (including state banks), improvements in swap mechanism, and elimination of swap premiums.
January 1992	Promulgation of new Banking, Insurance and Pension Fund Laws as well as the controversial worker social security law.
February 1993	A private owned and clearing and settlements firm, PT Kliring Deposit Efek Indonesia (KDEI) was established.
May 1993	Relaxation of prudential standard and extension of the deadline to meet the capital adequacy ratio of 8 per cent to December 1994.

In contrast to the "proper" sequence as suggested by some economic theoreticians (see Corbo and de Melo 1987; McKinnon 1991), Indonesia adopted a reverse course of deregulation. According to the theory, the proper sequence is as follows: First, commercial banks should be tightly monitored and regulated, a step which may include temporary credit rationing. The second step is to recapitalize the existing banks and their clientele. Third, during the transition period, the banking industry should be temporarily closed to new entrants, both domestic and foreign, since new entrants are not burdened with low-yield loans and can easily outcompete the banks that have been in existence during the pre-reform period.

The sequence of financial reforms in Indonesia was begun in October 1988 with the opening up of the banking industry to new entrants, the implementation of a liberal branching policy without adequate prudential rules, the liberalization of requirements to obtain licenses to operate in foreign exchange transactions, and sharp reductions in the legal reserve requirements. In May 1989, the authorities increased access of domestic banks to the international financial markets. In January 1990, Bank Indonesia announced the phasing out of many types of credit programs financed by its subsidized ("liquidity" credits); this was followed in June 1990 with measures to curtail the liquidity of the banking system. Nearly two years later, in February 1991, new prudential regulations were introduced. However, as many banks had difficulty in meeting these regulations, Bank Indonesia relaxed these prudential standards in May 1993.

The Indonesian style of financial sector reform did not provide for an adjustment period either for the "old" banks or their customers. The transition period has been most difficult for state-owned banks. This group of banks was undercapitalized and had inherited, from past financial repression, much larger problem loans, illiquid financial instruments which carry artificially low-yields, and a high risk of capital loss. In addition, at the time when they badly needed financial help, these banks faced the shocks of mid-1987 and February 1991.[3]

The nonfinancial firms have high debt/equity ratios because, during the past period of financial repression, borrowers adopted highly leveraged financing. This was due to the availability of the low, and even negative, real interest rates, particularly on various types of investment credit programs at state-owned banks. This included offshore loans fully guaranteed by these financial institutions. Part of the loans are on a "line of credit" basis which roll over short-term credit loans into long-term commitments (Cole and Slade 1993).

Coverage and Aspects of the Reforms

Types of Banks

The reform package removed the traditional functional specialization between various types of banks and eliminated the major areas of specialization for state-owned banks. There are now only two types of banks in Indonesia, namely, commercial banks and rural banks or *Badan Perkreditan Rakyat (BPR)*. With the 1988 reforms, the status of existing special purposes institutions changed and institutions such as development banks and savings banks were automatically converted into commercial banks. The nonbank financial institutions were given one year, beginning in December 1991, to become either security companies or commercial banks. The three development finance companies and the nine investment finance companies

---

[3] To suppress capital outflows, the Minister of Finance, Professor J.B. Sumarlin, instructed large state-owned enterprises to convert significant amounts of their deposits, mainly at state-owned banks, into certificates (SBI) issued by Bank Indonesia.

all opted, between January and March 1993, to became full-fledged joint-venture commercial banks, and were subsequently licensed to deal in foreign exchange transactions. The existing "secondary banks" were given an option either to convert their status into BPR or cease operations.

Scope of Bank Activities

Deregulation has provided banks with new opportunities: They can now expand their range of investment instruments and provide such services as innovative savings accounts and more extensive advisory services. The new rules and regulations, however, prohibit banks from engaging in securities underwriting, brokerage and trading, or giving loans for stock trading. From December 1991 on, such activities were to be transferred in total to their holding companies. The direct tapping of capital markets by companies does not necessarily totally exclude banks; instead of simply extending credit they can assist in the placement of financial titles. However, a special license is still required by commercial banks to deal in foreign exchange transactions, and the secondary or rural banks are not allowed to offer demand deposits.

Market Entry

Pakto, the October 1988 reform, also encouraged competition in the banking system by relaxing the barriers of entry into the banking industry. Prior to this reform, such entry had been closed since the early 1970s; since the reform, new licenses have been issued to allow new entrants, though no new entry is allowed for branching of foreign banks. The market opening includes further opening of domestic markets to foreign institutions. The mechanism by which foreign banks can currently penetrate the domestic markets is, however, through establishment of joint ventures with local banks.

To be eligible to establish a joint venture bank, both foreign and domestic partners must meet some requirements. The foreign bank must have representative offices in Jakarta, be reputable in the country of origin, and be from a country with a reciprocity agreement with Indonesia. The foreign partners' maximum share of ownership is 85 per cent. Domestic banks must be classified as "sound" for at least twenty of the preceding twenty-four months.

Requirements governing the opening of new branches have also been relaxed: Both national and foreign-owned banks must have been "sound" for twenty of the preceding twenty-four months. National private banks can open branch offices anywhere in Indonesia, while foreign and joint-venture banks are restricted to only one sub-branch each in seven major cities (Bandung, Denpasar, Jakarta, Medan, Semarang, Surabaya, and Ujung Pandang). The operations of BPR banks remain limited within their provinces. Pakto limits the operations of rural banks outside Jakarta, provincial capitals, and municipalities; if they do not wish to comply with these limitations, they must move or upgrade their status to become commercial banks. However, the May 1993 regulation re-allows operations of the BPR in large cities, including Jakarta and the provincial capitals.

Prudential Regulations

Pakto improved the prudential (or preventive) measures aimed at controlling the level of credit risks assumed by lending institutions, which in turn affected the probability of their failure. The measures include extension of Bank Indonesia's responsibility for setting a unified reserve requirement ratio and for supervising both banks and nonbank financial institutions to give them equal footing. The reserve ratio was reduced from 15 per cent to 2 per cent in October 1988 to reduce the cost of funds.

Bank Indonesia has adopted prudential regulations and a supervisory framework as outlined by the Committee on Banking Regulation and Supervisory Practices, under the auspices of the BIS in 1974. In line with the BIS recommendations, the new measures also raise the bank's capital adequacy ratio (CAR) which links the capital base of lending institutions to their risk-weighted assets.[4] The new regulation on Loan to Deposit Ratio (LDR), set at a maximum of 110 per cent, links new loans to the level of deposits.

The revised regulations of 29 May 1993 include capital in the calculation of the LDR. In addition, the bank is required to provide reserve for all loans. The revised regulation reduces the amount of this reserve from 1 to 0.5 per cent of total assets. Problem loans require additional reserves: 3 per cent for illiquid assets, 50 per cent for doubtful assets, and 100 per cent for bad assets. Assets in the form of SBI do not require reserves. These prudential standards, however, cover only credit risks; other market risks such as foreign exchange, interest rate, and position risks are not covered.

The capital adequacy ratio and new credit regulations required banks to strengthen their capital bases[5] and made extension of new loans more costly. Pakto set the minimum initial paid-in capital for a new bank at Rp 10 billion

---

[4] Bank Indonesia set a schedule for banks to raise the capital to 5 per cent of their risk-weighted assets by March 1992, 7 per cent by March 1993, and 8 per cent by the end of 1993. In May 1993, this schedule was extended to December 1994.

[5] Capital is defined as the sum of paid-in capital, general and specific reserves and retained earnings, plus subordinated and two-step loans. Both of these loans are basically government guaranteed foreign borrowings channeled mainly through state-owned banks. The 29 May 1993 regulations include 100 per cent of the previous years' profits, instead of 50 per cent in the calculation of the 8 per cent capital adequacy ratio.

or roughly $500,000, and for a new joint venture bank at Rp 50 billion or $2.5 million (assuming an exchange rate of Rp 2,000 per US dollar). The initial capital for a BPR is Rp 50 million. The minimum absolute capital requirement was raised following the promulgation of the Banking Law Nr. 7 of 25 March 1993 to Rp 50 billion for a domestic bank and to Rp 100 billion for a joint-venture bank.

The regulations on minimum initial capital requirements, capital adequacy ratio, LDR, and reserve provisions are designed to ensure that banks have an adequate capital base for operations and for the protection of depositors. Additional investment is also needed by the banks in the following areas: improvement in their software to modernize their systems; improvement in services, human resources, and management capabilities; introduction of new financial products and other forms of innovation to reduce marginal cost and spread; and expansion of branch networks to survive in a more competitive market.

Credit System and Access to Discount Windows

Though policy reforms have reduced distortions in the credit market, the credit policy in Indonesia remains segmented and pro-cyclical. In January 1990, Bank Indonesia narrowed the scope of the subsidized credit program to four areas, which include rice production, marketing, buffer stock, and investment financing in the eastern part of Indonesia.[6] Interest rates charged in these special credit programs are now much closer to market rates and their insurance is voluntary, with market-based premiums.

Three new credit regulations, related to the system of credit allocation, were introduced in 1989-1990. The first rule is implied in the legal lending limits regulations, which restricts the aggregate amount of loans and advances to insiders, whether to a single borrower (person or firm) or to a group of borrowers. The rule is not applicable to all forms of the subsidized credit program which are financed by Bank Indonesia's liquidity credit. The objective of the legal lending limits rule is to increase access to bank credits to inhibit concentration of financial power, protect the interests of uninformed depositors, and prevent misuse of funds by insiders. The second rule requires new joint-venture banks and branches of foreign banks outside Jakarta to allocate at least 50 per cent of their loan portfolios to export-related activities. The third rule mandates domestic private and state banks to allocate a minimum of 20 per cent of their loan portfolios to

---

[6] At the end of 1990, this concessionary credit was also made available for *Badan Penyangga dan Pemasaran Cengkeh* to finance its buffer stock of cloves, the main ingredient of clove cigarettes. The recipient of this credit is a consortium of private traders who have powerful backing and have been granted the exclusive rights to operate a buffer stock for that agricultural commodity.

small-scale enterprises and cooperatives (*Kredit Usaha Kecil*) or KUK.[7] These credit rules force lending institutions to operate with a portfolio that they might not otherwise voluntarily accept.

In addition, banks are competing with nonbank state-owned enterprises in providing credit to at least some segments of small- and medium-scale enterprises and cooperatives. Since 1989, all state-owned enterprises, including banks, have been mandated to channel between 1 and 5 per cent of their profits to help those sectors. Most of the funds have been distributed directly by the companies concerned to the target groups and through mechanisms of their own choosing, at very low or zero nominal interest rates.

Both the banks and the business sector faced difficulties in meeting the October 1990 Pakto deadline for implementation of the legal lending limits regulation. During the previous era of low, or even negative, interest rates and low credit risks, the business sector had become accustomed to high leveraged financing, mainly from bank credits. In the past, the main purpose of banks associated with big conglomerates was to mobilize funds for financing their sister companies. Partly because of an inward-looking development strategy, rarely is it possible to keep all the operations of the companies within a group to world standards. Often, inefficient subsidiaries are carried along by sales to other subsidiaries within the group.

To help both the banks and business sector adjust to the new rules, the deadline for meeting the regulations was extended to March 1997 by the Banking Law Nr 7 of 1992. The May 1993 regulations further clarified the schedule for meeting the legal lending limit regulations.

The reduction in the scope of the credit program automatically reduced Bank Indonesia's refinancing loans to banks. To assist financial institutions in coping with more competitive markets, Bank Indonesia, as the bankers' bank, established two rediscount facilities in February 1984. In contrast to the program-oriented liquidity credit, these facilities are for general purposes. Rediscount Facility I is designed to facilitate day-to-day reserve management by financial institutions, and Rediscount Facility II is intended to promote long-term lending by offering temporary relief to banks that face risks associated with maturity mismatches between assets and liabilities. The maximum duration of Rediscount Facility I is three working days and that of Facility II is 90 days.

---

[7] The subsidized credit program refinanced by Bank Indonesia's liquidity credit and credit denominated in foreign currencies are excluded from the calculation of the 20 per cent credit designated for the small-scale business sector (KUK). The May 1993 regulation raises the maximum size of KUK per customer from Rp 200 million to Rp 250 million and allows banks to use other financial institutions (including the BPR) to channel their mandatory KUK credit.

Access to Public Sector Deposits

The October 1988 deregulation (i.e., Pakto) removed the state-owned banks' exclusive access to public sector deposits by allowing state-owned enterprises to hold up to a maximum 50 per cent of their deposits with private banks. Implementation of this measure, however, has been slow, in part to prevent liquidity problems at state-owned banks. Because of the Sumarlin shocks of 1987 and 1991, however, there were rapid transfers of massive funds from the state-owned banks to Bank Indonesia.

Foreign Exposure

Prior to 1989, the only obstacle to capital inflows had been a system of complex ceilings on foreign borrowings. For a short period, beginning on 25 March 1989, the authorities replaced this system with the imposition of a daily net open position amounting to a total 20 per cent of the capital of the bank and 25 per cent to a single borrower and that borrower's group of companies.

To encourage inward foreign investment, from January 1979 to December 1991, a special effective exchange rate was made available to domestic borrowers by providing explicit subsidies on the exchange rate. The foreign exchange subsidy was extended through the exchange rate swap facility at Bank Indonesia. Under this facility, Bank Indonesia provided forward cover to foreign exchange borrowing contracts or swaps to banks and nonbank financial institutions as well as for their customers with a foreign-currency liability. The swap premium was set below the level of the realized depreciation of the rupiah. The size of the subsidy depends on the choice of interest rates used to calculate the interest rate differential. In reality, the swap facility was also used for liquidity purposes by the financial institutions either to fund themselves or to hedge or even speculate against a declining rupiah.

In the beginning Bank Indonesia set the amount of the swaps and allocated the premiums through a nonmarket mechanism. The ceiling was abolished in October 1986 as Bank Indonesia allowed the swap premium to be determined by market forces. In reality, the swap premium remained subsidized until December 1991. The subsidy came about because of the time lag in either an upward adjustment of the swap premium or a nominal depreciation of the rupiah, or a combination of both. Beginning from December 1991, Bank Indonesia, on average, set the swap premium above the market rate. This further induced speculation about depreciation of the rupiah.

As regards the control on offshore borrowings by the public sector, as a part of government policies to cool down the overheated economy, on 21 October 1991, the authorities reinstituted a complicated ceiling on foreign borrowings for various categories of borrowers (state and private banks and state and private companies). Offshore borrowings by state-owned banks and enterprises, including the private sector relying on public entities for their bankability, are now required to obtain approvals from a newly established Coordinating Team for Management of Offshore Loans. The Team is an interdepartmental body, chaired by the Coordinating Minister for Economy, Finance and Development Supervision. Ceilings on offshore borrowings of purely private sector enterprises are not binding, but these enterprises are required to report their borrowings to the authorities.

IMPACT OF THE BANKING REFORMS

The successive reforms introduced since 1988 have completely overhauled the function and structure of the banking system in Indonesia. The monopoly power of the state-owned banks has been eroded and private sector financial conglomerates are on the rise. Such adjustments have not been made without a price, however. The reforms have often given rise to acute short-run problems, especially for banks owned by or closely linked to business conglomerates. For example, the already undercapitalized state-owned banks were affected by the "Sumarlin" shocks in 1987 and 1991, and the rising interest rates and slowing down in the rate of economic growth, which declined from 7 per cent per annum in 1989 and 1990 to 6.5 per cent in 1991 and 1992, added to the problems.

Growing Market Competition

On the other hand, such reforms have increased competition. The relaxation of barriers to market entry, simplification of types of banks, and reduction in government controls strengthened competition in the banking industry and induced the formation of financial conglomerates owned by the private sector.

Number of Banks and their Branches

The numbers of banks and their branch offices grew rapidly following the October 1988 reform, increasing from 111 in 1988 to 192 in 1991 (excluding the 12 nonbank financial institutions which had become commercial banks between January and March 1993), while the number of bank branch offices (mostly of private banks) doubled from 1,728 to 3,563 during the same period. There were no changes in the numbers of state-owned banks, banks owned by the provincial governments, or branches of foreign banks.[8] Between 1988 and 1991, the number of domestic private banks doubled from 63 to 126, and the number of joint-venture banks/foreign banks[9] increased from 11 to 29 (Table 3.3). The competition became more intense as the number of BPR also rose from 65 in 1988 to 329 in 1991, especially as these banks compete with commercial banks in serving small- and medium-scale enterprises.

In terms of assets, the size of banks varies greatly as can be seen from Table 3.5 which shows the distribution for 40 major banks as of 1985 and for 44 banks in 1990. In 1985, the five state-owned banks had total assets in excess of Rp 1,100 billion ($670 million at the exchange rate of Rp 1,641 per one US dollar). The number of banks in this rank of assets increased to 18 in 1990, to include 12 national private banks and one foreign bank.

Conglomerations

Through networks of ownership, and business and management interlockings, nearly all of the domestic private banks were closely connected to large conglomerates in 1990. Close relationships with other companies (both financial and nonfinancial enterprises) in a group often provide new sources of direct capital and new business opportunities for the banks. Conglomeration also allows for improved information sharing and assists in dealing with scarce human resources. During the country's period of financial repression, the nonbank companies owned by the conglomerates grew rapidly because of the availability of subsidized and low-risk investment credits from state-owned banks. Also, nonfinancial companies within the groups obtained relatively cheap and low-risk medium- and long-term investment credits from state-owned banks; they channeled the proceeds to their own financial firms.

As long as all companies within the conglomerates operate according to international standards, conglomeration is synergistic. Problems arise if the less efficient subsidiaries are carried by sales to other companies in the group. Moreover, large firms may lose the flexibility necessary to continue to respond adequately to new challenges and changing business environments. The collapse of the large conglomerates—PT Astra, PT Bentoel, and PT Mantrust—in 1990–1991, indicated that certain sectors within the conglomerates could

---

[8] By nationality, four of the 10 foreign banks established between 1968 and 1971 are American institutions. The other six banks come from six countries, namely: England, Germany, Hong Kong, Japan, the Netherlands, and Thailand.

[9] These are joint ventures between domestic and foreign private banks. Prior to 1988, PT Bank Perdania was the only joint-venture bank in Indonesia. The foreign partners of the 19 new joint-venture banks are: 10 Japanese, 3 French, 2 Singaporean, 2 Dutch, 1 Korean, and 1 Australian.

become burdensome, in part because of their strategy of being highly leveraged, which may have been suitable in the era of subsidized interest rates and highly protected domestic markets but which is not suitable in the current period of deregulation.

Table 3.5 Distribution of Banking Assets

Asset Size (Rp billion)	Number of Banks	Per cent of Banks	Assets (Rp billion)	Per cent of Assets	Group of Banks
1985					
>1,100	5	12.5	28,701	79.7	5 State Banks
900–1,099	0	0.0	0	0.0	
700–899	1	2.5	708	2.0	1 Private Bank
500–699	1	2.5	560	1.6	1 Foreign Bank
300–499	6	15.0	2,402	6.7	5 Private Banks 1 Foreign Bank
100–299	16	40.0	2,877	8.0	8 Private Banks 8 Foreign Banks
<100	11	27.5	744	2.1	10 Private Banks 1 Foreign Bank
1990					
>1,100	18	40.9	125,528	90.7	5 State Banks 12 Private Banks 1 Foreign Bank
900–1,099	3	6.8	2,848	2.1	3 Private Banks
700–899	1	2.3	815	0.6	1 Foreign Bank
500–699	6	13.6	3,449	2.5	2 Foreign Banks 3 Private Banks 1 Joint-Venture Bank
300–499	13	29.6	5,036	3.6	5 Foreign Banks 5 Private Banks 3 Joint-Venture Banks
100–299	3	6.8	783	0.6	2 Foreign Banks 1 Private Bank
<100	0	0.0	0	0.0	

Note: Total assets of 44 banks in this sample are approximately 90 per cent of total assets of all banks in 1990.
Source: Bank Indonesia, *Annual Report* (various issues).

There has also been a rising trend in the formation of financial conglomerates following the October 1988 reform. Through a network of ownership, nearly all large domestic private banks are now linked to insurance, pension funds, and securities and finance companies.

Branch Network

Indonesia has a branch banking system. As a result of past discrimination, the state-owned banks have a much more extensive branch system than do other banks. As discussed earlier, prior to the October 1988 reform, foreign banks were not allowed to operate outside Jakarta and the number of their branches was limited to two. After 1988, foreign banks were allowed to open branch offices in six other major cities outside Jakarta, including Medan, Batam, Bandung, Semarang, Surabaya, and Ujung Pandang. In addition, joint-venture banks are allowed to operate in Denpasar, Bali—the hub of Indonesia's tourist industry. These groups of foreign banks specialize in providing services to multinational firms although some, particularly the American banks, have successfully expanded their bases to include local customers. The former regional development (now commercial) banks are restricted to operating within their respective provinces. Since 29 May 1993, BPR have been allowed to operate both in rural areas as well as in cities.

Geographic dispersion of commercial bank offices in Indonesia is strongly oriented toward the major cities. Table 3.6 shows that 72 per cent of the country's bank offices were located in large cities of the five major provinces (North Sumatra, Jakarta, West Java, Central Java, and East Java) with 30 per cent concentrated in Jakarta alone. To some extent the authorities have tried to redress the imbalances by encouraging national private banks to establish branches in small towns and the rural sector. The rapid expansion of branch offices of state-owned banks in the 1970s was related to this supply-led policy. To channel agricultural credits, for example, Bank Rakyat Indonesia (BRI) has a wide branch network of rural credit banks. The result of this policy, has been mixed, however, since there has not been a similar response from the real sector outside agriculture.

Following the branching liberalization policy, Bali, East Java, and Jakarta experienced a rapid rate of growth (over 20 per cent per annum) in the number of bank offices. The number in Jakarta and Bali nearly doubled in 1990 as compared to 1989.

Jakarta—the largest city in Indonesia with no surrounding rural area—has the lowest population per bank office (1991: 7,579) followed by Bali (1991: 21,194) and Central Java (1991: 55,054). The BPR in the area serve certain classes of customers, particularly medium- and small-scale firms.

Table 3.6 Number of Bank Offices and Population per Bank Office by Province

	1985	1986	1987[b]	1988	1989[b]	1990	1991
Bank Offices[a]							
North Sumatera	101	104	109	115	144	197	220
Jakarta	275	293	312	357	588	1,020	1,281
West Java	149	151	170	185	302	445	525
Central Java	144	159	167	181	252	356	412
East Java	159	168	179	189	297	490	609
Bali	43	45	48	50	62	102	134
South Sulawesi	71	72	73	75	91	103	113
Share (per cent)[c]							
North Sumatera	7.0	6.8	6.7	6.7	6.1	5.5	5.2
Jakarta	18.9	19.3	19.2	20.7	24.7	28.6	30.2
West Java	10.3	9.9	10.5	10.7	12.7	12.5	12.4
Central Java	9.9	10.5	10.3	10.5	10.6	10.0	9.7
East Java	10.9	11.1	11.0	10.9	12.5	13.8	14.3
Bali	3.0	3.0	3.0	2.9	2.6	2.9	3.2
South Sulawesi	4.9	4.7	4.5	4.3	3.8	2.9	2.7
Total	64.8	65.2	65.2	66.7	73.0	76.1	77.6

Annual growth (per cent)							
North Sumatera		3.0	4.8	5.5	25.2	26.9	11.7
Jakarta		6.6	6.5	14.4	64.7	42.4	25.6
West Java		1.3	12.6	8.8	63.2	32.1	18.0
Central Java		10.4	5.0	8.4	39.2	29.2	15.7
East Java		5.7	6.6	5.6	57.1	39.4	24.3
Bali		4.7	6.7	4.2	24.0	39.2	31.4
South Sulawesi		1.4	1.4	2.7	21.3	11.7	9.7
Population per Bank (thousands)[d]							
North Sumatera	93.6	93.0	90.8	87.9	71.7	53.5	48.9
Jakarta	28.8	28.0	27.2	24.6	15.5	9.2	7.6
West Java	207.6	209.8	190.6	178.9	111.8	77.4	66.8
Central Java	187.8	172.7	167.0	156.1	113.7	81.5	71.3
East Java	193.7	189.1	179.7	172.0	110.7	67.8	55.0
Bali	61.8	59.8	56.7	55.0	44.9	27.6	21.2
South Sulawesi	93.3	93.5	93.6	92.2	76.9	68.8	63.3

[a] Bank offices (state banks, private national banks, foreign/joint venture Banks) cover head offices, branch offices, and sub-branch offices (excluding rural banks and Bank Rakyat Indonesia Unit Desa).
[b] September 1987 and September 1989.
[c] Number of bank offices in each province is divided by total bank offices in Indonesia.
[d] Population divided by number of bank offices.

Sources: Bank Indonesia, *Indonesian Financial Statistics* (various issues); Central Bureau of Statistics, Population Projection by Province, 1985–1995.

Market Share

Increased competition in the financial industry has had an effect on the state-owned banks and non-foreign exchange domestic private banks. On the liability side, the state-owned banks are beginning to compete with nonbank financial institutions and private banks for lucrative pension funds and deposits of the nonbank state-owned enterprises. On the asset side, higher interest rates on bank credits have caused prime companies to sell securities in the newly liberalized stock exchanges. Some of these customers have turned to domestic and foreign banks for better services and lower interest rates. Erosion of the pool of creditworthy corporate clients creates pressures on all banks to replenish their portfolios with lower quality loans, to reduce lending rates, and to improve services so as to win new customers. The private banks have, in turn, suffered from the problems of lower asset quality, in part because of the slump in the real estate market.

Following the October 1988 reform, the state-owned banks experienced a particularly sharp decline in market share. Between 1988 and 1991, the market share of this group dropped sharply by 17.8 per cent in terms of assets, by 12.9 per cent in the loan market, by 16.1 per cent in savings and time deposits, by 16 per cent in demand deposits, and by 16 per cent in total deposits (Table 3.7). In contrast, the market shares of the private foreign exchange banks in all aspects grew rapidly.

The average ratio of assets, bank funds, demand deposits, saving deposits, and credit to the number of offices by ownership of the banks is calculated in Table 3.8. All indicators place the foreign and joint-venture banks far above the state-owned and private national banks. This indicates that, on average, the scale of business of foreign and joint-venture banks per office is much larger than that of other banks.

Internal Problems

During its transition period to a market-based system, the banking sector in Indonesia has faced a number of internal problems.

Rising Capital Requirements

The first challenge for the banking industry has been the need for increases in auto or internal financing to meet minimum capital and reserve requirements as well as other aspects of the prudential standards. Most lines of business face low profitability. Market competition is rising both on the asset and liability sides. Moreover, a number of assets seem to have questionable value, partly because of fraud, and partly due to underdevelopment in the legal and accounting systems. This gives rise to costly reform measures which reduce profitability.

Table 3.7 Market Shares of Banking Institutions by Ownership

	Assets		Loan		Savings and Time Deposits		Demand Deposits		Total Deposits	
	1988	1991	1988	1991	1988	1991	1988	1991	1988	1991
	(Billion rupiah)									
State banks	39,862	70,158	21,149	48,790	12,294	22,403	4,366	6,399	16,660	28,802
Private forex banks	10,189	45,654	7,629	36,844	4,463	19,639	1,403	5,777	5,866	25,416
Private nonforex banks	4,972	12,868	4,334	11,189	2,619	6,735	960	2,001	3,579	8,736
Foreign and joint-venture banks	3,215	12,070	2,109	9,181	763	1,538	342	645	1,105	2,183
Development banks	5,046	14,505	3,678	11,534	800	2,317	961	1,930	1,761	4,247
Total	63,284	155,255	38,899	117,538	20,939	52,632	8,032	16,752	28,971	69,384
	(As percentage of total)									
State banks	63.0	45.2	54.4	41.5	58.7	42.6	54.4	38.2	57.5	41.5
Private forex banks	16.1	29.4	19.6	31.3	21.3	37.3	17.5	34.5	20.2	36.6
Private nonforex banks	7.9	8.3	11.1	9.5	12.5	12.8	12.0	11.9	12.4	12.6
Foreign and joint-venture banks	5.1	7.8	5.4	7.8	3.6	2.9	4.3	3.9	3.8	3.1
Development banks	8.0	9.3	9.5	9.8	3.8	4.4	12.0	11.5	6.1	6.1

Source: Bank Indonesia, *Indonesian Financial Statistics* (various issues).

Table 3.8 Banking Indicators by Group of Banks in Rupiah and Foreign Exchange (Rp billion)

	1985	1986	1987	1988	1989	1990	1991
Assets/Bank offices							
State banks	34.9	41.5	46.7	57.8	67.8	77.2	87.9
Private national banks	14.3	17.3	20.7	27.0	230.5	23.1	21.9
Foreign/Joint-venture banks	108.3	128.2	132.1	156.7	143.2	209.6	243.9
Bank funds/Bank offices							
State banks	16.3	19.7	21.8	26.4	32.3	39.9	40.1
Private national banks	10.7	11.7	14.9	18.8	15.0	15.8	15.7
Foreign/Joint-venture banks	89.7	99.3	106.0	119.8	87.2	125.3	130.9
Demand deposits/Bank offices							
State banks	6.2	6.6	6.6	7.3	9.5	9.4	10.4
Private national banks	2.9	3.2	3.4	4.0	3.1	2.8	2.6
Foreign/Joint-venture banks	29.5	37.1	38.2	37.5	30.7	40.1	36.0

Time deposits/Bank offices							
State banks	9.0	10.8	13.7	17.1	20.0	26.4	23.2
Private national banks	7.5	8.1	11.1	14.2	9.9	10.6	10.1
Foreign/Joint-venture banks	60.2	62.2	67.8	82.3	56.3	85.0	97.2
Savings deposits/Bank offices							
State banks	1.1	1.4	1.6	1.9	2.7	4.1	6.5
Private national banks	0.3	0.4	0.5	0.7	1.9	2.4	3.0
Foreign/Joint-venture banks							
Credits/Bank offices							
State banks	13.4	22.0	28.1	33.6	42.9	52.6	57.3
Private national banks	9.6	11.9	13.9	18.1	14.2	16.3	15.3
Foreign/Joint-venture banks	51.1	57.3	67.0	91.1	82.0	128.7	160.6

Note: Bank offices cover head offices, branch offices, sub-branch offices (excluding rural banks and Bank Rakyat Indonesia Unit Desa).
Source: Bank Indonesia, *Indonesian Financial Statistics* (various issues).

The problems of meeting the new prudential regulations have been more difficult for the state-owned banks and private non-foreign exchange banks which are not licensed in dealing in foreign exchange. The latter are undercapitalized and their reserves and retained earnings are not sufficient to make up the deficit. In addition, the state-owned banks—already affected by the "Sumarlin shocks"—faced reduction in the scope of refinancing from Bank Indonesia in January 1990 which raised the interest costs of state-owned banks higher than that of private banks.

Further, capital inadequacy prevents state-owned banks from raising funds by selling shares in the capital market. In 1992, to raise the capital base of this group of banks, the Government obtained $307 million from a World Bank Financial Sector Development Project. The remainder of the capital requirements were generated from other sources such as asset revaluation, conversion of some Bank Indonesia's liquidity credits to equity, shifting the credit risks to government-sponsored insurance companies (such as PT Askrindo, PT Asuransi Ekspor Indonesia, and Perum Penjamin Kredit Koperasi), and taking over parts of the state-owned banks nonperforming loans by the Treasury.

Pressures for Banks to Adjust Portfolios

As a result of the changes in the capital adequacy ratio, the composition of banks' portfolios may need to be altered. Banks with substantial off-balance-sheet assets, which are used to calculate the minimum capital, are now required to raise new capital. Alternatively, these banks may try to lower their capital costs by either decreasing the size of their portfolio or by placing greater emphasis on those assets with a low risk weight. Banks with relatively low shareholder equity are then required to raise additional equity. The costs associated with doing so may be high, especially as the composite price index of securities traded on the Jakarta Stock Exchange has eroded nearly 50 per cent since 1989. The institution of capital adequacy requirements and regulations on legal lending limits, loan/deposit ratio, net open position, and back-up reserve have also made it more difficult for mergers and acquisitions to take place within the industry.

As was indicated earlier, portfolio adjustment is particularly difficult for the "older" banks, especially the state-owned banks which were established long before the current reforms. For example, the increase in interest rates on deposits raised the cost of funds, and the October 1988 rules and regulations relaxed the exclusive access to public sector deposits. On the other hand, loans extended by the state-owned banks prior to January 1990 were mainly program loans, which bear artificially low interest rates and have longer terms to maturity (particularly medium-term investment credits) than that of the bank's deposits.

Rough estimates of capital adequacy ratios from the published balance sheets of 44 banks show that only 32 had met the ratio requirements in December 1991. Of those 32, 18 were private national banks, 10 were foreign banks and four were joint-venture banks. According to these estimates, none of the state banks had met the capital adequacy ratio requirements. However, the balance sheets of the state banks for 1992, released in June 1993, indicated that all had met the prudential regulations.

Problem Loans

The third major challenge facing the banking sector during the country's transition to a market-based economy has been how to address the problem of the number of bad debts held by the "old banks," particularly the state-owned banks and the private non-foreign exchange banks.[10] Bad debts in the banking industry are financial tapeworms: they eat up capital and grow fat on capitalized interest that the banks never see. The exact size of problem debts is currently unknown in Indonesia, in part because they are difficult to calculate, partly because of the practice in Indonesia of refinancing bad debts, and because of the weak legal framework for debt collection and bankruptcy. Calculation of problem loans is made more difficult as credit is shifted from one bank to another. In addition, a portion of the credit risks are passed on to state-owned insurance companies.[11] Bank Indonesia estimated the amount of the banks' nonperforming loans—those on which no interest has been paid for more than three months—as about Rp 8,500 billion in 1992; the World Bank estimation was close to Rp 20,000 billion.[12] The high number of problem loans may be attributed to a combination of the following factors:

(i) careless lending, bad management, and foreign exchange and property speculation (a conclusion which can be drawn from evidence in the collapse of PT Bank Duta in 1990 and PT Bank Summa in 1990);

---

[10] This was noted in the statement by the Governor of Bank Indonesia before the VII Commission of the House of Representatives (Dewan Perwakilan Rakyat) on 15 September 1993.

[11] The credit risks were mainly insured by PT Askrindo, PT Asuransi Ekspor Indonesia (ASEI), and Perum Penjamin Kredit Koperasi (PKK). PT Askrindo is a state-owned company that insures banks' credit to small-scale and medium-scale companies. PT ASEI is a state-owned insurance company that covers export risks. Perum PKK is a state-owned credit insurance company that insures banks' credit to cooperatives.

[12] In a hearing before the House of Representatives on 15 September 1993, the Governor of Bank Indonesia noted that bad debts had increased from Rp 3.47 trillion, or 2.8 per cent of total credit, in March 1993 to Rp 3.9 trillion, or 2.9 per cent of total banks' credit, in June 1993 (*The Economist* 1993).

(ii) weak internal system, such as customer selection, credit administration and supervision, and recovery efforts;

(iii) elimination of credit subsidies and lower levels of protection following deregulation;

(iv) high real interest charged on bank credits since 1990, over 20 per cent per annum, have affected even the most efficient and feasible projects;

(v) cyclical factors and changing market conditions that are outside the control of either the banks, the borrowers, or the authorities, e.g., volatility in commodity prices and exchange rates fall into this category. Most of Indonesia's imports and its external debts are denominated in Japanese yen, which has been appreciating over the past few years;

(vi) rapid expansion of commercial real estate and consumer loans which turned sour;

(vii) financing of "mega projects" owned by the well-connected conglomerates in 1989–1991—these post reform loans were not market priced because they were subject to external political pressures. Many of the "mega projects" have been delayed or discontinued; and

(viii) weak legal system, particularly in the framework for debt collection and bankruptcy.

Part of the costs of bad loans can be measured in financial terms. Financial costs include: erosion in the allowance for loan and lease losses; cost of loans that are placed on a non-accrual status; costs attached to foreclosed assets; legal expenses; costs to build greater reserves to cover irretrievable loans; and opportunity costs of the bad loans. These financial costs will shrink capital and tighten the credit squeeze. Unquantifiable costs of problem loans include erosion in the bank's image because of the lack of confidence among investors and users of its services and loss of marketing impetus as the bank focuses its attention towards "cleaning up" the problem loans. In the long run, a low expansion of new loans may create more problems for the banks than the bad loans themselves.

Weak Internal Capacity

The fourth challenge faced by financial institutions is how to improve their internal capability as innovative financial intermediaries to mobilize and efficiently allocate domestic savings. The long period of financial repression, which created a seller's market operating within a target system and bureaucratic mechanism, drove banks to neglect the need to improve their internal capabilities, technology, and customer services. These issues will be discussed in more depth in a later section.

Monetary Policy

The financial problems in the banking system and the high debt/equity ratio of nonfinancial firms makes monetary policy in Indonesia—such as interest rate policy—less effective. A rise in interest rates, on the one hand, increases the cost of capital. On the other hand, there is a reduction in the cost of capital as the rise in interest rates induces banks to reduce the debt/equity ratio. In his study of the Korean banking system, Sundararajan (1987) found that the reduction in the cost of capital more than offset the increasing cost. The reduction in debt/equity ratio of firms, Sundararajan concludes, makes the interest rate policy more effective. At this early stage of capital market development in Indonesia, which will be discussed further in the following section, the adjustment lag to reduce debt-equity is going to be relatively long.

There are many factors which drive up lending interest rates, at least during the transition period, including (i) inadequate capital of the banks to back up their high nonperforming loans; (ii) high debt/equity ratio of nonfinancial firms; and (iii) a decrease in short-term capital inflows and cut in the subsidies both on interest rates and exchange rate "swap" premiums. The slowdown in capital inflows is partly related to the reinstitution of ceilings on foreign borrowings of state-owned banks. The successive failures of banks and nonbank firms in 1988–1992 also reduced access of Indonesia's firms to international financial markets.

Economic Overheating, 1989–1991

The combination of reverse course and different speed of deregulation and lack of policy coordination caused overheating in the Indonesian economy in 1989–1991. This was indicated by rising inflation and interest rates and accompanied by appreciation of the real effective exchange rate in 1989–1991 which eroded the international competitiveness of the economy. These problems made the adjustment process more difficult.

In addition, the economic ills in 1989–1991 were partly caused by the inability of Bank Indonesia to conduct market-based monetary policy and to coordinate policy instruments. When it liberalized access of commercial banks ("bank devisas") to international financial markets, the subsidies on foreign exchange "swap" premiums had not been removed. There were massive inflows of short-term capital in 1989–1991. Given the narrowness and shallowness of the money market, Bank Indonesia could not sterilize these inflows with open market operations. Thus, the central bank faced the difficult choice of either permitting the real value of the rupiah to appreciate or purchasing foreign exchange from the public. The first alternative would enlarge the external current account and complicate the structural adjustment process. The second option would rapidly expand the money supply to fuel inflation.

To suppress inflationary pressures, the authorities had to re-introduce stabilization measures using market-based and nonmarket instruments. The former included tightening Bank Indonesia's credit policy and eliminating the subsidy on foreign exchange "swap" premiums. The discretionary policies included the second "Sumarlin" shock and reinstituting the ceilings on public sector' foreign borrowings. These measures ignited an increase in interest rates and created liquidity problems for the commercial banks.

CAPITAL AND MONEY MARKETS

In general, the domination of credit or debt finance involves dangers both at the microeconomic and macroeconomic level. Japan's success of developing a credit-based and highly leveraged financial system may be the only exception to this general rule. A high debt/equity ratio and highly leveraged financial system leave enterprises and banks operating in a volatile market with a high degree of exposure to down-side risks. Economy-wide shocks affect the performance of a substantial number of enterprises and banks. Offshore borrowings through banking channels can have greater monetary effects than if borrowed through capital and money markets. Excessive reliance on debt, therefore, renders economies and enterprises vulnerable to internal and external shocks and contributes to financial instability.

In theory, well-developed money and capital markets contribute to the efficiency and solvency of the financial system, by enhancing competition, lowering intermediation costs, and providing savers and borrowers with an alternative to debt financing from the banking industry. From this point of view, the overall financial sector policy reforms in Indonesia may seem to have had two main objectives: (i) to pave the way for the corporate sector to switch from a reliance on bank loans to issuing securities directly in money and capital markets; and (ii) to move from a government-controlled and monopolistic financial system toward a market-based one.

Capital Market

To diversify sources of financing and channels of capital inflows, the authorities deregulated Indonesia's stock exchanges in 1988 and 1989. However, securities traded in the capital market have not been used by Bank Indonesia as instruments for open market operation.

The present Jakarta Stock Exchange (JSE) was established on 10 August 1977;[13] the Surabaya Stock Exchange (SSE) was established on 30 March 1989. Listing requirements in JSE are much more difficult than in SSE and by offering lower transaction costs, SSE has been able to attract Jakarta-based companies to issue bonds in Surabaya. Also, transactions at SSE are fully automated and, to further expand its area of operation, in September 1993, SSE introduced long-distance trading with the opening of a computerized gallery in Jakarta with direct links to the SSE. It is not clear, however, whether the benefits will justify the cost of this link, nor whether Indonesia's electric and telephone systems will support long-distance trading.

JSE and SSE are operated independently and trading in the two markets is not integrated. Of course, both trade the same products and serve the same customers and, because of this, they could be united as a cost-saving measure.

Since December 1978, the authorities have reduced control on the capital markets by reorganizing and privatizing them[14] and by simplifying the operating mechanism to reduce the cost of transactions. Prior to this, fiscal incentives for listing and buying securities on the stock exchanges had been abolished in the 1983–1985 tax reform.[15] The tax reforms did minimize the biases of the (corporate) income tax system on business financing, however.

---

[13] After having been closed at the beginning of World War II, the Jakarta Stock Exchange was reopened on 4 June 1952. However, due to economic and social instability, it was officially suspended again in 1968.

[14] The new organization and working mechanism of the capital markets are contained in Presidential Decree No. 53 of 10 November 1990 and in the Minister of Finance's Decisions No. 1548 of 4 December 1990 and No. 1199 of 30 November 1991.

[15] Prior to the 1983 income tax reforms, the authorities offered generous tax incentives for companies to sell their shares in JSE, including asset revaluation for the purpose of depreciation and other tax privileges. A company selling at least 15 per cent of its shares was eligible for tax relief on revaluation of its fixed assets, which was normally levied at the company tax rate. Tax relief on revaluation of fixed assets was as follows: (i) the difference between book value and acceptable revaluation was exempted from company tax; (ii) any capitalization following a revaluation was exempted from the stamp duty levied on the paid-up nominal value as a result of the revaluation of fixed assets of the company either in the form of bonus shares or the registration of additional capital without payments; and (iii) the value increase was exempted from assessment of company tax, personal income tax, and tax on interests, dividends, and royalties. Capitalization of reserves derived from retained or undistributed profits was also exempted from stamp duty if a company sold some of its share in the JSE. The buyers of securities at JSE were entitled to the privilege of no tax inquiries into the origin of a maximum of Rp 10 million of funds invested. The investment was tax deductible and free from property tax, and the yields as well as the capital gains were also exempted from tax. Shareholders and certificate holders of Indonesian nationality were granted full relief from corporation tax, income tax, and tax on interest, dividends, and royalties with respect to profits gained from the sale of shares, and to any dividends received within a period of 24 months after a company listed at JSE.

The first stage of reform in the capital market introduced in December 1987 relaxed initial listing requirements, streamlined procedures for new issues, and established the over-the-counter (OTC) market. Foreigners were allowed to own up to 49 per cent of the listed shares issued by Indonesian companies (except banks). However, this "foreign portion" rule is not only cumbersome to implement, it also limits the purchasing interest of foreign investors.

Also in this stage of reform, the special right of PT Danareksa to buy new issues was revoked in 1989; this made it a regular securities house and ended direct government controls on prices of securities, making them more volatile to produce capital gains or losses. As a regular securities house, PT Danareksa may conduct business as an underwriter, broker-dealer, investment manager, and investment consultant. Also, the Banking Reform of October 1988 removed the differential tax treatment by levying the same 15 per cent withholding tax rate on both interest income from bank deposits and yields from securities.

The second stage of reform introduced on 4 December 1991 privatized the JSE, separating it from BAPEPAM.[16] Since then, the role of BAPEPAM has been reduced to "capital market authority," the regulator and the supervisor of the stock exchanges. Though now directly under the control of the Minister of Finance,[17] there is no adequate legal basis for BAPEPAM to enforce the legislation and regulations governing the capital market.

The new rules define the key terms, such as "securities," "brokers-dealers," "underwriters," "custodians," and "securities administration bureaus;" identify prohibited activities, such as fraud, market manipulation, and insider trading; and contain sanctions for violations, both civil and criminal. To improve the efficiency of the capital markets, this set of regulations has improved disclosure rules, restrictions on insider trading, and other elements of the regulatory frameworks.

Since 1991, BAPEPAM has simplified the trading system and the listing procedures in Indonesia's stock exchanges. Crossing and rights issues were allowed on 20 July 1992 and, concurrently, automatic cross listing was discontinued, though as of November 1993, there were 10 companies automatically cross listed on both the JSE and SSE. The JSE and SSE have reduced their membership fees, initial listing fee, annual listing fee, and

---

[16] Before the December 1987 reforms, BAPEPAM set the operating procedures for the stock exchange and supervised the implementation of these procedures. As the operator of the JSE, BAPEPAM evaluates applicants for listing to ensure they meet the listing requirements and monitors the progress of the listed companies.

[17] Prior to 1988, Indonesia had a Capital Market Policy Council that assisted the Minister of Finance in drawing up policies and regulations for the operation of the stock exchange. In reality, the Council has never been inactive. Under Capital Market Law Nr. of 1952, the Minister of Finance is in charge of regulating the capital market.

commission and transaction fees. In 1990, the Minister of Finance set the minimum capital requirements for stock exchange and securities firms.

The number of intermediaries in both the JSE and SSE is large compared to the number of listed companies and volume of trading. At the end of 1992, there were 205 securities houses and stock brokers, 47 underwriters, 15 custodians, 16 securities administration bureaus, and eight trustees and guarantors operating in the capital markets.[18] Of the securities houses, 34 are affiliated with banks and 11 with nonbank financial institutions; the rest are independent firms engaged exclusively in the securities business. PT KDEI (Indonesia Capital Market Clearing Corporation) manages comprehensive clearing, settlement, and depository in the JSE.

Despite the reforms, however, the authorities have introduced two types of new distortions which may affect development of the capital market. The first distortion is the Decision of Minister of Finance Nr. 1232/KMK.013/ 1989 which requires state-owned companies to earmark between 1 and 5 per cent of their profits for helping cooperatives and small-scale firms. The second distortion is the use of the capital market for achieving equity objectives. On 4 March 1990, President Soeharto called for the redistribution of a maximum of 25 per cent of the shares of the domestic conglomerates to cooperatives. The objective is to redistribute wealth and income; the transfer is paid from dividends of the companies' shares. As of January 1993, the state-owned companies have accumulated funds amounting to Rp 216.2 billion for such purposes and 172 domestic private firms have transferred Rp 55.5 trillion of their shares to cooperatives. Such redistribution mechanisms have weakened the affected companies, but not necessarily benefited the cooperatives and small-scale private firms.

Despite government efforts to encourage Indonesian companies to sell securities on the stock exchanges, many factors prevent private and public enterprises from listing their bonds and equities. First among these is the predominance of family-owned companies. Second, most of the publicly traded firms typically list only small fractions of their shares, thereby keeping real control in the hands of majority share holders. This avoids pressures for radical cost-cutting and separation of ownership and company management. Third, most of the listed firms belong to private conglomerates. Fourth, there is a reluctance to disclose information to the tax authorities, competitors, and the public. Fifth, minority shareholders are not clearly protected in the currently outdated Indonesian company law.

A number of factors make securities traded in Indonesia unattractive as long-term investments, including (i) lack of stability and high mortality of companies, particularly the politically well-connected ones;

---

[18] BAPEPAM revoked the operational licenses of six securities houses in April 1993 and of another four firms in October 1993.

(ii) lack of confidence due to the absence of sufficient information on the part of the listed companies; (iii) lack of transparency with regard to accounting procedures and auditing standards; (iv) inadequate protection for minority shareholders; and (v) inadequate administration and enforcement of the legal system. Trading is conducted manually, with inadequate control and supervision. In addition, the counterfeit scandal of five listed companies in the first week of April 1993 further eroded public confidence in the country's stock markets. Losses were estimated at Rp 10 billion ($4.98 million).

In 1988, there were only 24 firms listed on JSE. Of these, 17 were joint-venture firms which sold equities in JSE to meet the divestment program and to take advantage of the generous special tax relief offered to listed companies. By December 1992, the number of companies listed increased to 162 and the number of firms floating bonds increased to 34. The number of companies selling shares in SSE was 148 and 28 companies were issuing bonds. Four companies were listed and 6 firms floated bonds in the OTC (parallel) market. The total value of shares and bonds traded in these two markets amounted to over Rp 7 trillion in 1992 or 40 per cent of total banks' credit or 10 per cent of GDP in the same year. In terms of new listings, 1990 was the busiest year at the JSE when there were 56 new companies. In December 1992, 49 per cent of the stock listed in Indonesian stock exchanges was owned by foreign investors.

On the other hand, the very high real interest rates on loans and the reduction in the levels of protection that followed the reforms have investors worried about the efficiency and competitiveness of the issuing banks and nonbank firms. A mix of market efficiency problems and high interest rates eroded the index of stock prices by 35 per cent between 1990 and 1992.

Bonds issued by the state-owned utility and toll road companies remain popular. Most of the bonds traded in the JSE and SSE are absorbed by foreign and domestic institutional investors, including state-owned institutions such as PT Taspen (a state-owned company which manages civil servants' pension funds) and pension funds owned by large state-owned companies and state-owned insurance firms. Participation of individual investors is absent. As the buyers keep the bonds to themselves, bond trading in the secondary market is inactive. At present, there is no bond rating agency to evaluate bonds or assist investors in making decisions. There is also no bond registration or book-entry clearing system or formal rate that can be made as a standard to price the bond.

The types of bonds listed in the markets are straight and convertible, mainly issued by regional development banks, mortgage banks (PT Perumnas and BTN), Bapindo (the state-owned development bank), PT Jasa Marga (state-owned toll-road company), and PLN (state-owned electric company). Twenty-three of the 38 bonds carry fixed coupons (interest rates), ranging from 15.5 per cent to 19.5 per cent per annum. The remaining 15 bonds carry floating rates which are often tied to the average interest rate on one-year time deposits at state-owned banks.

Money Market

Until 1983, the money market consisted mainly of the interbank money market. The short-term money market was inactive as the dominant state banks were in surplus. The Government has not floated treasury bonds in the domestic market. Only a small portion of the state-owned banks' excess liquidity was traded in rupiah in the domestic money market. To compensate small or even negative profits from domestic operations in channeling the credit programs, the state banks invested much of their excess reserves in foreign assets. During the "oil boom" period in the 1980s, the authorities also encouraged this practice as one of the ways to sterilize the monetary impact of the "oil money."[19]

Bank Indonesia developed its short-term money market by reactivating the use of central bank certificates (SBI–*Sertifikat* Bank Indonesia) as a money market instrument in 1984. This was followed by the introduction of private sector commercial paper (SBPU—*Surat Berharga Pasar Uang*) and large-denomination negotiable certificates of deposit (CD) in 1985. The SBPU consists of promissory notes issued by banks or nonbank financial institutions in connection with interbank borrowing and bills of exchange issued by third parties and endorsed by eligible financial institutions. All of the CDs are issued by domestic banks in rupiah. Other instruments being used are repurchase agreements and banker acceptances. There is no credit rating agency in Indonesia.

Interbank Money Market

Prior to the establishment of the SBI and SBPU markets, banks were heavily dependent on the segmented interbank money market to smooth out their liquidity problems. During the pre-reform period, the fund providers were mainly the overliquid state-owned institutions (state-owned banks and rural development banks) and the buyers were mainly private and foreign banks as well as nonbank financial institutions. Partly to popularize the SBI and SBPU markets, effective from 1 October 1984, Bank Indonesia set a limit on each bank's borrowing to 7.5 per cent of the bank's third-party liabilities. This upper limit was raised to 15 per cent on 7 August 1985. The value of the

---

[19] The other way to sterilize the monetary impact of the "balanced budget" was to increase government imports, accumulate primary or secondary foreign exchange reserves, and/or to retire external debt.

interbank call money market in May 1992 was 19.5 per cent of the outstanding SBI and SBPU traded in the money market.

SBI and SBPU Markets

Bank Indonesia introduced four measures to redevelop the SBI markets: (i) to make it more liquid by allowing the holders to trade SBIs before maturity date; (ii) to inject liquidity into the banking system by special credit facilities; (iii) to increase the number of participants in the primary markets who are allowed to make SBIs transactions with Bank Indonesia; and (iv) to raise the interest or "cut-off" rates of SBIs.

At first, Ficorinvest, a nonbank financial institution owned by Bank Indonesia, was the sole institution to act as a market maker of SBIs. Following Pakto, 15 banks and nonbank financial institutions have been authorized for this purpose. The number of participants in the primary market, which can trade SBI directly with the central bank, was raised to 21 in February 1993 (eight state banks and 13 private banks). The public was allowed to trade SBIs in the secondary market as well. Secondary market trading occurs in the OTC market generally for SBIs of maturities of 28 days and greater. Currently, SBIs are the most liquid domestic short-term instrument available in Indonesia.

Initially SBIs were auctioned on the basis of pre-announced rates. Daily auctions were then supplemented with a weekly auction. The cut-off rate in the auctions is set by Bank Indonesia, based on bids and offers received. Over 83 per cent of the outstanding money market instruments traded in May 1992 (Rp 17.2 trillion) were in SBI and the remaining 17 per cent were SBPUs. The increasing amount of SBI issuance and the rising cut-off rates of this instrument indicate the high price paid by Bank Indonesia to promote money market development. The amount of the subsidy paid by the central bank to operate the SBI market may in fact have been higher than its revenue from intervention in the foreign exchange market.

On 1 June 1993, the authorities allowed the market forces to determine the interest rate of SBI. Since then, Bank Indonesia's auction system of SBI has been changed from a cut-off-rate (COR) to a stop-out-rate (SOR) system. The former is equivalent to "price" control as the authorities predetermine the yield for each auction. In the latter, Bank Indonesia sets maximum and minimum yields for each auction. For daily auctions, the band is set at +/− 25 bps around the weighted average yield of the prior daily auction; for weekly auctions, the band is set at +/− 50 bps around the weighted average yield of the prior weekly auction. Bids outside the band are rejected and must be quoted on yield basis in the increments of one eight.

Following the second "Sumarlin shock" in February 1991, the outstanding SBI and SBPU has risen sharply from roughly 2 per cent to over

10 per cent of total assets of commercial banks or from 2 per cent to nearly 20 per cent of their total deposits.

Controlling Liquidity Through Nonmarket Mechanisms

The authorities use market and nonmarket policies to encourage money market development. Early in 1984, Bank Indonesia reduced the rate it was then paying on excess reserves in rupiah and stopped paying any interest on excess foreign exchange reserves deposited with the central bank. At the same time, Bank Indonesia announced that it would cease paying any interest on rupiah excess reserves after three months. To help sterilize the liquidity impact of the reduction in the required reserve ratio, Pakto required banks to temporarily use their excess liquidity to purchase SBIs with three and six-month maturities. As was discussed earlier, the Sumarlin shocks required the switching of deposits of nonbank state enterprises, mainly at state banks, into SBIs.

NONDEPOSITORY FINANCIAL INTERMEDIARIES

The nondepository or nonbank financial industry in Indonesia consists of insurance companies, pension funds, and other types of financial companies. Established mostly in the second half of the 1980s, the latter include multi-finance, leasing, factoring, consumer finance, venture capital, and credit card companies. A number of the nonbank financial institutions—including life insurance companies and, to a certain extent, property and casualty insurance firms and finance companies—generate income by the spread between the return that they earn on assets and the cost of their funds.

The nonbank financial industry in Indonesia is also currently centered around state-owned and quasi publicly owned enterprises. Despite deregulation, many of the public companies in this industry, such as insurance, pension funds, and other types of financial institutions, enjoy captive markets. As it did in banking, deregulation imposed capital requirements and solvency standards on this industry.

Insurance Industry

In terms of assets and number of branch offices, insurance is the second largest financial industry in Indonesia after banking. The industry is characterized by predominantly small-scale operations with small capital, a narrow range of products, outdated technology, and weak management, and because of these constraints, many of the financial institutions in the industry actually perform more brokerage functions rather than full-fledged insurance functions. As most of the risks are exported, the retention of business is low. This limits the role of this industry as a provider of long-term investment funds for financing economic development.

The second characteristic of the insurance industry is that most of the large companies are either state-owned or interlocked with the large business groups. Such interlocking management and ownership in a group may provide new sources of direct capital for the insurers and finance companies and offer new business opportunities. Where the groups include banks and other financial institutions, financial expertise and information can be shared. While the infusion of capital and financial expertise and information may help the industry to expand, the data on whether the economies of scale of the conglomeration have outweighed the diseconomies of scale are not yet conclusive.

The third characteristic of the insurance industry is that the market share of state-owned and quasi public enterprises has been eroded rapidly with the rapid growth of domestic private and joint ventures. The latter are equipped with a larger capital base, wider selection of products, and better technology.

Types of Business and Number of Insurance Companies

The insurance industry in Indonesia consists of five separate subsectors, namely, reinsurance, life insurance, general (or non-life or loss and casualty) insurance, social insurance, and insurance brokers. A company is licensed for each type of insurance activity. The brokering industry acts as an independent intermediary between consumers and the insurance companies, tailoring policies and seeking competitive prices.

Besides the 115 foreign-owned insurance companies operating in Indonesia in the mid-1950s, there were 20 private domestic insurance firms. As in the other areas of the financial sector, insurance companies were seriously threatened by monetary purges, high inflation, and political turbulence in the late 1950 and early 1960s. Dutch-owned companies were nationalized during this period and other foreign firms ceased operations in Indonesia. Following the transition to the New Order Government in the late 1960s, the Government reissued licenses to establish new private local companies as well as joint ventures and representative offices of foreign-based insurance firms in Jakarta.

By October 1992, there were 145 insurance companies in Indonesia, comprising four reinsurers, 46 life insurance companies, 90 general insurance companies, including brokers, and five social insurance companies. The state-owned PT Asuransi Eskpor Indonesia (ASEI) is the only export insurance company in Indonesia. PT Askrindo, a joint venture between Bank Indonesia and the Ministry of Finance, is the sole credit insurance company. Both PT ASEI and PT Askrindo are classified in the general insurance group. In addition, there were 70 insurance brokerage companies, 21 insurance adjuster firms, and 12 insurance consultant enterprises (Table 3.9).

Table 3.9 Number of Insurance Companies, 1987–1992

	1987	1988	1989	1990	1991	October 1992
Life insurance	26	30	31	37	41	46
State owned	1	1	1	1	1	1
Private owned	23	27	26	32	36	39
Joint venture	2	2	4	4	4	6
Social insurance	5	5	5	5	5	5
Casualty insurance	68	74	78	84	87	90
State owned	2	2	2	2	2	2
Private owned	54	60	64	69	72	75
Joint venture	12	12	12	13	13	13
Reinsurance	4	4	4	4	4	4
State owned	2	2	2	2	2	2
Private owned	2	2	2	2	2	2
Total	103	113	118	130	137	145
Insurance brokers	48	58	60	63	67	70
Insurance adjusters	12	13	14	17	18	21
Insurance consultants	5	8	14	17	18	19
Representative office of Foreign insurance	6	8	8	–	–	–

– nil.
Source: Department of Finance (1991).

Gross Premiums

The gross premiums of the insurance industry grew rapidly between 1987 and 1991, by 9.8 per cent in 1988, 19 per cent in 1989, 23 per cent in 1990 and 18 per cent in 1991. However, as a percentage of GDP, gross premiums remain relatively small at 1.17 per cent in 1991 as compared to 1.12 per cent in 1987. About 57 per cent of the gross premiums in the insurance industry in 1991 were generated by general insurers, 22 per cent by social insurance companies and 21 per cent by life insurance firms (Table 3.10).

Table 3.10 Gross Premiums by Type (Rp billion)

	1987	1988	1989	1990	1991
Total gross premium	1,400.4	1,537.5	1,832.4	2,254.7	2,655.6
Life insurance	243.8	298.7	346.7	455.4	562.1
Social insurance	316.4	350.2	391.9	458.1	588.8
Casualty insurance	840.2	888.6	1,093.8	1,341.2	1,504.7
Distribution in per cent					
Life insurance	17.4	19.4	18.9	20.2	21.2
Social insurance	22.6	22.8	21.4	20.3	22.2
Casualty insurance	60.0	57.8	59.7	59.5	56.7
Rate of growth (per cent)					
Total		9.8	19.2	23.0	17.8
Life insurance		22.5	16.1	31.4	23.4
Social insurance		10.7	11.9	16.9	28.5
Casualty insurance		5.8	23.1	22.6	12.2
Memo:					
Total gross premium as per cent of GDP	1.12	1.08	1.10	1.14	1.17

Source: Department of Finance (1991).

Regulations

Barriers to Market Entry. To protect the domestic insurance companies, those located in Indonesia are required to be insured by domestic insurance companies. Prior to the Insurance Law of 1992, export of risk was discouraged by requiring domestic insurance companies to retain 75 per cent of the premium for their business within the country. The objectives of such regulations have not been realized, however, mainly because of a lack of domestic insuring capacity. Risk is exported through multiple levels of reinsurance.

Traditionally, a foreign insurance company could penetrate the domestic market only through a joint-venture firm. The Insurance Law of 1992 relaxed this market barrier by allowing foreign firms to insure both foreign nationals resident in Indonesia and foreign companies operating there. Also, if the domestic insuring capacity is lacking, a special permit may be given by the Minister of Finance for others to be insured by foreign firms as well.

The decrees set forth by the Minister of Finance in 1988–1989 and the Insurance Law Nr. 2 of 1992 limit the percentage of foreign participation in an insurance company to a maximum 80 per cent share. They also require foreign partners to transfer their share ownerships to local investors within a 20-year period. These restrictions reduce the interest of foreign insurance firms in developing long-term capital investment in Indonesia. For foreign insurance companies, the establishment of subsidiaries may not be optimal either as there is always the possibility that the attention of the insurance management will be diverted from its primary tasks of developing an insurance market to the activities of the subsidiary.

Capital Requirements. Government Regulation Nr. 73 of 1992 raised the absolute minimum capital requirement.

Solvency Margin and Retention. Similar to the Capital Adequacy Regulation in banking industry, there is a solvency-margin regulation in insurance industry which directly links the required capital to risks. Reinsurance and general insurance companies are mandated by the Finance Minister's Decree Nr. 224 of February 26, 1993 to maintain their solvency margin at a minimum of 10 per cent of their net premium. The minimum ratio of solvency margin for a life insurer was set at 1 per cent of its reserve premium. For health and casualty, the minimum ratio solvency margin is 1 per cent of reserve premium plus 10 per cent of net premium. Solvency margin of an insurance company is defined as the difference between the value of the solid and liquid assets of the company and its debts and the required capital.

Protection of solvency requires that insurance investment be relatively secure and diverse. To pursue these objectives, investment by insurance companies is regulated in Indonesia: Investment in time deposits and certificate of deposits in each bank is limited to 5 per cent of its asset value; the same rate is applied for investment in shares and bonds. Investment in commercial paper is limited to those guaranteed by commercial banks with a maximum of 10 per cent of insurer assets. Investment in commercial papers for each issuer is limited to 2 per cent of insurer assets.

The Insurance Law of 1992 eliminated previous regulations on export of risk. The Law prescribes the maximum amount of own retention of the insurance and reinsurance companies at 10 per cent of their capital. The minimum amount of net premium is set at 30 per cent of their gross premium. The maximum amount of net premium retained by general insurance and reinsurance companies is set at 300 per cent of their capital. A life insurance company can retain net premium only for personal injury at a maximum amount of 150 per cent of its capital. The same limit of retaining net premium is applied for health insurance. The maximum limit for premium from indirect insurance is set at two thirds of the value of premium from direct insurance.

Because of the history of monetary purges, high inflation rates, deposit freezes, and rupiah devaluations, US dollar denominated policies represent a substantial share of the individual life policies of the large insurance companies. This requires them to maintain investment denominated in foreign currencies, mostly time deposits, for meeting current obligations, which ultimately limits the opportunity for insurance companies to make long-term investments in rupiah in Indonesia.

Insurance Premiums and Investment. Prior to 1987, insurance premiums were set by the Minister of Finance, and were calculated by the Team for Renewal of Insurance Rates at the Ministry of Finance in cooperation with the Board of Insurance. Members of the Board were the directors of state owned insurance companies. The Government's primary concern was that rates not be excessive for the coverage provided and that people be fairly classified for rating differentials. However, this concern had to be balanced with a concern that rates be high enough to maintain the insurers solvency, and the rates had to be determined in view of foreign competition and reinsurance.

Since 1988, insurance companies have been allowed to set their own rates, which has increased market competition and has put pressure on companies to reduce insurance premiums. On the other hand, cost of reinsurance is rising internationally while most lines of business face low profitability. Moreover, a number of assets seem to have questionable value, in part due to a combination of factors such as technical incompetence, falling prices, and weak market infrastructure. At the same time the need for auto-financing is rising as the Insurance Law of 1992 demands higher minimum capital and surplus requirements than before.

Reinsurance Industry

There are five reinsurers in Indonesia. The largest single company is PT Asuransi Kredit Indonesia (Askrindo), with an asset value of Rp 451 billion in 1991. Nearly 52 per cent of the share of PT Askrindo is owned by Bank Indonesia, and the other 48 per cent is owned by the Ministry of Finance. This company sells insurance policies to cover risks of certain types of bank credits. Past credit programs for small- and medium-scale industries were required to be insured by PT Askrindo. Other credits, including banks's credit beyond the legal lending limit regulations, are insured on a voluntary basis. Credit insurance for cooperatives is handled by the state-owned *Perum Penjamin Kredit Koperasi* (PKK). The second largest company is the state-owned PT *Reasuransi Umum* Indonesia (Indo Re) with assets totaling Rp 94 billion in 1991. The quasi state-owned PT *Tugu Jasatama Reinsurance* Indonesia, had assets totaling Rp 45 billion and *Maskapai Reasuransi* Indonesia (Marein), the only private domestic company, had assets totaling Rp 28 billion.

The ultimate retention of reinsurance and general insurance industries is low (roughly below 5 per cent in 1991). Most of the insurance premium is passed to international reinsurance companies overseas. This indicates that most of the reinsurers and general insurers are actually operating as agents or brokers rather than as full-fledged insurance companies. PT Askrindo, the credit insurance company, suffered from a deficit of Rp 390 billion ($187.52 million) in 1992, or more than nine times the amount of its paid-up capital at Rp 45 billion. Due to increasing problem debts, the value of its approved claims rose sharply over 55 per cent from Rp 160.85 billion in 1991 to Rp 250.46 per cent in 1992. This deepened the company deficit in insurance underwriting income, which totaled Rp 46.34 billion in December 1991 to reach over Rp 444.06 billion in December 1992.

Life Insurance

By all standards, Indonesia is still "underinsured" for life insurance. This is the result of a number of factors, including the following: (i) the long period of economic and political instabilities in the 1960s and 1970s and the series of the rupiah devaluations eroded people's faith in financial investments; (ii) the low income per capita at $650 in 1992 precludes such investments; (iii) until October 1988, investment in life insurance was subject to tax as compared to tax-free bank time deposits; and (iv) the reputation of many life insurance companies has suffered because of their small size, occasional failure, and solvency problems.

The ratio of life insurance policyholders to population in Indonesia remains low at less than 5 per cent in 1991. On an international basis, the value of total volume of life insurance premiums in Indonesia was ranked 34 in 1985. In terms of average per capita premium volume, which measures "insurance density," Indonesia's ranking slips even further to 52. The ratio of life insurance in force to national income was small at 6 per cent in 1988 as compared to 23 per cent in the Philippines, 16 per cent in Thailand, 165 per cent in the Republic of Korea and 392 per cent in Japan.

General (Non-Life) Insurance

As in the life insurance industry, the market in non-life insurance is relatively small in Indonesia. As a percentage of GDP, the gross premiums in Indonesia in 1985 (at 0.45 per cent) were the lowest in the ASEAN region, compared to 0.54 per cent in Thailand, 0.66 per cent in the Philippines, 1.61 per cent in Singapore, and 2.15 per cent in Malaysia.

Of the 90 general insurance companies in 1992, two were state-owned insurance firms, 13 were joint ventures and 75 were domestic private companies. The latter includes seven quasi state-owned firms.

As in life insurance, the market share of a small group of state-owned and quasi state-owned insurance companies has begun to decline. The largest state-owned PT Asuransi Ekspor Indonesia controlled 12 per cent of the industry assets in 1991 and the state-owned PT Jasa Indonesia had 9.7 per cent of the total industry assets. PT Tugu Pratama, which controlled 15.8 per cent of the industry assets, is partly owned by PT Pertamina, the state-owned oil company, and partly by PT Nusamba, a well-connected business group. PT Timur Jauh (formerly PT Berdikari Insurance) belongs to PT Berdikari group which, in turn, is controlled by the National Logistics Agency (BULOG).

Social Insurance

All of Indonesia's social insurance companies are state-owned. Participation is compulsory for the various client groups and each of the companies has monopoly power in their respective areas.

A controversial law (Jamostek) was passed on 17 February 1993 which regulates workers' compensation insurance, health insurance, life insurance, and pension benefits. The program is mandated for all workers, both in informal and formal sectors, and is exclusively mandated by PT Astek, a state-owned worker insurance company under the control of the Ministry of Labor. This mandatory regulation and the exclusive right of a state-owned company to administer the program go against the principles of competition and market-based deregulation programs. In addition, the Jamostek program is too administratively burdensome for a single entity. The potential costs are frighteningly high. It drives up costs of production and encourages substitution or displacement of labor for capital as the employers's contributions are equivalent to a tax on hiring workers.

Noninsurance and Nonbank Financial Companies

There are six types of noninsurance and nonbank financial institutions now licensed in Indonesia, namely, multi-finance, leasing, factoring, consumer finance, venture capital, and credit card companies. The number of finance companies by type and ownership is depicted in Table 3.11.

Ownership and Purposes

Established in 1975, PT PANN (Pengembangan Armada Niaga Nasional) Multi Finance was the country's first and only state-owned finance company. It is jointly owned by the state and PT Bapindo and handles leasing of ocean going vessels. PT Patra Nusa Lease Corporation, a private national company, and Perjahl Leasing Indonesia, a joint venture company, were established in 1976. Both are subsidiaries of Pertamina, the state-oil company. The other finance companies were established in the 1980s and early 1990s.

Table 3.11 Number of Finance Companies by Type and Ownership
(as of March 1993)

	Joint Venture	Private	State-owned	Total
Multi-finance	15	63	1	79
Leasing	26	34	–	60
Factoring	1	–	–	1
Consumer finance	1	3	–	4
Venture capital	3	4	–	7
Credit cards	–	2	–	2
Total	46	106	1	153

– nil.
Source: Ministry of Finance (1991).

Some of state-owned, quasi state-owned companies and their pension funds are shareholders of finance companies. The activities of such finance companies include leasing equipment for the parent companies as a way of escaping government control over their budget, procurement, salaries, and facilities.

PT Perjahl, partly owned by Pertamina, leases equipment for foreign oil contractors. PT Patra Nusa leases properties for employees of Pertamina and foreign oil contractors. Koperasi Pembiayaan Indonesia (Kopindo) is the only finance company fully owned by cooperatives with the main purpose of securing financing for the cooperatives.

The state-owned companies which are active shareholders of finance companies include Pertamina, Bank Rakyat Indonesia, Bank BNI, Bank Exim Bank Bumi Daya, Bapindo, Bank Dagang Negara and PT Danareksa. The quasi state-owned enterprises include PT Uppindo, PDFCI, Bank Indover, and PT Ficorinvest. The first three companies are subsidiaries of Bank Indonesia. Pension funds of state-owned companies, including those of Bank Dagang Negara and Bank BNI are also active shareholders in many banks, insurance, and finance companies. Some workers cooperatives of state-owned companies, such as that of BULOG, are active shareholders in banks and other finance companies, outside Kopindo. This group of quasi state companies and cooperatives are classified as national private companies.

The consumer finance companies consist mainly of captive finance companies and bank-related companies. The first category of firms are subsidiaries of sole agents and assemblers of automobiles and heavy equipment, and real estate companies. Their primary purpose is to secure financing for the customers of the parent companies. Nearly all of the large national and foreign banks have a finance company subsidiary that provides loans to individuals and businesses to acquire a wide range of products.

Size of Business and Sources of Financing

The years 1987–1990 witnessed an unprecedented surge in private sector investment, non-oil exports, private consumption, and growth performance. These factors induced rapid growth in the value of finance companies' lease contracts during this period. In 1991, however, the value contracted by 17 per cent with the slowdown in economic growth. As shown in Table 3.12, the manufacturing and transportation sectors absorbed over 62 per cent of the lease contracts: of this, over 37 per cent of the lease contracts value was on vehicles, over 28 per cent on machinery, 11 per cent on heavy equipment, 8 per cent on commercial buildings, and 4 per cent on computer equipment.

In 1990, 85 per cent of the contract value of the consumer finance companies was received by individuals and the rest by firms. Nearly 100 per cent of the funds were used for financing automobiles and only a small part was used for house financing. In 1991, individuals received 80 per cent of the financing and 98 per cent of the contracts were used for financing automobiles.

Table 3.12 Finance Companies' Lease Contracts by Economic Sector, 1986–1991 (Rp billion)

Economic Sector/ Year	1986	1987	1988	1989	1990	1991
Agriculture	–	–	1	12	25	35
Construction	65	115	257	208	386	308
Manufacturing	172	207	363	1,011	1,607	1,137
Mining	2	17	1	58	67	212
Transportation	296	555	689	793	1,373	1,087
Others	110	301	371	803	1,288	1,166

– nil.
Source: Ministry of Finance.

Areas of Operations and Market Share

Operations of the finance companies are concentrated in Jakarta and West Java. The distribution of lease contracts by province in 1990–1991 is given in Table 3.13. The table also indicates the predominant role of joint-venture companies which controlled over 70 per cent of lease contract values both in 1990 and 1991. The leasing, consumer financing, and factoring industries markets are also highly concentrated in the hands of a small number of companies. In 1991, 13 companies controlled 55 per cent of the total market in the leasing industry. Four firms (Adipura, Astra, Staco Tiga Berlian, and Swadharma), all subsidiaries of automobile assemblers, controlled 64 per cent of the market in consumer finance industry in 1991. PT Putra Surya Multi Finance alone controlled 26 per cent of the total factoring industry market.

Table 3.13 Distribution of Lease Contract Value by Province and Ownership of Finance Companies, 1990–1991 (per cent)

Province	Type of Companies					
	Joint Venture		National Private		State-Owned	
	1990	1991	1990	1991	1990	1991
Jakarta	48.8	42.6	17.1	15.3	1.5	0.0
West Java	7.9	12.1	3.1	3.8	–	0.0
Central Java	3.3	3.4	1.0	1.0	–	0.0
East Java	6.5	6.4	1.5	1.7	0.4	0.6
Off-Java	5.2	6.3	3.7	6.8	–	0.0
Total	71.7	70.8	26.4	28.6	1.9	0.6

– nil.
Source: Ministry of Finance (1991).

INFRASTRUCTURE OF THE FINANCIAL SECTOR

Prudential regulations and appropriate supervision and examination are important ingredients for developing a well-functioning financial system. The successive packages of deregulation issued since 1 June 1983 have improved Indonesia's prudential regulations. The new rules have minimized the overlap in regulatory responsibilities of various government bodies, regulated disclosure of information for financial institutions and listed companies, improved supervision procedures, expanded the developmental role of the regulatory bodies with regard to money and capital markets, and established a comprehensive and uniform system for public reporting of financial statements.

The progress of administering and implementing the new prudential framework, however, has not moved at the same pace. This is true for the financial institutions and regulators and supporting agencies, such as accounting and legal firms. The collapse of PT Bank Duta in 1990 and the bankruptcy of PT Bank Summa in 1992 indicates the weakness of corporate law, the limited credibility of audited financial statements, and the limited ability of the central bank and BAPEPAM to administer and enforce the new rules and regulations. Also, because of inadequate and unreliable information regarding solvency, banks rely on developing information through building long-term relationships with their clients (Caprio 1993), and demand high-value collateral to back up their credit. The financial institutions owned by the conglomerates channel most of the credit to their own subsidiaries.

The weak capability of financial institutions to supervise credit has led to a high number of nonperforming loans. Further, weaknesses in the legal framework for debt collection and bankruptcy procedure limit the recovery of bad debts. Financial institutions use private collectors for asset recovery, and the Government suspends the civil rights of debt defaulters by banning them from overseas travel.

Internal Capability of Financial Institutions to Supervise Credit

The allocation of capital and the monitoring of its use is the "brain" of the economic system, the central locus of decision-making (Stiglitz 1993). The large number of nonperforming loans within the domestic banks, particularly the state-owned institutions, is indicative of their weak internal capability to select customers, administer credit, monitor its use, and recover matured credits. While this weakness is due, in large part, to the country's long period of financial repression, the rapid growth of credit programs has also weakened the process of initiation and evaluation of credit. Banks have been concerned with targets and less concerned about credit analysis and profitability. The target markets for state-owned banks were determined by their main areas of specialization. Risk-asset-acceptance criteria and calling programs were determined by the central bank and specified in the criteria of the credit programs. In some sectors of the economy—such as agriculture, small-scale credit programs and credit for the cooperative sector—customers have been identified by the technical ministries. The chain of approval was centralized at Bank Indonesia which approved the ceilings, the use of the credits, and their modifications. As interest rates were subsidized, credit was allocated based on nonprice criteria. Collateral was not properly appraised and valued because the market infrastructure was unreliable and political considerations carried heavy weight in the decision-making process.

State-owned bank resources could not catch up with the complexity and high rate of growth of credit programs. This also weakened credit administration and monitoring of the ongoing credit. Customers performance was not properly monitored, collateral and securities were never reappraised, field inspection was lacking, personal assets of personal guarantors were never tracked, and insurance coverage was allowed to expire. This led to a delay in corrective actions.

Regulatory and Supervisory Agencies

Responsibility of supervision of financial institutions in Indonesia is divided between the Ministry of Finance and Bank Indonesia. The Ministry of Finance is responsible for supervision of insurance, pension funds, and securities industries and Bank Indonesia is responsible for bank supervision.

Ministry of Finance

The Ministry of Finance is the most powerful institution in the Government: the Minister is Chairman of the Monetary Board and as such is in charge of macroeconomic management. The Board coordinates the fiscal, monetary, and balance-of-payment policies and their implementation. Bank Indonesia deals with the day-to-day operational implementation and administration of the Government's monetary policy. Some of the Minister of Finance's power in budget allocation is shared with the Planning Agency (BAPPENAS), which controls allocation of the "development budget." Besides being in charge of financial management of the central government, the power of the Ministry of Finance is extended to the economy as a whole. The Ministry represents the Government as the owner of the state-owned enterprises, issues "guidance" to the financial institutions, and supervises the capital markets.

Further, the Ministry of Finance is the chartering agency for all types of financial institutions and public accountants in Indonesia. The Minister of Finance is also involved in the consolidation and bail outs of failing state-owned and private domestic companies through acquisitions, mergers, takeovers or capital injections undertaken under the Ministry's suggestions. Before the financial reforms, through PT Danareksa, the Ministry of Finance also oversaw price stabilization in the stock market.

The Ministry of Finance consists of six powerful Directorate Generals and Authorities, namely: Custom and Excise, Taxation, Budgeting and Treasury, State-Owned Enterprises, Financial Institutions, and the Capital Market Executive Agency (BAPEPAM). The Directorate General of Financial Institutions issues operating licenses for financial institutions and accountants and supervises nonbank financial institutions, particularly insurance and pension fund companies. With technical ministries, the Directorate General of State-Owned Enterprises controls operations of state-owned enterprises, including banks, insurance and finance companies.

Supervision of Insurance Companies and Pension Funds

The Directorate of Insurance issues licenses and sets operating standards, rules, and regulations for the insurance activities, administers their implementation, and makes regular inspections of insurance companies. Likewise, the Directorate of Pension Funds is in charge of pension funds. These two directorates are under the Directorate General of Financial Institutions. Prior to 1988, the insurance industry in Indonesia was regulated by an outdated and unclear legal framework.

Pension Funds companies were unregulated prior to the current Law Nr. 11 of 1992. As the regulations were subject to change by administrative fiat, they had became a source of uncertainty within the insurance and pension funds industries. During this era, the Minister of Finance had the right to revoke an operational license for failure to meet certain regulatory requirements. When this happened, it jeopardized the company's survival as well as placed the interests of its policyholders and pensioners at considerable risk. Such uncertainty has been corrected by the Decree of the Minister of Finance in 1988 and the Insurance Law and Pension Funds Law, both of 1992. These two regulations authorize the insurance and pension funds regulators to impose limited monetary penalties in lieu of license revocation, except in the most extreme circumstances.

As the insurance supervisor, the Directorate of Insurance is empowered with an internal audit function and field examination function at the offices of insurance companies. Likewise, the Directorate of Pension Funds supervises the pension funds companies.

BAPEPAM-The Capital Market Regulator and Supervisor

The new regulatory framework identified the tasks and responsibilities of accountants, legal advisors, and consultants in the capital markets. Like the insurance and pension funds supervisors, BAPEPAM has also increased the number of its professional staff. However, as a branch of the Ministry of Finance, BAPEPAM is facing deficiencies from government bureaucracy: lack of budget, low salary and working benefits, lack of skilled manpower, and lack of working facilities. As noted earlier, BAPEPAM revoked operational licenses of four securities houses in April 1993 and another six securities firms in October 1993. This demonstrates its seriousness to administer the new regulations.

Bank Indonesia-The Central Bank

As the central bank, Bank Indonesia is the sole issuer of currencies and coins, banker for the Government and other banks and, when a bank needs currency or needs to make payments, it can use its deposits with the central bank. It is the controller and administrator of gold and foreign exchange reserves of the country, and empowered to set up the level and structure of interest rates and control banks' credit, either quantitatively or qualitatively. As the lender of the last resort, Bank Indonesia can extend credit to bank and nonbank financial institutions and nonfinancial companies. Bank Indonesia is the regulator of all financial institutions, except the insurance industry and pension funds. It has the right to control, monitor, and supervise the soundness of the banking and payment systems. It sets, administers, and enforces the rules and regulations on safety standards for the banks and nonbank financial institutions, and makes regular inspections of them.

In addition, Bank Indonesia is an active promoter of the development of short-term money market and long-term financial markets. Its certificates (SBI) are the main instrument traded in the money market.

To encourage inflows of foreign direct investment, Bank Indonesia provides foreign exchange hedging, sometimes with subsidized premium. In many cases, such as the case of PT Bank Summa in 1992, it acts as a rescuer and catalyst for unwinding the effects of banks' failures and bankruptcies.[20] Before it collapsed, Bank Summa was a business partner of Bank Indonesia, through Indover Bank, in Summa International Finance of Hong Kong. Until 1 June 1983, the central bank guaranteed certain types of saving and time deposits and assumed parts of the risks of credit programs.

To support its wider activities, Bank Indonesia either fully or partly owns a number of financial institutions. These include Bank Indover, a commercial bank in the Netherlands, three development finance companies (PT Ficorinvest, PT Uppindo and PT Bahana), a credit insurance company for small business loans (PT Askrindo), and a housing finance company (PT Papan Sejahtera). In early 1993, PT Uppindo and PT Ficorinvest were converted into full-fledged commercial banks and licensed to deal in foreign exchange transactions. Its commercial department ceased to operate in April 1984.

As the central bank, Bank Indonesia is not an independent institution. Its authority and scope of action depend on the Government. Due to the absence of a well-developed money market and the lack of instruments, Bank Indonesia is powerless to sterilize the monetary impacts of the public budget. Also, because it intervenes so widely, Bank Indonesia is prone to political pressures. The Governor and Managing Directors of Bank Indonesia

---

[20] When Bank Summa was insolvent in 1991 because of mismanagement and poor investment decisions, Bank Indonesia and a group of private banks provided loans to it and its owners so that the owners could make repayments to their depositors and creditors. In the end, Bank Summa was closed and liquidated in 1992.

are appointed by the President and their terms in office are similar to the terms of the cabinet members. The Governor of Bank Indonesia is a member of the Monetary Board and attends regular meetings of the cabinet ministers. There is a Government Commissioner who supervises the operation of Bank Indonesia as an enterprise and Bank Indonesia's budget requires approval from the Government.

Along with the expansion of commercial banks and the BPR, Bank Indonesia also expanded its organization and number of bank supervisors. It now has 41 branches throughout the country. The department of bank supervision has been expanded to three departments, covering state-owned banks, private (domestic, joint venture and foreign) bank devisas—which are licensed to deal with foreign exchange transactions—and private banks which are not licensed to do so. To attract and keep professional staff, the salary and working benefits at Bank Indonesia have been raised comparable to the private sector.

Underdeveloped Legal and Accounting Systems[21]

The need to improve the antiquated legal and accounting systems in Indonesia is partly indicated by the recent failures and incidents of fraud that have occurred among banks (such as PT Bank Umum Majapahit, PT Bank Duta, PT Bank Putera Sampoerna, and PT Bank Summa) and nonfinancial companies (such as PT Bentoel, PT Mantrust, and PT Astra).

Weak Legal System

As the basis for securing contract and credit transactions is unclear in the Indonesian legal system, debtors enjoy strong protection. At present, there are no reliable information systems covering land registration, property and security, or credit information. Hypothecation is recognized only for removables and ships, not for a wide range of assets. Pledges are possible, but require physical possession, which is impractical. The transfer of ownership and mortgage of land (at present, the most important form of asset and collateral) are extremely difficult. Obtaining title documents is a time-consuming and costly procedure. This, and the absence of a registration system for collateral as well as the risks on leased property located on immovable property, makes leasing transactions nearly impossible. The present company law in Indonesia is based on the antiquated 21 rudimentary provisions of the 1987 colonial Commercial Code. A limited liability

---

[21] A number of the ideas in this section are drawn from Country Department III, East Asia and Pacific Regional Office of the World Bank. 26 May 1992. *Indonesia-Growth, Infrastructure and Human Resources*. Chapter 4. Report No. 10470-IND.

company requires approval of the Ministry of Justice, which is again a time-consuming and often costly procedure. Flexibility and mobility are limited as companies can be formed only for specific purposes and for defined periods. Voting rights are archaic and minority shareholders have little protection. The laws and procedures on exit and bankruptcy are unclear and untested because of the weakness of the court system.

Under Indonesian law, disputes of commercial interests, including disputes with government agencies and state-owned enterprises, can be solved by arbitration. Arbitration provides a much cheaper and more reliable alternative to the lengthy and often costly court system as the disputes can be settled quickly and fairly by technical specialists. Arbitration proceedings follow procedures similar to those in the Civil Procedure Court and the format of the arbitral award is simple and must be rendered within six months of submission. Though there is an arbitration board, *Badan Arbitrase Nasional*-BANI, established in Jakarta in 1977 by the Indonesian Chamber of Commerce, its services are not much utilized. It handles only five to ten cases a year.

Limited Credibility of Financial Statements

The present corporate and tax laws in Indonesia require that "adequate financial records" be kept, but do not impose accounting requirements and standards to ensure financial disclosure. Financial institutions are required to file audited financial statements to Bank Indonesia and to publish these quarterly. Public companies are also mandated by BAPEPAM to provide and publish regularly audited financial statements. For internal use, financial statements of state-owned companies are audited periodically by the Agency for Control of Finance and Development (BPKP) and the reports are submitted to regulating ministries. The State Auditory Agency, *Badan Pemeriksa Keuangan* (BPK), audits the public entities. Public accountants are licensed by the Ministry of Finance.

Public accountants have to be graduates of accounting from a state-owned or selected private university and have acquired three years of experience. There are no requirements to pass additional, rigorous examinations in accounting theory and practice, auditing practice or commercial laws before being licensed. There is an Association of Indonesian Accountants (*"Ikatan Akuntan Indonesia"*-IAI) which has the responsibility for setting the framework of accounting principles, auditing standards, and codes of ethics for accountants. IAI, however, neither has the statutory right nor the resources to conduct the accountancy examination or to enforce accounting standards. On a joint-venture basis, foreign accounting firms can penetrate domestic markets, but only a local accountant is authorized to sign audit reports.

REORGANIZATION AND RECAPITALIZATION OF FINANCIAL INSTITUTIONS

There is an immediate need to strengthen the viability of the financial institutions, to help them and the users of their services to adjust to the more competitive market environment. This includes recapitalization and improvement in knowledge of domestic banks and restructuring or corporatization or privatization of state-owned financial institutions. Further weakening of prudential standards and providing help for failed banks are not the correct solutions to the present problems. These measures only postpone the difficulties of the bigger problems which may result in more severe consequences and greater financial and social costs. Helping the economic agents in the process of transition from a formerly repressed environment to a market-based system also requires improvement in the market infrastructure. Among other things, this includes astute supervision by government regulatory bodies, such as Bank Indonesia for banks, the Ministry of Finance for insurance and pensions, and BAPEPAM for the capital market.

Corporatization of State-Owned Financial Institutions

Management of state-owned financial institutions must be made accountable and independent. While changing their legal status to *Persero* or limited company is a start, other measures are required to minimize the degree of political interference in their decision-making processes. Currently, state-owned financial institutions cannot approach risk objectively or with adequate information. To minimize political interference, the state banks should be de-linked from government bureaucracy, including the State Recovery Agency.

The ultimate objective of making state-owned enterprises, including financial institutions, independent is to encourage them to develop clear strategies for the future, reorient them towards market principles, strengthen their balance sheets, and generate clearer reporting and responsibility structures. There are several options for achieving these objectives, namely, to privatize, corporatize, merge the viable institutions and close down the ineffective ones. Given the shallowness and narrowness of the domestic capital market, selling off of the state-owned financial institutions' equities may be risky in terms of who the buyers will be and who will ultimately be in control. The best option may be to corporatize the viable state-owned enterprises and close down the most inefficient ones.

Bad Debts and Premium Collections

Reports indicate that most of the defaulted loans are concentrated at the state-owned banks and inherited from the previous system under which borrowers regularly failed to pay debts to banks or other companies. Even the state-banks' loans to the well-connected conglomerates in the post deregulation period were not market priced because they were subject to external political pressures. Bad debts of private banks were accumulated because some banks, such as PT Bank Duta and PT Bank Summa, proved to be carefree lenders and foreign exchange and property speculators, while taking high risks by offering attractive interest rates in bids to attract customers away from state-owned financial institutions.

Merger

One of the important functions of the financial institutions in an economy is to transfer funds for savers in the form of long-term investment funds to investing sectors of the economy.

The small size of the privately owned financial institutions in Indonesia limits their opportunity to perform this role. The combination of size, low capital ratios, and inefficient management of some of these financial institutions is a recipe for future problems. Mergers could prevent this scenario.

It may be possible to increase the capital base and volume of business of some financial institutions by encouraging the existing firms to merge together. For those financial institutions which have no significant amount of funds available for long-term investment, the best solution is either to seek merger or consolidation or exit. Some of the domestic private banks belong to this group. As in the early 1970s, instruments such as deferral of taxes and normally taxable events in the course of merger and/or temporary income tax relief could be used as incentives.

The first candidates of the merger program are the quasi state-owned financial institutions. These include the former nonbank financial institutions and those owned by pension funds of state-owned banks and nonbank companies. The former nonbank financial institutions have now legally become "private" commercial banks. Most of these new banks exist effectively as an external treasury department of their owner-customers and earn most of their profits not from interest on loans but from lucrative fees. These include fees for administering deposits, arrangement fees for loans and placement of surplus funds owned by their owners in the interbank credit market and SBI. The former nonbank financial institutions, now commercial banks, are competing with their owners in the same line of business.

Recapitalization of Banks

Recapitalization is the costliest component of the financial sector reforms. Traditionally Indonesian banks were undercapitalized. Bad debts appeared as assets on balance sheets of the financial institutions, particularly as Indonesia's banks kept financing the problem loans to make credit "evergreen." The value that is lost when loans are defaulted has to be replaced quickly to stop further deterioration in the portfolios of the financial institutions. The rising share of short-term deposits with higher interest costs in the banking industry coupled with the high cost of production severely limits the accumulation of investible funds.

To give banks a fresh start, Indonesia may well adopt a combination of methods, adopting the way Latin America bailed out their banks with the methods of Malaysia. In the former, banks were capitalized by removing nonperforming loans from the banks' portfolio and substituting them with a government-backed mechanism (e.g., treasury bonds) to inject additional capital (Larrain 1989). The banks' capital grew over time through the gradual elimination of bad loans and the positive net income flow from the government bonds which replaced the bad loans.

In contrast, Malaysia placed the burden of recapitalization of banks on both the owners and the central bank and less on the public budget (Sheng 1989). Shareholders of the ailing banks were required to inject additional capital through a rights issue. This new capital was supplemented by the central bank to meet the minimum 8 per cent capital adequacy ratio. The shares subscribed by the central bank were held under a buy-back scheme under which the shareholders who had participated in the bail-out program were allowed to buy back the unsubscribed shares at par plus holding costs.

The first mechanism could be used to bail out the state-owned banks and the second mechanism for strengthening the capital base of domestic private banks. Tax facilities alone, such as allowing banks to count provisions for bad loans against their tax liabilities, may not be enough. Conversion to capital of Bank Indonesia's credit liquidity may not help the state-owned banks if the credit refinanced by them turns out to be bad. Also, conversion of "two-step" offshore loans, mainly from Japan, requires approvals from the donors.

The downside of issuing government bonds to recapitalize financial institutions is, however, that this constitutes government spending. The public budget has to absorb losses on the nonperforming loans and transfer new resources to the banks through interest payments on the treasury bonds. This may help the economy in the long run, if it helps banks to start afresh with responsibility for their own actions. It will then make them more attractive for future buyers, and it will boost the banks' capital adequacy ratios without resorting to underprovisioning. To allow state-owned enterprises to strengthen their capital bases, other policies need to be reconsidered as well. This includes the policy which requires them to earmark between 1 and 5 per cent of their profits to help small- and medium-scale firms and cooperatives.

Operating System

To survive and stay healthy in a more competitive market, financial institutions should be able to maintain their financial integrity and performance. In addition, as the financial industry is a part of the service industry, to survive in a more competitive market, it must improve its service and quality for customer satisfaction through customer service. To improve service and quality of their products, financial institutions need to improve the way they deliver information about their products to the customers, to take interest in the customers, including their needs and purchasing power. The quality of service includes fast product delivery, and timely follow-up and claim settlements. Simply hiding the bad loans or passing them to state-owned insurance companies and the State-Recovery Agency will not solve the problems inherent in the system.

CONCLUSION

Bank-Oriented Financial System

The financial system in Indonesia is a bank-oriented system. The system itself is controlled predominantly by the state-owned institutions who provide financial intermediation and other services. Banks are involved in creating money and facilitating the payment system, transforming maturities, and diversifying the risk associated with credit. Banks not only provide short-term working capital, they also provide risk capital. However, they are not involved in taking up equity participation. Nonperforming loans are often hidden by transferring them to other banks and/or insured by state-owned insurance companies and simply passed to the State Recovery Agency.

In addition, The country's financial system is comprised of financial intermediaries and agents, such as insurance and finance companies which are also centered around state-owned or quasi publicly owned financial institutions. The insuring capacity of the insurance companies is still very limited and, in reality, these companies actually perform more brokerage functions than full-fledged insurance functions. As most of the risks are exported, retention business is low. This limits the role of insurance industry as a provider of long-term investment funds for financing economic development.

The country's money and capital markets are in the early stage of development. This is true for both the primary market, where new capital is raised, and secondary markets, where existing securities are traded. Information disclosure on behalf of the listed companies and candidates for listing could be improved if public confidence is to be strengthened. Remaining distortions, such as government "calls" for redistribution of wealth and income through capital markets and earmarking in the use of profits of state-owned enterprises, need to be eliminated. Because of these distortions, capital markets have only begun to provide financing substitute for bank credits.

Growing Market Competition

In general, the financial sector reforms begun in the early 1980s have strengthened market competition, and have sharply reduced distorted government controls, modified financial market segmentation, and induced formation of financial conglomerations owned by the private sector. The more competitive markets have allowed money and capital market instruments to compete with credit instruments and find areas of comparative advantage in financial systems to the benefit of savers, lenders, borrowers, and investors. Despite deregulation, however, growth-retarding government interventions remain extensive in some sectors of the economy. In some segments, "capitalism" has not improved market competition. The state-owned enterprises have been relatively untouched by deregulation, and some sectors of the economy remain highly protected. In the insurance industry, some segments of the market remain reserved for state-owned and quasi state-owned enterprises.

Savings Mobilization and Resources Allocation

At this point of the evolution of the financial system, the roles of financial intermediaries to provide long-term investment for economic development and to improve allocation of resources in Indonesia are still limited. Banks provide facilities for savers. Since the oil boom, with rising incomes per capita and increasing corporate profitability, along with improvements in operations of the financial institutions, the role of domestic savings has increased steadily. The role of the financial system in improving resource allocation has been mixed. On the one hand, Indonesia's economic growth and its success in promoting non-oil exports has been impressive. On the other hand, the high number of banks' nonperforming loans indicates their poor performance.

Despite deregulation, the credit policy remains segmented and procyclical. In addition, banks are now competing with nonfinancial state-owned enterprises in providing credit to small- and medium-scale enterprises and cooperatives. These firms also received transfer of shares from domestic conglomerates.

Risks

Risks arising from maturity transformation are increasing along with the reduction of the credit programs refinanced by liquidity credits from Bank Indonesia. Part of the commercial risks of the state-owned banks is assumed by the state-owned insurance companies and by the Treasury through the State Recovery Agency.

Weak Financial Institutions

In the short run, deregulation has created problems for Indonesia's financial system. It takes time for portfolio adjustment to accumulate information and human capital, to change incentive and promotion systems, and to improve market infrastructure.

Inadequate Capital

The need for auto-financing is rising because of the enforcement of the new risk-based capital requirement in the banking industry and the minimum absolute capital requirements in other branches of the financial industry. Inflow of offshore borrowings has begun to dry up, partly due to the re-imposition of ceilings on external borrowings of the public sector. The successive collapse of Indonesian banks and nonbank firms in 1988–1992 reduced their access to international capital markets. Concurrently the authorities are enforcing stricter prudential measures and solvency regulations. The quality of assets of the financial industry is deteriorating as shown by rising bad debts of the banking industry. This and rising market competition reduce profitability. Lower asset quality and declining profitability of the financial industry leads to a deterioration of the industry's balance sheets.

Hence, there is a need to strengthen the capital base of the domestic private and state-owned banks. This can be done by allowing banks to count provisions for bad loans against tax liabilities, merger, spin-off division of conglomerates and through temporary credit from the central banks. However, before injecting new capital into the state-owned banks, they must be reorganized, restructured, and consolidated, and their links to government bureaucracy cut.

Weak Human Resources and System of Operations

Market-oriented management, a sound system of operations and improvement in technical competence are needed along with new capital to turn the financial system around. The high ratio of bad debt of the banking industry may be related to widespread corruption and fraud as well as to technical incompetence of their management and staff.

Inadequate Market Infrastructure

Increasing competition and reducing direct government control of the decision-making processes are not enough to make the markets more effective. Improvements in market infrastructure, particularly the accounting, legal, and regulatory framework, including supervisory control, are also required. Improving this "software" of market institutions is a difficult and time-consuming aspect of deregulation.

Ineffective Monetary Policy

Finally, the fragility of the financial institutions and nonbank firms and the reduced access of Indonesian companies to international financial markets, at least during the transition period, have combined to make the country's monetary policy less effective. Interest rates have "overshot" both the international rate as well as the "long-term" equilibrium level.

Acknowledgments

The author gratefully acknowledges the contributions of Professors David C. Cole and Lawrence J. White, and other participants of the Conference on Financial Sector Development in Asia held at Asian Development Bank Headquarters in Manila on 1-3 September 1993, for improvements on the earlier versions of this report. Professor Betty F. Slade deserves special thanks for her contributions to the data, analysis, and style of this report. Mr. Edwin Syahruzad prepared the tables and graphs and the officials at the Ministry of Finance of the Republic Indonesia were helpful in supplying information on regulations. Last but not least, the author also gratefully acknowledges Drs. J. Malcolm Dowling and Shahid N. Zahid, both of ADB, for their assistance in this work.

Bibliography

Akerlof, G. 1970. "The Market for Lemons: Qualitative Uncertainty and the Market Mechanism." In *Quarterly Journal of Economics*. 84:288-300.

Bank Indonesia. (various issues). *Annual Report*. Jakarta: Bank Indonesia.

_____. *Indonesian Financial Statistics*. Jakarta: Bank Indonesia.

PT Barito Pacific Timber Company. 1993. *Prospektus*. Jakarta. 20 August.

Benston, George J. 1972. "Economies of Scale of Financial Institutions." In *Journal of Money, Credit and Banking*. VI(2).

Benston, George J., Robert Eisenbeis, Paul M. Horvits, Edward J. Kane, and George Kaufman. 1986. *Perspectives on Safe and Sound Banking: Past, Present, and Future*. Cambridge: MIT Press.

Binhadi, and Paul Meek. 1992. "Implementing Monetary Policy." In *The Oil Boom and After: Indonesian Economic Policy and Performance in the Soeharto Era*, edited by Anne Booth. Singapore: Oxford University Press.

Caprio, Gerard Jr. 1992. "Banking of Financial Reform? A Case of Sensitive Dependence on Initial Conditions." A paper presented at the Conference on the Impact of Financial Reform. Washington DC: World Bank. Mimeo.

Caprio Jr., Gerard, Izak Attiyas, and James Hanson. 1992. "Policy Issues in Reforming Financial: Lessons and Strategies." A paper presented at the Conference on the Impact of Financial Reform. Washington DC: World Bank. Mimeo.

Central Bureau of Statistics. *Population Projection by Province, 1985–1995.* Jakarta.

Chant, John, and Mari Pangestu. 1992. "An Assessment of Financial Reform in Indonesia, 1983-90." A paper prepared for the Conference on the Impact of Financial Reform. Washington, DC: World Bank. Mimeo.

Cole, David C. 1995. "Financial Sector Development in Southeast Asia." In *Financial Sector Development in Asia* edited by Shahid N. Zahid. Hong Kong: Oxford University Press for Asian Development Bank.

Cole, David C., and Betty F. Slade. 1991. "Development of Money Markets in Indonesia." Development Discussion Paper No. 371, Cambridge, MA: Harvard University.

———. 1992. "Financial Development in Indonesia." In *The Oil Boom and After: Indonesian Economic Policy and Performance in the Soeharto Era*, edited by Anne Booth. Singapore: Oxford University Press.

———. 1993. "How Bank Lending Practices Influence Resource Allocation and Monetary Policy in Indonesia." Development Discussion Paper No. 444. April. Cambridge: Harvard Institute for International Development.

———. Forthcoming. "Indonesia Financial Development: A Different Sequencing?" in *Financial Regulation: Changing the Rules of the Game,*" edited by Dimitri Vittas. Washington, DC: World Bank.

Corbo, V., and J. de Melo. 1987. "Lessons From the Southern Cone Policy Reforms." In *Research Observer 2.* No. 2:111–142.

Cukierman, Alex, Steven B. Webb, and Leila M. Webster. 1992. "Measuring the Independence of Central Bank and Its Effect on Policy Outcomes." *World Bank Economic Review.* 6(3): 353–98.

Department of Finance. 1991. Directorate General of Financial Institutions. "Report on Insurance Industry in Indonesia." Jakarta.

The Economist. 1993. "One Mountain Conquered." 23 October, pp. 83–84.

Edwards, Sebastian. 1986. "The Order of the Current Account and Capital Account of the Balance of Payments." In *Economic Liberalization in Developing Countries,* edited by Armaene Choski and Demetris Papageorgiu. New York: Basic Blackwell.

Edwards, Sebastian, and Mohsin S. Khan. "Interest Rate Determination in Developing Countries: A Conceptual Framework." In *IMF Staff Papers.* 32(3):377–403.

Gillis, Malcolm. 1989. "Comprehensive Tax Reform: The Indonesian Experience, 1981–1988." *Tax Reform in Developing Countries,* edited by Malcolm Gillis. Durham: Duke University Press.

Giovannini, Alberto. 1993. *Finance and Development: Issues and Experience.* Cambridge, UK: Cambridge University Press.

Greenwald, B.J., J.E. Stiglitz, and A.M. Weiss. 1984. "Informational Imperfections and Macroeconomic Fluctuations." *American Economic Review.* Papers and Proceedings. 74:194–99.

Hanna, Donald P. 1992. "Indonesia Experience with Financial Sector Reform." A paper prepared for the ECLAC/UNU-WIDER/UNCTAD Seminar on Savings and Financial Policy in Developing Countries. Santiago, Chile, October 5–6.

Harris, John, Fabio Schiantarelli, and Miranda Siregar. 1992. "Financial and Investment Behavior of the Indonesian Manufacturing Sector and the Effect of Liberalization: Evidence from Panel Data, 1981–1988." A paper prepared for the Conference on the Impact of Financial Reform. 2–3 April. Washington, DC: World Bank.

International Monetary Fund (IMF). (various issues). *International Financial Statistics.* Washington, DC: World Bank.

Larrain, M. 1989. "How the 1981–83 Chilean Banking Crises was Handled." Policy, Planning, and Research, Working Paper No. 300. Washington, DC: World Bank. Mimeo.

McKinnon, Ronald I. 1991. *The Order of Economic Liberalization: Financial Control in the Transition to a Market Economy.* Baltimore: Johns Hopkins University Press.

McLeod, Ross. 1993. "Labor: Sharing the Benefits of Growth?" A paper presented at the Conference on Indonesia Update 1993. Indonesia Project-Research School of Pacific Studies. Australian National University, Canberra.

Ministry of Finance. 1991. Directorate General of Financial Institutions, *Statistical Data of Finance Companies, 1990–1991.* Jakarta.

──────. (various issues). Directorate General of Financial Institutions, *Statistical Data of Finance Companies.* Jakarta.

Nasution, Anwar. 1983. *Financial Institutions and Policies in Indonesia.* Singapore: ISEAS.

──────. 1987. "Structural Adjustment for Sustainable Growth: The Case of Indonesia in the 1980s." A paper presented at The Conference on Structural Adjustment for Sustainable Growth in Asian Countries. Economic Planning Agency of Japan. Tokyo. 7–8 November. Mimeo.

──────. 1989. "Managing External Balances Under Global Economic Adjustment: Case of Indonesia, 1983–1988." A paper prepared for the Conference on the Future of Asia-Pacific Economies (FAPE III): Emerging Role of Asian NIEs and ASEAN, APDC-NESDB-TDRI. Bangkok. 8–10 November. Mimeo.

──────. 1990. "Recent Deregulation of the Banking Sector in Indonesia." In *Privatization and Deregulation in ASEAN and the EC: Making Markets More Effective*, edited by Jaques Pelkmans and Nobert Wagner. Singapore: ISEAS and European Institute of Public Administration.

──────. 1991. "Open Regionalism: The Case of Asean Free Trade Area." A paper presented at a seminar at the Department of Area Studies, University of California, Berkeley, 19 September. Mimeo.

──────. 1991. "The Adjustment Program in the Indonesian Economy Since the 1990s." In *Indonesia Assessment 1991*, edited Hal Hill. Canberra: Department of Political and Social Change, Research School of Pacific Studies, Australian National University.

──────. 1992. "The Years of Living Dangerously: The Impacts of Financial Sector Policy Reforms and Increasing Private Sector External Indebtedness in Indonesia, 1983–92." A paper presented at the Third Convention of the East Asian Economic Association (EAEA). Seoul, 20–21 August. Mimeo.

_____. 1993a. "Reforms of The Financial Sector in Indonesia, 1983–1991." PITO Business Environment in Asean. No. 7. Honolulu: East-West Center. May. Mimeo.

_____. 1993b. "Recent Developments and Future Prospects of Indonesian Economy." A paper presented at the Conference on Southeast Asia: Economic Experiences and Prospects. Organized by the American Council on Asia and Pacific Affairs, Inc. Washington DC. 29–30 November. Mimeo.

OECD. 1989. *Competition in Banking*. Paris: OECD.

_____. 1992. *Bank Under Stress*. Paris: OECD.

Portes, R., and A.K. Swoboda. eds. 1987. *Threat to International Financial Stability*. Cambridge, UK: Cambridge University Press.

Ravallion, Martin, and Monica Huppi. 1991. "Measuring Changes in Poverty: A Methodological Case Study of Indonesia During An Adjustment Period. *The World Bank Economic Review*. 5 No. 1 (January): 57–82.

Reisen, Helmut, and Bernard Fisher. eds. 1993. *Financial Opening—Policy Issues and Experiences in Developing Countries*. Paris: OECD.

Sheng, Andrew. 1989. "Bank Restructuring in Malaysia, 1985–88." Policy, Planning, and Research Working Papers, WPS 54. Washington, DC: World Bank. November. Mimeo.

Sundararajan, V. 1987. "The Debt-Equity Ratio of Firms and the Effectiveness of Interest Rate Policy: Analysis with a Dynamic Model of Saving, Investment, and Growth in Korea." In *IMF-Staff Papers*. June. 34(2):260–310.

Stiglitz, Joseph E., and Andrew Weiss. June 1981. "Credit Rationing in Markets with Imperfect Information." *American Economic Review*. 71:393–410.

_____. 1993. "The Role of the State in Financial Markets." A paper presented at the World Bank Annual Conference on Development Economics. May. Washington, DC: World Bank. Mimeo.

Warr, Peter G. 1992. "Exchange Rate Policy, Petroleum Prices, and the Balance of Payments." In *The Oil Boom and After: Indonesian*

Economic Policy and Performance in the Soeharto Era, edited by Anne Booth. Singapore: Oxford University Press.

Woo, W.T., and Anwar Nasution. 1989. "Indonesian Economic Policies and Their Relation to External Debt Management." In *Developing Country Debt and Economic Performance*. Vol. 3. Edited by J.D. Sachs and Susan M. Collins. Chicago: Chicago University Press.

Woo, W.T., Bruce Glassburner, and Anwar Nasution. 1991. *Macroeconomic Policies, Crises and Long-run Growth: The Case of Indonesia, 1965–1990*. Washington, DC: World Bank.

World Bank. 1990. *Adjustment Lending Policies For Sustainable Growth*. Country Economics Department of the World Bank. Washington DC: World Bank.

————. 1990 and various other issues. *World Development Report*. Washington, DC: World Bank.

————. 1992. "Indonesia—Growth, Infrastructure and Human Resources." Report No. 10470–IND. Country Department III–East Asia and Pacific Regional Office of the World Bank. 26 May.

————. 1993. *The East Asian Miracle: Economic Growth and Public Policy*. New York: Oxford University Press.

Zahid, Shahid N. 1995. *Financial Sector Development in Asia*. Hong Kong: Oxford University Press for Asian Development Bank.

CHAPTER FOUR
A Study of Financial Sector Policies: The Philippine Case

Mario B. Lamberte and Gilberto M. Llanto

OVERVIEW OF THE FINANCIAL SYSTEM

The domestic financial system of the Philippines consists of two major subsystems: the banking system and nonbank financial intermediaries. Institutions belonging to the banking system are authorized to accept traditional deposits (i.e., demand, savings, and time deposits), while institutions classified as nonbank financial intermediaries are not.

The banking system is composed of the commercial banks, thrift banks, rural banks, and specialized government banks. Nonbank financial intermediaries include insurance companies, investment institutions, fund managers, nonbank thrift institutions, and other financial intermediaries.

Table 4.1 shows the assets of the various types of financial institutions for the period 1986–1991. The total nominal assets of the financial sector, including those of the Central Bank of the Philippines, doubled during the period. The relative size of the financial system—measured here as the ratio of the total assets of the financial sector (including the Central Bank) to GDP—declined in 1987 and 1988, but recovered in subsequent years. As of 1991, total assets comprised 92 per cent of GNP, still slightly lower than the level attained in 1986.

The dominance of the banking system in the financial sector is clearly visible. Almost four fifths of the total assets of the sector belong to the banking system, and there is no recent indication that this share is declining. The dominance of the banking system in the financial system will likely remain in the medium term especially with the relaxation of bank entry and branching regulations recently adopted by the Central Bank of the Philippines and the relatively successful rehabilitation of several failed banks.

Within the banking system, commercial banks are the largest group, with assets comprising about 90 per cent of the total assets of the system. The private commercial banking system has been mainly in the hands of the private sector. The relative size of government-owned commercial banks appears to be significant and even larger than that of foreign banks.

Thrift banks, which include savings and mortgage banks, private development banks, and savings and loan associations, are the second largest group of banks. Though their assets comprise only between 6 per cent and 7 per cent of the total assets of the banking system, this does not mean that thrift banks are not dynamic. In fact, two fast-growing thrift banks recently obtained commercial bank status.

Table 4.1 Assets of the Domestic Financial System, 1986–1991
(Pesos billion)

	1986	1987	1988	1989	1990	1991
Central Bank	156.959	139.326	140.169	153.061	164.032	226.616
Banking System	289.000	320.278	360.100	465.357	579.783	682.949
Commercial banks	252.257	278.439	312.349	409.182	510.349	591.339
Private	164.400	179.400	224.600	296.134	365.700	406.637
Government	50.757	49.939	51.849	71.219	88.949	120.880
Foreign	37.100	49.100	35.900	41.829	55.700	63.822
Thrift banks	17.600	19.500	24.900	32.204	37.294	47.044
Savings and mortgage banks	8.100	10.600	14.200	19.601	21.721	29.628
Private development banks	5.600	5.400	6.700	8.346	11.180	12.166
Stock savings and loan associations	3.900	3.500	4.000	4.257	4.393	5.250
Rural banks	9.100	9.700	10.700	12.160	13.459	15.488
Specialized government banks	10.043	12.639	12.151	11.811	18.681	29.078
Nonbank financial intermediaries	118.497	130.137	147.986	172.009	172.703	245.836
Insurance companies	77.490	90.100	106.100	125.630	121.470	181.700
Government	57.190	64.728	76.426	89.420	76.320	130.320
Private	20.300	25.372	29.674	36.210	45.150	51.380
Investment institutions	23.300	20.800	21.400	20.985	21.621	25.627
Investment houses	7.500	9.000	8.400	6.749	6.053	6.972
Finance companies	5.600	7.000	7.400	4.439	4.611	5.942
Investment companies	10.200	4.800	5.600	9.797	10.957	12.713
Trust Operations (Fund managers)	1.300	1.600	1.800	2.590	2.869	3.325

Other financial intermediaries	16.407	17.637	18.686	22.804	26.743	35.184
Securities dealers/brokers	0.945	2.072	1.706	2.725	2.437	2.924
Pawnshops	1.011	1.318	1.666	2.170	2.664	3.451
Lending investors	0.237	0.715	0.636	1.015	1.303	2.034
Venture capital corp.	0.126	0.129	0.100	0.090	0.102	0.101
Specialized government nonbanks	14.088	13.403	14.578	16.804	20.237	26.674
Nonbank thrift institutions	1.194	1.830	2.620	3.926	4.771	5.318
Mutual building and loan association	0.016	0.014	0.014	0.016	0.018	0.019
Nonstock savings and loan associations	1.178	1.816	2.606	3.910	4.753	5.299
Total (excluding Central Bank)	408.691	452.245	510.706	641.292	757.257	934.103
Per cent of GNP	68.541	67.416	64.482	70.315	69.951	73.780
Per cent of GDP	67.121	66.237	63.904	69.296	70.296	74.847
Total (including Central Bank)	565.650	591.571	650.875	794.353	921.289	1,160.719
Per cent of GNP	94.864	88.185	82.180	87.098	85.103	91.679
Per cent of GDP	92.899	86.644	81.443	85.835	85.523	93.006
Memo item:						
GNP (nominal)	596.276	670.826	792.012	912.027	1,082.557	1,266.070
GDP (nominal)	608.887	682.764	799.182	925.444	1,077.237	1,248.011

Note: Assets of the Central Bank exclude three nonperforming assets: exchange stabilization adjustment account, monetary adjustment account, and revaluation of international reserves.

Source: Central Bank of the Philippines (1991).

The three specialized government banks are the third largest group of banks in terms of assets. The smallest group are the rural banks, a great majority of which are unit banks. Because of the failure of many rural banks, the relative size of the rural banking system has declined in the recent past.

MICROECONOMIC EFFICIENCY ISSUES OF THE FINANCIAL SYSTEM

The financial sector is the vehicle which mobilizes and transfers the economy's surplus to net users. If this vehicle is efficient and healthy, then the economy gains: the lack of confidence between net savers and net users is eliminated in the financial markets and resources flow from lower-return uses to higher-return uses. The coincidence of net savers is brought about by financial intermediaries that (i) create financial assets and issue liabilities, and (ii) serve to facilitate direct transfers from surplus to deficit units. A typical example of the first type is a commercial bank which provides loans to firms and individuals (i.e., creates assets) from the deposit liabilities they make (i.e, issues its own liabilities). Other institutions in the financial markets either facilitate such direct transfer of resources to net users or create substitute instruments with essentially the same objective of moving financial resources to their most efficient alternative use. Credit information and rating agencies, stock brokers, securities underwriters, finance companies, leasing companies, and pawnshops arise to enhance and complement the operations of the first type of financial intermediaries.

Financial transactions, however, do not occur in a frictionless financial market where transaction costs are either negligible or nil and information is perfect. The reality is that in most developing countries, huge transaction costs, risk, uncertainty, and severe information problems affect the efficiency of these transactions. Information asymmetry in the financial market gives rise to screening, incentive, and enforcement issues (Llanto 1989).

The financial market in the Philippines is highly segmented. The largest and single most important traditional lender is the commercial bank whose immediate clients are the larger and more established commercial, industrial, and service enterprises. The commercial banks have always financed trade and commerce and large manufacturing establishments. Priority is given to the profitable and established enterprises to ensure lowered transaction costs and reduced risks of loan default. Other financial intermediaries have developed their own clientele base by creating their own "market niches" and by developing credit programs which employ various screening devices in the hopes of attracting borrowers who will make good the loan contract. This sorting behavior is a response to information asymmetry and the transaction costs surrounding heterogenous borrowers.[1] Those

---

[1] See Lamberte (1992) for a review of policy-based lending programs in the Philippines.

borrowers not accommodated by the larger formal financial institutions are served by the smaller financial intermediaries such as finance companies, credit unions, lending investors, and pawnshops. Thus, in the segmented financial market, there exist particular types of institutions to serve: specific clientele (industrial versus agricultural); size of loan (depending on type of project and capacity to repay the loan); needs of clientele (short versus medium to long term); and purpose of loan (consumption versus capital formation).

CREDIT RATIONING IN A LIBERALIZED FINANCIAL MARKET

The Philippines embarked on financial liberalization in the 1980s. By the late 1980s, many restrictions imposed on financial intermediaries had been lifted and monetary and credit policies had become market oriented. Credit subsidies in agricultural credit programs were eliminated and the banks were encouraged to mobilize savings and lend out of their loanable funds instead of depending on the Government for such funds.

The naive expectation was that in a deregulated environment, access to credit would be easier for all types of borrowers. Recent studies (Lamberte 1987; Lapar 1988; Llanto and Magno 1992) have shown that microeconomic considerations—such as risk and uncertainty, information problems, and transaction costs—affect a banks' willingness to lend; their rational response is to employ tight screening and credit rationing. Lapar, for instance, showed that banks go through three stages in lending: the screening stage, the acceptance/rejection rationing stage, and the quantity rationing stage. During the screening stage, the bank manager determines the eligibility of the prospective borrower in an interview. The loan application is given to this borrower upon success of the screening stage. The loan application is then evaluated to determine the creditworthiness of the borrower. The bank makes use of the information submitted by the borrower as well as by facts generated from a credit investigation. Once a calculated judgment on creditworthiness has been made, the bank decides how much loan will be provided. This is the quantity rationing stage in which the bank strives to determine the optimal combination of loan size, interest rate, and maturity.

Field survey data from a sample of 65 banks and 344 bank clients from private development banks and rural banks show that:

— activity one—banks engage in initial screening to determine initial creditworthiness of loan applicants—was practiced more often by commercial banks and private development banks;

— activity two—banks accept or reject loan applications—was higher among private development banks and commercial banks than rural banks; and

— activity three—loans granted are much lower than what was applied for—was more commonly practiced by rural banks than either commercial banks or private development banks.

This suggests that in the financial market there is a complex interaction between the microeconomic realities and the macroeconomic and policy environment. While a liberal banking policy environment is a necessary condition for efficient transactions, it is not a sufficient condition. It is true that a great deal of the microeconomic behavior is motivated by the macroeconomic and policy environment surrounding financial markets. Financial intermediaries are among the most highly regulated institutions in the economy. Capital and lending portfolios are heavily regulated. Capitalization of banks and other financial intermediaries is not uniform. The single-borrower limit constrains lending to a single individual up to a certain maximum loan size. Mandatory loan allocations[2] seem to be increasing instead of decreasing, as financial liberalization would normally require. The distortions created by government intervention and the microeconomic inefficiencies hinder efficient contracting and transformation of surplus resources into their most profitable alternative use.

FINANCIAL INTERMEDIARIES

Banking System

There are four major types of banks in the Philippines: commercial banks; thrift banks; rural banks; and specialized government banks. Commercial banks are further divided into two groups: universal banks or commercial banks with expanded commercial banking functions and ordinary commercial banks.

The 1980 financial reforms reduced differentiation among the different bank categories as far as function is concerned. For instance, prior to the reforms, the term "commercial bank" applied to a financial institution that accepted demand deposits subject to withdrawal by check. This definition is no longer completely valid in the Philippine context since other types of banks may now be authorized by the Central Bank to accept demand deposits provided they satisfy certain prerequisites. In addition, enforced specialization has been eliminated. For instance, rural banks, which before the reforms were allowed to lend only to small-scale farmers, may now lend to medium-size farm and nonfarm enterprises.

---

[2] This refers to regulations that require banks to allocate a certain portion of their total loan portfolio to a particular sector. This is different from special credit programs in which the government lends directly or indirectly (through banks) to target groups or sectors.

The creation of commercial banks with expanded functions (or universal banks) was one of the important aspects of the 1980 financial reforms. The intent was to create "one-stop banking facilities" which offered clients a broad range of financial services. Universal banks, therefore, have been authorized to perform some functions, such as securities underwriting and syndication activities, which before were reserved for investment houses. In addition, universal banks are also now allowed to have direct equity investments in allied and nonallied undertakings, with some restrictions, to ensure the flow of long-term funds into the economy.

Among the different types of banks, the commercial banking system appears to be the largest group. As of December 1991, it accounted for 82 per cent of total bank assets, 80 per cent of total loans, and 86 per cent of total deposits of the banking system. Further, commercial banks have the most extensive branching network in the country. There are 30 commercial banks with 1,892 branches, 101 thrift banks with 562 branches, 784 rural banks with 279 branches, and 3 specialized government banks with 139 branches. About 32 per cent of the total number of banking offices are located in the National Capital Region (Metro Manila) and 25 per cent in the two adjoining regions, i.e., Regions III (Central Luzon) and IV (Southern Tagalog). All commercial banks have their headquarters in Metro Manila.

Rural banks are basically privately-owned banks. However, to encourage the private sector to invest in rural banks, the Government instituted a policy of matching every peso infused by private investors into a rural bank, with the Government's participation of nonvoting preferred shares. The same policy was instituted for private development banks, although its equity exposure in each private development bank was much less than its exposure in the rural banks. In the 1980s, the Government ended this policy, in keeping with its newer policy of withdrawing subsidies to private banks.

Few thrift banks have accepted foreign equity participation, with the exception of Planters Bank and the Northern Mindanao Development Bank.

The Philippine National Bank (PNB) was, prior to 1989, a wholly government-owned institution. In 1989, the Government began the process of privatizing PNB, and now 43 per cent of the total outstanding shares of PNB are owned by the private sector. The Land Bank of the Philippines and the Development Bank of the Philippines are wholly owned by the Government, and will likely remain that way because of their special role: the Land Bank is involved in the implementation of the Government's agrarian reform program, and the Development Bank provides long-term funds to the industrial sector.

After the establishment of the Central Bank in 1949, only four branches of foreign banks were allowed to continue their operations. These four are authorized to operate as an ordinary commercial bank and may establish branches in the country with prior approval by the Central Bank. However, new branches may not accept demand deposits. Equity participation of foreign entities in domestic banks up to 40 per cent (30 per cent voting; 10 per cent nonvoting) has been allowed by the Monetary Board since 1972. Several commercial banks have received substantial foreign capital infusions, and more recently, a number of foreign entities (mostly foreign banks) have bought shares of several domestic banks via the "debt-equity conversion scheme" that was introduced in 1986 to help alleviate the country's foreign debt burden. To date, 10 of the 26 domestic commercial banks have foreign equity participation.

In the early 1980s, the Philippine financial system experienced a severe liquidity crisis. Several banks became insolvent. To prevent a crisis of confidence in the entire banking system, the Government intervened by infusing capital into the ailing banks, which led to government ownership of six of the ailing banks. All, except one, have since been sold to the private sector.

Sources of Funds

Banks have four major sources of funds, namely, deposits, borrowings, other liabilities, and capital and retained earnings. Table 4.2 shows the volume and relative importance of the sources of funds of the different bank categories. With the exception of the specialized government banks, banks have relied mainly on mobilized deposits to finance their lending and investment operations. Thrift banks depend most on mobilized deposits, which account for at least 70 per cent of their total resources. Commercial and ru-ral banks' reliance on deposits has increased significantly from 60 per cent to 65 per cent and from 41 per cent to 55 per cent, respectively, during the period 1986–1991. For specialized government banks, deposits contributed only 28 per cent on the average to their total resources and much of that came from the national government and government-owned corporations.[3]

Saving deposits are the most important source of deposits for all bank categories, though the degree of dependence on this type of deposit varies across bank categories. Commercial banks are the least dependent on saving deposits among the different bank categories for the simple reason that they offer a wider array of attractive deposit instruments than other banks. The recent competition in the saving deposit market has become more intense across different bank categories, and even among banks within the same category, especially with the introduction of automatic teller machines. Commercial banks have a decided advantage in this area, as most have automatic teller machines already and have long experience in managing branches.

---

[3] All government agencies and government-owned corporations are required to bank only with government financial institutions.

Table 4.2 Sources of Funds of the Banking System
(Pesos million)

Sources	1986	1987	1988	1989	1990	1991
Commercial banks						
Deposits	141,158	153,336	195,898	253,368	312,979	364,158
	(59.5)	(62.0)	(65.4)	(64.2)	(62.5)	(65.0)
Borrowings	33,772	23,911	27,038	39,982	51,675	56,913
	(14.2)	(9.7)	(9.0)	(10.1)	(10.3)	(10.2)
Other liabilities	32,473	39,473	40,317	58,348	82,196	70,189
	(13.7)	(16.0)	(13.5)	(14.8)	(16.4)	(12.5)
Capital accounts	29,652	30,516	36,365	42,668	54,290	69,048
	(12.5)	(12.3)	(12.1)	(10.8)	(10.8)	(12.3)
Total	237,055	247,236	299,618	394,366	501,140	560,308
	(100.0)	(100.0)	(100.0)	(100.0)	(100.0)	(100.0)
Thrift banks						
Deposits	12,773	14,822	18,635	24,097	26,839	33,658
	(72.8)	(75.9)	(74.9)	(74.8)	(72.0)	(71.5)
Borrowings	1,831	1,482	1,882	2,511	3,120	4,377
	(10.4)	(7.6)	(7.6)	(7.8)	(8.4)	(9.3)
Other liabilities	1,279	1,275	2,041	2,493	2,889	3,631
	(7.3)	(6.5)	(8.2)	(7.7)	(7.7)	(7.7)
Capital accounts	1,664	1,943	2,319	3,103	4,446	5,378
	(9.5)	(10.0)	(9.3)	(9.6)	(11.9)	(11.4)
Total	17,547	19,522	24,877	32,204	37,294	47,044
	(100.0)	(100.0)	(100.0)	(100.0)	(100.0)	(100.0)
Rural banks						
Deposits	3,767	4,516	5,218	6,200	7,010	8,491
	(41.4)	(48.3)	(48.8)	(51.0)	(52.1)	(54.8)
Borrowings	3,082	2,495	2,682	2,495	2,525	2,553
	(33.9)	(26.7)	(25.1)	(20.5)	(18.8)	(16.5)
Other liabilities	802	773	1,080	1,526	1,644	1,770
	(8.8)	(8.3)	(10.1)	(12.5)	(12.2)	(11.4)
Capital accounts	1,452	1,575	1,713	1,939	2,280	2,674
	(16.0)	(16.8)	(16.0)	(15.9)	(16.9)	(17.3)
Total	9,103	9,359	10,693	12,160	13,459	15,488
	(100.0)	(100.0)	(100.0)	(100.0)	(100.0)	(100.0)
Specialized government banks						
Deposits	8,577	5,307	6,847	6,182	12,189	19,017
	(33.3)	(21.9)	(27.2)	(23.1)	(30.4)	(32.5)
Borrowings	2,866	2,202	1,224	1,213	3,817	8,888
	(11.1)	(9.1)	(4.9)	(4.5)	(9.5)	(15.2)
Other liabilities	7,904	9,418	7,877	8,395	10,909	16,074
	(30.7)	(38.9)	(31.3)	(31.4)	(27.2)	(27.5)
Capital accounts	6,405	7,263	9,241	10,944	13,215	14,551
	(24.9)	(30.0)	(36.7)	(40.9)	(32.9)	(24.9)
Total	25,752	24,190	25,189	26,734	40,130	58,530
	(100.0)	(100.0)	(100.0)	(100.0)	(100.0)	(100.0)

Note: Figures in parentheses are percentage of total.
Source: Central Bank of the Philippines (1991).

Thrift and rural banks have responded to this competition by offering higher rates on saving deposits and requiring lower deposit balances than commercial banks. Some commercial banks acquired thrift and rural banks to make their presence felt in the retail market as well.

Time deposits are the next most important type of deposits for banks. Such deposits have the same relative importance as a source of funds to commercial, thrift, and rural banks.

Demand deposits are the sole domain of commercial banks. Prior to the 1980 financial reforms, only commercial banks were allowed to accept demand deposits, and because of their long experience in offering demand deposits, commercial banks have retained their advantage over other types of banks in this instrument.

The rural and specialized government banks' low level of deposit mobilization in the 1980s was partially compensated for by subsidized borrowings, mainly from the Central Bank. However, the market orientation of the rediscounting policy of the Central Bank toward the second half of the 1980s has prompted both rural and specialized banks to reduce their dependence on the Central Bank and to start mobilizing deposits.

Structure of Interest Rates on Deposits

The Philippines had, until recently, a regulation prohibiting banks from giving interest on demand deposits.[4] While in effect, banks could circumvent this regulation by offering clients, especially large depositors, an automatic transfer service from interest bearing deposit accounts to demand deposits.

Data on interest rates on savings and time deposits are collected regularly by the Central Bank from a sample of banks. Prior to 1989, these banks reported only the minimum interest rates on regular savings deposits. Nominal interest has ranged between 4 and 5 per cent since 1987, well below the inflation rates for most of the years. However, these interest rates are not reflective of the interest rates on total saving deposits since banks usually pay higher interest rates on saving deposit accounts with higher average daily balances and on special saving accounts. The interest rates approximate those on short-term time deposits. The weighted average interest rate on total saving deposits is higher than that on regular saving deposits alone: these stood at close to 6 per cent in 1989 and were greater than 10 per cent in the last three years. However, the weighted nominal interest rates on total saving deposits has been below the rate of inflation with the exception of 1992.

Time deposits can be as short as 30 days and as long as five years. Unlike the interest rates on saving deposits, interest rates on time deposits were

---

[4] This regulation was lifted recently.

volatile during the period 1986–1992. For short-term time deposits (i.e., one year and below), the weighted interest rates ranged between 8.6 per cent and 20 per cent: except in 1991, they were above inflation rates during the indicated period. For long-term time deposits (i.e., above one year), the weighted interest rates ranged between 10.2 per cent and 21.2 per cent: except in 1991, they were above inflation rates.

Uses of Funds

The major uses of funds of the various bank categories are: loans net of valuation reserves; investments; cash; and others. Since 1986, banks in the Philippines have been generally liquid, particularly commercial banks where the loan portfolio has been less than 50 per cent of their total assets. This high liquidity during the period 1986–1991 can be attributed to several factors. First, there was during this period a general feeling of uncertainty about the country's political and economic situation. Between 1986 and 1989, seven coup attempts were staged against the Government; and in 1990, the economy suffered a balance-of-payments crisis. Second, short-term, high-yielding treasury bills had provided banks with alternative investment opportunities, and a significant amount of banks' funds went into this investment instrument. Even specialized government banks, which were under pressure to become self-reliant, invested heavily in treasury bills. Third, the reserve requirement on all deposit liabilities had been raised to 25 per cent in 1990 and had been maintained until December 1992.

About 70 per cent of the commercial banks' loan portfolio are short-term instruments (i.e., either demand or short-term loans with maturity of one year or less). The objective of the 1980 financial reforms—to encourage commercial banks to lend long-term—seems not to have been achieved. Thrift banks, on the other hand, have a more balanced loan portfolio comprised of from 35 to 54 per cent medium-term (with maturity of more than one year but less than three years) and long-term loans (with maturity of more than three years). Of course, banks under the thrift bank category have had a long history of providing medium- and long-term loans. Specifically, private development banks have been engaged in the provision of long-term funds to small- and medium-scale industries, while savings and mortgage banks and stocks savings and loan associations have been more active in the home mortgage market.

Structure of Interest Rates on Loans

Data on the structure of interest rates on loans are available only for secured loans. These are obtained by the Central Bank from a sample of commercial banks. From this data it can be seen that nominal interest rates fluctuated during the period 1986–1991. The weighted interest rate on short-term loans peaked in 1990 at 24.2 per cent, and medium- and long-term loans, in 1991 at 24.3 per cent. The rise in the nominal interest rates in 1988 and 1989 was due mainly to the strong recovery of the economy. The continuing rise in interest rates in 1990 could be attributed to inflationary expectation as the economy underwent another crisis.

As expected, the interest rate on short-term loans was below that on medium- and long-term loans except in 1990. However, it had been well above the inflation rate in all years during the period 1986–1991.

Government Policies and Regulations

Interest Rate Policy. Though the anti-usury law was abolished *de facto* in the early 1970s, the Central Bank administratively set all interest rates until 1981. Interest rate liberalization was accomplished in several stages: in 1981, interest rate ceilings on all types of deposits and loans, except on short-term loans, were lifted. The interest rate on short-term loans was subsequently lifted in 1983.

Prior to 1985, the rediscount window was used by the Central Bank to direct the flow of credit to priority sectors by giving these sectors preferential rediscount rates, which could be as low as 1 per cent, and rediscounting value, which could be as high as 100 per cent. This policy was changed in 1985 when the Central Bank began setting one rediscounting value equivalent to 80 per cent of the value of the original loan and one rediscount rate for all eligible papers aligned with the market rate. The basis used now for determining the rediscount rate is the 90-day Manila Reference Rate (MRR90).[5] The rediscount rate is re-evaluated and, if necessary, adjusted every quarter to reflect the prevailing cost of funds.

Government's Direct Participation in the Banking System. Toward the second half of the 1980s, the Government adopted a policy of reducing its direct participation in the banking system. Except for one commercial bank, all banks acquired by the Government in the early 1980s have already been privatized.[6] The PNB, which is engaged only in commercial banking, is now partly owned by the private sector. Plans to fully transfer ownership of this bank to the private sector are underway.[7]

---

[5] The MRR90 is based on the weighted average of the interest rates on promissory notes and time deposits with a 90-day maturity.

[6] This bank cannot be privatized pending resolution of the legal disputes among previous owners.

[7] There is a regulation in the Philippines that only government-owned banks can become depository institutions of the national government: PNB is one of these banks. Hence, if PNB's ownership were to be transferred to the private sector, the national government would be obliged to transfer its deposits with PNB to other government-owned banks. Since the national government has at the moment huge deposits with PNB, a gradual privatization of PNB is necessary to avoid a run on the bank.

Because they perform special functions, the Land Bank of the Philippines and the Development Bank of the Philippines remain in the hands of the Government. They are the main conduits of special credit programs for the agriculture and industrial sector, mostly funded by multilateral agencies.[8] Both banks have reduced their retail lending activities substantially, except in areas not adequately served by the private sector, and have instead concentrated on wholesale lending so as not to compete with private banks. With the newly opened rediscounting facility for rural banks, the Land Bank has become a *de facto* apex bank for the agriculture sector. The Development Bank is far from this status because of the very active participation of commercial and thrift banks in industrial lending.

Although there are only three government-owned banks, their presence in the banking system is highly visible. As of December 1991, their combined assets comprised nearly one fifth of the total assets of the banking system. The PNB is the largest commercial bank in the country with assets of P92 billion as of December 1991. Despite their size, the government-owned banks have not been used by the Government to influence interest rates in the last six years. Instead, the banks follow the market rates of interest.

Ownership Restrictions. Foreign equity participation is limited to 30 per cent of the voting stock of any banking institution. On a case-by-case basis, foreign equity may go up to 40 per cent of the outstanding shares of any institution, provided that equity participation in excess of 30 per cent is placed in nonvoting stocks. Though well received by foreign investors when first introduced in the 1970s, the experiences of some joint ventures have made foreign investors conscious of the need to have greater control of the operations of the bank to protect their investments.

There are, however, a number of successful joint ventures that have attracted foreign investors to further increase their investment. Since much of the foreign equity participation in banks has already reached the equity ceiling, foreign investors can only increase their investment in a bank if their local partners increase their investment proportionately. Unfortunately, the lack of capital of their Filipino partners has constrained them from doing so.

For democratic ownership of a bank, the maximum ownership share of an individual is set at 20 per cent and of a corporation at 30 per cent.

Bank Entry and Branching. Beginning in 1989, the regulations on bank entry have been relaxed. Since then, the Central Bank approved the conversion of two thrift banks into commercial banks. When examined closely though, the bank entry policy still appears restrictive: First, the Central Bank has continued to increase the minimum capital requirement for banks,

---

[8] This is discussed in greater detail in Lamberte (1992). Interest rates on almost all of these special credit programs have been aligned with the market rates.

making it difficult for potential entrants to enter the market. Second, rules on bank entry are not transparent. Although the Central Bank has issued guidelines for the establishment of a bank, it continues to base its decision to grant a banking license to an applicant on criteria other than those spelled out in the guidelines. Third, the existing merger or consolidation incentives clearly suggest that the Central Bank prefers fewer, but large banks in the financial system. Fourth, the limit on the foreign equity investment in domestic banks discourages foreign banks from entering the domestic financial system. Finally, although the General Banking Act does not prohibit the Central Bank from allowing foreign banks to establish branches in the Philippines, the Central Bank has never granted a new license for a branch of a foreign bank since its creation in 1949. Also, if allowed, a new branch could not accept deposits as provided for under the General Banking Act, which would make it a less competitive banking institution. Hence, no foreign bank has applied for the establishment of a branch in the Philippines since the passage of the General Banking Act.

The regulations on bank branching have been relaxed substantially in the last three years. Unlike before when the Central Bank designated areas which were overbanked and prohibited the opening of branches based on this criteria, recent rules allow commercial and thrift banks to open branches without restriction as long as they meet certain capital requirements.

Loan Portfolio Regulations. There are three loan portfolio regulations that affect the operations and profitability of banks: the deposit retention scheme, the agri/agra law, and mandatory credit to small-scale enterprises.

Under the deposit retention scheme, at least 75 per cent of the total deposits, net of required reserves against deposit liabilities and total amount of cash in vault, accumulated by branches, agencies, extension offices, units and/or head offices of specialized government banks in a particular regional grouping outside the National Capital Region, must be invested in that region as a means to develop that region. For purposes of this regulation, the country was divided into thirteen regions. In 1990, the thirteen regional groupings were reduced to three, which effectively relaxed the regulation since banks can now transfer funds to their branches in a much wider geographical area.

The second is the agri/agra law that mandates all banking institutions to set aside 25 per cent of their net incremental loanable funds for agricultural lending, 10 per cent of which is to be lent to agrarian reform beneficiaries, and 15 per cent for general agricultural lending. Banks have an easier time complying with the latter regulation because of the large number of well-performing agri-business corporations, some of which are multinational firms and corporate giants. However, they have difficulty complying with the former simply because their operations are not structured to provide credit to agrarian reform beneficiaries: banks face severe information problems aside from high transaction costs when it comes to lending to agrarian reform beneficiaries. Banks are therefore compelled to invest the money allocated for the agrarian reform beneficiaries in alternative eligible government securities whose yields are well below the market rate. The losses from such investments are ultimately passed on to borrowers.

The third and most recent loan portfolio regulation is the mandatory credit to small enterprises. Under the Magna Carta for Small Enterprises, all lending institutions are mandated to lend at least 10 per cent of their total loan portfolio to small enterprises whose total assets, inclusive of those arising from loans but exclusive of the land on which the particular business entity's office, plant and equipment are situated, amount to five million pesos and below. Commercial and large thrift banks are adversely affected by this law, since they do not possess adequate information about the creditworthiness of small enterprises and their structure is not designed to service small borrowers.

Intermediation Taxes. The financial system is heavily taxed. The current reserve requirement on all deposit liabilities including deposit substitutes is 23 per cent, down from 25 per cent six months ago. This is expected to be reduced further to 21 per cent in the next six months. Even at this rate, however, the reserve requirement remains the highest in Asia, ostensibly meant to finance the losses of the Central Bank. Bank reserves deposited with the Central Bank earn only 4 per cent, while the interest rate on Central Bank bills is around 12 per cent. As interest rates on savings and time deposits are well in excess of 4 per cent, banks pass on the additional costs to their borrowers.

Another intermediation tax is the 5 per cent gross receipts tax imposed on interest income of banks. This tax is also passed on by banks to borrowers.

Safety and Soundness Regulation. In the wake of several bank failures during the first half of the 1980s, the Central Bank made an effort to improve its prudential regulations. The changes were aimed at strengthening the financial position of banks and minimizing insider abuse. The major changes in prudential regulations are discussed below.

(i) *Capital Adequacy.* The Central Bank imposes different minimum capital requirements on different types of banks: higher for those with expanded functions and lower for those with limited functions. Since 1980, the Central Bank raised the minimum capital requirements for various bank categories three times. The latest schedule of minimum capital requirement for the banks ranges from P2 million to P1.5 billion depending on types and location of banks. The net-worth-to-risk assets ratio is set at 10 per cent for all banks. However, the Monetary Board is empowered to authorize a bank to maintain a ratio lower than 10 per cent under certain conditions. Universal banks with a minimum capital requirement of P1.5 billion are allowed a ratio of 8 per cent. In addition, a tighter definition of risk assets was made to reduce discretionary actions of both the banks and regulators.

(ii) *Single Borrower's Limit.* The single borrower's limit of 15 per cent of the unimpaired capital and surplus of banks has long been imposed on banks. In response to a series of bank failures that were partly caused by the concentration of loans to a few large borrowers, this regulation was strengthened by including contingent liabilities in the determination of the limit to which banks can lend to a single borrower or a group of affiliated borrowers. The regulation limits the aggregate ceiling of guarantee outstanding to 50 per cent of a bank's unimpaired capital and surplus standby letters of credit, foreign and domestic, including guarantees, except those fully secured by cash, hold-out deposit/deposit substitutes, or government securities.

As a result of this regulation, corporations wanting to expand their production have had to secure credit from several banks, which has increased the transaction costs for both banks and borrowers. More recently, the single borrower's limit was increased to 25 per cent.

(iii) *Loans to Banks' Directors, officers, stockholders, and related interests (DOSRI Loans).* DOSRI loans should not exceed, at any one time, an amount equivalent to his/her outstanding deposits and book value of the paid-in capital contribution in the lending bank. Further, unsecured credit accommodations to each of the bank's directors, officers, or stockholders should not exceed 30 per cent of his/her total credit accommodations. Though Central Bank examiners have noted violations of DOSRI loan ceilings, in cases where dummies are used, proving such violations are difficult.[9] The passage of New Central Bank Act in June 1993 addresses this problem.

---

[9] The Secrecy of Bank Deposits Law states that: "All deposits of whatever nature with banks or banking institutions in the Philippines... are hereby considered as of an absolutely confidential nature and may not be examined, inquired or looked into by any person, government official, bureau or office, except when the examination is made in the course of a special or general examination of a bank and is specifically authorized by the Monetary Board after being satisfied that there is a reasonable ground to believe that a bank fraud or serious irregularity has been or is being committed and that it is necessary to look into the deposit to establish such fraud or irregularity..."

Depositor Protection

Membership in the Philippine deposit insurance program administered by the government-owned Philippine Deposit Insurance Corporation (PDIC) is compulsory for all banks. The maximum coverage per depositor was raised from P40,000 to P100,000 in 1992. The total risk exposure of PDIC is projected to increase to 39 per cent from 26 per cent of the banking system's total deposit liabilities with the recent increase in maximum coverage per depositor.

The credibility of the deposit insurance system depends on the speed of paying claims, which, in turn, depends on its financial and human resources. The PDIC used to be severely undercapitalized and understaffed, and its available funds are still insufficient to meet payments of insured deposits of failed banks. Also, it is behind in its payment of claims. In 1988 alone, PDIC paid a total of P368 million to insured deposits of failed banks that had been closed more than two years earlier.

Despite the inadequacies of the deposit insurance system, deposits continued to rise even in those years when PDIC experienced difficulty in paying claims, which seems to suggest that the deposit insurance system does not play a significant factor in inducing people to place their savings in deposits. The health of the financial institutions could be a more important factor.

Efficiency of the Banking System in Mobilizing and Allocating Savings

Deposit Mobilization

There are several indicators of the extent of the efficiency of the banking system in mobilizing deposits. One is the volume of deposits in real terms. As Table 4.3 shows, the volume of real deposits mobilized by the banking system increased from P162 billion in 1986 to P243 billion in 1991, or an average annual increase of 8.3 per cent. Thrift banks obtained the highest annual growth rate in deposits at 9.2 per cent, closely followed by commercial banks at 8.7 per cent. Rural and specialized government banks had lower growth rates at 5.4 per cent and 5 per cent, respectively.

Though the growth in deposits will be affected by the growth in the number of branches of banks, new branches may not be as efficient as the old branches in mobilizing deposits. It is therefore important to measure the volume of deposits per branch through time. The data in Table 4.3 show that real deposits per bank increased consistently during the period 1986–1990, declining slightly in 1991, which suggests that the increase in the volume of real deposits of banks noted earlier is not only brought about by the increase in the number of branches but also by the rising productivity of a bank branch. This supposition holds true for the branches of commercial and thrift banks, while real deposits per branch of rural banks only marginally increased and that of a specialized government banks was erratic during the period 1986–1991.

Another measure of the efficiency of banking system in mobilizing deposits is the ratio of loans and investment to deposits. A ratio of less than one indicates that the banking system relies more on mobilized deposits to finance the acquisition of earning assets, whereas a ratio of more than one suggests that the banking system relies less on mobilized deposits and more on borrowed funds. The Central Bank, the interbank market, and special credit programs of the government are the three main sources of borrowed funds. Table 4.3 shows that the ratio of loans and investments to deposits of the banking system was close to one during the period 1986–1991. Again the performance of the different types of banks with respect to this indicator varies significantly. Commercial and thrift banks tend to rely more on mobilized deposits to finance their earning assets, while rural and specialized government banks tend to depend heavily on borrowed funds. Interestingly, the degree of reliance of rural banks on borrowed funds has declined rapidly during the period. The change in the rediscounting policy of the Central Bank and the withdrawal of several subsidized credit programs for the agriculture sector that were coursed through the rural banks have compelled these banks to improve their deposit mobilization. As regards specialized government banks, the degree of reliance on borrowed funds has remained high mainly because they have been designated as the main conduits of special credit programs for the agriculture and industrial sector.

Credit Allocation

The ratio of past due loans and discounts (i.e., past due loans divided by loans outstanding) was high for all types of banks in 1986 (Table 4.4). Of course, this may be a reflection of the 1984–1985 balance-of-payments crisis during which many firms collapsed and failed to repay their loans.

Past due loan ratios in all types of banks declined during the period as indicated in the Table, suggesting an improvement in loan allocation. However, the rate of decline varies greatly among banks: commercial banks were able to reduce their past due ratio drastically in 1987 and since then have been able to maintain a low past due ratio. As of 1992, their past due ratio was only 5.62 per cent. In contrast, the other types of banks did reduce their past due ratios but in a more gradual manner. Interestingly, specialized government banks obtained the lowest past due ratio in 1992. This was because the nonperforming assets of the DBP were transferred to the national government. Also, as part of the recent restructuring of the government-owned banks, both DBP and the Land Bank of the Philippines recently changed themselves from retail lending institutions to wholesale

Table 4.3 Indicators of Deposit Mobilization, 1986–1991

	1986	1987	1988	1989	1990	1991
Commercial banks						
Loans/Deposit (per cent)	69.30	77.28	72.71	72.52	76.40	72.52
Real deposits (Pesos million, 1985=100)	137,113.16	138,665.22	160,703.86	191,221.13	208,388.71	208,090.29
Real deposits/Number of banking offices (Pesos million)	77.64	80.29	92.04	108.34	114.94	108.21
Thrift banks						
Loans/Deposit (per cent)	73.60	79.91	77.30	84.54	85.89	83.44
Real deposits (Pesos million, 1985=100)	12,406.99	13,403.87	15,287.12	18,186.42	17,870.03	19,233.14
Real deposits/Number of banking offices (Pesos million)	18.66	20.37	23.02	26.94	27.37	29.01
Rural banks						
Loans/Deposit (per cent)	173.69	153.32	146.51	137.05	133.02	121.27
Real deposits (Pesos million, 1985=100)	3,659.06	4,083.92	4,280.56	4,679.25	4,667.42	4,852.00
Real deposits/Number of banking offices (Pesos million)	3.38	3.86	4.08	4.49	4.47	4.56
Specialized government banks						
Loans/Deposit (per cent)	136.38	163.56	153.53	208.46	156.40	156.76
Real deposits (Pesos million, 1985=100)	8,331.23	4,799.24	5,616.90	4,665.66	8,115.72	10,866.86
Real deposits/Number of banking offices (Pesos million)	83.31	46.15	54.01	44.43	63.90	76.53
Total						
Loans/Deposit (per cent)	75.46	82.00	77.23	77.80	80.93	78.12
Real deposits (pesos million, 1985=100)	161,510.44	160,952.25	185,888.43	218,752.45	239,041.88	243,042.29
Real deposits/Number of banking offices (Pesos million)	44.69	45.38	52.19	60.97	65.71	64.11
Total deposits/GNP (Pesos million)	0.28	0.26	0.28	0.32	0.33	0.34

Source: Central Bank of the Philippines (1991).

Table 4.4 Ratio of Past Due Loans

	1986	1987	1988	1989	1990	1991	1992
Rural banks	36.70	31.70	29.50	26.30	25.00	23.20	24.13
Commercial banks	23.87	13.59	9.63	7.14	6.77	9.36	5.62
Private development banks	...	...	43.94	36.98	24.22	25.68	15.61
Specialized government banks	...	...	22.33	19.61	12.92	8.35	3.73

Note: ... data not available.
Past due ratios = Past due loans/Loans outstanding.
Among thrift banks, only data for private development banks were made available to the research team.
Source: Central Bank of the Philippines (1991).

lending institutions; thus, they are able to significantly reduce their risk exposure by lending only to conduit private banks.

Bank Spread and Concentration

One way of determining the extent of competition in the banking system is to examine bank spread, i.e., the difference between lending and deposit rates after removing the effects of the reserve requirement and the gross receipts' tax. Greater competition can lead to lower bank spread; conversely, less competition can lead to greater bank spread.

As shown in Table 4.5, bank spread in the Philippines declined in the early 1980s. However, this was not caused by increased competition but rather by the losses incurred by some banks as a result of the liquidity crisis in 1981. Interestingly, bank spread tended to increase after 1983 despite the economic crisis in 1984–1985 and the deceleration in the growth of the economy in 1990. There are a number of explanations for this.

One of the objectives of the 1980 financial reforms was to improve competition in the financial system. However, the policy of not allowing new entrants into the banking system and of encouraging mergers and consolidation conflicted with this objective. Since monetary authorities gave primary consideration to strengthening the financial position of banks by increasing their capital and encouraging the formation of universal banks, several bank mergers, consolidations, and acquisitions of small banks and other financial institutions by large banks occurred. This had an unequivocal effect on market concentration.

Banking concentration based on deposits has increased since 1981 (Table 4.5). The share of the five largest banks in total deposits increased from 31 per cent in 1981 to 52 per cent in 1990. Likewise, the Herfindahl concentration index almost doubled from 0.044 to .074 in the same period.

The result of regressing bank spread on the Herfindahl concentration index shows that 65 per cent of the total variation of the former can be explained by variations in the latter. This clearly suggests that the increasing banking concentration has led to larger bank spreads.

The increasing bank spread of commercial banks translates into increasing rates of return on assets. As Table 4.6 shows, rates of return on assets of all types of commercial banks generally increased during the period 1981–1990, with the exception of a government bank that incurred huge losses between 1984 and 1986.

The increasingly large bank spread could not have been due to increasing risk. Although the country suffered political and economic instability until 1990, political and economic risk was definitely highest during the 1984–1985 balance of payments crisis, which occurred shortly before the change in government. However, as has been shown in Table 4.5, both bank spread and rates of return on assets of all types of commercial banks continued to increase after the 1984–1985 balance-of-payments crisis.

Table 4.5 Interest Rates on Loans and Deposits, Bank Spread Consolidation of all Commercial, Government and Foreign Banks, Herfindahl Concentration Index and Share of Banks in Total Deposits

Year	Interest Rates on Loans	Interest Rates on Deposits	Bank Spread	Herfindahl Concentration Index	Share of 5 Largest Banks in Total Deposits
1981	17.58	10.73	1.11	0.044	30.7
1982	17.05	10.49	0.83	0.038	30.1
1983	15.41	10.18	0.11	0.048	36.4
1984	19.74	10.95	2.81	0.058	43.3
1985	19.86	12.20	1.31	0.061	46.6
1986	21.09	9.26	5.14	0.068	45.9
1987	15.00	5.74	4.93	0.066	50.0
1988	17.18	6.66	5.77	0.070	51.8
1989	21.16	8.26	7.57	0.073	52.4
1990	24.66	10.53	7.82	0.074	51.9

Sources: Central Bank of the Philippines, *Selected Philippine Indicators* (various years); Central Bank, *Fact Book of the Philippine Financial System* (various years); Business Day's *Top 1,000 Corporations* (1975, 1983, and 1984); and *Business Star*, 5 August 1991 (for year 1990).

Table 4.6 Rates of Return on Assets by Type of Commercial Banks

	Universal Banks	Ordinary Commercial Banks	Branches of Foreign Banks	Government Banks
1981	1.3	0.9	1.2	0.7
1982	1.3	0.6	1.6	0.5
1983	1.5	0.6	1.5	0.5
1984	1.1	0.3	2.3	-0.8
1985	1.5	0.4	1.8	-7.6
1986	1.6	0.6	1.7	-10.2
1987	1.8	0.8	1.3	2.4
1988	1.8	1.4	3.5	4.0
1989	2.0	1.5	3.0	3.4
1990	2.3	2.2	2.9	4.7

Source: Central Bank of the Philippines.

Finally, interest rate liberalization without being accompanied by a liberal bank entry policy can lead to the widening of bank spread. This hurts both borrowers and depositors since banks tend to increase lending rates and depress deposit rates.

Accounting System

Unlike firms, banks are heavily regulated. The Central Bank prescribes a detailed accounting system for banks which is embodied in the *Manual of Accounts*. Though there is a separate *Manual of Accounts* for each type of bank, the contents of these manuals are similar.

The *Manual of Accounts* prescribes the form of the balance sheet, off-balance sheet, and income and expense accounts of banks. Each item is described in a fairly detailed manner. Banks are required to submit financial reports to the Central Bank on a regular basis. For example, a consolidated statement of condition is to be submitted to the Central Bank 7–15 banking days after the end of each month; a consolidated statement of income, expenses, and surplus, 20 banking days after the end of every quarter; and a consolidated statement of condition by banking unit, 10–15 banking days after the end of the reference quarter.

In addition, the Central Bank requires that banks keep records of renewed loans, restructured loans, secured and unsecured loans to directors, officers, stockholders and related interests of banks (DOSRI), time loans payable on amortization, large loans to a single borrower, and required reserves.

Besides the external auditors hired by the banks, the Central Bank examines the books, although the regularity of this examination varies by the size and type of bank. For example, the books of all commercial banks and large thrift and rural banks are examined by the Central Bank at least once a year, while examination of small thrift and rural banks is less frequent: once every two years. Understandably, the financial reports of commercial banks and large thrift and rural banks reflect more closely their financial health than small thrift and rural banks.

Despite examination, however, the financial reports of large banks cannot be taken at face value. Some commercial banks have, in the past, presented erroneous reports which were discovered in the course of examination. These included fictitious/indirect/questionable DOSRI loans and nonreporting of past-due DOSRI loans. Central Bank examiners watch for under-reporting of DOSRI loans and past due loans and inadequate allowance for probable losses,[10] which are key problem areas insofar as

---

[10] This represents the amount set up against current operations to provide for losses which may arise from noncollection of loans and discounts. This account is also known as "valuation reserve" in existing rules and regulations.

financial reports of banks are concerned. Also, despite the effort to make the banks' accounting systems transparent, and despite the on-site and off-site examinations, and sanctions imposed for fraud, irregularities exist. This problem will be discussed in greater depth in a later section of this chapter.

NONBANK FINANCIAL INTERMEDIARIES

As of December 1991, there were a total of 4,064 nonbank financial intermediaries in the Philippines: 2,423 pawnshops, 1,093 lending investors, 191 finance companies, 64 investment companies, 33 investment houses, 123 securities dealers/brokers, 10 venture capital corporations, 6 mutual building and loans associations, 13 fund managers, 104 nonstock savings and loan associations, and 4 government nonbank financial institutions. Seventy-five per cent of these institutions are located in the National Capital Region and the adjoining regions.

Nonbank financial intermediaries had total resources of P77.4 billion as of December 1991. For the last five years their total resources grew at an average 11.4 per cent compared to the liabilities' average growth of 8.8 per cent and 16.0 per cent growth in capital. The bulk of these assets are in the form of loans and discounts (P40.5 billion or 52.4 per cent of total resources) and investments in bond and securities (P21.7 billion or 28 per cent of total resources) as of December 1991. Increased business activity has been made possible by an increase in paid-up capital and borrowings.

Finance Companies

There are two types of finance companies: those with quasi-banking functions and those without.[11] Finance companies have emerged as an alternative source of credit facilities for consumers and agricultural, commercial, and industrial enterprises. As of December 1991, there were 191 finance companies composed of 119 head offices and 72 branches. These companies are heavily involved in consumer credit which enables households to acquire houses and lots, appliances, cars, and other consumer durables on an installment basis. Through lease financing, commercial enterprises and

---

[11] Quasi-banking refers to the borrowing of funds, for the borrower's own account, through the issuance, endorsement, or acceptance of debt instruments or any other kind of deposits, or through the issuance of participations, certificates of assignment or similar instruments with recourse, or of repurchase agreements, from 20 or more lenders at any time, for the purpose of relending or purchasing receivables and other obligations. The Central Bank issues a certificate of authority to engage in quasi-banking which allows the nonbank financial intermediary to borrow substantial amounts from the public, subject to the Central Bank's rules and regulations, and to relend those borrowed funds to deficit units. The activity has a striking similarity to banking, hence the term, quasi-banking.

producers are given the opportunity to use equipment, business and office machines, and other fixed assets and at the same time to have more cash flow for other purposes. Trading of securities and residential mortgage finance are two recent activities where finance companies are active.

Short-term borrowings constituted the major source of funds of finance companies. In 1986–1991, short-term borrowings averaged 41 per cent of total sources of funds. Finance companies had to compete with short-term instruments of other financial institutions to mobilize their loanable funds. The cost of the borrowed funds figures heavily in the terms and conditions for receivables financing and direct lending: the cost of borrowed funds is high, especially when the national government has a large budget deficit. Though the Financing Company Act allows the finance companies to offer high-yielding instruments, the high cost of mobilizing loanable funds is ultimately passed on to the consumers. Own capital and retained earnings account for 23 per cent of finance companies' total resources. The rest are other types of liabilities.

Finance companies are heavily involved in receivables financing, direct lending, and trading of securities. Receivables financing constituted on average about 34.1 per cent of the total business of finance companies in 1986–1991, and has grown steadily in importance. Trading of securities was at its peak in 1986 when the Government had to float high-yielding securities to stabilize the economy; trading has since become secondary to receivables financing because of the declining trend of treasury bill rates. The amount of securities traded by finance companies in 1986 was P13.95 billion. By 1991, the volume declined to P2.2 billion. Direct lending is another important activity of finance companies, and was second only to receivables financing in 1991. It represented a 19.5 per cent share in the total business of finance companies in 1986–1991 and will continue to be a major source of profits for finance companies given the growth in consumer income and the development of small- and medium-scale enterprises.

Government Policies and Regulations Affecting Finance Companies

The Financing Company Act (RA 5980), as amended, governs the operations of the finance companies. An allied law is the General Banking Act. Finance companies are regulated by the Central Bank and the Securities and Exchange Commission. They are allowed to engage in receivables financing and financial leasing. There are two types of lease transactions: (i) the "financial lease" which is a noncancelable contractual arrangement whereby the lessor provides the lessee use of an asset, usually fixed equipment, building or land for the consideration of periodic rental payments and (ii) the "operating lease" which is a cancelable contractual arrangement whereby the lessee makes periodic payments to the lessor over a number of years for the use of an asset. Under the latter lease, the usable life of the asset is longer than the term of the lease but because of technological obsolescence, the asset is not leased for a period which is commensurate to its usable life. In the Philippines, leasing is mostly of the financial type, hence this section discusses only this type of lease.

The Financing Company Act allowed these companies to charge any rates even before the *de facto* abolition of the Anti-Usury Law. A finance company performing quasi-banking functions (i.e., borrowing from 20 or more lenders for relending or purchase of receivables) is subject to the Central Bank's rules and regulations on quasi-banking functions. More recently, the single borrower's limit and ceilings on DOSRI loans of finance companies with quasi-banking functions have been imposed while selling of receivables of finance companies has been confined to banks, investment houses, and other finance companies. The SEC regulates the issuance of commercial papers.

The regulatory and policy environment and competition from other financial institutions for loanable funds determine to a great extent the performance of the finance companies. The competition for funds is stiff but nonbank financial institutions develop market niches which enable them to reach particular types of clientele not ordinarily served by the commercial banks, such as entrepreneurs who use the financial lease as a strategy for effective use of capital equipment and to free their cash flow for alternative uses.

Other Nonbank Financial Institutions

Other nonbank finance intermediaries that provide short-term retail credit are pawnshops, credit unions, lending investors, and institutions involved in capital formation, namely, investment houses, securities/dealers, and venture capital corporations.[12] These institutions provide an alternative financing mechanism for borrowers unable to access loans from the financial institutions.

Examples of their credit accommodations are: (1) the Market Vendor Loan, which is extended to stallholders in public markets or private commercial establishment; (2) Salary Loan, which is granted to bona fide employees of business establishments; and (3) Real Estate or Chattel Mortgage Loan, a loan extended for the fulfillment of an obligation or for the payment of debt. In the case of credit unions, the borrowers are the members themselves, hence, in most cases collateral is not required.

---

[12] While fund managers are sometimes included as nonbank financial intermediaries, in actual practice a fund manager may be a bank, an investment house, a finance company, or a securities dealer or broker, who has been tasked to manage assets belonging to other parties. Fund managers are not included in the discussion in this section.

Pawnshops and lending investors represent the most numerous of the nonbank financial institutions in the Philippines. As of December 1991, there were 2,423 pawnshop offices consisting of 1,829 head offices and 594 branches. This represented a 77.2 per cent increase from the number of pawnshop offices in 1986. In the same year, there were a total of 1,093 lending investors. Though there were no data on the total number of credit cooperatives or cooperative credit unions in the country, their importance in the rural areas is evident.[13]

Investment houses/companies, securities dealers/brokers, money brokers and fund managers cater mostly to individuals and institutions seeking an alternative investment outlet outside the banking system. The activities of these institutions range from the trading and sale of securities to financial consultancy and portfolio management. Venture capital corporations are intended to be joint undertakings between the private entrepreneur, a financing source such as a bank, and a government development agency to provide equity to promising small- and medium-scale enterprises.

As of December 1991 there were 123 securities dealers/brokers, mainly head offices, and some investment houses/companies. Venture capital corporations numbered 17 in 1986 but had declined to 10 companies in 1991.

Sources and Uses of Funds

These nonbank financial intermediaries, other than the finance companies, are not engaged in deposit-taking activities.[14] Their capital plus borrowings are the major sources of funds. Those involved in lending activities rely heavily on their own funds to support their operations. As of December 1991, about 64 per cent of the total sources of funds of pawnshops were obtained from capital accounts; 49 per cent for lending investors and 94 per cent for venture capital corporations. The amount of capitalization required by these institutions is low relative to the capital levels required of banking institutions[15]—in part because the operations of the nonbank financial intermediaries are less complex and not as diverse

---

[13] Credit unions are not supervised by any government agency; cooperative credit unions are supervised by the Cooperatives Development Authority.

[14] However, credit unions do require members to make deposits. In general, a fixed amount of deposit per month for every member is imposed. Lending investors are also authorized by the Central Bank to accept deposits but this is limited to a maximum of 19 lenders at one time, and these deposits are not covered by the Philippine Deposit Insurance Corporation. Nonbank finance institutions mobilize funds from the public by offering high-yielding instruments which are not strictly categorized as "deposit liabilities."

[15] For example, the minimum paid-up capital of pawnshops and lending investors is P100,000. Financing companies have a minimum paid-up capital of P500,000.

as those of commercial banks. Pawnshops are also primarily single proprietorship; lending investors are mostly partnerships while venture capital corporations are most often organized by private banks (e.g., Metrobank) generally in conjunction with other government agencies such as the National Development Corporation or the Technology Livelihood and Resource Center. The nature of venture capital corporations' ownership may also be the reason why they rely less on borrowings and why a significant percentage (32 per cent) of profits obtained from their operations are retained in the business. In contrast, the dependence of pawnshops and lending investors on borrowing as a source of funds is higher at 28 per cent and 36 per cent, respectively.

Nonbank financial intermediaries engaged mainly in trading (purchase and sale) of securities also rely on own capital and borrowings, while investment companies and fund managers, in contrast, rely to a great extent on capital and profits obtained from their business.

With the exception of investment companies, nonbank financial intermediaries use most of their funds for direct lending. For pawnshops, in particular, 87 per cent of total funds were lent. Investment in securities is practically nonexistent. A similar trend is observed for lending investors but the volume of loans granted by lending investors is lower. As of 1991, these entities had generated P1.32 billion in loans. Venture capital corporations made significant investments (other than loans) especially in the years 1988 to 1991, investing most often in the equities of small- and medium-scale enterprises.[16] There was, in fact, a decrease in loans granted by venture capital corporations during the period 1986–1991, from P87 million to only P9 million.

By comparison, during the same period, investment houses and companies as well as securities dealers and fund managers granted a larger volume of loans than pawnshops, lending investors, and venture capital corporations. Investment houses alone generated P7.31 million in loans in 1991. Even investment companies with only 14 per cent of their funds in loans granted P1.84 billion loans in 1991.

Regulatory Environment

The nonbank financial intermediaries, other than the credit unions, are considered to be formal financial institutions because their activities are supervised by the Central Bank. Cooperative Credit Unions are categorized as informal institutions but are supervised by the Cooperative Development

---

[16] By virtue of Central Bank Circular No.733, venture capital corporations are organized for the purpose of developing, promoting, and assisting small- and medium-scale enterprises through debt or equity financing.

Authority. In general, the Central Bank's regulation helps stabilize the financial system and gives it a better handle for monetary and credit policy. Regulations of the nonbank financial institutions are, however, less strict than those applied to banks.

More specifically, branching policies do not directly affect the nonbank financial institutions. Pawnshops and lending investors, in particular, need not secure special licenses usually required of banks. The only major requirement is to secure a license from the city or municipality where operations will be conducted and to register with the Bureau of Domestic Trade for single proprietorship or with the SEC for corporations or partnerships. Pawnshops are restricted from operating or establishing branches in Metro Manila, however. Lending investors are not similarly constrained.

Capitalization requirements for nonbank financial institutions are also much lower than those of banks. Entry into the system is less cumbersome, which has created some advantages. The increase in the number of pawnshops, for instance, has created the potential for accessible credit and lower interest rates for small- and medium-scale borrowers. The average interest rate now charged by pawnshop is 5 per cent per month. This rate represents a significant decrease from the 10 per cent rate charged in the 1970s. Stiff competition has also improved the services of pawnshops. Some pawnshops, for instance, have gone into computerized operations and use modern technology to appraise jewelry and gems submitted as collateral.

Nonbank financial intermediaries (in particular, pawnshops) service small loans and incur higher transactions costs per peso lent than do banks. In addition, the former cater to riskier clients. In general, the growth of the nonbank financial institutions is due to the failure of the banking institutions to respond to the credit demands of particular types of clientele and conversely, their ability to tailor fit the financial instruments to the needs of these clientele. The segmentation of the country's credit markets, however, has not made possible the mobilization and integration of funds for the larger capital needs of small- and medium-scale enterprises.

Insurance Companies

Structure of the Industry

Insurance companies in the Philippines are principally private sector companies, with the exception of the five government-owned and operated corporations, namely: Government Service Insurance System (GSIS); Social Security System (SSS); Philippine Crop Insurance Corporation (PCIC); Home Insurance and Guarantee Corporation (HIGC); and Philippine Deposit Insurance Corporation (PDIC). Only the private insurance companies, including the 12 Mutual Benefits Associations and 39 Trusts for

Charitable Uses, are regulated by the Insurance Commission. The government-owned insurance companies are governed by their respective charters.

There are 127 private sector insurance companies in the country, of which 23 are life, 98 non-life, two composite, and four reinsurance companies. There are 12 foreign-owned companies, of which two are life insurance, nine are non-life insurance, and one is a professional reinsurer. Foreign registered companies and foreign-owned but locally registered companies hold 55 per cent of total assets of the life insurance companies and 42 per cent of the assets of the non-life companies.

Among the government-owned companies, the GSIS and SSS offer life insurance for government and private employees, respectively. These corporations are also involved in lending. The PCIC, on the other hand, is the only agricultural insurance in the country which insures rice and corn crops of small farmers. The PCIC is also engaged in livestock insurance and credit guarantee for small farmers. The HIGC is involved in financing and insurance of home-related projects, while PDIC insures the deposits of banking institutions.

Growth of the Industry

There are three general measures of growth: assets, premium income, and the amount of insurance in force. Overall, these measures of growth have kept pace with inflation and developments in the financial sector.

(i) *Private Insurers.* As of 31 December 1991, the combined total assets of the private insurance companies were P50 billion, of which P31.6 billion were life insurance assets and P18.4 billion were non-life. Between 1986 and 1991, assets of life companies grew by 150 per cent while those of non-life companies grew by only 98 per cent.

Life insurance companies also had a larger volume of business. The amount of life insurance in force in 1986 was P282 billion from about 3 million life insurance policies. Policies come largely from ordinary policies (69.7 per cent), group policies (15.2 per cent), and industrial policies (15.0 per cent). The total premiums obtained by life insurance companies is also higher than for the non-life. Gross and net premiums[17] of life companies also grew at a faster rate than non-life. In particular, net premiums of non-life companies displayed, on average, a 30 per cent difference with gross premiums. This reflects a relatively high dependence on reinsurance. Further,

---

[17] Net premiums are adjusted for the cost of reinsurance.

the extent of reinsurance for non-life companies has been increasing. In 1991, for instance, net premiums were 36 per cent less than gross premiums compared to a 1986 level of 28 per cent.

In terms of profits, combined operations of life and non-life insurance exhibited a 38 per cent increase in profits (excluding capital gains) from 1990. Gains are mainly attributed to non-life operations whose profits increased from P75 million in 1990 to P408 million in 1991. Profits of life insurance companies slightly improved from P896 million in 1986 to P929 million in 1991. Income derived from investments has been the major and stable source of profits for the insurance companies. Investments provided 80 per cent of the insurance companies' profits; 95 per cent for life and 46 per cent for non-life. The high returns obtained by life insurance companies from investments stem from their ability to invest in longer-term investments since they are engaged in long-term contracts compared to the shorter time frame of the non-life companies. About 56 per cent of the assets of life and non-life companies are invested in stocks, government securities, and bonds.

(ii) *Public Insurers.* The asset base of government insurance companies is almost three times that of private insurance companies. As of 1991, combined assets of the five government insurance corporations was P130.3 billion. The Social Security System has the largest asset base of P76.4 billion, which is twice that of the Government Service Insurance System with P43.6 billion. This is because the former extends insurance to nearly the entire work force and, as such, has 10 million members. The latter offers insurance only to government employees and presently has 1.4 million members. The Philippine Crop Insurance Corporation has the lowest asset base (P648 million) among the government corporations. It is also the only insurance corporation whose operations are subsidized. The Government's subsidy in premiums alone is at 60 per cent. Subsidy in crop insurance is justified on the basis of its clientele: the small-scale farmers who are unable to afford the full premium that results from the high risk of crop failure.

The investments of these government corporations consist of government bonds and short-term treasury bills. In 1991, of the combined investment portfolio of P53.3 billion, 45.9 per cent was invested in securities and about 50 per cent was used for lending operations. The Government Service Insurance System and the Social Security System, in particular, have been major sources of long-term loans especially with regard to home development. The former has also been involved in financing projects of local government units. The huge resource base of the two and the minimal supervision they enjoy allow them flexibility in investments.

Capitalization of Private Insurance Companies

The composition of an insurance company's net worth indicates its strength in supporting insurance policies generated. As a whole, the country's life insurance companies are well capitalized. Their strength lies in their high reserves and unassigned surplus which together cover 60 per cent of liabilities. By comparison, the capitalization of non-life insurance is inadequate. Reserves and surplus are only able to support 32 per cent of liabilities.

At present, the minimum paid-up capital for life and non-life companies is P75 million inclusive of the P25 million contributed surplus. The Insurance Code also requires that domestic insurers deposit 25 per cent of paid-up capital with the Insurance Commission in the form of government securities or associated securities. Foreign insurers are required to deposit with the Insurance Commission securities satisfactory to the Commission and equal to the minimum paid-up capital; and 50 per cent of these securities must be in bonds or other forms of indebtedness of the government.

Annual valuations are to be done for life insurance companies and these must be certified by an actuary. The allowable bases are specified and interest rate assumption cannot exceed 6 per cent. This specified limit (of 6 per cent) results in a conservative asset valuation which constrains the offering of competitive financial products vis-á-vis alternative saving instruments. Finally, all insurance companies are required to contribute a *pro rata* share into a security fund which is used to pay outstanding claims against any insolvent insurance company. This fund is administered by the Insurance Commission.

Ownership of Private Insurance Companies and Barriers to Entry

Most Philippine insurance companies are owned by small groups of individuals, though there is potential for more ownership by commercial banks following a June 1991 policy announcement by the Central Bank allowing up to 35 per cent ownership by banks. Such ownership would allow banks to tap the long-term savings generated by the insurance companies and their distribution network. The banks could also place additional capital which the insurance companies need for long-term growth and viability perspective. The immediate issue for policy is the impact of increasing bank ownership on the concentration of the insurance industry and on the prudential supervision of both the insurance and banking industry.

With respect to foreign investment, the Insurance Commission encourages foreign participation in the inadequately capitalized non-life sector. The Omnibus Investment Code regulates foreign investments in the country. With this Code as its framework, the Insurance Commission has since 1987 allowed foreign equity participation up to a maximum of 40 per cent without the need to seek prior authority from the Board of Investments. Before 1987, 30 per cent maximum foreign equity participation was allowed. The attitude toward the life insurance sector is in contrast to the above. The 1990 Insurance Commissioner's report states that "the policy is not to allow foreign equity participation except for those foreign companies which until the present have already been allowed to transact business in the Philippines."

The result of disallowing foreign investment in the life insurance sector is a lack of access to foreign capital, expertise, experience, and innovative products. A policy issue, therefore, which merits closer examination is the challenge posed by the availability of foreign capital and expertise without necessarily ceding ownership and control of life insurance companies. One area of policy is to encourage the growth of mutual benefit societies—such as that currently pursued by the second largest domestic life insurance company—and the entry of international mutual societies. Policyholders own such societies. The broad ownership seems to be a good antidote for concentration of control over the company.

Government Policies and Regulations Affecting Insurance Companies

The Insurance Code treats the insurance industry as a monolithic industry, basically as a risk-spreading, non-life industry. The reality is that there are important differences between the two principal segments of the insurance industry. The main differences are as follows:

(i) The life insurance sector is concerned with the long term while the non-life sector has a relatively shorter time frame because of the nature of the activities it covers.

(ii) The life insurance sector is involved with long-term contractual savings while the non-life is involved with short-term renewable contracts.

(iii) The life insurance sector has a savings focus and contributes to the availability of funds for long-term investments while the non-life, which has a risk-spreading orientation, helps secure capital investments.

Another issue relates to the tax treatment of insurance policies vis-á-vis alternative savings instruments. Premiums are levied both a 5 per cent premium tax and a documentary stamp tax of 50 centavos on every P200 or any fraction of the amount of policy. The premium tax is a direct tax on long-term contractual savings. The insurance companies pay the normal corporate income taxes; the 20 per cent withholding tax on interest earnings; real estate tax; transfer taxes on securities and the customary examination and license fees. The structure of taxation imposes a huge tax burden on the insurance industry relative to other sectors and, in particular, to the less regulated pre-need industry. This tax treatment puts the insurance industry at a disadvantage and acts as a disincentive to the growth of long-term savings and of the industry.

A fast-emerging segment of the insurance industry is the pre-need sector which provides benefits relating to education, pension fund accumulation, internment and death-related services. The pre-need industry enjoys tax concessions which are not given to the insurance industry. In addition, the SEC imposes different standards of regulation to the pre-need and the insurance companies. The Insurance Commission has no jurisdiction over the former.

The products of the pre-need industry are similar to insurance products although the former have a shorter time frame (one to five years). Pre-need plans are classified by law as securities. Because of the advantages of the relative taxation, investment controls, and financial and operational aspects of pre-need companies over the more regulated insurance companies, more domestic savings are placed in the shorter-term products of pre-need companies than in long-term savings.

The regulations on insurance companies tend to be conservative as insurance companies are enjoined to have "investments which are safe and free from excessive market price fluctuations since a relatively small shrinkage in asset values could endanger their insolvency." Thus, the Insurance Code contains provisions on allowable investments and rules for their valuation. Non-admitted assets such as intangible assets, shares in own company or a company in which the insurer owns an interest, prepaid or deferred expenses, unpaid checks, advances to officers, furniture and fixtures and any excess of aggregate book value of assets over market value, may not be used to cover policy reserves or meet the solvency requirements. The Code also requires that the investments be valued at the lesser of book value or market value.

The result of these regulations is a conservative investment portfolio. As of December 1991, a high percentage of assets were in short-term assets and in government paper. Investment in stocks and property were only 28.5 per cent of total assets.

There is ample room for reform of existing regulations which date back to the 1970s. The emphasis then was prudential control; this is still the thinking today.

It is recommended that the Insurance Code be amended to provide for a differential regulatory treatment of life and non-life insurance companies; clear guidelines for prudential control; increased flexibility for the Insurance Commission to act within the legal guidelines; focus on regulation and control of excesses rather than control of specific activities; and a developmental role for the Insurance Commission.

The potential for further growth of the industry is limited by the slow growth of the economy, the volatility of interest rates and inflation, the competition given by less-controlled alternative savings instruments and inadequate capitalization. The importance of the insurance industry, particularly, the life insurance component, stems from its role in accumulating long-term domestic savings and in matching these savings against long-term assets such as commercial or infrastructure investments.

Private Pension Funds

Types of Pension Funds and Investment Portfolio

The pension system is composed of occupational and personal retirement schemes that are administered privately; these are in contrast to the public sector social security system (GSIS and SSS). Included in the private schemes are the Military Pension Fund and other pension funds of government entities. The occupational retirement schemes are set up by the individual's employer on a voluntary basis. These are usually established by private companies which offer the pension schemes as part of the compensation package. Personal retirement plans are set up by the individual and are rapidly becoming an attractive, although still considered as supplementary, source of income upon retirement.

There are two types of retirement schemes, differentiated according to the benefits that will eventually be received. Most of the funds are of the defined benefit type and thus, employers enjoy flexibility in determining contribution levels in the absence of guidelines by regulatory agencies. The defined benefit plans have fixed benefits, and contributions are computed to meet the required benefits when due. The other type of pension fund is the defined contribution plan. In this, the amount of contribution is fixed and benefits accrue based on the actual contribution and investment earnings of these contributions. The defined benefit plan is the most common type of plan. It pays lump sum benefits upon retirement or separation.

Tax concessions given to pension funds are enjoyed by both employer and plan members. For income tax computation purposes, the employer can deduct the full amount of the contribution to the pension fund; the employee cannot. Benefits are exempt from taxes, and some fund earnings may be tax exempt. These tax advantages are a major reason for the growth of private pension schemes. As of June 1991, there are 1,500 approved private occupational pension plans.

There are no comprehensive data on pension fund assets. However, pension assets have been estimated to be as much as P20 billion or almost one fourth of the financial assets held in trust. The 1990 pension assets are estimated to equal roughly 12.5 per cent of the total market capitalization of publicly traded companies and about 1.8 per cent of GNP. It is also estimated that 25 per cent of the labor force is presently participating in a pension plan.

The average pension fund has about 50 per cent in corporate loans; 24 per cent in other fixed interest instruments; 12 per cent in stocks and the rest mostly in treasury bills. Investments in real estate and equity comprise a small percentage of assets.

In the absence of any formal regulatory agency to monitor and supervise pension plans, an informal self-policing system has been set up by participants. This system provides informal standards and procedures for preventing abuse. The participant follows more or less its own sectoral professional standards, e.g., the banks are regulated by the Central Bank Act, the General Banking Act, and similar laws. A Board of Trustees usually administers the funds which are established as trusts. The Board selects a fund manager, usually a commercial bank, to invest and manage the funds. The informal system of monitoring and supervising seems to work so far as there has been no report of any abuse. However, the moral hazard problems of such a system which does not impose credible sanctions on erring participants would suggest the need for a formal system of monitoring and supervision.

Government Policies and Regulations Affecting Pension Funds

Pension funds are important sources of long-term funds for the capital markets. However, the coverage of the labor force is still small because of the weak economy. The average retirement benefits are not adequate to meet the needs of the retiring employee. There is no single government agency that monitors and supervises the pension funds. The Bureau of Internal Revenue is concerned solely with the approval of pension plans for the tax concessions that the law provides to pension funds.

The self-employed pension plans are discriminated against by taxation policy relative to occupational plans. Only the employers are allowed to deduct their contributions to pension plans. On the other hand, retirement benefits are fully tax exempt, though earnings of retirement accounts from investments in deposit liabilities, money market instruments, commercial paper, and government securities are subject to the 20 per cent withholding tax. The taxation policy also does not allow the portability of benefits. Income derived from the benefits the employee receives upon his resignation and made prior to his retirement, is subject to tax and cannot be rolled over into a new plan. The tax code allows the exemption from tax of benefits only once during the individual's lifetime. Hence, a review of the differential tax treatment of personal and occupational pension schemes is necessary for consistency and growth of the system.

Investor confidence can be strengthened by imposing rules on financial disclosure and by using generally accepted accounting and auditing principles for pension funds. There are three areas where a regulatory framework will be necessary: (1) the funding of the pension plan and the required contribution; (2) the administrative expenses of the plan; and (3) the management of the investment portfolio.

Nonbank Government-Owned Credit Guarantee Institutions

Besides the nonbank financial intermediaries discussed above, there are nonbank government-owned credit guarantee institutions that can be classified as special development finance institutions. These consist of the Guarantee Fund for Small and Medium Enterprises (GFSME); the QUEDAN Guarantee Corporation; the Small Business Guarantee and Finance Corporation (SBGFC); and the Philippine Export and Foreign Loan Guarantee Corporation (Philguarantee). In addition, the Bankers Association of the Philippines recently established the Bankers Association of the Philippines Credit and Guarantee Corporation (BCGC) which to date has been an inactive credit guarantee institution.

The credit guarantee institutions are not yet significant contributors to the financial system. They are highly fragmented, undercapitalized, and overstaffed. In addition, the credit guarantee they provide has failed to act as a collateral substitute.

FINANCIAL FIRMS AND MARKETS

Money Markets

Types of Money Market Instruments

The money market is a market for short-term debt instruments, most of which have maturities of seven days or less. There are four major instruments in the Philippine money markets, namely: interbank call loans;

deposit substitutes; commercial papers; and government securities. Deposit substitutes dominated the money market until 1986; however, its share in the market has declined since. By 1992, the money market became almost synonymous with interbank call loans and government securities markets, accounting for 43 per cent and 49 per cent, respectively, of the total volume of money market transactions. Among government securities, treasury bills have been the dominant instrument. These changes in the relative shares of the various money market instruments can be better appreciated by an examination of the major participants in each submarket.

Participants in Each Submarket

Interbank Call Loans Market. Participants in this market are exclusively banks and nonbank quasi-banks. Commercial banks are the largest group of borrowers, accounting for more than 85 per cent of the total volume of transactions. Although still small, the volume of interbank borrowings of investment houses and savings banks have increased in 1992.

Commercial banks are also the largest group of lenders in this market, though its share in the total volume of loans did decline markedly from 96 per cent in 1986 to 76 per cent in 1992 as other banking institutions and rural/thrift banks increased their share.

It appears that banks use this market not only to adjust their liquidity position but also to finance their regular lending operations. With the lowering of the reserve requirement of interbank call loans from 5 per cent to 1 per cent in 1980, they became a cheaper instrument, especially compared to deposit substitutes where reserve requirement have been aligned with those of traditional deposits, at 23 per cent. The growth in the interbank call loans market, especially in the recent past, can also be attributed to higher demand for funds to cover reserve deficiencies as a result of the increase in the reserve requirement.

Deposit Substitute Markets. Commercial banks used to be the single largest group of borrowers in this market, however, their share declined from 55 per cent in 1986 to only 40 per cent in 1988. The main reason for this is that with the interest rate liberalization, commercial banks were able to raise funds through the traditional deposit market by making their deposit instruments as competitive as money market instruments. Investment houses and financing companies caught up fast as borrowers in this market. In 1992, commercial banks increased their borrowing from this market while investment and financing companies reduced theirs.

Lenders in this market are quite diverse. Commercial banks were the largest group of lenders in 1986; their share has since declined. Replacing the commercial banks as large lenders in this market are individuals and other banking institutions, which together account for more than half the total volume of transactions.

Commercial Paper Market. This market has been clearly dominated by the nonfinancial corporate sector since 1986.

The major suppliers of funds in this market are individuals and other private corporations, which together accounted for more than 90 per cent of the total volume of commercial paper transactions in 1992.

Government Securities Market. As a borrower in this market, the National Government in 1983 contributed only 33 per cent of the total volume of transactions. Its share rose sharply to 60 per cent in 1986. Since then, treasury bills have become the primary instrument in the government securities markets, accounting for more than 80 per cent of the total volume of transactions. The floating of high-yielding marketable treasury bills also provided monetary authorities an important mechanism through which they could exercise an effective open market operation.

Private corporations have been the largest investors in this market, although their share has declined since 1986. Other large investors are commercial banks, trust/pension fund, and individuals. These four groups of investors account for about 70 per cent of the total volume of transactions.

Government Supervision and Regulations of the Money Market

The Philippine money markets have undergone several regulatory environments. Until 1973, they were virtually unregulated, which caused a significant shift of funds from regulated traditional deposits to money market instruments.

Between 1973 and 1979, money markets became regulated. The Central Bank Act was amended to place all banks and nonbank financial institutions under the supervision of the Central Bank. Stricter measures were imposed on nonbank financial institutions with quasi-banking licenses. Among the major regulations imposed on the money markets were: requiring firms to present authority to issue debt instruments; prescribing reserve requirements on interbank loans and deposit substitutes; imposing a 35 per cent transaction tax on all primary money market borrowings; imposing interest rate ceilings on deposit substitutes; and imposing minimum lot size.

The Securities Act was amended in 1975 to place commercial paper under the responsibility of the SEC. The SEC required all issuers of short- and long-term commercial papers to obtain the Commission's approval for the maximum amount of commercial paper they could issue. Among the important criteria used to determine approvals were the prescribed maximum debt/equity ratios ranging from 10:1 for financial institutions without quasi-banking licenses to 70:30 for firms engaged in agriculture and the minimum current ratio of 1:1.

From 1981 onward, the money markets were placed in another policy environment. Interest rate ceilings on deposits and money market instruments were removed. However, the liquidity crisis in 1981 dealt a heavy blow to the money markets. The SEC imposed stricter rules on commercial paper registration, requiring that the prospective commercial paper issuer obtain a credit line from a bank or other authorized financial institutions earmarked for the repayment, on a *pro rata* basis, of the aggregate outstanding commercial paper issued. The credit line should cover at least 20 per cent of the aggregate commercial paper outstanding at any time.[18]

Despite these regulations, the money markets have remained buoyant, with the volume of money market transactions increasing from P14.2 billion in 1975 to P60 billion in 1983. Though it declined in the next four years during the period of economic instability, in 1988 the volume of money market transactions recovered and reached P780 billion. It further increased to P2 trillion in 1992. These increases were mainly in the volume of transactions in the interbank call loans and the government securities markets, as these submarkets are less regulated than the deposit substitute and the commercial paper markets.

Capital Markets

There are two major segments of the capital markets which exist today at varying degrees of development in the Philippines: the non-securities and the securities markets.

Non-securities Markets

These markets provide non-negotiable medium- and long-term debt to enterprises. The instruments include loans, mortgages, leases, and sale and lease back arrangements. The major institutions participating in these markets are banks, insurance companies, and leasing companies.

Banks have long dominated the non-securities markets. The share of medium- and long-term loans increased during the period 1980–1985, due in part to the financial reforms initiated in 1980 that encouraged banks to

---

[18] The Monetary Board has allowed exemptions from this requirement subject to certain prerequisites, such as: assets should not be less than total liabilities for the previous three years; average annual return on equity of at least 8 per cent; a three-year average acid test ratio of at least 0.5:1; average interest service coverage ratio for the previous three years should be at least 1.2:1; and a maximum debt/equity ratio of 2.5:1.

increase the proportion of their medium- and long-term loans in their total loan portfolio. However, there was a reversal in the trend after 1985 as a reaction of banks to the 1984–1985 crisis and to the generally unstable political atmosphere that prevailed since 1986 when a new Government took over. Interest rate volatility also made banks more conservative in lending. To protect themselves from losses arising from sudden changes in interest rates, banks have applied floating rates on their medium- and long-term loans.

As a method of providing long-term capital to enterprises, leasing appears to be an insignificant market. The highest leasing volume was achieved in 1989 at P1.2 billion. The succeeding years saw much lower levels.

There are four major government-owned financial institutions that play a key role in the non-securities markets: the Development Bank of the Philippines; the Government Social Insurance System (GSIS); the Social Security System (SSS); and the Home Mutual Development Fund (HMDF).

The Development Bank has long been the supplier of medium- and long-term capital to the private industrial sector. Its outstanding loans and advances as of December 1991 amounted to P14 billion, and its equity investments stood at P142 million. It is a conduit of several medium- and long-term loan programs funded by multilateral agencies and bilateral grants for small, medium and large industries.

The GSIS and SSS provide a stable source of long-term funds to the domestic economy. As of December 1991, their combined assets stood at P119.6 billion. Together their loans and investment, including housing loans to their members, equity investments, and investments in long-term financial instruments were P97.5 billion.

The HMDF is a provident fund among employees of private and government agencies. It collects contributions from its P1.2 million members at the rate of 80 million per month. It provides housing and consumer loans to its members. As of 1991, HMDF had a total resource base of P17 billion, of which P9 billion were housing loans to members. It is one of the largest institutional investors in securities, with investments in 1991 totaling P4.8 billion in securities.

Securities Market

The securities market provides enterprises with medium- and long-term debt through negotiable instruments. These also provide liquidity to investors. The securities market is comprised of primary and secondary markets.

Primary Markets. In the primary markets, enterprises raise long-term funds by selling new issues of securities which include equity, equity equivalents, and debt securities. The institutions involved in these markets are investment banks, brokers/dealers, venture capitalists, issuers of securities (private corporations and government agencies), and the SEC.

The debt securities or bond markets are still underdeveloped. Very few private companies have issued long-term bonds. This market has long been dominated by government securities.

The equities market had been thin, with total equities raised in the exchanges below P10 billion. However, in the last three years—spurred partly by the liberalization of foreign investment and deregulation of the foreign exchange markets and partly by improvement in political stability—the equities market soared to new heights. Total equity capital raised through the stock exchanges was P11.3 billion in 1990 and P26 billion in 1991. In 1992, this declined to P22.8 billion. There seems to be a shift in emphasis in the recent past from debt financing to equity financing due to massive corporate projects, expansion plans, and capital requirements, as well as the high cost of borrowing.

A change in the character of the equities market is noticeable. First, there was an improvement in the quality of new listings as more issues came from the commercial and industrial group. Second, there is now more representation from sectors not traditionally listed. Third, the peso value of the issues has been large, thereby increasing the depth and breadth of the market.

Secondary Market. The secondary market provides opportunities for trading in already issued securities, usually through the exchanges. The institutions involved are the stock exchanges, securities brokers/dealers, over-the-counter markets, clearance and settlement agencies, and mutual funds.

Under the newly established Philippine Stock Exchange (PSE) foreign investors may buy shares of stocks of domestic firms, though ownership is limited to 40 per cent. This has necessitated the creation of two classes of shares to more conveniently monitor the limit on foreign ownership of firms. Class "A" shares are limited to Filipino nationals, while Class "B" shares are open to any nationality. Both classes of shares have the same voting and dividend rights.

There are at present 164 listed firms, and more are expected to list in the next two years. Of the total listed firms, only 62 belong to the top 1,000 corporations in the country. In 1992, only 35 stocks were actively traded.

The combined turnover of the two exchanges displayed a cyclical pattern during the period 1986–1992, reaching a high of P77 billion 1992. Market capitalization rose dramatically from only P40 billion in 1985 to P391.24 billion in 1992. As a ratio to GNP, market capitalization stood at 28.8 per cent in 1992, the highest ever achieved by the market. There are indications that this trend will continue in the next few years as large firms implement their expansion plans.

The increase in market capitalization can be attributed to three factors. On the supply side, high grade stocks of well-known corporations and the partially privatized government-owned commercial bank were offered to the public and listed in the stock exchanges. On the demand side, the debt conversion scheme and the creation of four Philippine country funds brought in foreign investment. It is estimated that of the total equity transactions in 1989, 40 per cent could be accounted for by debt equity swaps. In addition, foreign exchange liberalization has encouraged foreign investment.

Given the high interest rates on treasury bills, the return on stocks must be much higher to induce investors to place their money in those instruments. During the period November 1986–March 1992, common stocks yielded the highest after-tax real rates of return in the Philippine financial markets at an average of 21.5 per cent per annum, followed by treasury bills at 4.1 per cent, time deposits at 2 per cent, and savings deposits at –5.6 per cent. Common stocks also obtained the highest risk as measured by the standard deviation.

What is notable though is the large spread between the average rate of return on common stocks and treasury bills, which means that those who have access to the stock market have the opportunity to earn more than those who have access only to the treasury bills and bank deposits.

Regulatory Bodies. There are three government agencies involved in regulating and monitoring the exchanges, namely the Central Bank, the Bureau of Internal Revenue and the Securities and Exchange Commission. The role of the first two are minor compared to the last one.

The Central Bank regulates and monitors all direct foreign investments. The law guarantees full repatriation of principal, profits, and dividends. With the recent liberalization of foreign investment and foreign exchange deregulation, the Central Bank's regulatory function with respect to foreign portfolio investment has been greatly reduced.

The Bureau of Internal Revenue (BIR) is the regulatory body as far as taxes on transactions are concerned. There is no capital gains tax on transactions involving listed securities. Instead, a transactions tax of one-quarter of 1 per cent is levied on gross sales. In addition, a documentary stamp tax is applied on the sale or transfer of debt or equity securities.

The Securities and Exchange Commission (SEC) is the lead agency that regulates the securities market. In particular, it oversees the activities of the exchanges and the stockbrokers and security dealers. Securities have to be registered with the SEC. Issuing companies are required to periodically submit corporate information and financial statements. The SEC reviews all candidates for listing approved by the listing committees of the stock exchanges.

Recent Reforms. There were three major policy reforms effected in the last six years that have had a direct bearing on capital market development. First, as part of the 1986 tax policy package, the double taxation of dividend income was eliminated through the abolition of the tax on inter-corporate dividends and gradual phase out of the tax on shareholder's dividend income. In the case of domestic and resident foreign corporations, the abolition of the tax on dividends was immediate, while in the case of shareholders, the double taxation on dividends was removed gradually over a period of three years.

Second, the SEC formally signed on 31 October 1989 the "Rules and Regulations Governing Investment Companies under Republic Act No. 2629" signaling the revival of mutual funds. One open-end mutual fund company, i.e., Magellan Fund, with foreign participation as minority shareholders, was approved under the new rules and regulations, bringing the total number of mutual funds to three.

Third, rules and regulations covering foreign investments in Central Bank-approved securities have been relaxed.[19] Under the new rules, some of the functions of the Central Bank have been downloaded to the custodian banks to reduce red tape.

Transactions related to foreign investment in Central Bank-approved securities can now be settled in three to four days compared to four to six months under the old rules and regulations, which has resulted in a significant increase in portfolio investment inflow since the last quarter of 1991. Portfolio inflows in January 1992 amounted to $84.2 million compared to $32.5 million for the same month in 1991.

To boost the capital market even further, the SEC recently issued rules and regulations covering warrants and asset-backed securities.

The SEC has yet to change its regulatory approach to a developmental approach insofar as the capital market is concerned. A clear example is how it deals with insider trading. Saldaña (1989) reviewed four cases in which the SEC had raised the question of insider trading: Engineering Equipment, Inc. (EEI), San Miguel Corporation (SMC), Benguet Corporation, and PNB. Under its regulatory approach, SEC was unable to find any violations.

Saldaña therefore suggested a developmental approach to insider trading, i.e., push for the immediate dissemination of material facts and interim reports submitted to SEC by listed companies to reduce potential gains from insider trading. In opposition to SEC's plan to revise the regulation on disclosure of insider trading to catch broker-directors who have already traded based on inside information, Saldaña suggested the prohibition of trading

---

[19] This was promulgated in Circular No. 1284, 1 June 1991, later amended by Circular 1318, 3 January 1992, which has since been consolidated in Central Bank Circular No. 1389 (April 1993).

by brokers who are also directors/officers of a listed company to prevent them from gaining access to internal information rather than to catch them after the fact. This developmental approach would give a "level playing field" to all brokers.

More reforms are expected to be implemented in the near term. However, a number of these require legislative action, including the installation of an efficient clearing and settlement system; the redefinition of the maturities of various instruments; the gradual phasing out of the gross receipts tax; the elimination of the documentary stamp tax on collateral instruments for bonds, secondary trading of bonds, and on inward and outward remittances of foreign exchange for CB-registered foreign investments in long-term domestic securities; the allowance of a nonresidents tax exemption on dividend income from listed shares of stock; and the lowering of the transaction tax on sales of SEC-registered over-the-counter (OTC) shares.

Foreign Exchange Market

The foreign exchange market used to be highly regulated. The rigid regulations effectively segmented the market into three submarkets: interbank market, customer market, and parallel market.

In the interbank market, banks were allowed to trade among themselves on-floor at the trading floor of the Bankers Association of the Philippines (BAP) from 4:30 pm to 5:00 pm daily on weekdays. The Central Bank also participated in the trading. Off-floor trading was prohibited.

Trading in the interbank market was mainly between the Central Bank and the banks for the purpose of correcting banks' position or for the account of their clients. Little trading occurred among banks and hardly any market-making was conducted by banks. The official exchange rate was based on the rates determined at the BAP's trading floor.

The customer market was much larger than the interbank market but was highly segmented. Banks tried to match their sources and uses of foreign exchange to maintain their customers.

The parallel market, whose transactions did not pass through the formal banking system, was estimated to be as large as the customer market. A substantial amount of the funds transacted came from remittances of Filipino overseas workers. In the 1980s, the premium in the parallel market was as high as 20 per cent, especially during the severe balance-of-payments crisis.

The process of deregulating the foreign exchange market began in mid-1991 and has continued to the present day. The Central Bank initially relaxed some rules to test the reaction of the market. For example, it initially increased the retention limit for the export proceeds of exporters from

2 per cent to 40 per cent to be deposited in special foreign currency deposit accounts with authorized banks, and then later raised this to 100 per cent with no restrictions. In addition, the Central Bank removed the limits on foreign exchange purchases for such things as travel, educational expenses, and medical expenses. The ceilings were later completely removed. Some relaxation of the rules did not require any gradual phase-out. For instance, the requirement that Filipino nationals working overseas remit specified minimum shares of their earnings was removed.

In sum, trade and nontrade related foreign exchange transactions have been liberalized. As for the capital account, restrictive regulations remain in effect with respect to foreign borrowing by the private and public sectors, especially that guaranteed by the National Government or government financial institutions.

An important measure instituted by the Central Bank was the lifting of the prohibition on off-floor trading. The creation of the Philippine Dealing System, which began operations in April 1992, was the Bankers Association of the Philippines' response to the lifting of the ban on off-floor trading. The system links participants through an electronic screen-based network for sharing information and undertaking transactions. All participants are required to submit two-way quotes. This provides participants with quick and sufficient information about the foreign exchange market at a low transaction cost.

While the Philippine Dealing System is limited to members of the BAP and the Central Bank, there are a number of corporate subscribers. However, they are not allowed to trade in the system.

In a recent assessment of the Philippine Dealing System, Saldaña (1993) found that it:

(i) has led to an increase in the average daily volume of trading from $9.4 million to $18 million;

(ii) has reduced the dominance of the Central Bank which characterized the previous on-floor trading arrangement;

(iii) has increased competition among participants (as evidenced by the bid-offer spreads of transaction at the Philippine Dealing System, ranging from 10 to 12 centavos);

(iv) has led to a more volatile exchange rates (with standard deviations of the daily exchange rates ranging from 7 to 63 centavos as contrasted with standard deviations ranging from 1 to 25 centavos before), indicating that the exchange rate is no longer managed by the Central Bank; in addition, the exchange rate differential between the formal and informal markets has on average been lower.[20]

Foreign exchange market deregulation has opened the way for banks to transact in financial derivatives. Some banks have offered new financial products, such as the peso-dollar convertible in which borrowers have the option to pay loans in foreign or domestic currency.

Interestingly, many anticipated that foreign exchange deregulation would lead to a larger depreciation of the peso. On the contrary, the peso has appreciated. Indeed, the stabilization program implemented by the Philippine Government between 1990 and 1992 led to high interest rates on peso-denominated assets when converted to dollar terms. This induced substantial capital inflows facilitated by the foreign exchange liberalization, which in turn caused the appreciation of the peso. The Central Bank entered the foreign exchange market to purchase foreign currency to arrest the further appreciation of the peso. However, since it had to meet its monetary target, it sterilized its purchases of foreign exchange through open market operations, which led to a high interest rate. This resulted in a dramatic increase in the gross international reserves of the Central Bank from $2 billion in 1990 to $6.7 billion in April 1993. This leads to the conclusion that the microeconomic gains from the foreign exchange reforms have been negated by inappropriate macroeconomic policy.

INFRASTRUCTURE SUPPORT SYSTEM

The country's infrastructure support system determines to a great extent the efficacy of its financial markets. As noted earlier, there are information asymmetries both from the borrowers' and the lenders' perspectives. The absence of a well-developed and uniformly applied accounting system makes the information gathering task of lenders difficult. The financial information about the borrower that is extracted from formal financial statements is important in determining the likelihood of success of the venture or project to be financed by the loan. It will also provide vital information on the ability of the borrower to repay the loan. Savers and investors must be able to gauge the integrity of deficit units and the relative riskiness of alternative financial instruments. Hence, an accounting system that is reflective of the true financial condition of agents in the financial markets is important. The other side of the information problem is the need to have a reliable and fair yardstick with which to measure the accuracy and

---

[20] Saldaña (1993) argued that the informal market is not a parallel market in that it is basically a retail market while the Philippine Dealing System is a market for wholesale transactions.

reliability of the accounting information presented to interested parties. An independent assessment or audit of the accounting information that is fair and consistent is a necessary institution in financial markets. The existence of independent credit rating agencies can reduce the transaction costs of both lenders and borrowers.

On the other hand, the existence of a legal framework which will assure the lender prompt and just repayment of the loan will give the confidence needed to extend more loans. With efficient legal remedies to assure debt recovery, there might be less emphasis on the collateral requirements and more attention to the viability prospects of the project to be financed. Khanna and White (1993) observed that the recourse available to lenders in the event that the borrowers have provided misleading information and the ease of loan recovery in the event of a loan default can be important in the lender's decision to provide loans.

Accounting and Auditing Systems[21]

There are 206 accounting and auditing firms registered with the SEC. Although nearly all are located in the National Capital Region, auditing services are provided to firms located outside the region by using auditing teams.

Philippine firms and banks are required to maintain appropriate financial records which observe standard and generally accepted accounting principles and procedures (GAAP). Banks and firms (or enterprises, in general) prepare financial statements that follow the legal and regulatory framework and the accounting policies and procedures set by the supervisors (Central Bank for financial institutions and SEC for non-financial firms). Deliberate misstatement of information or misrepresentation in financial statements is a criminal offense under Philippine law.

Accounting System

The Accounting Standards Council (ASC) was organized in 1981 by the Philippine Institute of Certified Public Accountants (PICPA) to provide suitable financial accounting standards for the Philippines. The Central Bank, the Professional Regulation Commission through the Board of Accountancy, the SEC, and the Financial Executives Institute of the Philippines (FINEX) endorsed the creation of the ASC. Its main function is to establish and improve accounting standards that will be generally

---

[21] This discussion is drawn from the *Compilation of Statements of Financial Accounting Standards Nos. 1-18* and the *Statement of Auditing Standards of the Philippines* issued by the Accounting Standards Council and the Auditing Standards and Practices Council, respectively.

accepted in the Philippines. The ASC uses the accounting pronouncements issued by PICPA, the International Accounting Standards Committee, and the Financial Accounting Standards Board. There are eight members distributed as follows: four from PICPA; one from SEC; one from Central Bank; one from the Board of Accountancy; and one from FINEX.

The Professional Regulation Commission must give its approval to the Statements of Financial Accounting Standards and Related Interpretations that the ASC will issue. The Statements of Financial Accounting Standards are applicable to the financial statements of any commercial, industrial, or business enterprise which must be prepared according to the generally accepted accounting principles. The ASC has issued nineteen Statements of Financial Accounting Standards, of which eighteen are generally applicable to commercial and industrial enterprises; one is for banks and related institutions.

The generally accepted accounting principles encompass the conventions, rules, and procedures necessary to define accepted accounting practices at a particular time, and include both the broad guidelines of general application and detailed accounting practices and procedures. For instance, the principles define the nature and valuation of the account and how it is presented in the financial statement. An example is the treatment of the Interbank Loans Receivable account. According to the GAAP, it covers call or other loans extended in the interbank market to local banks and non-bank financial intermediaries engaged in quasi-banking functions. Interbank money comes from banks with excess reserves and is borrowed by banks with deficiency in reserves and other liquidity requirements. Interbank loans are stated at the outstanding principal amount. If the balance of this account is material, it should be shown as a separate line item in the statement of conditions. Otherwise, it is categorized under the heading of "Loans." Enterprises are duty bound to adopt sound accounting policies and practices, maintain an adequate and effective system of accounts, safeguard the integrity of the assets, and devise a system of internal control that will help establish the viability of the business.

Financial Reports on Cash versus Accrual Basis

Accrual Basis. This method of accounting is used by almost all financial institutions in the Philippines. The method measures income for the period as the difference between revenues recognized in that period and the expenses matched with those revenues. In the case of revenues, amounts earned rather than amounts collected are recorded; in the case of expenses, it is amounts incurred rather than amounts paid. The period's revenues are not necessarily the same as the period's cash disbursements. The income statement of a corporation would normally be measured on an accrual basis, and would include in its measure current costs such as depreciation allowance or depletion allowance.

Cash Basis. The cash basis of accounting approximates the flow of funds and resources, avoids evaluation problems, corresponds closely with other financial statistics and is, therefore, eminently useful for policy and financial programming.

Money receipts and payments are recorded at the time they are settled in cash, sales are recorded for the period in which they are received in cash, and costs are subtracted from sales in the period in which they are paid for by cash disbursements. The acquisition of inventories are recorded as a reduction in profit when the acquisition costs are paid, rather than when the inventories are sold. The cost of acquiring plant, property and equipment is treated as a profit reduction when paid in cash, rather than in the later period when these long-lived assets are used. There is no recognition of bad debts since revenue is not recognized unless cash is received. There is no recognition of depreciation since the entire cost of the equipment is recognized as an expense at the time of payment.

The Bureau of Internal Revenue permits the filing of income tax returns on either the accrual or cash basis. However, the cash basis for tax purposes is actually a combination cash-accrual basis, since it is recognized that the application of a strictly cash approach could result in serious distortions in net income measurement. Further, a strictly cash approach could offer a means of shifting significant amounts of revenues and expenses from one year to another by control of cash receipts and disbursements.

Use of the cash basis, then, generally means the use of a hybrid system, with sales, purchases, depreciation, and doubtful accounts being reported as on the accrual basis, but with the remaining revenue and expense items being measured by cash receipts and disbursements. This method offers a simple and more economical accounting.

A summary of operations prepared on the cash basis may be acceptable when failure to recognize accruals and prepayments results in relatively minor misstatements that are largely counterbalanced in periodic accounting. However, when accruals and prepayments are material in amount and vary significantly from period to period, satisfactory net income measurement would call for the adoption of the accrual basis of accounting.

These accounting principles are generally applicable to commercial, industrial, business enterprises, banks, and related financial institutions, though bank accounting does have some major differences from conventional double-entry bookkeeping. Significant differences are as follows:

(i) Banks have daily posting of transactions and trial balancing of the general ledger. A statement of condition is prepared daily.

(ii) Banks are governed by special rules and regulations issued by the Central Bank.

(iii) A large portion of assets held by banks consists of cash and negotiable items which may be kept both for its own account and in trust for others.

(iv) Loans constitute a major group of assets. Their fair presentation involves an evaluation of the adequacy of the reserve for losses.

(v) Good internal control is essential.

(vi) Trust functions constitute a major activity of commercial banks. Trust activities are performed by a separate department within the bank and are governed by the Trust Regulations promulgated by the Central Bank.

Under current regulations, banks submit monthly a set of financial statements to the Central Bank which should conform to the prescribed Regulatory Accounting Policies of the Central Bank Manual of Accounts. In accordance with the generally accepted accounting principles, audited financial statements have to be submitted at the end of the fiscal year. The audited financial statements carry a reconciliation of the monthly financial statements and the audited financial statements. While most of the differences between the GAAP and Regulatory Accounting Policies have been removed, three major points of difference with respect to the application of the equity method in accounting for investments remain:

(i) Under the *Regulatory Accounting Policies*: Evidence of significant influence is ownership of more than 50 per cent; for ownership between 20 per cent and 50 per cent, a determination of the existence of significant influence is made by the Central Bank before the equity method may be applied.

Under *GAAP*: Twenty per cent ownership presumes the ability to exercise significant influence. There is, however, a need to evaluate the facts and circumstances relating to the investments which will confirm the existence or absence of such influence by the investor.

(ii) Under *Regulatory Accounting Policies*: The excess of cost over book value of the investment is charged to operations over a maximum period of five years.

Under *GAAP*: The excess of cost over book value is amortized over the economic life of specifically identifiable undervalued assets and/or the estimated economic life of the goodwill, which period should not exceed 40 years. The Regulatory Accounting Policies is a stricter application of GAAP. The Central Bank will permit the amortization beyond the five-year period only with prior approval from the Monetary Board.

(iii) Under the *Regulatory Accounting Policies*: In applying the equity method on equity investments in foreign subsidiaries, the parent's share in the income/loss of the subsidiary is valued at the foreign currency amount converted into pesos at the exchange rate existing as of the balance sheet date of the investee. The equity investment in foreign subsidiaries are considered as nonmonetary accounts and are not revalued for fluctuations in exchange rates. Instead, they remain revalued at historical cost.

Under *GAAP*: The financial statements of foreign subsidiaries should first be translated into pesos. The translated financial statements shall then be the basis of applying the equity method.

Auditing System

The Philippines has an Auditing and Standards Council (ASPC) established in 1987 which issues statements on "Generally Accepted Auditing Standards." These standards are approved by the Board of Accountancy and the Professional Regulation Commission. The standards require the independent auditor to indicate whether the financial statements presented conform with generally accepted accounting principles. The general standards for the audit are as follows:

— The examination is to be performed by a person or persons having adequate technical training and proficiency as an auditor(s).

— In all matters relating to the assignment, an independence in mental attitude is to be maintained by the auditor.

— Due professional care is to be exercised in the performance of the examination and the preparation of the report.

— The work is to be adequately planned and assistants, if any, are to be properly supervised.

— There is to be a proper study and evaluation of the existing internal control as basis for reliance thereon and for the determination of the resultant extent of the tests to which auditing procedures are to be restricted.

— Sufficient competent evidential matter is to be obtained through inspection, observation, inquiries, and confirmations to afford a reasonable basis for an opinion regarding the financial statements under examination.

— The report must state whether the financial statements are presented in accordance with generally accepted accounting principles.

— The report must state whether such principles have been observed consistently in the current period in relation to the preceding period.

— Information disclosures in the financial statements are to be regarded as reasonably adequate.

— The report must contain either an expression of opinion regarding the financial statements, taken as a whole, or an assertion to the effect that an opinion cannot be expressed. When an opinion cannot be expressed, the reasons for this should be stated. In all cases where an auditor's name is associated with financial statements, the report should contain a clear-cut indication of the character of the auditor's examination, if any, and the degree of responsibility the auditor is taking.

The generally accepted auditing standards are applicable in the audit of the financial statements of banks, related financial institutions, and business enterprises. Besides the audit, management letters prepared by the auditors give important information on various facets of the enterprise's operations. For example, deficiencies in internal control, or inadequate allowance for losses may be brought to the attention of interested parties by such management letters. The audit is undertaken by a cadre of qualified professional auditors who have passed stringent qualification standards and procedures laid down by the Professional Regulation Commission and the Auditing Standards Council of the Philippines. Although the audit is not a statement of the future viability of the enterprise, it is important in that it indicates the credibility and fairness of the financial statements.

Legal Framework for Debt Recovery

The existence of a functioning legal structure that will guarantee ease of loan recovery in the event of a loan default is critical because it will help assure the lender that, in the face of credit risks brought about by information asymmetry, the lender can be protected from severe financial losses brought about by defaulting borrowers. By the same token, the borrower is aware that the legal structure will compensate the lender in the event of a loan default. A functioning legal structure protects property rights and permits the transfer of financial claims, and should have rules for the issuance and trading of those claims. The first requirement is the recognition of an enforceable loan contract. The Civil Code of the Philippines governs the enforcement of credit transactions.

Real Property Mortgage

Upon loan default, the lender can foreclose on the mortgaged property. A mortgage may be foreclosed judicially by bringing an action for that purpose in the Regional Trial Court of the province or city where the real property is located. The proceeds of the sale are applied to the (i) payment of the costs of the sale; (ii) the amount due the mortgagee; (iii) claims of persons holding subsequent mortgages in the order of their priority; and (iv) the balance, if any, is paid to the mortgagor (i.e., the borrower). Under a judicial foreclosure, the sale of the mortgaged and foreclosed real property is not subject to the one year redemption period except when the mortgagee is a bank or credit institution where the foreclosure process (whether judicial or extrajudicial) is subject to special law. In this case, the sold real property can be redeemed within one year after the date of the sale. This implies that the law on statutory redemption has an asymmetric application on the mortgagees whereas there is a good argument for applying the law uniformly to bank and nonbank mortgagees.

In the case of extrajudicial foreclosure, the loan contract must stipulate that the mortgagee (the lender) has the power to foreclose the mortgage by an extrajudicial sale of the mortgaged property upon default of the debtor. The sale of real property which was the subject of extrajudicial foreclosure is subject to a one year redemption period. The extrajudicial sale must be done through a public auction conducted by a judge, notary public, or sheriff after posting advance public notice of the sale. Publication once a week for at least three consecutive weeks in a newspaper of general circulation in the municipality or city where the foreclosed real property is located is required. Philippine experience shows that there are attendant long delays and expenses in the usual judicial foreclosure process. There is, thus, a preference for extrajudicial foreclosure but the reality is that under Philippine practice, the extrajudicial foreclosure may become the subject of judicial process at the instance of the mortgagor or any other interested party. The realization of the mortgagee's debt will, therefore, take some time.

On the other hand, the borrower has some recourse with respect to the mortgage that would be auctioned. The borrower has what is called "equity of redemption" which is the right of the mortgagor to redeem his mortgaged property after foreclosure but before the sale of same. In judicial foreclosure, the borrower-mortgagor can exercise his equity of redemption before but not after the sale is confirmed by the court of law. The "right of redemption" is the right of the borrower-mortgagor to redeem the mortgaged property within a certain period after it was sold to satisfy the debt. In extrajudicial sale, the borrower-mortgagor may redeem his property at any time within one year from and after the date of the registration of the sale. In judicial foreclosure, the general rule is that the mortgagor cannot exercise his right of redemption after the sale is confirmed.

Chattel Mortgage

The chattel mortgage also acts as a security for the performance of the loan contract. From the point of view of the Civil Code, a chattel mortgage is both a formal contract—since it is registered in the Chattel Mortgage Register—and a unilateral contract—because it obliges the creditor to lift the encumbrance on the personal or movable property upon repayment of the loan. Certain laws such as the Chattel Mortgage Law, as amended and the Civil Code and the Revised Penal Code, among others, govern chattel mortgages. Upon loan default, the mortgagee has the right to sell the property at public auction through a public officer. To be valid, the chattel mortgaged must have been registered with the Register of Deeds. The mortgagor has a right of redemption which is exercised by paying or delivering to the mortgagee the amount of the loan and the costs and expenses that have been incurred because of the breach of the loan contract. The redemption must be made before the sale of the property or within the 30 day grace period.

Some Implications of the Insolvency Law

The country's insolvency law has a direct bearing on the ability of banks and other creditors to recover debt. They may end up with unsecured claims for deficiencies after foreclosing on their security because of certain provisions of the insolvency law. The Insolvency Law of the Philippines, which was enacted on 20 May 1909 as Act. No. 1956, has remained substantially the same for the last nine decades. It seeks to distribute equitably the property of debtors in financial distress and to free debtors from the burden of indebtedness. This law lists 14 categories of preferences that must be satisfied before the rule of *pro rata* distribution is implemented.

This particular rule of law seems burdensome to creditors of insolvent debtors. The number of preferences should be limited and/or a ceiling should be put on the amounts that can be claimed by category.

In conclusion, there are concrete efforts to professionalize and upgrade the standards of accounting and auditing in the Philippines and to encourage individuals and firms alike to maintain those standards. The existing accounting and auditing systems are comparable with those of more developed countries. However the actual accounting and auditing practices are different from their respective standards. In particular, small- and medium-scale firms seldom maintain good accounting systems; this is one of the reasons why banks seldom lend to this group of borrowers. Also, it has been a general practice among firms to keep two books of accounts, one for their creditors (this is called in-house financial statement) and one for the BIR for tax purposes; the differences between the two are glaring. If Executive Order No. 10 mentioned earlier in this chapter is implemented, many firms may not qualify for a loan from a bank, or if they do, they may obtain a much lower amount of loan than if they presented the real financial status of their firms to the BIR at the start.[22]

With respect to the legal framework for debt recovery and insolvency, the process of judicial and extrajudicial foreclosures is lengthy and time-consuming. Delays negate the potential of judicial and extrajudicial foreclosure as remedies that can be easily availed of by creditors in the face of loan defaults. In addition, because of the delays and the uncertainty created by the long litigation process, banks will not be encouraged to provide secured long-term credits.

Credit Rating Institution

The Credit Information Bureau, Inc. (CIBI)—the only business information and credit rating institution in the Philippines—was organized as a non-stock, nonprofit private corporation by the Central Bank of the Philippines, the Securities and Exchange Commission, and the Financial Executives Institute of the Philippines on 14 April 1982 to develop and maintain an efficient mechanism for gathering and analyzing credit information for the use of its subscribers.

---

[22] Under the recently issued Executive Order No. 10, banks are required to accept only the financial statements that have been submitted to the BIR by firms when applying for a bank loan.

The CIBI accumulates, organizes, and makes available on a continuing basis, accurate and reliable data on the financial status and performance of businesses operating in the Philippines. Its charter also allows it to render credit evaluation services to interested parties including government regulatory agencies. The SEC has in fact asked CIBI to evaluate several commercial paper issues for investor protection.

Presently, the CIBI has 1,077 local and 70 foreign subscribers. Its international correspondence network covers 262 countries. The CIBI provides a variety of services, which include among others, the following: information on commercial establishments, individual, provincial or foreign businesses; verification of property, court cases, addresses of potential creditors/investors and database reports—portions of the business information report which are provided upon request by the client. It also gives information on the corporate family tree which traces the relationships among individuals, companies, and business groups.

The CIBI's assessment of commercial papers has been well-received by investors and the business community because of the acknowledged independence of CIBI credit ratings. The CIBI publishes a quarterly *Commercial Paper Ratings Guide* which is a comprehensive report on the credit rating of the commercial papers as well as the financial condition and operating trends of each corporate issuer.

Commercial paper issuers agree to pay a fee to CIBI for the evaluation and rating services prior to receiving the rating. For short-term commercial paper issue (i.e., those with maturity of 365 days or less), the CIBI has six ratings to indicate the relative repayment capacities, ranging from strongest capability to expected default. For long-term commercial paper (i.e., those with maturity of more than 365 days), nine ratings are used, ranging from smallest degree of investment risk to extremely poor prospects of ever attaining any real investment standing.

For the past ten years of its existence, the CIBI has shown the advantage of having an independent credit rating and business information institution. However, its coverage is still limited because of the cost involved in obtaining and updating information, the limited public listing in the Philippine Stock Exchange; and the small number of companies that can issue quality commercial papers. A principal problem (but prospective growth and development area) is the limitation of the automated information network to the National Capital Region. Nationwide automated on-line access to CIBI information will give investors in other cities of the country the same advantage as a Manila-based user. On the demand side, Philippine business— while appreciative of the significant role that objective credit evaluation brings—is aware of the high cost of obtaining information.

REVIEW AND ASSESSMENT OF GOVERNMENT POLICIES AND RECOMMENDATIONS

Government policies affecting the financial sector have changed considerably over the last decade. What follows is a general assessment of those policies.

Stabilization Policies

Borrowing and lending involve a contract in which the borrower promises to pay the lender in the future the principal and a certain interest rate. A highly unstable economy reduces the profitability of the business and increases the probability of default by borrowers. Even if default is unlikely, an unstable economy reduces the real return on lending. It, therefore, makes sense to institute prudent macroeconomic policies to achieve a certain degree of economic stability necessary to enhance debt contracting.

The Philippine economy in the last decade was highly unstable, which contributed to the weak financial intermediation process. Measures to stabilize the economy will likely preoccupy policymakers in the next decade. A rapid decline in the rate of inflation and in interest rates in the last six months can be sustained if prudent macroeconomic policies are maintained. In this regard, managing the consolidated public sector deficit is crucial over the next five years. More emphasis should be given to increasing tax revenues through improvement in tax administration, broadening of the tax base, and reduction in the losses incurred by government corporations.

The move toward improving efficiency in financial intermediation has been hampered by a series of crises that struck the economy over the last fifteen years. However, as a result of these crises, weaker financial institutions have been purged and the weaknesses of prudential regulations have been exposed. This section makes some assessments of key microeconomic reforms undertaken by the Government since the mid-1980s and recommends some policies.

Pricing Policies

Interest rate liberalization was supposed to have been completed in 1983 (i.e., banks could freely set their lending and deposit rates). However, special credit programs of the Government continued to carry substantially subsidized interest rates, which distorted the structure of interest rates in the market. This distortion was greatly reduced when the Government in 1986 began to align the interest rates on its special credit programs with the market rates. Since then, banks, especially rural banks, have been compelled to mobilize more deposits which are much cheaper than funds from the credit programs.

Interest rate liberalization came at a time when the monetary authorities were trying to strengthen the financial position of institutions (i.e., closing down weak financial institutions and increasing the capitalization requirement of the remaining ones). The stricter policy on bank entry and branching prevailing at the time when interest rates were liberalized provided incumbents with opportunities to exploit their monopoly power by increasing bank spread.

Taxation of Financial Intermediation

The explicit and implicit tax rates on financial intermediation are not insignificant. They drive a wedge between lending and deposit rates. Moreover, the differential tax treatment on various instruments have distorted the rates of return on financial assets, causing some shifts from highly taxed financial instruments to less taxed instruments. This would not have occurred had there been no differential tax treatment between the instruments. Some extra costs must have been involved in such shifts, both on the part of depositors and financial institutions. A case in point is the differential reserve requirement on traditional deposits and trust funds that favors the latter over the former. The same is true of pension funds. In particular, self-employed pension plans are discriminated against by taxation relative to occupational pension plans. The structure of taxation imposes a heavy burden on the insurance industry relative to other sectors and, in particular, to the less regulated pre-need industry. Some taxes do not have any economic justification. One example is the documentary stamp tax on secondary trading which obviously discourages secondary trading of financial instruments.

There is indeed a need to do a comprehensive review of the tax policy on financial institutions, transactions, and instruments.

Entry and Branching

Entry and branching policy is applied more restrictively on the banking system than on the nonbank financial system because the former accepts deposits from the public. This is possibly one of the reasons why some nonbank financial institutions, particularly pawnshops and lending investors, attained a rapid high growth in terms of number of offices/branches during the period 1986–1991.

The rules and regulations on bank entry and branching have been relaxed significantly since 1989. This has paved the way for the entry of new banks and has led to significant growth in bank branches. However, within the banking system, entry into a more sophisticated type of bank is more restrictive than entry into the less sophisticated one. In recent years, it has taken the Central Bank less than one year to approve an application for a thrift bank, whereas it took about three years to approve an application for a universal bank despite the fact that the applicant had already complied with all the necessary requirements to operate as a universal bank.

Entry of foreign banks into the domestic banking system either as a subsidiary or a branch is still prohibited.[23] Joint ventures are encouraged but foreign banks are limited to being minority stockholders. In the insurance industry, foreign equity participation up to a maximum of 40 per cent (the same requirement for a bank) is encouraged in the inadequately capitalized non-life sector, but is not allowed in the life sector except for those foreign companies which have already been allowed to transact business in the Philippines.

An abundance of suppliers of credit does not necessarily lead to more competition.[24] Because of asymmetric information, there is already less competition in the loans market even with the presence of a number of financial institutions, and the merger/consolidation policy will further reduce that competition.

Prudential Regulations

The near collapse of the money markets and the failure of several banks and nonbank institutions, particularly investment houses and finance companies, during the 1981 liquidity crisis, the closure of several weak banks during and in the aftermath of the 1984–1985 balance-of-payments crisis, and the sporadic scandals in the stock markets which left many potential investors suspicious of the said markets have convinced the regulatory authorities of the need to strengthen prudential regulations. The measures taken by the Government to improve prudential regulations took several forms. The minimum capital requirements for different types of financial institutions have been adjusted upward three times since 1980 and a clearer definition of risk assets has been made.

The single borrower's limit that applies to all lending institutions, including government financial institutions, is another prudential measure. A much broader definition of liabilities has effectively reduced the ceiling that any one borrower can borrow from a bank. The chances that banks will lend more to individuals with uncorrelated risks are greater under this regulation.

---

[23] The bill liberalizing the entry and scope of operations of foreign banks is expected to be passed by Congress in 1994.

[24] Customers cannot move from one lender to another and get the same services at the same cost.

The recent deregulation of the foreign exchange markets and the establishment of off-floor trading have provided banks with profit opportunities through direct position-taking or service to clients. To prevent excessive risk-taking on the part of banks under this deregulated environment, the Central Bank has imposed limits on the open foreign exchange position of banks. Specifically, banks' long and short foreign exchange positions are not allowed to go beyond 25 per cent and 15 per cent, respectively, of their unimpaired capital.

In some areas, major efforts in improving prudential regulations are to be exerted before problems arising from abuses and excessive risk-taking by unscrupulous players reach crisis proportions. In the stock markets, stronger regulatory measures are needed, such as prohibition of trading by brokers who are also directors/officers of a listed company, and immediate dissemination of material facts and interim reports submitted to SEC by listed companies to reduce potential gains from having access to inside information. In the insurance industry, there is a need to subject to prudential controls the investments, reserves, and liability valuation of pre-need companies and to impose actuarial supervision of the risks associated with the pre-need products.

Government's Direct Participation in the Financial System

The Government has reduced its direct participation in the financial markets by selling to the private sector several financial institutions it acquired during the 1980s. It has since focused its intervention in strategic areas not being served by the private sector. Since the securities market is relatively underdeveloped, the long-term debt market could not be relied upon by the private firms for their long-term capital requirements. On the other hand, private banks have concentrated on the provision of short-term loans. Given the historically unstable economy, term transformation might be a very risky proposition for banks. This plus the desire to respond to the demands of private firms for long-term capital to expand their capacity leaves the Government with no choice but to be actively involved in the long-term debt market through its government financial institutions.

These institutions mobilize long-term funds mainly from multilateral and bilateral donors. They lend wholesale funds to private financial institutions, which, in turn, retail them together with some funds of their own to ultimate users. Thus, the government financial institutions and the private one complement each other in the long-term debt market. The best way to increase the share of the private financial institutions in this market is to maintain a stable economic environment and financial system.

Accounting, Auditing, Legal, and Institutional Framework

The accounting and auditing standards in the Philippines are fairly developed with some degree of sophistication. However, small- and medium-scale firms seldom keep a good accounting system and generally do not use the services of well-established external auditors, which are sufficiently available in the Philippines. Therefore, banks have serious difficulty in evaluating their financial performance and potential when they apply for a loan.

Banks generally require financial statements of their clients to be audited by external auditors. However, auditors certify only the accuracy of the accounting information and methods used (i.e., the firm follows the generally accepted accounting principles), not necessarily the veracity of the information. This is because under Philippine practice, it is up to the firm to disclose pertinent information about itself and the project. Therefore, audited financial statements are as good as the amount and quality of information provided by the firm. In this environment, the problem of asymmetric information persists.

Banks try to remedy the situation by maintaining an in-house cadre of credit investigators that do both off-site and on-site verifications and investigations, supplemented by corollary information from the CIBI. However, because the cost of obtaining information through this approach is high, banks are not encouraged to cater to small- and medium-scale firms. This suggests that the existing accounting and auditing framework must be accompanied by a good incentive structure for firms to use the appropriate accounting method and to fully disclose information.

With respect to the legal framework for debt recovery and insolvency, the existing laws and procedures are not conducive to lending. For instance, judicial and extra-judicial procedures for disposition of foreclosed collateral involve a long and tedious process generally in favor of the debtor. Partly because of this, banks tend to demand highly liquid collateral, such as government securities, hold-out deposits, and shares of stocks. Under the existing Insolvency Law, there are many claims senior to debt that must be satisfied first.

The discussions above suggest that the existing legal framework must be reviewed thoroughly with the objective of simplifying procedures for quick settlement of loan obligations.

Supervisory Bodies

Improvement in prudential regulations does not guarantee the stability and efficiency of financial intermediation unless the supervisory bodies have the capability to process information and enforce regulations. This means that these bodies must have the necessary number of skilled personnel, which unfortunately is not present in government agencies. For instance, the Central Bank examines large banks at least once a year, but does the same for small banks once every two years due to lack of personnel.[25] Because of their incentive structure, Government supervisory bodies find it difficult to maintain an adequate cadre of qualified personnel. The incentive structure must, therefore, be redesigned in such a way that compensation is tied to performance.

A comprehensive review of the mandates of supervisory bodies, particularly the SEC and the Insurance Commission, is necessary to ensure their effectiveness in supervising and regulating financial institutions and markets.[26]

---

[25] The staff complement of the Central Bank is approximately 316 persons per $1 billion, compared with the Bank of England with 443 persons, Singapore with 9 persons, and Australia with 79 persons per $1 billion (Bertrand et al. 1992).

[26] This is no longer necessary for the Central Bank because of the passage of a recent law creating the New Central Bank Act. This law provides the new Central Bank with strong powers of intervention, such as issuing an order requiring the bank or officers of the bank to cease and desist from doing something that endangers the financial status of the bank concerned.

Bibliography

Aiyagari, S. Rao. 1993. "Explaining Financial Market Facts: The Importance of Incomplete Markets and Transaction Costs." *Quarterly Review*. Winter. Federal Reserve Bank of Minneapolis.

Aries Group. 1990. "A Study of Securities Market Institutions in the Philippines." Manila: Asian Development Bank.

Bautista, Ernesto D. 1992. "A Study of the Philippine Monetary and Banking Policies." Philippine Institute for Development Studies (PIDS). Unpublished Paper.

Bertrand, Trent, Compton Bourne, Ponciano S. Intal, Jr., Mario B. Lamberte, Juan Carlos Protasi, and Eli Remolona. 1992. "Foreign Exchange Liberalization in the Philippine Economy." Unpublished.

Business Day's *Top 1,000 Corporations*. 1975, 1983, 1984.

Business Star. 5 August 1991.

Central Bank of the Philippines. 1991 and various years. "Philippine Financial System." *Fact Book of the Philippine Financial System*. Manila: Central Bank of the Philippines.

_____. (various years). *Selected Philippine Indicators*. Manila: Central Bank of the Philippines.

Diokno, Benjamin, Mario B. Lamberte, and Rudolph Penner. 1992. "Philippine Public Debt Management." Unpublished.

Estanislao, Basilio, and Gilberto M. Llanto. 1993. "The Comprehensive Agrarian Reform Program and the Collateral Value of Agricultural Lands." Unpublished.

Flood, Mark D. 1991. "An Introduction to Complete Markets." *Review* Vol. 73. No. 2. St. Louis, Missouri: The Federal Reserve Bank of St. Louis.

Government Service Insurance System. 1990. *1990 Annual Report*.

_____. 1991. *1991 Annual Report*.

Home Insurance and Guaranty Corporation. 1991. *1990–1991 Performance Report.*

Khanna, Ashok, and Lawrence J. White. 1993. "The Role of the Financial Sector in Developing Countries: A Methodological Framework." Manila: Asian Development Bank. Mimeo.

Lamberte, Mario B. 1992. "Review and Assessment of Policy-Based Lending Programs in the Philippines." Unpublished.

_____. 1990. "Credit Unions: An Underrated Mode of Mobilizing and Allocating Resources in Rural Areas." *Working Paper Series No. 90–21.* Manila: PIDS.

Lamberte, Mario B. 1982. "Behavior of Commercial Banks: A Multiproduct Joint Cost Function Approach." Ph.D. Dissertation submitted to the School of Economics, University of the Philippines. Diliman, Quezon City.

_____. 1985. "Financial Liberalization and the Internal Structure of Capital Markets: The Philippine Case." *Staff Paper Series No. 85–07.* Manila: PIDS.

_____. 1987. "Comparative Bank Study: A Background Paper." *Working Paper Series No. 87–04.* Manila: PIDS.

_____. 1989. "Assessment of the Problems of the Financial System: The Philippine Case." *Working Paper Series No. 89–18.* Manila: PIDS.

_____. 1990. "An Assessment of Policies Affecting the Financial Sector, 1986–1988." *Working Paper Series No. 90–05.* Manila: PIDS.

_____. 1991. "Private Financial Savings and Direct Foreign Investment." Unpublished.

_____. 1992. "Assessment of the Financial Market Reforms in the Philippines, 1980–1992." Paper presented during the Third Convention of the East Asian Economic Association, Seoul, Korea.

Lamberte, Mario B. et al. 1985. "A Review and Appraisal of the Government Response to the 1983–84 Balance-of-Payments Crisis." *Monograph Series No. 8.* Manila: PIDS.

Lamberte, Mario B., and Eli M. Remolona. 1992. "Deregulation of the Foreign Exchange Market." Unpublished.

Land Bank of the Philippines. *Annual Report 1991.*

Lapar, Ma. Lucila A. 1988. "An Empirical Analysis of Credit Rationing in the Rural Financial Markets of the Philippines." Masters Thesis submitted to the School of Economics, University of the Philippines. Diliman, Quezon City.

Licuanan, Victoria S. 1986. *An Analysis of the Institutional Framework of the Philippine Short-term Financial Markets.* Manila: PIDS.

Lin, See Yan. 1992. "The Savings-Investment Gap, Financing Needs, and Capital Market Development." *Malaysia's Economic Vision.* Malaysia: Pelanduk Publications.

Llanto, Gilberto M., 1989. "Asymmetric Information in Rural Financial Markets and Interlinking of Transactions Through Self-Help Groups." *Journal of Philippine Development*, No. 28, Vol. XVI.

Llanto, Gilberto M., and Marife Magno. 1992. "Agrarian Reform and Credit Markets in the Aquaculture Sector: Analytical Framework and Some Preliminary Findings." Paper presented in the Workshop on the Analytical Frameworks. Dynamics of Rural Development Project Phase II. Imus Sports Complex, 29-30 October 1992.

Llanto, Gilberto M., and Aniceto C. Orbeta. 1992. "Identification and Classification of Problem Banks." *Journal of Philippine Development.* No. 34, Vol. XIX.

National Statistical Coordination Board. 1993. "The National Accounts of the Philippines for CY1990–1992." Manila.

PAG-IBIG. 1991. *1991 Annual Report.*

Philippine National Bank. 1990. *Annual Report 1990.*

_____. 1991. *Annual Report 1991.*

Saldaña, Cesar G. 1993. "An Assessment of the Philippine Dealing System." Report prepared for USAID-Manila Mission.

Social Security System. 1991. *Annual Report 1991.*

Tan, Edita A. 1991. "Interlocking Directorates, Commercial Banks, and Other Financial Institutions and Nonfinancial Corporations." *Discussion Paper No. 9110.* Quezon City: UPSE.

_____. 1989. "Bank Concentration and the Structure of Interest." *Discussion Paper No. 8915.* Quezon City: UPSE.

Torres, Onofre, Robert C. Vogel, Dale Adams, Mario B. Lamberte, and Gilberto M. Llanto. 1989. "Philippine Financial Markets Development Strategy." Unpublished.

United States Agency for International Development/Philippines. 1991. "Capital Market Development Project." Unpublished.

Yap, Josef T., Mario B. Lamberte, Teodoro S. Untalan, and Ma. Socorro V. Zingapan. 1990. "Off-Balance Sheet Activities of Commercial Banks in the Philippines." *Working Paper Series No. 90–25.* Manila: PIDS.

Zingapan, Ma. Socorro V., Mario B. Lamberte, and Josef T. Yap. 1990. "Central Bank Policies and the Behavior of the Money Market." *Working Paper Series No. 90–24.* Manila: PIDS.

CHAPTER FIVE

FINANCIAL SECTOR DEVELOPMENT IN THAILAND

PAKORN VICHYANOND

The beginning of the 1990s saw crucial changes in most financial markets around the world. Numerous types of financial institutions were liberalized to conduct business transactions never before undertaken. Stringent rules and regulations once imposed by the monetary authorities on financial practices were relaxed, allowing greater flexibility as well as more business opportunities. One motivation for the deregulation policies was to strengthen the degree of competition among the financial institutions, a move which should have ultimately benefited the general public. However, there are no guarantees that strengthened competition or improved efficiency will leave financial stability unshaken. Financial stability depends on both the tactics adopted by the authorities during the different stages of liberalization and the readiness to adopt or adjust to those tactics on the part of the financial institutions.

This trend toward liberalization of the financial markets was also felt in Thailand. Not only did the Bank of Thailand allow more room for financial institutions to compete, but those institutions adopted both new technology and know-how to offer several new services and products to their customers.

Along with other financial markets in the world, Thailand addressed the question of efficiency versus stability. When gains in the former come at the expense of the latter, the monetary authorities may hesitate to take the next step; in Thailand, the authorities are already concerned about how to preserve financial stability given that a large number of financial institutions are exercising their new rights in a seemingly formidable fashion. This issue is addressed throughout this chapter.

This chapter first provides a broad perspective of the financial institutions in Thailand. Given the dominance of commercial banks and finance companies, a section of the chapter addresses these units specifically. Included as well is a review of Thailand's Financial Institutions Development Fund, its stock exchange, the country's accounting standards, and the recently adopted value-added tax (VAT). The chapter also examines some macroeconomic issues, such as the objectives and instruments of monetary, fiscal, and exchange rate policies, and traces the development of various aspects of the country's financial markets. Recent reforms, such as the dismantling of exchange controls, freer interest rate specification, more flexible portfolio management, less rigid reserve requirements, broader scope of operations, globalized standards of capital adequacy, and the opening of Bangkok International Banking Facilities (BIBF) are then discussed.

OVERVIEW AND EVOLUTION OF THE FINANCIAL SECTOR

The financial system in Thailand consists of a wide variety of institutions, the majority of which are privately owned. Commercial banks are predominant. In the past two decades, commercial banks accounted for 70 per cent of all financial institutions' total assets (Tables 5.1 and 5.2).

The commercial banking business first came to Thailand in 1888 with the establishment of a foreign bank branch. The first Thai bank was set up in 1906. Initially the branch offices of the foreign banks were the most active in the country, but the Thai banks soon became firmly established, especially under the continual protection provided by the Government. By the end of 1992, the circuit of 15 Thai commercial banks commanded an extensive network of 2,566 branches while foreign banks' activities were restricted to only 15 local branches. In terms of assets, advances, and deposits, Thai banks controlled roughly 97 per cent of the total share while the foreign banks controlled only 3 per cent.

Commercial banks mobilize funds by accepting time, savings, and demand deposits and by borrowing from domestic as well as foreign sources. They play an important role in the development of Thailand by fostering rapid growth, as is evident by their 73 per cent share of household savings captured and credits extended by all financial institutions in the past decade. Within the commercial banking sector, business is concentrated among a few banks. For example, assets of the largest three banks altogether made up about 50 per cent of the entire banking system while those of the smallest five totaled only 6 per cent in 1992.

Finance and securities companies represent the second largest group, with their assets representing 14 per cent of all financial institutions' assets in the regulated market. The first full-fledged finance companies began operations in 1969; many more have since been established. They obtain funds mostly through the issuance of promissory notes and borrowing from commercial banks and other countries. Securities companies, on the other hand, engage primarily in securities issuance and trading business. At the end of 1992, there were 92 finance and securities companies operating in Thailand. Similar to finance and securities companies in capturing savings are the credit foncier companies, but the latter specialize in immovable properties. Credit foncier companies are allowed to tap funds by issuing medium-term notes (with maturities longer than one year). They may then use such funds in the following schemes: lending money on the security of mortgage on immovable properties, buying immovable properties under sale contract with a right of redemption, and others as specified in the ministerial regulations. In December 1992, there were 18 credit foncier companies in Thailand.

An important rival in long-term finance is life insurance. Most of the life insurance companies in Thailand (10 out of 12) are local. They mobilize funds through sales of insurance packages, the most popular of which is an endowment policy; this represents roughly 60 per cent of all sales. Life insurance policies in Thailand may be broadly classified as ordinary, industrial, and group policies. Among these three types, ordinary policies represent nearly three quarters of the total.

Besides the powerful or specialized financial units mentioned above, cooperatives are active in the country as well. The cooperative movement in Thailand began in 1916, and has since grown extensively. By the end of 1992, there were 1,797 agricultural cooperatives and 898 savings cooperatives. Agricultural cooperatives are organized by farmers for the purpose of making credits available to members at low interest rates specifically for farm activities. Their principal sources of funds are borrowings and capital accounts. Savings cooperatives, on the other hand, are formed mostly on the basis of occupation, e.g., school teachers, university employees, policemen, and public employees. Most members of savings cooperatives are salary earners. The dominant source of funds is paid-up capital from members, with subordinate sources from reserves and deposits.

Pawnshops are institutions which lend money against a variety of personal articles such as gold ornaments, jewelry, wrist watches, sewing machines, and electrical appliances. They typically operate as small units without branches. At the end of 1992, there were 367 pawnshops, some of which were privately run while others were government owned. Private pawnshops rely on their own resources for loanable funds, although they occasionally borrow from commercial banks and other sources. Municipal and government pawnshops obtain initial capital from municipal and central government budgets, and they may also borrow from other sources as well. Most loans extended by pawnshops are small and are usually meant for household consumption.

Among government-owned financial institutions, the Government Savings Bank outranks others in a number of aspects. For instance, it had an extensive network of 525 branches throughout the country in 1992. Thus it is not surprising that its deposits reached 6 per cent of the entire financial system in the past two decades, and its total assets reached 11 per cent. The Government Savings Bank dates to the year 1913, when King Rama VI set up the Savings Office attached to the Royal Treasury. In 1929, the Savings Office was transferred to the Post and Telegraph Department and remained there until 1946 when its assets and liabilities were transferred to the newly created Government Savings Bank.

Table 5.1 Salient Features of Financial Institutions in Thailand at the end of 1992
(Baht billion)

	Operations Began	Number	Number of Branches	Household Savings Captured	Capital Account	Credits Extended	Total Assets
Commercial banks	1888	30	2,581	1,552.2	170.2	2,161.9	2,528.1
Finance companies	1969	92	42	293.7	66.9	547.7	671.0
Credit foncier companies	1969	18	—	3.4	1.2	5.1	6.2
Life insurance companies	1929	12	1,016	53.0	12.5	19.7	66.9
Agricultural cooperatives	1916	1,797	—	6.9	7.3	13.8	17.8
Savings cooperatives	1946	898	—	52.3	51.9	69.2	75.0
Pawnshops	1866	367	—	—	2.0	8.5	9.2
Government Savings Bank	1913	1	525	130.4	11.1	18.5	150.9
BAAC	1966	1	266	19.0	6.1	62.0	76.9
Government Housing Bank	1953	1	18	38.3	3.5	53.5	57.1
Industrial Finance Corporation of Thailand	1959	1	7	—	7.7	35.9	57.1
Small Industrial Finance Corporation	1992	1	—	—	0.3	0.1	0.6
Total		3,219	4,455	2,149.0	340.7	2,996.1	3,716.7

— nil.

Source: Bank of Thailand.

Table 5.2 Total Assets of Financial Institutions
(Baht billion)

	1977	1978	1979	1980	1981	1982	1983	1984	1985	1986	1987	1988	1989	1990	1991	1992
Commercial banks	173.3	218.2	259.9	300.0	353.2	423.7	528.2	639.4	701.9	767.6	920.1	1,126.1	1,406.5	1,789.6	2,147.6	2,528.1
Finance companies	38.3	56.7	58.3	66.2	76.9	94.9	105.4	118.1	131.0	139.8	160.0	195.7	281.7	365.6	481.0	671.0
Life insurance companies	3.3	4.0	5.1	6.5	8.1	10.1	12.2	15.2	18.4	21.2	25.0	29.3	37.7	47.4	54.7	66.9
Agricultural cooperatives	4.1	4.7	5.2	5.7	6.7	8.5	9.7	7.2	7.8	8.5	11.2	14.0	17.7	12.7	15.4	17.8
Savings cooperatives	2.1	2.7	3.4	4.4	5.6	7.4	8.0	12.1	15.7	19.6	24.4	30.9	39.1	56.8	65.0	75.0
Pawnshops	1.8	2.1	2.8	3.2	3.3	3.8	3.8	4.1	5.1	5.1	5.4	6.5	7.3	9.7	8.5	9.2
Credit foncier companies	1.7	2.8	3.3	3.9	5.1	5.6	5.3	4.4	4.2	4.3	3.8	3.9	4.1	4.4	5.0	6.2
Government Savings Bank	18.2	19.8	24.2	28.0	31.2	37.0	45.2	54.2	64.9	94.1	111.5	124.3	125.7	138.8	140.2	150.9
BAAC	9.6	12.6	14.4	17.3	19.3	20.9	22.4	25.2	28.5	30.2	33.6	37.3	45.0	54.8	67.2	76.9
Government Housing Bank	3.1	5.0	7.4	10.1	11.7	11.5	10.8	12.9	13.7	14.7	15.4	19.5	27.8	35.8	45.4	57.1
Industrial Finance Corporation of Thailand	2.0	2.5	3.2	4.2	5.5	6.6	7.4	11.1	15.0	17.9	22.6	24.1	29.6	38.0	47.8	57.1
Small Industrial Finance Corporation	0.1	0.1	0.1	0.1	0.1	0.1	0.1	0.1	0.1	0.1	0.1	0.1	0.1	0.2	0.2	0.6
Total	257.6	331.2	387.2	449.6	526.8	630.1	758.5	903.9	1,006.0	1,123.2	1,333.0	1,611.6	2,022.3	2,553.8	3,078.0	3,716.7

Source: Bank of Thailand.

The Bank's main function is to mobilize small savings through acceptance of demand, savings, and time deposits. It also issues savings certificates and premium bonds. In the past, a large portion of its funds were invested in government securities, but in recent years, as the Government's fiscal status has improved, the Government Savings Bank has been allowed to extend more credits to state enterprises as well as to private entrepreneurs.

The Bank for Agriculture and Agricultural Cooperatives (BAAC), established in 1966 to take over the Bank for Cooperatives, provides credits for farmers and cooperatives at low interest rates. Recently its scope has been enlarged to cover both technology and personnel. Its resources are appreciably drawn from commercial banks which lend to the BAAC as one means of satisfying their compulsory rural credit extension. The BAAC also accepts time and savings deposits from the private sector. In emergencies, the central bank supplies the BAAC with additional credit facilities.

The Government Housing Bank was established in 1953 to assist people of moderate income in obtaining houses of their own. It accepts deposits of any type and maturity from the public, and can also on-lend funds borrowed from abroad.

Two other specialized financial institutions are industrial financing units: the Industrial Finance Corporation of Thailand (IFCT) and the Small Industrial Finance Corporation of Thailand (SIFCT). Established in 1959 to supersede the Industrial Bank of Thailand, the IFCT assumed some of its predecessor's assets which were in the form of an interest free 50-year loan from the Government. The pattern of ownership has changed over the years, however, to include more private entities, principally the Thai commercial banks, while the remainder is owned by the Ministry of Finance. The IFCT operates along the lines of a private development bank or a development finance company. It focuses on financing fixed assets through the extension of medium- and long-term loans to different industries. The IFCT also participates in its customers' equity. It is not allowed to extend credit to any company which is more than one-third government owned. As it has no foreign exchange license, IFCT may issue guarantees to commercial banks on behalf of its customers to obtain letters of credit. In the past, IFCT was principally funded by foreign borrowing under government guarantees and its own capital. In recent years it has depended more upon borrowings in local currency due to rapid and wide exchange rate fluctuations.

The SIFCT is the counterpart of IFCT for small-scale industrial undertakings, including cottage and handicraft industries. Ordinarily, SIFCT could not extend credit to any enterprise which has total assets worth more than B10 million. Originally, SIFCT was under the Department of Industrial Promotion in the Ministry of Industry; more recently, its status has been changed to a juristic entity which can mobilize funds from the public. The SIFCT's paid-in capital and scope of operation has been expanded as well.

As regards Thai financial markets, the money market is an arena for short-term dealings in such financial instruments as treasury bills, interbank loans, repurchase agreements, certificates of deposits, and commercial bills. The capital market, on the other hand, trades stock issues, long-term bonds or debentures, and floats rate notes. The credit market is distinguishable from the money and capital markets in that it consists of both short- and long-term credits, including direct loans, overdrafts (which may or may not be rolled over), bills discounted, and leasing, extended by different financial institutions. The financial markets will be discussed in some depth in a later section of this chapter.

Relative Sizes of the Financial Institutions

Table 5.3 reveals the size of the financial system relative to that of the economy. Broad money supply or M2 represents the banking sector, and total assets of financial institutions (TAFI) represents the financial system. Two points are immediately evident. First, the banking sector and the financial system grew steadily between 1977 and 1986, as M2/GDP rose from 37 per cent to 59 per cent while TAFI/GDP from 64 per cent to 99 per cent. Six years later these two ratios grew at a much faster pace to 79 per cent and 139 per cent, respectively. Rapid expansion of GDP was a primary factor inducing financial institutions to enlarge their assets; gradual financial liberalization undertaken since the late 1980s also encouraged new activities, which led ultimately to the expansion of assets. Second, the banking sector was able to retain its predominance over other financial institutions, as the M2/TAFI remained constant at around 58 per cent throughout the period. This indicates that while the banking circuit scored an impressive record plus improvements in numerous aspects during the period 1987–1992, other parts of the financial system experienced much less development.

The data in Table 5.4 compare the performance of the credit market to that of the capital market. Between 1975 and 1986, though the stock market was already established, its role was insignificant. Annual changes of credit outstanding relative to GDP averaged 7 per cent while capital funds tapped from the stock market to GDP stayed at only 0.3 per cent. The main reason for this imbalance was that the stock market was in its infancy; its regulations were excessively intricate and burdensome. The period of 1987–1992 saw not only exuberant strides of the Thai economy, with real GDP growing at the pace of 7–13 per cent per annum but also much greater contributions from both the credit and capital markets. The credit flows/GDP ratio surged to 21 per cent while the capital tapped from the stock market/GDP ratio rose to 2.1 per cent in 1992. The more active role of the capital market can be attributed to familiarity as well as appropriate legal amendments.

Table 5.3 M2, TAFI, and GDP
(Baht billion)

	1977	1978	1979	1980	1981	1982	1983	1984	1985	1986	1987	1988	1989	1990	1991	1992
M2	151.1	180.3	205.5	251.8	292.9	363.8	450.5	537.9	593.5	672.8	808.6	956.1	1,207.1	1,529.1	1,832.4	2,123.0
TAFI	257.6	331.2	387.2	449.6	526.8	630.1	758.5	903.9	1,006.0	1,123.2	1,333.0	1,611.6	2,022.3	2,553.8	3,078.0	3,716.7
GDP	403.5	488.2	558.9	662.5	760.4	841.6	921.0	988.1	1,056.5	1,133.4	1,299.9	1,559.8	1,856.5	2,182.1	2,509.4	2,671.4
M2/GDP (per cent)	37.4	36.9	36.8	38.0	38.5	43.2	48.9	54.4	56.2	59.4	62.2	61.3	65.0	70.1	73.0	79.5
TAFI/GDP (per cent)	63.8	67.8	69.3	67.9	69.3	74.9	82.4	91.5	95.2	99.1	102.5	103.3	108.9	117.0	122.7	139.1
M2/TAFI (per cent)	58.7	54.4	53.1	56.0	55.6	57.7	59.4	59.5	59.0	59.9	60.7	59.3	59.7	59.9	59.5	57.1

Source: Bank of Thailand, National Economic and Social Development Board.

Table 5.4 Credit Extension and Capital Funds Tapped from SET
(Baht billion)

	1975	1978	1979	1980	1986	1987	1988	1989	1990	1991	1992
Credit flows from 12 financial institutions	21.8	55.8	43.2	39.3	33.5	152.4	238.4	370.2	494.5	442.8	569.3
Capital mobilized by quoted companies in SET	0.1	0.4	2.9	0.7	2.2	14.5	10.9	25.3	46.6	55.6	55.3
GDP	303.3	488.2	558.9	662.5	1,133.4	1,299.9	1,559.8	1,856.5	2,182.1	2,509.4	2,671.4
Credit flows/GDP (per cent)	7.2	11.4	7.7	5.9	3.0	11.7	15.3	19.9	22.7	17.6	21.3
Capital tapped/GDP (per cent)	0.0	0.8	0.5	0.1	0.2	1.1	0.7	1.4	2.1	2.2	2.1

Source: Bank of Thailand, Securities Exchange of Thailand (SET).

Since there is a close link between the real and financial sectors, tracing the evolution of the former will help develop a better understanding of the latter. Therefore, the following discussion summarizes economic development in Thailand during the past three decades.

Evolution

Overall, the Thai economy underwent three periods of economic development. The first came in the 1960s when the Government emphasized two targets, i.e., developing necessary economic infrastructure (such as roads, dams, reservoirs) and diversifying agricultural production (which had largely hinged upon rice). Efforts in this wave were quite successful, as supported by agricultural statistics. In the early 1960s, paddy represented 50 per cent of all crop production, though it declined to 35 per cent in 1970. All crops constituted 74 per cent of agricultural output in the early sixties. This share fell to 70 per cent in 1970.

The second wave occurred in the 1970s, when the focus shifted toward import substitution. The Government felt that those industries were readily equipped with market, labor, and raw materials, and thus deserved some tariff protection. The protective measures granted were effective, as evidenced by import figures. The consumer goods portion of total imports dropped from 19 per cent in 1970 to only 10 per cent in 1980. Even the capital goods portion sank from 35 per cent to 24 per cent in the same period. By the end of the 1970s, the Government realized that the formidable constraint in import substitution strategy was limited market potential.

It is thus not surprising that the third wave, which began at the beginning of the 1980s, moved toward exports instead of import substitution. As the Government aimed to achieve a higher degree of competitiveness and efficiency, it dismantled tariff walls one after the other. In addition, the Board of Investment geared its policy toward export promotion, which proved beneficial as the ratio of exports to GDP grew from 22 per cent in the first half to 32 per cent in the second half of the 1980s. Most of the macroeconomic policies implemented were market-oriented, such as flexible exchange rates and fewer constraints on interest rates. Also, the timing of third wave in the 1980s corresponded to the period when several industrial countries were trying to relocate their industries abroad.

As regards the financial sector, the first and second waves of economic development did not generate much financial pressure. In the first wave, agricultural diversification was not a capital intensive endeavor whereas the Government resorted to loans from international organizations for the building-up of infrastructure. In the second wave, the scale of involved production was not large because import substitution had limited markets. Therefore, during the first and second waves, the primary task assigned to the domestic financial system was the allocation of financial resources, which explains why the 1960s and 1970s witnessed the imposition of related regulations on commercial banking. These regulations included the following: each year banks had to extend agricultural credits up to a certain portion of their deposit outstanding. Further, if they wished to open new branches, they had to meet the prerequisite of holding government securities of at least 16 per cent of their deposits outstanding, a rule which was meant to give financial support to the Government. Once new branches became operative, they were obligated to extend credits to customers located in the same vicinity as the bank branches by at least 60 per cent of the deposits tapped. Further, for branches in small regional districts, one third of the required local credits had to be for agricultural purposes. Though these regulations may be perceived as market interference, they were designed to accommodate the underlying objectives of economic development policies, even at the expense of some efficiency.

The third wave (export promotion) has had a strong impact on the financial system because the target is the world market. The support of export industries leads to a large demand for funds to handle related investment, personnel, and machinery. Domestic financial institutions need to upgrade themselves with respect to both efficiency and fund mobilization to handle the demand, or they risk losing their market shares to their foreign rivals. On the part of the Government, it is widely believed that the final outcome of international trade negotiations (e.g., the Uruguay Round) will be in favor of liberalization in the fields of both goods and services. The entry as well as the branching of foreign banks is inevitable. Thus, Thai banks must adjust or improve. Their survival to date has been partly attributed to government protection. In anticipation of successful international trade negotiations, the authorities have begun to relax regulations to encourage more competition among commercial banks. It is hoped that a larger degree of competition will strengthen commercial banks. In short, the trend toward globalization has forced Thailand to liberalize its financial system so as to enable it to compete on equal terms. Consequently, in this third wave of economic development, the Thai financial arena has undergone several structural changes.

The two cores of financial policy direction during the third wave are to favor the market mechanism and to develop the financial infrastructure. Examples of market-oriented policies are relaxation of exchange controls, liberalization of interest rate specification, and a wider scope of financial institutions' businesses. Examples of financial infrastructure which have been institutionalized are establishment of the Securities and Exchange Commission (1992), adoption of the Bank for International Settlements' rule on capital adequacy (1993), operation of the Bangkok International Banking Facilities (1993), and most recently, establishment of the EXIM Bank as well as a few credit rating agencies. These will be discussed in greater depth in a later section.

FINANCIAL INFRASTRUCTURE

Among the various types of financial institutions now operating in Thailand, commercial banks and finance companies command predominant shares, i.e. 68 per cent and 18 per cent of all financial institutions' assets by the end of 1992. Thus, this chapter focuses primarily on how these two groups of financial institutions are regulated.

Commercial Banks

Commercial banks in Thailand are governed by the Commercial Banking Acts of 1962 and 1979, as amended in 1985 and 1992. Under the Acts, the term "commercial banking" is defined as "the business of accepting deposits of money subject to withdrawal on demand or at the end of a specified period, and of employing such money in one or several ways such as (i) granting of credits, (ii) buying and selling of bills of exchange or any other negotiable instrument, and (iii) buying and selling of foreign exchange." In addition, commercial banks may undertake related businesses such as collection of payments against bills, giving avals to bills, accepting bills, issuing letters of credit, providing guarantees, or any other business of a similar nature. A commercial bank can be established only in the form of a limited public company with a license granted by the Ministry of Finance.

Restricted Shareholding

To divest shareholding in commercial banks to the public, the Acts require that a commercial bank have at least 250 individual shareholders. Such shareholders must together hold shares of not less than 50 per cent of the total shares issued, and each shareholder cannot hold shares of more than 0.5 per cent of the said total shares. A commercial bank must also have among its shareholders Thai nationals who hold not less than three fourths of the total shares issued. The Acts do not allow any "person" to hold any commercial bank's shares in excess of 5 per cent of the total shares issued. Included in the definition of "person" are (i) the shareholder's spouse or offspring; or (ii) partner in any ordinary or limited partnership or limited company holding more than 30 per cent of total shares issued by such partnership or company.

Adequate Assets

Each bank must maintain a minimum "liquidity reserve ratio" of 7 per cent of total deposits to ensure there is always enough cash to meet withdrawals. At least 2 per cent of this 7 per cent must be held at the Bank of Thailand without interest, at most 2.5 per cent as cash in hand, and the rest as eligible securities. Further, provisions must be made for the possibility of losses occurring as a result of bad debts.

Commercial banks must maintain capital funds at not less than 8 per cent of risk assets, i.e., credits cannot be extended beyond 12.5 times their capital base. The amendment in 1992 lowered this ratio to 7 per cent, effective 1 January 1993. Nevertheless, the adjustment was made in consonance with the Bank for International Settlements (BIS) guideline whereby two tiers of capital are recognized and different assets are given different weights according to their associated risks. It is expected that the BIS standard of 8 per cent will have been attained in January 1994 and be maintained from then on.

In addition, commercial banks must cover their contingent liabilities as well as foreign exchange exposure with sufficient capital funds: banks' capital should total at least 20 per cent of their total contingent liabilities while their net foreign exchange liabilities or assets must remain within 20 per cent or 25 per cent of their capital accounts, respectively.

Risk Diversification

As a means to achieve a good degree of financial stability, the Acts require that commercial banks diversify their customer profiles. A bank cannot extend credits to any individual client by more than 25 per cent of the bank's capital funds. Neither can it offer off-balance-sheet obligations to any customer by more than 50 per cent of its capital funds. Moreover, each bank may neither hold shares/debentures of other incorporated companies totaling beyond 20 per cent of the bank's capital funds nor hold more than 10 per cent share in any company's equity. These rules help commercial banks to avoid the clustering of risks, regardless of the prospects of any specific client.

Branching

Prior to 1993, a bank wishing to open a branch was required to hold government bonds of at least 16 per cent of its total deposits outstanding, a requirement which was meant to help finance budgetary deficits or bond issuances. However, since the government's fiscal balance improved markedly in the late 1980s, no government bonds were issued after 1992, and the 16 per cent ratio on branching was lowered gradually in 1990–1992 until it was eradicated in the middle of 1993.

Three additional requirements regulating new bank branches in suburban districts are still in effect: first, credits extended by those branches to community clients must add up to at least 60 per cent of their deposit outstanding. Second, as for small districts in particular, a third of the community credits must be agriculturally related. Third, if bank branches cannot fulfill these community credit targets, they must deposit the residual at the central bank, or buy government/state enterprise bonds to four and a half times the residual, or Government Housing Bank bonds by three times the residual.

Compulsory Credits

While the central bank does not explicitly require that commercial banks extend any special credits to specific customers under particular terms, it has formally requested collaborative efforts in two respects. First, in 1975 commercial banks "should" allocate credits totaling at least 5 per cent of their deposit outstanding to agriculture. The ratio was increased from 7 per cent to 13 per cent during the period 1976–1986 while its 2 per cent constituent belonged to agro-businesses. In 1987, these requested credits reached 20 per cent of deposits outstanding; their coverage was also broadened to cover credits offered to small-scale industrial enterprises.

Further, the more credits that commercial banks extend to the so-called priority sectors (consisting of agriculture, mining, manufacturing, exports, and wholesale purchases from farmers), the more privileges those creditor banks will be awarded in the forms of rediscount facilities from the central bank or partial exemption from the minimum capital/risk asset requirement.

Compulsory credit extension allows credit to be circulated to vital rural economic sectors which tend to be rather weak, sectors which if left alone may be ignored by commercial banks. As investment in weak economic sectors will not yield steady and satisfactory financial returns, such investment needed to be mandated.

Intra-affiliate Services

In contrast to compulsory credits, commercial banks are often tempted to offer excessive credits and financial services to family members of their executives as well as to their affiliated firms. It is thus not surprising to find the following restrictions in the Commercial Banking Acts: Section 12(2) prohibits commercial banks to extend credits or obligations to their executives; and Section 12 BIS expands the coverage of Section 12(2) to cover family members of bank executives and affiliated firms in which bank executives hold more than 30 per cent.

Management Control

The Commercial Banking Acts stipulate that the following persons are prohibited to be appointed as a commercial bank's director, manager, deputy manager, assistant manager or adviser: a bankrupt person; a person who was sentenced by final judgement to imprisonment for an offense related to property committed with dishonest intent; persons who had been dismissed or discharged from the government service or other government organizations or agencies for fraudulent conduct in performance of his/her duties; a person who had been a director, manager, deputy manager, assistant manager of a commercial bank that had its license revoked; a person who had been removed from a position in a commercial bank by the Minister of Finance; a political civil servant; a government official with duty to control the operations of commercial banks or an officer of the Bank of Thailand; and, a manager, deputy manager, or assistant manager of the partnership of the company in which the person (or the person pursuant to Section 12 BIS) has held shares, except as a director or adviser of the commercial bank who does not have power to sign on behalf of the bank.

Three fourths of the total number of directors of the commercial bank must be of Thai nationality.

Finance Companies

Finance companies are financial institutions that borrow funds from the public through issuance of promissory notes or similar instruments and lend those funds to other persons. Finance companies are regulated under the Act on the Undertaking of Finance Business, Securities Business and Credit Foncier Business, B.E. 2522 (1979). This Act was amended in 1983 and again in 1985.

Under this Act, finance business is defined as the business of procuring funds and using such funds for any type of business operations which may be categorized as follows: business of finance for commerce; business of finance for development; business of finance for disposition and consumption; business of finance for housing; and, other types of finance business as prescribed in ministerial regulation.

The finance business may be undertaken only by a limited public company and only after a license has been obtained from the Minister of Finance. At the end of 1992, there were 92 finance companies authorized by the Ministry of Finance. In addition, a finance company may be granted a license for securities business under the same Act, in which case the company will be both a finance and securities company. Finance companies are regulated in a fashion similar to commercial banks.

Restricted Shareholding

A finance company must have at least one hundred shareholders who are of Thai origin. Shareholders must hold shares altogether of not less than 50 per cent of the total number of shares sold. Each shareholder must not hold shares more than 0.6 per cent of the total number of shares sold, while the number of shares held by persons of Thai nationality must not be less than three fourths.

Adequate Assets

Each finance company must maintain liquid assets of at least 7 per cent of the total borrowings from the public. Liquid assets can be held in the form of balance at the Bank of Thailand (no less than 0.5 per cent), government securities, and other unobligated bonds and debentures guaranteed by the Thai government (no less than 5.5 per cent), and the rest as deposits or call loans at domestic commercial banks. As regards capital adequacy, finance companies are required to satisfy the minimum capital funds to risk assets ratio of 6 per cent. The registered and paid-up capital of each company must be at least B60 million.

Risk Diversification

Finance companies cannot offer loans to any one person of more than 30 per cent of its capital funds. Neither can it invest in securities by more then 60 per cent of its capital funds. Avals and guarantees are subject to similar limits, i.e., those given to any one person must not exceed 40 per cent of capital funds and the aggregate of avals plus guarantees should be within a quadruple of capital funds. Collectively, all financial commitments given to any person must not be above 40 per cent of capital funds.

Finance companies are also prohibited from concentrating their holdings in any one enterprise. The maximum shareholding in any limited company is 10 per cent of such company's shares sold.

Branching

A finance company may set up a branch office only with authorization by the Minister of Finance who may attach any condition to the authorization.

Controls on Credits and Borrowings

The central bank may, at times, specify some restrictions on credits and borrowings, such as for the terms of car hire-purchase financing (minimum down payment, maximum hire-purchase period), maximum interest rates on loans, minimum amounts of borrowing from the public, and prohibition on discount payments.

Company Linkages

A finance company cannot purchase or hold shares in other finance companies except (i) those acquired as a result of a debt settlement or a guarantee in respect of a loan granted; in this case, such shares must be disposed of within six months of the date of acquisition; (ii) those acquired as a result of the conduct of other business activities as authorized by the Minister of Finance; and (iii) those acquired as permitted by the Minister upon recommendation of the Bank of Thailand.

Financial Institutions Development Fund

In November 1985, after a series of crises and bankruptcies in the circle of finance companies, the Financial Institutions Development Fund was established within the Bank of Thailand. The Fund has separate legal entity status for the purpose of reconstructing and developing the financial institution system to accord it strength and stability. The establishment of the Fund was part of the Government's measures to provide financial aid for financial firms having difficulties and to restore public confidence after the collapse of several firms in 1983.

Each financial institution remits to the Fund a sum of money at a rate prescribed by the Fund Management Committee. The rate may not exceed 0.5 per cent of the total amount of deposits, borrowings, or funds received by that financial institution from the public or what is outstanding at the end of the financial year preceding the year in which the remittance is to be made. In addition, the Bank of Thailand may from time to time allocate to the Fund any suitable amount of its reserves.

The Fund is empowered to do the following:

(i) hold ownership or possessory rights or any real rights to build, buy, acquire, sell, dispose of, hire, lease, buy on hire-purchase, extend hire-purchase, borrow, lend, accept pledges, accept mortgages, exchange, transfer, accept transfer of or engage in any act concerning properties, within or outside the country, including accepting properties from donors;

(ii) lend money to a financial institution against reasonable security;

(iii) guarantee or certify, accept, give aval to, or intervene to honor bills;

(iv) provide financial assistance in reasonable amounts to those depositors or lenders of a financial institution who have sustained losses as a consequence of such financial institution having suffered a serious financial crisis;

(v) hold deposits with such financial institutions as the Fund Management Committee deems necessary and appropriate;

(vi) purchase or hold shares of any financial institution;

(vii) purchase, discount or re-discount debt instruments or accept transfers of claims from any financial institution;

(viii) borrow money with or without interest, issue bills and bonds;

(ix) make investments for the purpose of earning income as permitted by the Fund Management Committee; and,

(x) undertake all business in connection with or incidental to the attainment of the purposes of the Fund.

The Fund Management Committee has the authority and duty to lay down policies and take general control and supervision of the affairs of the Fund. The Committee consists of the Governor of the Bank of Thailand as Chairman, the Permanent-Secretary for Finance as Deputy Chairman and not fewer than five, but not more than nine, other committee members appointed by the Minister of Finance. The Fund's Manager is an officer of the Bank of Thailand, appointed by the Committee. The Manager holds the dual role as the Secretary of the Fund Management Committee.

Stock Exchange

Following the enactment of the Public Company Act and the Amendments to the Civil and Commercial Codes in 1973, limited companies were not allowed to make any initial public offerings of their shares, nor were they allowed to offer debentures to the public. Initially, new shares were to be offered only to existing shareholders. These provisions significantly slackened the mobilization of funds in the primary market, and also had an adverse effect on the secondary market.

According to the provisions in the Public Company Act, only public companies can make initial public offerings of their shares and debentures. In addition, the Act provides for severe criminal as well as civil liabilities on guilty companies' directors, requires that directors be shareholders, limits the entire board of directors to a one-year term, and requires that any company having more than 99 shareholders must be converted into a public company regardless of its intention to raise funds from the public or not. Due to its strict bindings and other weaknesses, in over 10 years after the Act had taken effect, only 33 companies had become public.

In 1984, an attempt was made to reduce these obstacles by using the Securities Exchange of Thailand Act to overrule some of the provisions in the Public Company Act and the Amendments to the Civil and Commercial Codes. The Securities Exchange of Thailand Act allows listed and authorized companies in the Securities Exchange of Thailand—as well as those companies which have received tentative approval to be listed and authorized in the Securities Exchange who are waiting for final approval from the Finance Minister—to offer their shares and debentures to the public. However, the Securities Exchange of Thailand Act was designed primarily to regulate the trade of securities in the secondary market, not the primary market. Hence, most of the companies listed in the Securities Exchange were limited companies according to the Civil and Commercial Codes, and were, therefore, not regulated by the Public Company Act. Consequently, there was no effective legal framework from which to supervise the primary market.

Another weakness in the previous legal system is the number of laws that securities market participants must observe, namely, the Securities Exchange of Thailand Act, B.E.2517 governing the activities of the Securities Exchange; the Act on the Undertaking of Finance Business, Securities Business and Credit Foncier Business, B.E.2522 governing the business of securities companies; the Public Company Act, B.E.2521 governing the public offering of shares and debentures; and the Civil and Commercial Codes specifying general provisions regarding the civil and commercial practices in Thailand such as the setting up of limited companies. The greater the number of laws, the greater the chance of inconsistency among them, and the greater the burden upon market participants.

Besides the different laws that needed to be observed, there were also various supervisory agencies in charge of the securities businesses, namely, the Ministry of Finance, the Securities Exchange of Thailand, the Bank of Thailand, and the Ministry of Commerce. There was no single supervisory agency: i.e., the Bank of Thailand supervised the securities companies but did not supervise the activities in the Exchange; the Securities Exchange supervised brokers but did not have the authority to regulate the subbrokers. The enforcement of securities regulations was, thus, inefficient.

The Securities and Exchange Act was enacted in March 1992 to address this problem. Its objectives were (i) to set up the framework for the development of financial instruments; (ii) to provide greater investors' protection; (iii) to make the supervisory systems for securities companies and other related institutions more transparent and unified; and (iv) to facilitate the development of securities businesses and the Securities Exchange.

The Act includes provisions for the following:

(i) *public offering of securities*: An eligible issuer of shares and equity-related securities will be restricted to a promoter of a public limited company or a limited company, while an issuer of debt instruments can be both a public limited company and a limited company. The Act also emphasizes the disclosure of information by issuers. For this purpose, the issuers must disclose as much reliable information as possible to investors by filing registration statements and draft prospectuses with the Securities and Exchange Commission (SEC). The contents of the registration statement, as stipulated by the SEC, include registered capital, nature of business, financial condition, important business information, management team, major shareholders, and other information necessary for investors. Under the Act, it is both a criminal and civil offense for anyone involved in the preparation of the registration statement and draft prospectus to mislead the public by making a false disclosure.

(ii) *securities business*: Securities companies are supervised by SEC while the Bank of Thailand supervises the finance companies. The Act also provides an opportunity for other types of financial institutions (e.g., commercial banks and finance companies) to engage in securities businesses with a license from the Minister of Finance upon the recommendation of the SEC.

Under the Act, the following are considered to be securities businesses and therefore require a license: securities brokerage; securities dealing; investment advisory service; securities underwriting; mutual fund management; and private fund management. The rules and conditions regulating each are similar to those discussed earlier pertaining to the undertaking of finance, securities, and credit foncier businesses.

(iii) *securities exchange*: The Stock Exchange of Thailand (SET) was set up in 1974 under the Securities and Exchange of Thailand Act. Amendments to the Act in 1984 allowed for better control of the listed companies and trading activities. In 1992, the Securities and Exchange of Thailand Act of 1974 was repealed, and most of its provisions are now incorporated into the 1992 Securities and Exchange Act. Under the law, SET is a center for the trading of listed securities, while matters concerning primary issues are the responsibility of SEC. In addition, the power to grant listing approval, which once belonged to the Finance Minister, will now be vested in the SEC. The SET's Board of Directors is comprised of 11 persons: a president who is appointed by the Board, five members appointed by SEC, and five elected by the members.

(iv) *over-the-counter center*: The Act provides for the setting up of an over-the-counter center to facilitate the trading of unlisted securities. Under the Act, public offering of securities must proceed through SEC, after which issuers can apply to SET for a listing. If the securities cannot be listed in SET, they can be traded over the counter. An over-the-counter center can be formed by at least 15 securities companies with permission from SEC. The center would be managed and operated by its own board of directors which would be wholly elected by its members.

(v) *unfair securities trading practices*: Unfair securities trading practices are criminal offenses with severe penalties under the Act. Although the legal framework differs only slightly from the previous statute, the power of investigation is now transferred to SEC which is an independent regulator. This is a crucial change since the effectiveness of the law largely depends on the effectiveness of enforcement.

(vi) *takeover*: The SEC Act requires the disclosure of every 5 per cent holding of securities by an individual and associates. Once the holding reaches 25 per cent of all securities sold, the person holding such securities may be required to make a tender offer to the existing shareholders. In certain cases, the law requires that the person making a tender offer shall have to purchase all of the securities being offered.

Accounting Standards

Ordinarily, all business enterprises in the country are required by the Institute of Certified Accountants and Auditors in Thailand to complete and submit financial statements in accordance with generally accepted accounting principles. Such financial statements serve as public information and consist of the following four components: balance sheet, statement of income, changes of financial position, and notes to financial statements. As for financial institutions in particular, most have to supply additional information for the purpose of specific disclosure. Other government-controlled businesses, such as insurance companies, also have to furnish responsible agencies with additional information.

Value-Added Tax

The value-added tax (VAT) was enacted in January 1992 to replace the business tax. Based on the value of the goods or services provided, the VAT is collected at each stage of production and distribution.

Entities required to pay the VAT include traders, whether as persons or juristic entities, providers of services, importers, and others deemed by law to be traders. The sale of goods and services must be professional in nature, serve in furthering the sellers' business interests, and be undertaken in the regular course of business. A one-time sale of a personal item would not subject the seller to the VAT requirements.

Parties exempt from the VAT are businesses subject to the provisions of the specific business tax laws (e.g., commercial banks, financial institutions); businesses expressly exempted from VAT liability; businesses having an annual income not exceeding B600,000; and businesses exempt by law, including sellers of farm products, newspapers, and magazines.

The obligation to pay VAT for the sale of goods arises at the earliest occurrence of the following events: upon delivery, when title is transferred, when the price has been paid, or when the related tax invoice has been issued. For the sale of services, the obligation occurs when the service provider receives compensation. However, the VAT obligation will also arise upon the issuance of a related tax invoice.

All taxable revenue from the sale of goods and services, defined as the tax base, is calculated as the sum of the sales value and the applicable excise tax, if any. Calculation of the tax base differs for activities other than the sale of goods and services which are subject to VAT. The tax rate is then applied to the tax base. The standard VAT rate of 7 per cent applies to the sale of goods and services, though there are many exceptions to this rate. Small-scale traders having an annual gross sales revenue exceeding B600,000 but less than B1,200,000 may, instead of paying 1.5 per cent tax on the gross receipts, elect to pay 7 per cent VAT. There is no tax rate applicable to certain business activities, including export of goods or services, international transportation by air or sea, and the sale of goods and services to United Nations (UN) related organizations.

The amount of VAT either payable by, or refundable to, a party is calculated monthly by subtracting the total input VAT from the total output VAT. If the output VAT exceeds the input VAT, the business remits the excess to the tax authorities. If the input VAT exceeds the output VAT, the excess is either refunded or credited to the business. VAT must be paid by the taxpaying entity within the first 15 days of the month following the month of taxation. The VAT law contains detailed provisions for reporting and record-keeping, and these regulations dictate the required type, format, and contents of the taxation report to be filed by the tax paying entity.

Various penalties and surcharges are levied on tax-paying entities for their failure to undergo VAT registration or for failing to comply with its regulations. These penalties range from nominal fees to imprisonment. Individuals acting for juristic parties such as directors and managers are personally liable for the said penalties if there is no proof that the individuals had no knowledge of, or involvement in, the offense committed.

COMMERCIAL BANKING SYSTEM

The commercial banking system was begun in Thailand during the reign of King Rama V. The first bank was a branch of the Hong Kong and Shanghai Bank opened in 1888. Its main objective was to facilitate international trade. In its early stages, the banking system was heavily influenced by the British banking tradition. Thus, it was a branch banking system with a network of branches throughout the country. Between 1888 and 1941, the banking business was confined largely to trade financing and remittance of funds to China. At the end of the period there were 12 banks operating in Thailand; five were local and seven were foreign affiliates.

Later, during World War II, as Thailand was forced to enter into an alliance with Japan, trade with Western nations ended and many foreign bank branches had to close. Those units were then replaced with five new local banks. From the end of World War II until 1962, foreign banks resumed their role in the system, and more locally incorporated banks opened as well. At the end of 1992, Thailand had 29 commercial banks: 15 locally incorporated and 14 foreign owned. There were 2,566 (excluding head offices) Thai bank branches and 15 foreign.

The Thai commercial banking system developed steadily, particularly in the three decades after the Economic and Social Development Plan was introduced in 1961. During the 1960s and early 1970s, operations of the banking system proceeded smoothly and in line with economic expansion. In that period, commercial banks encountered no rivals in the business of finance; they easily exercised their clear-cut oligopolistic edge. Later, commercial banks were considerably affected by two major factors: (i) the emergence of finance companies brought about a greater degree of competition; and (ii) the world markets became more volatile in terms of commodity prices, interest rates, and exchange rates. Consequently, the health of commercial banking in Thailand deteriorated markedly and its development came to a standstill.

In 1979, the central bank revised the Commercial Banking Act to include the following multiple objectives: increase share divestiture; prevent commercial banks from exposing themselves to the businesses of their executives or related persons; limit exposure on contingent liabilities; and improve the flexibility as well as effectiveness of bank supervision. However, in the first half of the 1980s, the commercial banks were severely affected by global recession and volatile exchange and interest rates, and the revision of the Act did little to improve the soundness of the Thai banking system.

In addition, at the height of the finance company crisis in 1984, the financial position of some Thai commercial banks was further weakened by mismanagement and fraud. For example, Asia Trust Bank suffered from imprudent management, e.g., maturity mismatching and excessive exchange risk. Its financial position deteriorated to the extent that foreign creditors lost confidence and stopped lending. The authorities intervened and offered assistance similar to the "April 4 Lifeboat Scheme" which had been given to rehabilitate troubled finance and securities companies in 1984. A new management team was sent in; organizational and operational systems were revamped; and soft loans and new equity were injected. In essence, the bank was effectively nationalized. It was then renamed the Sayam Bank. These measures were, to a certain degree, successful. Sayam Bank eventually merged with Krung Thai Bank in 1987, as it was considered unnecessary to have two state-owned banks.

During 1985–1987, critical problems occurred at two other commercial banks, one caused by excessive speculation in foreign exchange, the other by unscrupulous lending practices. In both cases, the authorities ordered the banks to upgrade their management system and operational efficiency. Their capital base was to be enlarged and strengthened as well. After having written off losses and injected new equity, shareholders were encouraged to take up new shares and new investors were encouraged. Shareholding by the authorities and the granting of soft loans were restricted. In cases where soft loans were necessary, stringent covenants were attached.

Improving the Effectiveness of Bank Supervision

The Commercial Banking Act was amended again in 1985, together with the Bank of Thailand Act and the Act on the Undertaking of Finance Business, Securities Business and Credit Foncier Business. These amendments were aimed at enabling the authorities to take action on troubled financial institutions in a more flexible, effective, and timely fashion. For example, the amendments included provisions under which the Bank of Thailand would be legally deemed as the damaged party entitled to press criminal charges against fraudulent managers of financial institutions. The accused managers could be prevented from leaving the country and their assets frozen pending trial. The amendment to the Bank of Thailand Act also gave rise to the Financial Institutions Development Fund established as a separate legal entity to rehabilitate ailing financial institutions and safeguard depositors.

Lines of Business

Thai commercial banks adopted electronic banking in 1970 by providing deposit and withdrawal services through an on-line system in Bangkok. By 1987, this system was extended to branches in all provinces. It is now possible for commercial banks to communicate throughout the country via satellite using the Very Small Aperture Terminal (VSAT) system. Since 1985, the transfer of funds among banks throughout the world has been possible through the SWIFT system. In addition, a central credit center allows for data exchanges among banks, and check clearing was recently computerized. Over a thousand automated teller machines (ATMs) in two pools, BANKNET and SIAMNET, now provide after-hours deposit and withdrawal services nationwide. Commercial banks are also moving more toward fee-based activities, offering investment advice, private and office banking services, and credit card administration.

In 1992, commercial banks were allowed to undertake the following businesses: underwriting government and state enterprises' debt instruments; managing, issuing, underwriting, and trading of private debt instruments; providing economic, financial, and investment information as well as financial advisory services; and acting as agents for the sale of mutual funds, representatives of holders of secured debentures, mutual fund supervisors, and securities registrars.

At the end of 1992, assets of the commercial banking system stood at B2,555.6 billion. Deposits amounted to B2,010.7 billion or 75.2 per cent of GDP. Thai commercial banks held a market share in deposits of 98 per cent while the remaining 2 per cent belonged to foreign bank branches. Credits totaled B2,183.4 billion or 81.7 per cent of GDP, with Thai commercial banks commanding a 95 per cent market share and foreign bank branches 5 per cent. Capital funds amounted to B170.2 billion and net profits B31.1 billion in 1992.

Also in 1992, there were 722 bank branch offices in Bangkok and 1,860 in the provinces, compared to 133 in Bangkok and 202 in provinces in 1960. In the last three decades, expansion of the branching network in the provinces reduced the proportion of population per branch from 101,250 to 22,266.

Banking Structure and Current Issues

From a microeconomic perspective, commercial banks in Thailand are highly concentrated. The total assets of the largest three Thai commercial banks account for one half of the entire banking system; thus the behavior and policies of the three greatly influence the banking atmosphere at large.

For example, a decision by the three to adjust interest rates was followed, to some extent, by the medium- and small-scale banks, especially when the interest rate ceilings proved effective.

However, as a result of financial liberalization, competition among the commercial banks has intensified with respect to both prices and services offered. Unfortunately this competition benefits the larger customers—major depositors and borrowers—to the greatest degree. From the viewpoint of large depositors, the benefits include more investment alternatives such as equities, private debt instruments, and real estate. To continue to attract these large depositors, commercial banks must offer higher deposit rates; conversely, deposit rates offered to small depositors are 2–3 per cent lower. On the lending front, large-scale borrowers can currently borrow abroad or directly from the equity market. As a result, lending rates to prime customers, i.e., the minimum lending rate, have been reduced steadily while the lending rates to general customers have hardly moved. This has created a situation whereby new borrowers with high credit risks have inadequate access to commercial banks' credits and/or commercial banks are able to exercise oligopolistic power over small-scale borrowers.

The several phases of exchange rate liberalization have encouraged commercial banks to rely on foreign borrowing as another main source of funds, besides domestic deposits. Consequently, net foreign liabilities of the commercial banking system increased steadily from B11.4 billion in 1989 to B94.4 billion in 1992. This trend may jeopardize the attempts by the central authorities to encourage domestic savings mobilization. Nevertheless, the Thai financial system, of which the banking sector is the backbone, has been able to capture a growing portion of gross national savings, and has, at the same time, financed a large part of gross domestic investment. This will be discussed at greater length in a later section of this chapter, but for now it can be concluded that the Thai commercial banking sector has contributed much to the "financialization" process along the savings-investment growth path.

Bank Stability

The Government has long been concerned about the stability of the country's commercial banks. It is not surprising, therefore, that whenever any bank faces financial problems of any kind, the authorities are ready to extend assistance. This raises the question as to whether, if left to their own devices, these banks could take care of their own safety.

In that context, the loan-loss reserves and capital funds maintained by commercial banks as a buffer against risks should be examined. In the early 1980s commercial banks held loan-loss reserves at 3.6 per cent of their capital while their capital/risk assets ratio was at 9 per cent. When the growth momentum of the Thai economy accelerated in the late 1980s and the early 1990s, commercial banks were adequately prudent, as witnessed in the rise of their loan-loss reserves/capital ratio to 6.4 per cent in 1986–1989 and 7.9 per cent in 1990–1992. During the two periods, the capital/risk asset ratio also grew to 9.8 per cent and 9.2 per cent, respectively. The commercial banking system in Thailand is endogenously stable to a satisfactory degree.

NONBANK SAVINGS INSTITUTIONS

Though nonbank savings institutions in Thailand consist of various units—including finance companies, credit foncier companies, life insurance companies, agricultural cooperatives, and savings cooperatives—finance companies outrank all others in most respects. In terms of total assets, at the end of 1992 finance companies commanded 18 per cent share of all financial institutions' assets, second only to commercial banks with 68 per cent. Four other types of nonbank savings institutions together controlled less than 4.5 per cent share of the total assets. Therefore, primary attention in the following discussion will be focussed on finance companies, including their evolution, performance, and development.

Finance Companies[1]

Origin and Rules

The emergence of finance companies in the late 1960s occurred at a time when the banking industry needed competition. In the beginning, finance companies were allowed to operate loosely with neither specific licenses nor supervision. In fact, prior to 1969, private limited companies were able to mobilize funds from the public without Ministry of Finance approval. On 26 January 1972, the Revolutionary Council Announcement No. 58 was issued to regulate business undertakings which affected the safety or wellbeing of the public; finance and securities businesses were included within this regulation, and the Ministry of Finance, authorized by Section 14 of the Announcement, delegated to the Bank of Thailand the regulatory authority over finance and securities businesses after 19 September 1972. The businesses included acceptances or purchases of bills, mobilization of funds for the purpose of onlending or discounting bills and other negotiable instruments, trading of debt instruments and securities (such as stocks, bonds, debentures, and commercial papers), brokerage, management, advisory services relating to the trading of debt instruments, and securities.

---

[1] For details and a comprehensive analysis on this topic, see Trairatvorakil and Punyashthiti (1992).

Finance companies may not accept deposits in the same form as commercial banks, and are prohibited from engaging in foreign exchange trading activities. Rather, these companies borrow funds from the public through the issuance of promissory notes or similar instruments. Ordinarily, finance companies tap funds from medium-sized savers; each promissory note issued must be worth at least B10,000 in Bangkok Metropolis and adjacent provinces, and at least B5,000 in the other provinces. Though promissory notes carry higher interest rates than bank deposits, their maturities are rarely beyond one year. Another important source of funds is borrowing from other financial institutions.

Finance companies use funds for a variety of lending activities, such as acting as acceptors and givers of avals, providing medium- and long-term credits to industrial, agricultural or commercial undertakings, hire-purchase or installment lending, and housing credits. It is notable that discounting of post-dated checks and issuance of bills of exchange has traditionally been the main short-term activity of finance companies while consumer financing and hire-purchases serve as medium-term lending channels. A wide variety of goods are eligible for loans under hire-purchase agreements, ranging from automobiles and motorcycles to computers and electrical appliances. Under this type of loan, down payment must be at least 25 per cent of the merchandise value with the maximum repayment period of 48 months.

Consolidation Period

Prior to 1979, the growth of finance and securities companies was high, but the quality was questionable. One major finance company faced liquidity shortages due to mismanagement and widespread speculation together with manipulation of stock prices, further compounded by the tight money markets and higher interest rates during the period. Fearing that the entire system would suffer, the Ministry of Finance and the Bank of Thailand took control of the company. Nevertheless, the company closed in August 1979, and despite government intervention, many companies survived the crisis but suffered from its aftermath. These companies included those having a direct connection to the failed company, such as having invested in that company's stocks, having made loans to that company, or whose borrowers incurred securities losses during the crisis.

The economy during 1981 to 1983 was also weakened by major changes in the world economy, namely, the slowing down of global demand, the increase in foreign interest rates relative to domestic ones, and the larger extent of exchange rate volatility. In an attempt to capture larger market shares, a number of finance companies failed to exercise sufficient care when extending credits; fraud and mismanagement were rampant in many others. These factors led to the second crisis in 1983.

The magnitude and severity of the second crisis was much greater than the first one, and it led to a series of revocations of licenses and mergers as well as acquisitions of companies. This is illustrated in Table 5.5, which shows that in 1982, before the second crisis, there were 127 companies in the finance and securities industry. This number declined to 105 in 1987 due to the above-mentioned revocations and rescue measures adopted between 1983 and 1987.

Such rescue measures included amendments to the Act on the Undertakings of Finance Business, Securities Business and Credit Foncier Business, B.E. 2522 (1979) made under Royal Decrees in 1983 and 1985, and the 1984 establishment of the "April 4 Lifeboat Scheme" by the Ministry of Finance and the Bank of Thailand (to rehabilitate 25 finance and securities companies). The conditions of ailing finance companies improved markedly.

Table 5.5 Revocation of Licenses, Mergers, and Acquisition of Finance and Securities Companies

Year	Finance Companies/ Finance and Securities Companies	Securities Companies	Total
1982	112	15	127
1983	109[a]	15	124
1984	104[b]	15	119
1985	100[c]	15	115
1986	98[d]	11	109
1987	94[e]	11	105
1988	94	11	105
1989	94	11	105
1990	94	11	105

(Number of Companies)

[a] Three revocations of licenses.
[b] Five revocations of licenses.
[c] Four revocations of licenses.
[d] One finance company and one finance and securities company merged together to become a new finance and securities company; one revocation of license.
[e] Two finance companies, three finance and securities companies and one credit foncier company merged together to become a new finance and securities company; one finance company and two credit foncier companies merged together to become a new finance and securities company.

Source: Bank of Thailand.

Pluralistic Character

Though many small and weak companies' licenses were revoked and a number of small companies were merged, the character of finance and securities industry remained pluralistic. Market participants consisted of various companies at quite different stages of development. The level of market concentration is one indicator of this pluralistic character. Table 5.6 illustrates this: in 1986, assets of the top five companies accounted for 24.9 per cent of the total; by 1990, the assets represented 28.9 per cent. Nevertheless, the business concentration in Thailand's finance and securities industry was less than that in the commercial banking circuit: in 1990, assets of the top three commercial banks were 56.3 per cent of the entire industry.

Table 5.6 Market Share of Top Five Companies in Finance and Securities Industry in Terms of Total Assets
(Baht million)

Year	Amount	Percentage of Industry's Total Assets
1986	35,362	24.9
1987	45,041	28.2
1988	51,526	26.3
1989	75,747	26.9
1990	105,773	28.9

Source: Bank of Thailand.

Soundness

In the post-crisis period, the country's finance and securities companies improved their status and sustained their buffer stocks to a satisfactory degree. This is evident from the data on capitalization and capital/risk assets ratios (Tables 5.7 and 5.8). As shown in Table 5.8, the capital/risk assets ratio of the industry has exceeded the legal requirement of 6 per cent in all the years covered.

Another measure of a finance company's soundness is its level of required reserves for doubtful debts and its ability to meet such requirements. In the second half of the 1980s, the level of required reserves held for doubtful debts of the entire industry remained nearly the same (Table 5.9), and in April 1991 only 16 of 105 companies in the industry could not meet their reserve requirements. Of those 16 companies, 10 were in the Lifeboat Scheme.

Table 5.7 Capital Increase of Finance and Securities Industry and that of the Listed Finance and Securities Companies
(Baht million)

	1986	1987	1988	1989	1990
Capital increase of the industry					
– Finance Companies and Securities Companies	1,472	1,126	3,496	1,726	3,756
– Securities Companies	(4)[a]	15	260	30	553
– Total	1,468	1,141	3,756	1,756	4,309
Capital increase of listed companies (in the Stock Exchange of Thailand) in finance and securities industry	277	275	966.5	986.5	1,135
As a percentage of total increase of the industry	18.9	24.1	25.7	56.2	26.3
Number of listed companies in finance and securities industry	12	15	20	22	25

[a] In 1986, the number of securities companies decreased from 15 to 11 companies.
Source: Bank of Thailand.

Table 5.8 Capital to Risk Assets Ratio of the Industry
(per cent)

Year	Capital to Risk Assets Ratio	Growth Rate	
		Capital	Risk Assets
1986	9.52	11.19	4.48
1987	8.63	5.04	15.96
1988	9.51	43.07	29.72
1989	7.63	22.02	52.11
1990	9.43	66.14[a]	34.48

Note: The legal requirement of capital to risk assets ratio is 6 per cent.
[a] The reason for high capital growth rate in 1990 was a high capital injection (over Baht 800 million) by a new investor group which took over Dynamic Eastern Finance Thailand Co., Ltd., a firm in the Lifeboat Scheme.
Source: Bank of Thailand.

Table 5.9 Industry's Required Reserves for Doubtful Debts
(Baht million)

	1986	1987	1988	1989	1990
Finance Companies and Securities Companies	2.60	3.21	2.87	2.34	2.66

Source: Bank of Thailand.

Credit Foncier Companies

Regulations concerning credit foncier or mortgage lending were contained in Article 104 of the Civil and Commercial Codes dated January 1924, although the name "credit foncier" did not appear until Article 104 was replaced by the Act on Controlling Trading Activities which Affect the Welfare of the General Public in 1928. In 1942, the Ministry of Finance officially specified the definition and allowable terms of credit foncier businesses. Then in January 1972, the business was subject to Announcement No. 58 of the Revolutionary Council. Under the Act on the Undertaking of Finance Business, Securities Business and Credit Foncier Business, 1979 (amended in 1983, 1985, and 1992), credit foncier companies are examined, supervised, and controlled by the Bank of Thailand.

In the first half of the 1980s, when several finance and securities companies encountered crises, a number of credit foncier companies experienced severe distress as well. However, though the problems resulted from similar causes, the effects were not the same: credit foncier companies were faced with the difficulties of liquidating abundant immovable properties, which resulted in inadequate liquidity and accumulated losses. Thus, the licenses of eight credit foncier companies were revoked, and six others joined the Lifeboat Scheme.

It is expected that credit foncier businesses will soon face more competition from both commercial banks and finance companies for two reasons. First, these institutions are expected to give more attention to the field of mortgage loans, as these loans carry low risks or weights according to the newly adopted BIS rule on capital adequacy. Second, commercial banks have more experience and expertise in liquidity management than credit foncier companies. Hence, if the Government wishes to maintain the current credit foncier companies, these companies will have to be allowed to tap more short-term funds and issue mortgage-backed securities so as to improve their asset management ability and efficiency.

Life Insurance Companies

Though it is not known when the life insurance business was first undertaken in Thailand, during the reign of King Rama VI insurance matters were referred to in the January 1924 Civil and Commercial Codes. In 1929, life insurance became regulated by the Act on Controlling Trading Businesses Which Affect Safety of the Public. Initially, in 1930–1938, all five life insurance companies operating in Thailand belonged to foreign corporations. After World War II, these companies closed and the first Thai company emerged in 1942. There are now 10 Thai and two foreign companies operating in the country.

Most life insurance companies in Thailand depend on policy reserves and capital funds as their major sources of funds. In 1992, life insurance policy reserves accounted for 75 per cent of total liabilities and capital funds accounted for 18.7 per cent. Nevertheless, these policy reserves remained small in size relative to all private savings placed at financial institutions in Thailand. The ratio of the former to the latter fluctuated between 2.09 per cent (in 1977) and 2.77 per cent (in 1971). As for 1992, this ratio stayed at 2.47 per cent, demonstrating a minor role of life insurance companies in tapping savings.

Regarding the uses of funds, investments in government and private securities attract strong attention from life insurance companies. In addition, the companies extend credit to customers with collateral or guarantees. In 1992, these two channels of fund allocations absorbed 42 per cent and 29.5 per cent of life insurance companies' total assets, respectively. However, credits released to the private sector (including those via securities) by life insurance companies represent only a small portion of total private credits extended by all financial institutions, i.e., 1.4 per cent in 1992.

All life insurance companies in Thailand now operate under the Life Insurance Act, B.E. 2510 (1967). They are regulated and supervised by the Ministry of Commerce. The Act requires that each life insurance company make a security deposit of B2 million, maintain minimum capital funds of B5 million, and deposit a minimum portion of its insurance reserves with the official insurance registrar. The Act also empowers the Ministry of Commerce to specify various forms of investment in which life insurance companies can engage. Any change in insurance premium rates must be approved by the Ministry of Commerce.

Growth prospects of the life insurance business depend on several important factors both within and outside the industry, such as personnel ability, average income level, population growth, and cultural values. Thailand lacks personnel trained in actuarial science. Further, the majority of the population are farmers in the low-income brackets who believe that buying life insurance is neither affordable nor worthwhile, and most lack an understanding concerning the objectives and mechanism of a life insurance policy. In addition, current regulations are stringent and outdated. Given these problems, it is not surprising that only 5 per cent of the population hold life insurance policies; for comparison, the proportion reaches 90 per cent in industrial countries such as Japan and the US.

Agricultural Cooperatives

Despite their large number, agricultural cooperatives are beset by a series of difficulties. First, the size of each agricultural cooperative is small and there is little cooperation among them, hence possible economy-of-scale benefits from numerous and widespread cooperatives are lost. Second, agricultural cooperatives have to rely on the BAAC as a crucial source of funds, because most members of the cooperatives belong to the agriculture sector where incomes as well as savings are too low to serve as a reliable source of funds. Therefore, the operation of the cooperatives largely depends on the BAAC's policies. Third, the staff of most agricultural cooperatives lack the capability to analyze creditworthiness or feasibility of business undertakings, and therefore contribute little to the growth of their organizations. Finally, the status as well as potential of agricultural cooperatives essentially fluctuates in consonance with how agricultural output fares in the economy. For example, recently (1990–1992), the Thai agriculture sector suffered a lengthy and miserable downturn. The incidence of bad debts increased; the BAAC and the central authorities intervened and provided assistance. All cooperatives currently operate under the Cooperative Act of 1968. The Department of Cooperatives Promotion and Department of Cooperative Auditing, both of which are under the Ministry of Agriculture and Cooperatives, are empowered to regulate cooperatives and are in charge of their organization, registration, liquidation, supervision, and auditing.

Savings Cooperatives

In principle, a savings cooperative is an institution which ought to be most able to tap savings from community members, as not only should it have close contact with its members, but its members hold joint ownership and share in its immediate benefits. In practice, however, savings cooperatives have failed to prosper, mainly because they lack capable managers who can handle both credit extension and long-term planning. Another hindrance to their success may be the regulating agencies. Savings cooperatives are now subject to the same government units as agricultural cooperatives. Should those controlling units devote financial support and attention to updating technology, effective management techniques, and unconventional yet worthwhile business opportunities, savings cooperatives could easily be upgraded to become competitive and viable members of the financial system.

GOVERNMENT FINANCE INSTITUTIONS

Government finance institutions in Thailand comprise the Government Savings Bank, Bank for Agriculture and Agricultural Cooperatives, Government Housing Bank, and Industrial Finance Corporation of Thailand.

Government Savings Bank

The Government Savings Bank was established to encourage the general public to save. It therefore offers various formats of savings to the general public ranging from ordinary deposits to premium bonds and savings certificates for particular purposes such as housing and education. Normally, the Bank allocates most of the tapped funds to government agencies and state enterprises by way of notes and bonds. Occasionally, it extends short-term credits to private traders, and concessional lending to civil servants.

The majority of its funds (87 per cent) come from deposits. Notably, however, in the past two decades, the Government Savings Bank has captured a declining portion of savings placed at all financial institutions, i.e., 17.3 per cent in 1971, 9.5 per cent in 1981, and 6.1 per cent in 1992. On the asset side, the recent trend is in favor of the private sector. Since 1988, the Government has consistently achieved a surplus on its cash balance, thus leaving the Bank with no burden on financing. Consequently, the Bank was able to allot more credits to private customers.

There are both positive and negative aspects regarding the internal structure and operations of the Government Savings Bank. On the negative side, its operations are subject to numerous regulations, as stipulated in the Government Savings Bank Act, B.E. 2489 (1946), which constrain management. Decision-making is hindered by lengthy bureaucratic processes. Updated technology or instruments such as ATMs are absent, and capable personnel who have expertise in handling flows of funds, especially credit extension, in an efficient manner is lacking. These problems help to explain the Bank's low and declining market share in deposits captured by all financial institutions. At the same time, the Bank's average cost of funds remains high.

On a more positive note, however, the Government Savings Bank does have some advantages over the commercial banks. First, as the Bank has had a long history in savings mobilization, many generations, including a portion of the current ones, associate savings with the Government Savings Bank. Second, the Government Savings Bank has a large number of branches or agencies and they are in a number of locations throughout the country. For example, in 1992, the Government Savings Bank consisted of 525 branches while the two largest commercial banks—Bangkok Bank and Krung Thai Bank—had only 386 and 401 branches, respectively. Moreover, the branches of the Government Savings Bank are not as clustered in any region or city as some commercial banks (shown in Table 5.10) because the Bank does not give as high a priority to credit extension or business opportunities as commercial banks normally do. For these reasons, the prospect of the Government Savings Bank becoming a major vehicle for the promotion of domestic savings is promising. With improved access to updated technology, management know-how, and capable staff, the prospect is even greater.

Table 5.10 Bank Branches in 1992

	GSB	Bangkok Bank	Krung Thai Bank	Other Commercial Bank
Number of branches	525	386	401	1,794
Percentage share of number of branches				
Bangkok	8.5	12.4	8.4	70.7
Central	16.2	11.2	12.8	59.8
North	21.7	13.1	15.5	49.7
Northeast	24.7	13.1	15.1	47.1
South	20.4	13.3	16.0	50.3
Whole country	16.9	12.4	12.9	57.8

Source: Bank of Thailand.

One area which could be developed by the Government Savings Bank to further enhance its deposit share, is penetration into the informal lending circuits or "pie-share" by offering credit lines to clients who do not possess collateral or guarantees but who have maintained deposits at its Bank branches over a long period. Commercial banks rarely provide credit lines to clients who do not have collateral or guarantees. Further, such clients view credit lines as highly valuable. The resulting credit risk is minimal, given close monitoring and a high degree of client enthusiasm for preserving credit worthiness. The extent of credit lines should vary in accordance with the extent of deposits, the steadiness of past deposits, and the length of time that depositors have been in the program.

The widespread network of Government Savings Bank branches enables it to coordinate flows of funds from different branches or stations, and arrange volume together with terms of credit lines in a way that will keep the Bank from encountering liquidity shortages, and will keep the nation as a whole from being flooded with excessive net spending. Further, though this program of tapping deposits via credits may seem expensive, at a time when the Bank is equipped with excess liquidity, such as now, the program is practical and worthwhile. Finally, this program would, in the longer term, strengthen the Government Savings Bank's competitive position.

Bank for Agriculture and Agricultural Cooperatives

Established in 1966 for the purpose of taking over the Bank for Cooperatives, the Bank for Agriculture and Agricultural Cooperatives (BAAC) is geared toward furnishing farmers and cooperatives with concessional credits. Since 1977, BAAC has cooperated with both the Government and private corporations in supplying credits, technology, and personnel to participating farmers and cooperatives under a number of agricultural development projects, such as the Rice Production Promotion, the Para-Rubber Production Promotion, and the Agricultural Development Project in Land Reform Areas. Under current regulations, BAAC is prohibited from extending direct credit to sectors other than the primary agricultural one, which involves only the production of primary agricultural products which are not processed. The Ministry of Finance owns 99 per cent of BAAC's shares; the remaining shares belong to agricultural cooperatives, farmers' groups, and individuals.

The BAAC's funding comes, for the most part, from deposits from the public, and from commercial banks in accordance with regulations on compulsory rural credit extension, totaling roughly 58 per cent of all liabilities. Other domestic borrowings account for 14 per cent. The BAAC's credits granted to agriculture absorb 81 per cent of total assets. Such concessional lending is managed through an extensive network of 266 branches. Most loans of the BAAC are short and medium term. Very few have maturities over three years.

When deposits and credits of BAAC are compared to those of all financial institutions, it is evident that BAAC has not been energetic in tapping deposits, and its role on credit extension appears ambiguous. The BAAC's deposit share of the total has been less than 1 per cent throughout, whereas its credit share has fluctuated between 2 per cent and 4.9 per cent. Such variation can be attributed largely to volatility in prices and market conditions of agricultural products.

Problems facing BAAC are similar to those faced by the Government Savings Bank—lack of proficient staff, stringent ceilings on interest rates, high operating costs, and an unsatisfactory financial performance. However, in one area, the problem faced by BAAC is unique. While the Government Savings Bank has experienced growing liquidity since 1988, BAAC has experienced a continually inadequate inflow of funds. For long-term correction of this problem, BAAC must revise its fundamental structure and management policies, especially in the areas of manpower and interest rate specification. Increased cooperation between the Government Savings Bank and BAAC (e.g., through securities purchases), would also benefit both banks.

Government Housing Bank

The Government Housing Bank was established in 1953 under the Government Housing Act, B.E. 2496. It is wholly owned by the Government and under the supervision and control of the Ministry of Finance. The basic function of the Bank is to offer mortgage lending to moderate income earners. It once supplied land and houses to medium-income groups under a long-term hire-purchase scheme. However, such operations were transferred to the newly established National Housing Authority in 1972. Since then, the Government Housing Bank has been delegated the responsibility of providing financial services or assistance to low-income and middle-income people for housing and real estate purposes.

Deposits represent a major source of funds, representing 68.7 per cent of the Bank's total liabilities while 21.2 per cent belong to local and foreign creditors. Most of the Bank's funds, or 94 per cent of total assets, are dedicated to housing.

The Government Housing Bank faces two main problems. First, because it has few branches, the Bank is often troubled by insufficient fund inflows. It thus resorts to foreign debt from time to time. Second, the regulations under which it functions entail some rigidities in funds management (e.g., all uses of funds must be housing-related), which restrict the Bank's competitive edge. In this respect, the Government Housing Bank faces strong competition from commercial banks due to the newly adopted BIS rule on capital adequacy, which assigns a low weight to risk from housing credit.

Nonetheless, the Bank has improved and upgraded its operations to some extent by offering loans for real estate, hiring outsiders to appraise collateral of clients, and adopting flexible interest rates in order to compete more evenly with other financial institutions. Owing to these efforts, its credits extended have grown relative to those granted by all financial institutions, i.e., 1.4 per cent in 1987 and 1.8 per cent in 1992.

Industrial Finance Corporation of Thailand

Classified as a private limited company whose main purpose is to promote domestic industries, the Industrial Finance Corporation of Thailand (IFCT) comprises six groups of shareholders: commercial banks, financial companies, life insurance companies, private corporations, ordinary persons, and the Ministry of Finance. It operates in the following three areas.

(i) *development of industries*: Examples of projects in this area are export industries, energy saving programs, programs which use domestic raw materials, those which are located in rural areas, and those which rely heavily upon labor input. In these areas, IFCT supplies customers with credits, financial advice, merchant banking facilities, and investment data.

(ii) *capital market development*: Efforts in this area are exerted through three units attached to IFCT, i.e., the Mutual Fund Company, the Para Pattana Finance and Securities Company, and the Capital Market Research Institute. The Mutual Fund Company specializes in the issuance and trading of unit trusts, which represent one channel of IFCT's fund mobilization. Para Pattana focuses on securities businesses as well as tapping short-term funds, while the Capital Market Research Institute studies various facets of capital markets and offers information to investors.

(iii) *operations concerning affiliated firms*: Tasks in this area stem from the companies in which IFCT holds equity and bears partial responsibility for laying down management policies.

Because of legal constraints, and unlike commercial banks and finance companies, IFCT cannot tap funds directly from savers. It therefore focuses on capital market development, and borrows funds from the capital and money markets. In this way the IFCT functions practically as a private development bank.

Before 1985, the IFCT counted on foreign debt as an important source of funds, which amounted to 64.4 per cent of total liabilities in 1984. After the devaluation of the baht and the adoption of a flexible exchange rate regime in Thailand in November 1984 (this will be discussed in a later section of this chapter), IFCT was beset with colossal exchange losses and its debt obligations fluctuated to a frightening degree due to exchange rate volatility in the world markets. It therefore tapped funds from domestic markets by issuing bonds and debentures, an action which is certified by the changing composition of its total liabilities. The portion under the format of baht bonds and debentures grew from 2 per cent in 1984 to 36 per cent in 1992, while the portion due abroad fell to 23 per cent.

As regards the use of funds, most, or 63 per cent of IFCT's total assets, are loans to private industries while 17.5 per cent are held as bank deposits and short-term notes. Its lending to private enterprises constituted 1.2 per cent of credits released by all financial institutions to the private sector in 1992.

NONINTERMEDIARY FINANCIAL FIRMS

Securities Companies

Private entrepreneurs first entered the securities businesses in Thailand during the period between 1959 and 1961. Led by foreign partners, several companies were established to promote securities trading. In 1962, these companies jointly founded an organized stock exchange: the Bangkok Stock Exchange. Unfortunately, local private corporations were not convinced of the benefits of being listed on the Exchange, and, at the same time, potential investors were hesitant about the degree of risk involved as well as possible returns that corporate securities could yield. Therefore, only a small number of securities were listed in the Exchange, and only a quarter of those were traded while the rest stood idle. In the early 1970s, partly due to the economic slowdown, most securities companies in the Exchange turned their attention to other areas of financing, such as automobile hire-purchase and housing loans.

The securities business was revived when the Stock Exchange of Thailand (SET) became operative in 1975.

In the 1970s, when the stock market was in its infancy, the central authorities allowed finance companies to pursue securities businesses simultaneously so that the securities companies would be firmly backed financially. After a crisis occurred in some finance and securities companies, it was discovered that one cause was the intermingling of finance and securities affairs. The authorities recognized then that the two should be separated in a manner similar to the stipulation in the Glass-Steagall Act of the US. Such separation would preserve the safety of deposits at finance companies, ensuring that they were not used for speculative investment in the stock market. However, as the link between the two activities is tight in many of the country's finance and securities companies, and as compulsory separation may lead to tension in both the money market and the securities market, regulations ensuring their separation have not yet been undertaken. During this transitional stage, the authorities have allowed each finance and securities company to regulate its own activities in this regard; a number have already opted to divide their businesses to be handled by different entities.

Mutual Fund Companies

Mutual fund companies mobilize funds from the general public by issuing unit trusts which are sold by underwriters in the stock market. The acquired funds are then invested in various firms' equity together with fixed-income securities as allowed by the central authorities. Subsequently, mutual fund companies recycle returns from those investments to unit trust holders in the form of dividends. Unit trust holders may decide to re-invest those returns in whatever options proposed by the mutual fund company. Mutual fund companies offer the following favorable benefits.

(i) Because of the small denominations of unit trusts, people from the low-income and medium-income brackets have the opportunity to place their savings in the capital market.

(ii) Managers in mutual fund companies have investment expertise and skills which allows laymen or outsiders without that knowledge or time to be linked with stock and securities investment.

(iii) Due to economies of scale, the mutual fund companies can diversify risks in stock and securities investment more efficiently than each small investor could.

(iv) The demand in the stock and securities market is strengthened and that widens the chance to develop the capital market further.

From an investors' point of view, however, there are certain constraints or limitations attached to an investment via the mutual fund company.

(i) Returns that investors expect from unit trusts are typically not high as a result of the adopted risk diversification tactics.

(ii) Prices of unit trusts tend to fluctuate in accordance with the general sentiment in the stock market. Since unit trust holders are ordinarily uninformed about particular securities in the SET, they can be panicked by rapid fluctuations of stock prices, and may liquidate their accounts prematurely, even though the mutual fund company has not allotted funds to the falling stocks or securities.

(iii) As of 1993, all mutual funds offered in SET are close-end which means that investors have to bear the long-term risks that the capital market may experience severe or lengthy recession, resulting in unsatisfactory returns when those unit trusts fall due. This drawback can be averted by issuance of open-end unit trusts which allow investors to liquidate them with the mutual fund company before they are due. Though investors are likely to have more confidence in these open-end unit trusts, the open-end quality does impinge on the mutual fund company as a binding constraint in investment strategy. Consequently, returns to open-end unit trusts may not equal or exceed those of the close-end trusts.

Prior to 1992, there was only one mutual fund company authorized by the Bank of Thailand to mobilize onshore funds from the general public for investment in both listed and unlisted securities in SET. In 1992, financial institutions were allowed to jointly establish mutual fund management companies and apply for such licenses. However, regulations require that shareholding in these companies be diversified to avoid biased decision-making. The maximum equity holding is specified at 25 per cent for commercial banks, 50 per cent for finance companies and securities companies combined, and 25 per cent for foreign juristic with expertise in the business. People eligible to apply for licenses must hold at least 25 per cent of the company's total shares, have a sound financial position, and present plans to operate mutual funds. Currently, there are eight mutual fund companies operating in Thailand.

As the unit trusts managed by mutual fund companies are designed to function as a bridge between the capital market and low-income earners whose locations are widely scattered, one means to reach those people is to allow branches of banks or finance and securities companies in remote areas to serve order placement and processing. Without this service, people in the rural districts would be unable to participate in the capital market, and their potential to speed up development of the capital market remains unquestionably strong.

Another area where the central authorities can encourage unit trust investment is to differentiate between equity-related and fixed-income unit trusts. Equity-related unit trusts are those whose proceeds are invested in equities of firms listed in the stock market; proceeds from fixed-income trusts are allotted to securities with fixed interest rates. Compensation or dividends from the former vary from year to year depending on the firms' performance, while returns from the latter are fixed or assured. This differentiation tempts different investors, and helps circumvent or lessen the adverse side-effects that plummeting share prices may have on confidence of unit trust holders.

MACROECONOMIC ISSUES

Monetary Policy

The Bank of Thailand, as other central banks, is the sole agent responsible for the formulation and conduct of monetary policy. Its main objective is to maintain stability of the economy and financial system although it also performs other roles such as fiscal agent and provider of development financing. The latter are sometimes in conflict with the Bank's main objective.

In formulating monetary policy, the Bank first needs to find appropriate monetary aggregates with specific characteristics as follows:

(i) *stability*: Good monetary aggregates should have stable relationships with their ultimate economic targets, i.e., economic growth, inflation, and unemployment.

(ii) *predictability*: The relationship between selected monetary aggregates and the ultimate economic targets should be predictable so that the impact of monetary policy will be positive.

(iii) *controllability*: The selected monetary aggregates should be under the full command of the Bank through available monetary instruments.

(iv) *simplicity*: The aggregates should be simple and easily understood by the public, especially in countries which announce monetary targets.

The Bank of Thailand currently monitors the expansion of narrowed money (M1), broad money (M2), and private credits, although they are not treated as monetary targets in the conventional sense, but rather as monetary indicators which are subject to economic condition and policy direction. Under certain circumstances, one monetary aggregate may receive more attention than another. For example, during the period of severe external instability, 1983–1984, the Bank exercised stringent credit policy by setting the growth target of credit expansion at 18 per cent (1984). Thereafter, the Bank gave priority to exchange rate policy to facilitate international trade while forgoing, to some extent, its autonomy in controlling monetary aggregates. Presently, with the need to mobilize domestic savings to contain the widening savings-investment gap, the Bank pays more attention to the growth of M2 and implements various measures to boost financial savings.

In practice, the Bank does not command full control of the abovementioned monetary aggregates. Therefore, the Bank selects another variable which can be closely managed on a day-to-day basis and which has a stable and predictable relationship with other monetary aggregates. For that purpose, a monetary base has been adopted which is essentially part of the balance sheet of the Bank and is linked with other monetary aggregates through multipliers. The effectiveness of the monetary base in terms of control procedure depends on the ability of the Bank to handle its own balance sheet and predict movements of the multipliers. In Thailand's case, the central bank can only exert control on parts of the monetary base. Under the present exchange rate arrangement, net foreign assets of the Bank can vary considerably, depending on how the Bank sets daily exchange rates and the exchange rate fluctuations of Thailand's major trading partners.

Regarding net claims on government by the Bank, it is virtually preset and largely hinges upon fiscal position. The Bank has a better grip on net claims on financial institutions. Nevertheless, the Bank's financial assistance to ailing banks and finance companies during the 1980s, and its development financing to priority sectors has sometimes complicated its operations and, in turn, reduced its ability to control the monetary base. Further, under the relatively stable exchange rate mechanism and diminution of exchange control measures, attempts to sterilize capital inflows partly undermine the effectiveness of monetary base control.

The factors affecting money multipliers act in accordance with theory. For example, the fall in cash reserve ratio will increase both multipliers of M1 and M2. Development of the payment system as well as an increased use of credit cards will strengthen both multipliers, while the rise in interest rates will increase M2 multipliers but reduce M1 multipliers. In practice, these relationships are weak in the short run due to various types of friction in the financial system, making it difficult for the Bank to fine-tune its monetary operations.

The central bank has a number of monetary instruments to control the monetary base. These are normally used in daily and medium-term operations, and include:

(i) *liquidity ratio*, which is currently set at 7 per cent of total deposits at commercial banks. The ratio comprises at least 2 per cent in the form of deposits at the central bank, at most 2.5 per cent as cash in hand, and the rest as eligible securities. The Bank of Thailand does not usually use this ratio for monetary control but rather for prudential purposes.

(ii) *credits to priority sectors*, which are given to commercial banks on a semi-annual basis. Under the present format, once the commercial banks commit themselves to 50 per cent of the credits requested by priority sectors, the central bank will furnish the remainder. Although this facility charges priority sectors at concessional interest rates, demand is constrained by limited credit lines and qualifications of eligible borrowers pre-specified by the Bank.

(iii) *loan window,* which is the central bank's lending facility, available to commercial banks when they face short-term liquidity shortages. The borrowing must be pledged by government bonds and has a maximum maturity of 7 days. The Bank can influence demand for this type of borrowing by varying the bank rate or the credit ceiling set for each commercial bank.

(iv) *repurchase market,* which was established in 1979 to add liquidity to government bonds held by financial institutions and to reduce the oligopolistic practice of large commercial banks in the interbank market. The central bank can exercise its monetary policy by intervening in the repurchase market but the principal discipline of anonymity in this market is always maintained. Normally the Bank's intervention aims at reducing the degree of volatility in money market rates. This intention is sometimes in conflict with the objective of monetary control.

Fiscal Policy

The Government normally formulates budget expenditures in April each year for the following fiscal year starting in October. The expenditures depend on the projection of government revenue, financing need of each ministry and, most importantly, the fiscal stance of the Government. During the first half of the 1980s, when the Thai economy was in difficulty, the Government adopted a disciplinary fiscal policy with a package of austerity measures. Although such fiscal policy was perceived as inevitable at that time, it was also the intention of the authorities to reform the government sector and reduce its role as the main economic locomotive. As a result of the continued disciplinary fiscal stance together with an extraordinary pace of economic growth, the Government experienced a fiscal surplus in 1989 for the first time in 30 years; that surplus continued into the 1990s. However, the current fiscal stance is aimed toward more economic stimulation, particularly in the area of rural development and income distribution. Thus, the government increased investment expenditures in FY1993, causing a small fiscal deficit that year. Based on past experiences, the prolonged use of a fiscal stimulus may not be effective, since government revenue usually varies in line with economic conditions. When the economy encounters a slowdown, government revenue tends to fall, leaving little room for expenditure increases without jeopardizing economic stability. In contrast, when the country experiences a boom, revenue tends to rise and there will be no need for further stimulation.

The following briefly demonstrates essential features of the tax structure in Thailand.

(i) *personal income tax*: As of 1 January 1992, there were five income tax brackets; the minimum tax rate was 5 per cent while the maximum tax rate was 37 per cent. The maximum expenditure exempted for income taxes was fixed at B60,000 or 40 per cent of taxable income, whichever is lower.

(ii) *corporate income tax*: As of 1 January 1992, the corporate income tax rates imposed on listed companies on SET and others were unified at the rate of 30 per cent.

(iii) *tax on interest income and dividends*: Taxpayers have the following two options: paying a flat rate of 15 per cent, or adding the income from interest and dividends to other types of income which are subject to progressive personal income tax.

(iv) *tax on capital gains*: Taxpayers have the same options in paying taxes owed on capital gains as they have on interest income and dividends. In addition, for common stocks issued by listed companies, investors who are juristic entities must add the received capital gains to other types of income and be subject to progressive income tax. Capital gains received by ordinary persons are exempt, as are capital gains from the sale of government bonds and debentures.

(v) *sales tax*: In 1992, the sales tax was replaced by the VAT of 7 per cent which was lower than the revenue neutral rate. The objectives of the introduction of the VAT were to reduce tax evasion and to increase efficiency in tax collection.

(vi) *tariffs*: In 1990, the tariffs on imported capital and raw materials of 419 items were harmonized and reduced to 5 per cent so as to increase efficiency and competitiveness of domestic industries.

Exchange Rate Policy

Thailand abolished the par value of the baht vis-á-vis the US dollar on 8 March 1978 and adopted the system which pegged the baht value to a basket of currencies of its major trading partners. With regard to the relevant legislative procedures, the Currency Act B.E.2501 (1958) was amended to enable Thailand to choose any exchange rate regime which is more flexible and consistent with the IMF's Agreement. Other than following the exchange rate regime adopted by most countries at the time, the change helped liberate the baht value from movements of any particular currency and allowed it to reflect more accurately the prevailing economic and monetary situation, particularly those related to Thailand's external position. To render the exchange rate determination process more flexible and amenable to timely adjustment in line with domestic and international monetary conditions, the Exchange Equalization Fund (EEF), which is the sole agent attached to the central bank responsible for the exchange rate policy, ceased to announce bid and offered exchange rates in dealing with commercial banks. Instead, it allowed commercial banks to participate in the process of determining exchange rates on a daily basis, called "daily fixing." This method took effect on 1 November 1978.

In 1981, the waning confidence in the value of the baht was evident as the public perceived that EEF was deliberately overvaluing the baht in relation to the basket of currencies so as to stabilize the baht/dollar exchange rate regardless of the rapidly appreciating US dollar. Meanwhile, the balance-of-payments position was deteriorating rapidly. Speculative activities in the foreign exchange market and prepayments of foreign borrowings were widespread, leading to continuous depletion of foreign exchange reserves. On 12 May 1981, EEF intervened in the foreign exchange market at the exchange rate of 21 baht per US dollar as against 20.775 baht per US dollar on the previous day, equivalent to 1.07 per cent devaluation of the baht. This devaluation did not revive public confidence, however, because the US dollar continued to surge and the balance of payments showed no sign of improvement. In addition, the "daily fixing" system clearly revealed daily transactions of EEF to commercial banks (including sustained sales of US dollars in large amounts by EEF), thus inducing bankers to engage in speculative transactions. In this light, EEF devalued the baht vis-á-vis the US dollar for the second time by 8.7 per cent or from 21 baht per US dollar to 23 baht per US dollar, and replaced the "daily fixing" system by the system that entitled EEF to fix exchange rates independently as from 15 July 1981.

The growth-oriented economic policies adopted in the following years and the rapidly appreciating US dollar caused by high US interest rates led to serious economic problems in Thailand in 1984, especially in the external accounts. Thus, the baht value, which had been firmly tied with the US dollar since 1981, began to appreciate at an alarming rate relative to major currencies such as the deutsche mark and pound sterling. To counteract the adverse impact on the economy, the authorities announced that the exchange rate system would be modified in two major aspects effective 2 November 1984. First, while EEF continues to be the central agency in specifying exchange rates of the baht, those exchange rate specifications vary on a daily basis depending on fluctuations of value of major currencies selected to constitute a reference basket for Thai baht. Second, an adjustment in the exchange rate between the baht and the US dollar was considered necessary for successful transition to the new exchange rate mechanism. In this connection, EEF on 5 November 1984 set an initial midrate of 27 baht per US dollar, equivalent to a 15 per cent devaluation of the baht against the US dollar from 23 baht per US dollar. Along with the new exchange rate regime, supplementary measures were launched to prevent unfair exchange gains and to closely monitor commercial bank activities in foreign exchange markets. To comply with these measures, commercial banks were required to submit their windfall profits from exchange rate differentials, if any, to EEF and report their daily foreign exchange positions to the Bank of Thailand.

The exchange rate arrangement adopted in 1984 is still in effect in 1994. In conducting the exchange rate policy, EEF works to achieve three objectives corresponding to its short-term, medium-term, and long-term goals. In the short run, EEF adjusts the exchange rate according to the basket of currencies and day-to-day developments in foreign exchange markets abroad; the amount of foreign exchange traded with commercial banks is taken into account in the medium term. As regards the long-term goal, EEF tries to minimize exchange rate disturbances to exports, since the overall performance of the Thai economy heavily relies upon this sector.

FINANCIAL MARKETS

Financial markets in Thailand comprise money markets, the foreign exchange market, government securities and commercial paper markets, and the stock market.

Money Markets

The range of money markets in Thailand includes those for interbank, government bond repurchase, treasury bills, and Bank of Thailand bonds.

The interbank market, the oldest among all formal money markets, is relatively well developed with lending at call and for fixed terms. Transactions are normally unsecured. Because of their widespread branches and abundant deposits, a number of large Thai banks have traditionally been influential lenders in the market, while small Thai banks, foreign banks, and finance companies are perpetual borrowers. In March 1985, commercial banks introduced the Baht Interbank Offered Rates (BIBORs)—the average rates at which prime banks lend to one another—to function as reference rates for floating rate loans. However, lending committed at BIBORs is inactive because prime banks tend to adjust their liquidity positions through the repurchase or foreign exchange markets. Interbank rates have been rather volatile and more variable than Eurodollar rates. To some extent this may reflect the seasonal pattern of cash demands during crop financing and tax payment seasons. The variability also reflects the thinness and shallowness of the market as well as the limited day-to-day liquidity management by the Bank of Thailand. The efficiency of the market is low as a result of a small number of participants and oligopolistic practice of lenders, which allows them to discern borrowers' positions and gives rise to multiple interest rates.

The government bond repurchase market was established on 9 April 1979 with four main objectives: (i) to increase liquidity of government bonds held by commercial banks to satisfy the branch opening requirement; (ii) to reduce the oligopolistic edge of large commercial banks in the interbank market; (iii) to introduce an impersonal money market which reveals neither identities nor liquidity positions of lenders and borrowers as is typically evident in the interbank market; and (iv) to create a new channel for the Bank of Thailand to conduct monetary policy. The Bank of Thailand acts as a principal on one side of the repurchase transactions and matches the demand for and supply of repurchases. Nevertheless, it can intervene in the repurchase market by purchasing or selling from its own account. Presently, there are seven maturities in the market, namely, 1, 7, 15, 30, 60, 90, 180 days. However, most of the activities are concentrated in the short end, i.e., one and seven days. The repurchase market has an advantage in that transactions are secured against government bonds; therefore rates can be more uniform, although they appear to be as volatile over time as in the interbank market.

The treasury bills market in Thailand is very restricted. Bills are auctioned on a weekly basis and purchased mainly by commercial banks and other financial institutions to satisfy certain requirements, and by the Bank of Thailand. The secondary market is small. A captive market and limited supply normally give rise to treasury bill rates which are below other market rates. The Bank of Thailand is usually ready to sell bills from its portfolios and has occasionally purchased bills before maturity. However, the treasury bill market has disappeared in the past few years due to the Government's persistent fiscal surplus. In the future, the Bank can reactivate treasury bills in its day-to-day management of money market liquidity as a supplementary instrument to its repurchase market transactions.

The market for Bank of Thailand bonds was created in May 1987 when the Bank for the first time issued its bonds to absorb excess liquidity from the financial system. The bonds were worth B2 billion with a coupon rate of 6 per cent per annum and 180-day maturity. The bonds were sold to commercial banks, and could be used by them to satisfy the branch opening requirement. Foreign bank branches could use them to meet the rule on capital adequacy. The enduring excess liquidity generated pressure upon the Bank of Thailand to issue a second batch of its bonds in 1988. The B2 billion bonds, carrying 6 per cent annual interest with one-year maturity, were sold to commercial banks and the Government Savings Bank. In March 1990, to slow down the growth of domestic credits, the Bank sold its bonds by auction, amounting to B13,485 million with a coupon rate of 9.125 per cent per annum and one-year maturity. The infrequent issuance of the Bank of Thailand bonds led to a limited supply, but notably did not facilitate an establishment of its secondary market since most of the bonds were short-dated and meant for particular purposes or held to satisfy certain requirements.

Foreign Exchange Market

Thailand has two institutions which can accommodate foreign exchange transactions, namely, the Exchange Equalization Fund and commercial banks (including the recently established offshore units). Major suppliers of foreign exchange in the Thai market are exporters, foreign investors, and large corporations which rely on foreign borrowing to finance their businesses; major buyers of foreign exchange are importers and large Thai corporations investing abroad. Recently, due to exchange control deregulation, commercial banks have tended to rely more on foreign borrowings. As a result, commercial banks have also become one of the major providers of foreign exchange. The recent liberalization allows residents to have foreign currency accounts and nonresidents to hold baht accounts, leading to more international capital movement and more transactions between baht and foreign currencies. However, the forward market between the baht and the US dollar or the dominant foreign currency has not been very active since the exchange rate between the baht and the US dollar is stable and there have yet to be benchmark interest rates for determination of appropriate premiums or discounts.

Cross-currency transactions are normally done in the interbank foreign exchange market so as to hedge against risks, and to seek profits. According to banking regulations, certain ceilings are imposed on commercial banks' foreign exchange positions (covering both spot and forward status): net foreign assets not exceeding 25 per cent of capital funds or net foreign liabilities not exceeding 20 per cent of capital funds. This regulation was designed to limit the extent of foreign exchange exposure; it does not contribute much to monetary control as commercial banks can cover their spot positions in the forward market.

Government Bond Market

The government bond market in Thailand has been dormant since its inception. The captive nature of government bonds was initiated by the requirement imposed on commercial banks to hold government bonds to satisfy the branch opening requirement. As a result, coupon rates did not reflect market conditions. In setting coupon rates, the Government usually fixed the rates at 1–1.5 per cent above the time deposit rates of commercial banks.

The captive market also hindered development of the secondary market because commercial banks could not release their bonds, and if they did, they would realize losses due to thin market status. The Government has run a fiscal surplus in the past five years (1988–1992), and there is no need to issue new government bonds. Consequently, the government bond market has steadily diminished. After the Bank of Thailand abolished the branch opening requirements on 17 May 1993, a large number of government bonds were released, though some must be maintained under liquidity requirement for prudential reasons. It is expected that if the Government continues to run a fiscal surplus in the next few years, there will be an insufficient amount of government bonds for commercial banks to satisfy the liquidity requirement. The central bank may need to allow commercial banks to hold other types of securities, or issue its own bonds for such objectives.

Commercial Papers

In Thailand, commercial papers serve two main purposes: for short-term borrowing and for debt payments. The transactions of commercial papers issued for debt payments have risen fairly rapidly, especially those without guarantee from commercial banks or finance companies. With respect to the first category, commercial papers are normally issued by financial institutions or large corporations, e.g., Citinotes by Citibank and IFCT-notes by the Industrial Finance Corporation of Thailand.

Stock Market

Evolution

In 1953, Houseman & Co., Ltd., Siamerican Securities Ltd., and Z & R Investment and Consultants began to act as intermediaries for securities transactions, however, their volume of business was negligible and prices did not truly reflect demand and supply. The Bangkok Stock Exchange was set up as a limited company in 1962 and later registered with B250,000 as capital funds. However, only seven or eight of the 35 listed securities were frequently traded, demonstrating tenuous attention from the general public. The annual market turnover during 1964–1973 was only B50 million. In addition, fees for different securities differed from one another.

In 1969, the central bank hired Professor S.M. Robbins to study the establishment of a stock market in Thailand; this led to the enactment of the Securities Exchange of Thailand Act in May 1974 and the Stock Exchange of Thailand (SET) in April 1975.

Initial period (1975–1978)

During its first two years, SET did not see much trading and, although activities picked up considerably in 1977 (trading volume: B26.6 billion; index: 181), most transactions were short-term speculations. In May 1978, the authorities decided to impose a 10 per cent tax penalty on capital gains realized from securities held no longer than six months. As a result, both prices and business volume in SET plunged. The Government, therefore, replaced the capital gains tax on short-term trading with a much lower sales tax; as a result, in 1978, the trading volume increased to B57.3 billion and the price index to 258.

Recession (1979–1983)

The atmosphere in SET deteriorated markedly as a consequence of the second oil price crisis and the surge of interest rates worldwide. In 1979, the performances of listed companies declined, as did the trading volume to B22.5 billion and the SET index to 149. The authority implemented rescuing measures such as setting up the Capital Market Development Fund in August 1979 with a contribution of B1 billion from the Central Bank, Government Savings Bank, and Thai Bankers Association. This Fund, operated by the IFCT, intervened in SET whenever share prices were too low. Moreover, the Thai Bankers Association lined up two pooled funds (B1 billion each) to support liquidity positions of SET members, while the Government organized the Krung Thai Fund of B3 billion for a program to buy securities with repurchasing rights left with sellers.

In 1980, several rounds of oil price and interest rate increases in the domestic market worsened the performances of listed companies, leading to further downfalls of both trading volume and SET index to the level of B6.6 billion and 125, respectively. The period between 1981 and 1983 corresponded to the duration that the Thai economy suffered from various strokes. For example, deficits on the country's foreign trade and current account rose precipitously while that on the Government's fiscal balance was equally threatening. Volatility of exchange rates and interest rates also climbed to a formidable extent. The widespread lack of confidence in the value of the baht and collapses of a few finance companies at the time thus came as no surprise. Trading volume and the SET index sank to only B2.9 billion and 106, respectively in 1981. In response to such tension, the Government organized a special committee to study and suggest suitable methods of resolution.

Recuperation (1984–1986)

Although Thailand encountered the second round of crisis among financial firms together with the breakdown of several chit funds in 1983–1984, the central authorities intervened and solved the problems, through such efforts as adopting a new scheme of exchange rate specification in November 1984 (as discussed earlier), amending the SET Act, and authorizing the emergency decree on illegal borrowings in 1984. Further, some recovery in the world economy was evident as oil prices and interest rates started to decline. Consequently, the stock market was revived with trading volume rising to B10.9 billion in 1984 and B16.5 billion in 1985.

In 1986, high liquidity persisted to such an extent that commercial banks lowered their interest rates five times. In addition, the central authorities amended the exchange control regulations so that foreign investors could easily transfer returns abroad. The Bangkok Fund and Thailand Fund were created to attract foreign investment. Trading volume thus grew to B29.8 billion while the SET index closed at 207.

Expansion (1987–1989)

Despite the Black Monday event on 19 October 1987 which pulled down the SET index by 229 points within two months, the stock market experienced appreciable growth, with trading volume reaching 123.4 billion and the index at 285 in that year. An unusually high pace of economic expansion in 1988 and 1989 (13.2 per cent and 12.0 per cent) enabled listed firms in SET to perform remarkably well. In addition, six more foreign funds were formed. These factors, together with favorable trends in the world stock markets, tripled the trading volume in SET to B377.1 billion and the index to 879 in 1989.

As regards the policy reaction to the Black Monday event, finance and securities companies were allowed to invest more in stocks, from 60 per cent to 100 per cent of their capital funds. Members of SET constructed a close-end six-year Ruam Pattana Fund of B1 billion in November 1987. On the part of SET, the limit on daily fluctuations of stock prices was decreased from 10 per cent to 5 per cent between 30 October and 26 November 1987.

Adjustment (1990-now)

The SET was healthy in the first half of 1990, as a result of higher confidence partly attributed to the Central Bank's declaration to accept Article 8 obligations of IMF and the creation of three more foreign funds totaling B8.6 billion. The operating hours of SET were lengthened from two to three per day. On the other hand, the Gulf Crisis in August 1990 generated adverse impact worldwide, and the impact on SET was no exception: the index fell to 613 in 1990 while trading volume totaled B627.3 billion. Though the general sentiment was not aided by the political takeover by the national peace-keeping force in February 1991, trading volume in 1991 remained heavy at B793 billion while the SET index recovered to 711. In addition, in 1992, despite some negative events such as the unrest in Thailand in May, price rigging in the stock market, and psychological effects of subsequent prosecution, trading activities doubled to B1.83 trillion and the SET index rose to 893. One technical reason for the increase was the increase in trading hours from three to four hours in two sessions since July 1992.

As for measures by the Government to cope with the Gulf Crisis, the stock purchasing margin was lowered from 100 per cent to 70 per cent on 10 August 1990 and 50 per cent on 20 September 1990. Finance and securities companies were permitted to invest more in stocks, an increase from 60 per cent to 100 per cent of their capital funds. Finally, the second close-end five-year Ruam Pattana Fund of B5 billion was set up to supply SET with needed liquidity. In 1992, more funds were established for the purpose of maintaining stability in SET in the presence of some disturbances to investors' sentiment.

RECENT REFORM EXPERIENCES

Historically, the Thai financial system was highly regulated and shielded from both domestic and foreign competition. Financial operations were subject to interest rate ceilings on both deposits and lending, as well as portfolio and branching restrictions, and various types of compulsory credits. Even so, in the 1960s and 1970s, with the pace of economic development toward industrialization relatively slow and the degree of openness of the economy still small, the financial system could manage to cope with prevailing activities. During the 1980s, in the presence of rapidly advancing computer technology and telecommunications and increasing volatility in the world's interest rates as well as foreign exchange markets, numerous financial innovations in terms of instruments, procedures, and techniques have been developed. These help institutions to hedge against risks and to circumvent financial regulations to create greater profit opportunities or to survive in the highly competitive and volatile financial markets. Financial institutions must compete with each other, and also with nonfinancial companies in areas such as consumer credits and housing loans.

During the period 1980–1987, though the competitive scenario described above prevailed in Thailand, financial liberalization received little attention. That was due to several factors. First, the Thai economy during that period was experiencing a slowdown and was plagued by frequent internal and external instabilities. Second, the financial system itself was confronted with a few problem banks, a score of ailing finance companies, and failures of various chit funds. Therefore, the authorities gave top priority to retrieval of economic and financial stabilities.

After economic and financial stability had been restored, Thailand underwent a process of economic expansion and recorded balance-of-payments surpluses as well as persistent positive cash balance on the fiscal budget as a result of increased revenue collection. The need to balance growth in the manufacturing and financial sectors, enhance competitiveness of domestic financial institutions, and restructure the financial system thus became more pressing policy issues.

In this light, medium- and long-term plans were formulated with a view to address domestic problems and to cope with competition on the external front, i.e., the outcome of rapid modernization of technology and integration of international money and capital markets. The plans focused on the development and modernization of the financial system and on the diminution of government intervention. Toward this end, efforts were directed at encouraging savings mobilization to support future economic development, boosting competitiveness of the financial institutions, promoting development of new financial instruments and services, and turning Thailand into a regional financial center.

In this connection, the Bank of Thailand implemented the three year financial reform plan in 1990–1992. The first component of the plan included deregulation of interest rates and financial institutions' portfolio management as well as relaxation of exchange controls. The second involved development of the supervision and examination system, the monitoring and analytical procedure, adjustment of the capital to risk asset ratio in line with the guideline set by BIS, training of bank examiners, and development of an information system. The third promotes financial innovations in both primary and secondary markets. The fourth, on improvement of the payment system, has as targets reform of the interbank clearing system and establishment of the second note-printing works.

Deregulation and Relaxation

Financial deregulation is normally related to relaxation of supervisory constraints or intervention by the authorities, thereby allowing the market mechanism to function more freely. In Thailand, deregulation has been undertaken in a gradual manner, beginning with interest rate reform after the second oil shock (1979–1980). Traditionally, Section 654 of the Civil and Commercial Codes restricted general interest rates to within 15 per cent since 1924. Wide variations in international interest rates in 1979–1980 induced the promulgation of the Financial Institutions Lending Rate Act, B.E.2523 (1980), which helped cushion the domestic financial system from the impact of volatile world interest rates. The ceiling on financial institutions' lending rates was raised from 15 per cent to a level deemed appropriate by the Minister of Finance.

Further, the government bond repurchase market was established in 1979 to create a new channel for financial institutions to adjust their liquidity positions and to provide an additional tool for the conduct of monetary policy. The exchange rate system was revamped in 1984, by replacing the previous practice of pegging the baht solely to the US dollar with the basket system linked to the currencies of major trading partners to better reflect their genuine economic and financial relationship.

Deregulation measures implemented under the three-year plan adhere to one principle originated in the previous decade—to reduce the role of, and intervention by, monetary authorities in the financial system. Financial institutions have the opportunity to adjust their operations in preparation for greater competition in the future when the financial liberalization process takes its full course. In addition, the financial system must be geared toward interest rate liberalization, relaxation of constraints on financial institutions' portfolio management, an expanded scope of operations of financial institutions, and future development as a regional financial center.

Dismantling of Exchange Controls

Acceptance of the obligations under Article VIII of the International Monetary Fund's Articles of Agreement and implementation of the first phase of exchange control relaxation on 21 May 1990 were important steps in the process of exchange control deregulation. The aim was to liberalize the foreign exchange regime in line with globalization of the economic and financial systems, and greater freedom of international capital movement. Phase I of the exchange control deregulation allowed commercial banks to process customers' applications for the purchase of foreign exchange without prior approval from the Bank of Thailand. The Bank also raised the limit on foreign exchange transactions for services and allowed commercial banks to process applications for purchase of foreign exchange in small amounts, remittances of loan repayments, sale of securities, or liquidation of companies.

The deregulation was expected to benefit the public at large as it streamlined the approval process without altering major elements of exchange controls. Preparations for further liberalization of exchange controls were already under way. However, the second phase required the amendment and the change in principle of exchange controls as embodied in Ministerial Regulation No. 13, B.E. 2497 (1954) pursuant to the Exchange control Act, B.E. 2485 (1942). The amendment to the legislation involved a lengthy process and commercial banks also needed a transitional period to prepare and make necessary adjustments. In the following year, on 1 April 1991, the Bank of Thailand announced Phase II of the exchange control deregulation. The second phase allowed greater flexibility to private businesses and the general public in the purchase and sale of foreign exchange while retaining minimum controls on some instruments for the purpose of monitoring and safeguarding the system. Under Phase II, the public could freely purchase foreign exchange from commercial banks for current account transactions.

The role of the authorities was reduced to the adjustment of rules, regulations, and operational guidelines to ensure greater convenience for the public. The measures designed to streamline procedures for export proceeds to provide more convenience for the public were implemented on 1 May 1992. Exporters were allowed to receive payments from nonresident baht accounts and to transfer foreign currency receipts to pay for imports or to debit foreign liabilities to nonresidents. With regard to foreign currency deposit accounts, commercial banks were permitted to withdraw funds freely from their accounts and/or accept deposits from government departments and state enterprises to facilitate foreign currency transfers.

The only remaining restrictions concern purchases of foreign currencies for the acquisition of real estate abroad and foreign equities.

Interest Rate Liberalization

The three-year plan of the Bank of Thailand aimed to fully liberalize the interest rate structure, thereby enabling the system to adjust to fluctuating demand and supply conditions, both domestically and externally. With continuous economic expansion since 1987, there has been a growing need to mobilize long-term funds for economic development. When it became evident that long-term deposits had not expanded in line with borrowing needs, the Bank of Thailand lifted the ceiling rate on term deposits with maturity greater than one year in June 1989 so as to accelerate the process of savings mobilization. Thereafter, interest rates on deposits with maturities longer than one year moved from the previous ceiling of 9.5 per cent to 10.5–11.0 per cent per annum.

Ceilings on time deposits of all maturities were abolished on 16 March 1990. On 8 January 1992, the cap on saving deposits was removed. In June 1992, the interest rate ceilings on finance companies' and credit foncier companies' borrowings, deposits, and lending, together with the ceiling on commercial banks' lending, were terminated, thereby effectively completing the liberalization of all types of interest rates. However, the interest rate ceilings on mortgage lending to low-income individuals were maintained, provided the loan contracts were made before 1 June 1992. For mortgage lending from commercial banks, the ceiling was fixed at their minimum lending rate (MLR), and for finance companies and credit foncier companies, it was fixed at their MLR less 1.5 per cent.

Fewer Constraints on Financial Institutions' Portfolio Management

In the past, and in a manner similar to other developing countries, the central bank required financial institutions to contribute to rural and overall development of the economy. Commercial banks were required to fulfill the agricultural credit target and the branch opening requirement. However, during the past few years, the structure and economic environment of the country have changed markedly. The Bank of Thailand has, thus, reduced its intervention in the decision-making process of the country's financial institutions. In retrospect, the intervention policy could have been perceived as unfair to the financial institutions concerned as well as to customers and other economic sectors. Also, inefficient allocation of resources must have arisen as the market fell short of free competition.

The central bank's three-year plan was designed to streamline and eliminate certain requirements imposed on financial institutions while retaining those needed to maintain stability and solvency of the financial system. This was intended to provide greater flexibility in management and operation, while reducing operation costs.

Rural Credit Policy

In the past, the Bank of Thailand set targets for commercial banks to allocate credits to the agriculture sector in proportion to total deposits (compulsory credit was discussed in an earlier section of this chapter). In 1987, however, because the structure of the country's agricultural production had shifted toward small-scale industries and the services sector, a new definition and ratio were introduced under the "rural credit policy." To give commercial banks greater flexibility in asset management and in accordance with the Government's policy to support regional small-scale industries, the rural credit policy was modified to cover a wider scope of activities besides the narrowly defined agricultural activities. In January 1991, the coverage of rural credits was broadened to include credits for wholesale trading of agricultural produce and regional industrial estates. The definition was further expanded in 1992 to incorporate credits for secondary occupations of farmers and exports of farm products. Computation of the deposit base was also adjusted to exclude interbank deposits to facilitate compliance with this requirement.

Branch Opening Requirements

The mandatory requirement for commercial banks to hold adequate government and other eligible bonds before opening branches came about when the Government needed to issue a large number of bonds to finance budget deficits which had arisen from huge expenditures devoted to economic development. However, now that the Government's fiscal position has improved, no government bonds have been issued since 1991. The central bank, therefore, progressively relaxed this requirement to allow commercial banks more flexibility in dealing with their tapped resources. The ratio was reduced progressively from 16 per cent of total deposits to 5.5 per cent between 13 November 1990 and 14 February 1993, until it was finally abolished on 17 May 1993.

Modification of the Reserve Requirement

Effective 23 June 1991, the Bank of Thailand relaxed the constraints on commercial banks' portfolio management by replacing the reserve requirement ratio with the liquidity ratio. Although in practice commercial banks must maintain the minimum of 7 per cent of deposits in securities and cash in accordance with the Bank's conditions, the liquidity ratio permits commercial banks to substitute other securities for government securities, namely, Bank of Thailand's bonds, debentures, and bonds of government organizations and state enterprises. This provides commercial banks with greater opportunities in relation to investment options and asset management. As the Bank had not used the liquidity ratio as an instrument to conduct monetary policy, the change, in effect, had no significant policy implications.

Besides the above measures, the central bank revamped a regulation on compulsory credits, which formerly required each bank branch to lend an amount in proportion to its total deposits to customers in the area where it was located. Under the revised regulation, bank branches situated in each region are required to collectively extend credits, in proportion to total deposits, to customers in their respective region.

Expansion of Financial Institutions' Scope of Operations

Financial institutions were permitted to broaden their scope of activities and to better use their resources as well as expertise. It was expected that consumers of financial services would benefit from a wider variety of financial services together with greater competition among financial institutions. The customary practice of receiving deposits and extending loans has not been as profitable recently as before because clients have more options for saving and borrowing. Financial institutions, therefore, must seek more revenue from fee-based income. Nevertheless, the policy to widen the scope of operations of financial institutions has been undertaken in gradual steps, with regard to the efficiency and readiness as well as available expertise of financial institutions. In the initial stages, the Bank of Thailand expanded the scope of activities of financial institutions within the framework of relevant legislation.

As regards commercial banks, at present, they are allowed to conduct businesses related to banking practice, such as loan syndication, recommending insurance companies to customers, completing feasibility studies for projects or investment options, providing advisory services in merger and acquisition, and offering safekeeping or custodian services. In 1992, commercial banks were permitted to do more business related to financial instruments, including managing the issuance, underwriting and distribution of securities, trading of debt instruments, acting as supervisors as well as selling agents for mutual funds, and securities registrars.

Other financial institutions are also entitled to enlarge their scope of activities. Since the end of 1991, finance companies can conduct leasing businesses. In March 1992, they were authorized to act as selling agents for government bonds, to provide economic, financial, and investment information service, and to advise companies seeking listings in SET. Securities companies are allowed to carry out those same businesses as well as safekeeping and custodian services, and acting as registrars or selling agents of securities.

Improvement of Supervision and Examination of Financial Institutions

The Bank of Thailand, as the supervisory authority of the financial system, needs to closely monitor the operations of financial institutions to ensure the stability and soundness of the financial system and, thus, the viability of market participants. In this regard, the central bank amended the relevant legislation to suit specific situations. The Commercial Banking Act, B.E. 2505 (1962) was revised in 1979, 1985, and 1992, while the Act on the Undertaking of Finance Business, Securities Business and Credit Foncier Business, B.E. 2522 (1979) was adjusted in 1983, 1985, and 1992.

Besides amendments to relevant legislation in line with current conditions, the Bank of Thailand has also focused on the development of supervision and examination procedures. However, as plans to improve the supervision and examination process consisted of several stages, the Bank first emphasized the adequacy of financial institutions' capital funds in line with the international standard. The conventional capital risk asset ratio which had been in use since 1962 was replaced by the BIS guideline on capital adequacy, modified in line with local conditions.

In this respect, commercial banks' capital funds are divided into first-tier and second-tier capital. First-tier capital comprises paid-up capital, retained earnings, reserves appropriated from net profits, and statutory reserves. Second-tier capital includes revaluation of land and buildings and certain types of instruments issued by banks such as hybrid debt capital instruments and subordinated term debts. Moreover, risk assets are weighted by the degree of their risks according to the BIS guideline. Effective 1 January 1993, Thai commercial banks are required to maintain the capital risk asset ratio of not less than 7 per cent; this ratio will be adjusted to 8 per cent by the end of 1994. A minimum of 5 per cent must be maintained as first-tier capital. Due to the difference in the structure of their capital funds, foreign bank branches will be required to maintain a minimum of 6.25 per cent capital risk asset ratio, to be adjusted to 6.5 per cent by the end of 1994.

Given the recent popularity of contingent liabilities or off-balance-sheet items, the Bank of Thailand finds it proper to add those items to risk assets after appropriate credit conversion factors and risk weighing methodology are applied. In other words, the new 8 per cent rule by 1994 will take into account the correct extent of risks from both credits and off-balance-sheet obligations. Besides, normally the traditional rule on capital adequacy restricted the amount of credits as well as obligations that banks can extend to any single entity. After the evolution toward the BIS standard occurred as mentioned above, the single-entity limit is similarly adjusted by weighing both credits and obligations by their associated risks before summing them up and comparing them with lenders' capital funds.

Bangkok International Banking Facilities

The Bank of Thailand has recently decided that the Thai financial system was ready for development toward a regional funding center, taking into account stable economic conditions, liberal exchange controls, and a high level of international borrowing transactions. Thus, the Bank proposed the establishment of Bangkok International Banking Facilities (BIBF) to facilitate and reduce the cost of international borrowing, while encouraging fund inflows to finance the current account deficit. The BIBF is also designed to link the financial systems in the region.

Local banks, existing foreign bank branches, and new banks which the Bank of Thailand and the Ministry of Finance have deemed ready and qualified, were granted licenses for international banking facilities. The license permits the following activities: acceptance of deposits in foreign currencies; lending in foreign currencies to both residents and nonresidents; and foreign exchange transactions (cross-currency only), with relevant tax concessions.

In 1993, there were 47 commercial banks granted BIBF's licenses. Of those, 15 are domestic commercial banks, 12 are branches of foreign banks already established in Thailand, and 20 are new foreign banks with no branches in Thailand.

The major competitors of BIBF are the offshore markets in Hong Kong and Singapore. As the main objective of BIBF is to provide financing for regional development, particularly in Indochina, comparable tax incentives to the above centers are given to BIBF. However, there are some types of taxes—such as withholding tax on interest payments abroad and the permanent establishment tax—which have yet to be exempted or reduced.

The second three-year plan of the Bank of Thailand, which began in 1993, consists of two parts. The first is the continuation of the 1990–1992 plan, comprising deregulation and financial development, improvement of supervision and examination, and development of the payment system. The second deals with savings mobilization, development of Thailand into a regional financial center, and improvement of the central bank's operations with respect to monetary policy and internal organization.

After the Bank of Thailand implemented its first three-year plan and while it pursues its second, the ability of the Bank to directly influence monetary conditions and operations of financial institutions is weakened to a notable degree: this may have an adverse effect on monetary policy. The Bank, therefore, has begun to develop new financial instruments and methods of monetary controls which are market-oriented. For example, the Bank is now considering use of its own bonds and state enterprise bonds to actively conduct open market operations, especially as the secondary market for government bonds is virtually nonexistent. However, before the successful beginning of the open market operations, i.e., buying or selling securities outright, the Bank must be able to forecast short-term liquidity conditions in the money market and establish secondary markets for those securities. To forecast short-term liquidity correctly, the Bank must be aware of the timing and extent of major inflows or outflows of funds from money markets and understand behavior of money and credit demand. To establish active secondary markets, the returns to those securities must be market-determined and there should be a wide range of maturities.

Meanwhile, with inadequate indirect monetary instruments, the Bank sometimes turns to less market-oriented methods to influence monetary conditions, such as market intervention. For example, in 1993, after financial liberalization had begun, the Bank noticed that the gap between lending rates for general customers and those for prime customers, and the gap between the effective lending rate and deposit rate were widening. To solve this problem for the long term requires that competition in banking industry be strengthened by allowing new entries and continuing the deregulation process. However, as a short-term solution, the Bank tried to convince bankers to adjust their interest rates on credits down to a reasonable level. In addition, the Bank decided to supply more liquidity to the market by offering concessionary financing to new factories in the rural area which received promotion from the Board of Investment. Under this concessionary scheme, commercial banks have to extend 70 per cent of the credits to clients at the minimum lending rate while the central bank covers the rest. This places commercial banks in a dilemma because sharing the involved risks with the central bank means less profits than if they assumed all requested credits solely by themselves.

From a broader perspective, the Bank of Thailand appears to be entangled in a self-inflicted dilemma. The wider interest rate margin or profit that commercial banks now enjoy is primarily due to interest rate liberalization. Exchange control deregulation, however, does not equip all borrowers with equal access to credits abroad. Therefore, in the initial period after liberalization, commercial banks gain from an inadequate degree of competition in the market. The central authorities could counteract this bias by increasing the degree of competition in money markets by, for example, allowing finance companies to perform commercial banks' functions or granting local licenses to more foreign bank branches. Both would lead to quicker adjustment or narrower spreads than would be possible by waiting for general customers' improved credit ratings or smooth access to funds abroad.

Other Recent Moves

As was discussed earlier in this chapter, in the early 1990s, most government policies on the financial system were designed in favor of two underlying principles, i.e., market mechanisms and improving the financial infrastructure. Examples of the former have been described above, e.g., exchange control deregulation, interest rate liberalization, and wider scope of financial institutions' businesses. Those of the latter are adoption of the BIS rule on capital adequacy and establishment of the BIBF. Three others concern the capital market.

The Securities and Exchange Act was enacted in March 1992 for two purposes: (i) to unravel the legal intricacies and drawbacks of previous laws; and (ii) to reinvigorate the local capital market, thus decreasing the dependency on the local money market. In other words, support of the stock market strengthens market competition in the financial market and, consequently, diminishes to some extent the oligopoly long held by powerful commercial banks. Functioning as an overseer of the Act is the newly established SEC (discussed in some detail earlier in this chapter).

Another institution recently established is the credit rating agency which should serve as a bridge linking investors with borrowers in the capital market with respect to the disclosure and evaluation of information. At the end of 1993, there were two credit rating agencies; both were in their infancy but had shown some promise. Given a strong potential of the Thai capital market and the high likelihood that Bangkok will emerge as a regional financial center, their role may be strengthened in the future.

Finally, the Export-Import (EXIM) Bank was recently organized to promote international trade-related sectors of the economy by offering various financial arrangements and risk sharing schemes. From a broader perspective, the EXIM Bank represents a new but crucial element of the Thai financial infrastructure which is truly essential amid the current trend of globalization or trade liberalization.

CONCLUSION

While the previous sections of this chapter provide a clear picture of the evolution, function, and recent reforms of different financial institutions and their markets in Thailand, there is no indication of future trends. That is provided in this final section. First, however, the efficiency of the sector will be examined, because, other than events in international markets, the direction of the Thai financial sector also depends on actions of both commercial bankers and the central authorities.

"Financialization"

In any theoretical economic growth model, the size of national savings is compared to that of domestic investment. What is often neglected is how much or how quickly those savings resources are transferred to investment use. Without adequate and efficient intermediation, too few resources will be channeled to optimal use, or, conversely, too many to misuse. The following discussion examines the relative extent of national savings and domestic investment which are "processed" through the financial sector, which should provide some indication of how well the financial system in Thailand has functioned in funneling resources to proper uses.

First, gross national savings can be compared to changes in M2, or broadly defined money supply, selected because it represents financial liabilities of both the monetary authority and commercial banks, or currency plus deposits. In other words, the ratio of changes in M2 to gross national savings should reflect how much the financial system has been able to capture savings. Data suggest that the Thai financial sector has been increasingly efficient in financializing savings in the past three decades. The ratio of changes in M2 to gross national saving, on average, grew from 27 per cent in the 1970s to 38 per cent in the 1980s, and 44 per cent in 1990–1992.

The same conclusion is reached when attention is turned to investment financing. Gross domestic investment is compared to changes in domestic credits extended by commercial banks and the monetary authority. The ratio of changes in domestic credits to gross domestic investment signifies the portion of domestic investment that is financed via the financial system. The trend of that ratio confirms a higher degree of the Thai financial sector's efficiency in financializing investment, since that ratio on average rose from 30 per cent in 1970s to 33 per cent in 1980s and 43 per cent in 1990–1992.

However, efficiency in domestic financialization should not be confused with efficiency in market competition. Further, conclusions about the former have neither implication nor connotation upon the latter, and vice versa. Before discussing the sequencing of financial liberalization measures and the degree of efficiency in market competition, it is valuable to examine the fundamental characteristics of the Thai banking system and the corporate culture as well as the dogma evolved therein.

Fundamentals

The banking industry in Thailand has been less than competitive. For decades, there has been no entry or exit of commercial banks, since the central authorities rank stability over all other objectives, and have maintained that stability by limiting the number of operating banks and by helping ailing banks to survive by rehabilitating their financial positions. Even in the case of fraud or mismanagement, the troubled banks or financial firms were rarely closed. Neither were unlucky depositors left uncompensated. Most problem banks or finance companies were taken over by healthy banks or by state banks or companies which were jointly owned by a group of financial institutions. In short, the industry has been well protected for the purpose of preserving financial stability as well as confidence of the general public.

Because of this protection, the number of powerful banks is limited and price competition scarcely exists. Interest rate changes are always spelled out in an oligopolistic fashion by the Thai Bankers' Association. Other aspects of cartel also exist due to the fact that few suppliers of funds command a great portion of the market shares in all regards, for example, deposits, credits, number of branches, variety of services, and technology. Most of the above characteristics are applicable to the circuit of finance and securities companies as well after a series of crises were experienced in the early 1980s.

Sequencing

Theoretically, efforts to liberalize should be exerted upon elements in the domestic market before being directed to the external frontier. Such sequencing should minimize disruptions and repercussions upon the domestic economy. Thailand's sequencing of liberalization measures complied with how policymakers derived the suitable timing and direction of changes from prevailing political sentiments.

Though the Thai sequence of financial liberalization did not correspond to theory, the economy was not critically affected by the following series of policies: exchange control deregulation, phase 1 (1990), phase 2 (1991), and phase 3 (1992); interest rate liberalization, long-term deposits (1989), time deposits (1990), savings deposits (1992), and lending rates (1992); broader scope of operations (1992), for example, for banks, underwriting and distributing securities, trading debt instruments, selling mutual funds, and for finance and securities companies, leasing, selling government bonds and financial information, and serving as securities custodians and registrars. In part, the sequencing of these policy changes did not have an adverse impact on the economy because Thai society is predominated by conservatism, an inclination which prevails in the financial circuit as well as macroeconomic management, and which favors gradualism.

Gradualism is evident in the stages of financial liberalization already pursued in Thailand. The stages correspond to the first two components of the series of growing degrees of intensity in financial liberalization, as follows:

(i) relaxing the stringency of previous rules step by step, e.g., branch opening requirement, mandatory rural credits, foreign exchange exposure, exchange controls, and interest rates;

(ii) broadening the scope of operations;

(iii) allowing financial institutions to branch out in both metropolitan and provincial or rural areas, and allowing those branches to function to an equal extent to headquarters; and

(iv) authorizing the establishment of new members of each type of financial institutions.

These different degrees of intensity in financial liberalization apply to all types of financial institutions and services, such as commercial banks, finance and securities companies, credit fonciers, mutual fund companies, and BIBF. Should financial liberalization be viewed in this manner, one can easily foresee formats of liberalization that the monetary authorities in Thailand will choose to authorize in the future.

Consequences and Outlook

The effects that financial liberalization has had on the economy or the financial system cannot be concluded with any certainty for two main reasons. First, the move toward liberalization is a recent event, so conclusions would be premature. Second, according to theory, the atmosphere which suits liberalization should contain no barriers or rigidities. Only in such a scenario can liberalization bring about more efficient allocation of resources. However, as has been shown in this chapter, Thailand has numerous barriers and rigidities, such as restriction on entry and exit of commercial banks and finance companies, and some portfolio restrictions still in effect. In addition, some strategic moves have already been institutionalized such as the dominant role of the Thai Bankers' Association in specifying interest rate changes. Therefore, it is not surprising to discover that after one to two years of financial liberalization, the effective interest rate spread attained by commercial banks grew from 3 per cent in 1988 to 3.5 per cent in 1990 and 4 per cent in 1992. Also, some cartel pricing persists and a large number of small clients, especially those in the rural areas, remain neglected. The Government has to devote more effort to reducing these market barriers and rigidities further to provide customers with more mobility and to achieve greater competition and efficiency. Examples of possible measures in the future are the following. Regarding entry, the government may have to further broaden the role of finance companies so they can compete with commercial banks to a greater extent. More foreign banks and more Thai banks may be allowed subsequently.

Some exits may have to be approved as well. In this respect, an explicit Deposit Insurance Institute should be established so as to protect depositors and examine financial institutions. The results of such examinations should be disclosed to the general public so that they can help function as a safety valve. On the contrary, if the Government insists on no exit for the sake of stability, customers will be more tempted to take risks, or moral hazard is encouraged. Financial institutions can very well fall into the same trap. In other words, they may adopt a riskier strategy of financial management or the so-called adverse selection.

From another perspective, there are costs to pursuing financial liberalization. The Government may seem to have forgone its interest rate and exchange control policies as instruments to control the temperature of the economy. However, benefits of such sacrifice have not emerged yet, so the question arises as to whether it was worth it. Careful consideration indicates that the Government has not fully lost 100 per cent of its control. It can still use indirect means to influence the market, for example, through intervention in the repurchase market and moral suasion. Those means may prove effective but they are distinctly different from the previous system, which was full of directives or explicit interference, in that current indirect instruments lead to smoother adjustment or transition.

In the midst of the present competitive global climate, financial liberalization is the only viable means for Thailand to cope with rapid market fluctuations. It is hoped that the final outcome of such liberalization will provide better results than those derived from the previous directive system.

Bibliography

Bank of Thailand (various issues). Bangkok: National Economic and Social Development Board.

_____ (various issues). Securities Exchange of Thailand.

Trairatvorakil, Prasarn, and Prakid Punyashthiti. 1992. "Some Structural Changes and Performances of Finance and Securities Companies in Thailand during 1981–1990." In *Pacific-Basin Capital Markets Research*. Vol. III. edited by S.G. Rhee and R.P. Chang. North Holland: pp. 487–505.

CHAPTER SIX

REFORM OF INDIA'S FINANCIAL SYSTEM

P. JAYENDRA NAYAK

This chapter provides an interpretation of India's financial system, with implications for its redesign and reform, based on the economics of imperfect information. It focuses attention on the microeconomics of financial markets, characterizes the formation of incentives for lenders in the presence of endemic information asymmetries, situates the rationale for bank regulation and supervision in the need to mitigate moral hazard and adverse selection, and argues that risk diversification is crucial to enhancing the competitive vitality of the financial system.

India has now embarked on a comprehensive reform of its formal financial system comprising commercial banks and development finance institutions (DFIs), bringing the sector to the forefront of the Government's wider economic reform agenda. Banks and DFIs are predominantly in the public sector, and a recognition of their state of distress has been slow in coming, with weak accounting conventions successfully camouflaging for several years the extent of their bad debts. However, India also has a diversified semi-formal and informal financial sector, large segments of which report virtually no bad debts, are growing rapidly, and have become crucial to satisfying the credit needs of the country's production and commercial activity.[1] Although relatively under-researched, the dominance of the informal sector is not in doubt, and a study in 1989 revealed that over 56 per cent of total outstanding urban debt (inclusive of enterprise debt) was owed to the informal sector (Das-Gupta et al. 1989). Although the share of informal rural debt has been contracting, nevertheless the latest comprehensive 1982 estimates indicate that about 40 per cent of outstanding rural household debt is owed to the informal sector (RBI 1988). Clearly, informal finance is flourishing.

Hence, the nature of credit contracts supported by the informal sector merits scrutiny, so as to establish where the comparative advantage of this sector lies. Can the formal financial sector mimic the manner in which informal credit institutions identify their borrowers and structure loans to them? If they cannot, would it be preferable for formal financial intermediaries to phase out loans to such borrowers? These queries and the

---

[1] The categorization into formal, semi-formal, and informal can depend on the extent of regulation. For convenience, semi-formal and informal will henceforth be categorized as informal.

microeconomics underpinning them, have relevance to the design of financial sector reform, but have typically not been addressed in the extensive recent literature on this subject.[2] Unless asset portfolios and the borrower composition of banks can be rapidly restructured by identifying new ways of doing banking business, the fragility of most banks will continue. These issues are examined in the next section of this chapter, which explores the linkages between formal and informal finance.

MICROECONOMICS: SOME THEORETICAL ISSUES

The investigative logic of this chapter is fundamentally microeconomic. The analysis of the financial sector is anchored to a set of theoretical issues which are invoked repeatedly throughout the discussion.

Incentive Compatibility of Financial Sector Contracts

Financial sector contracts are intertemporal and not contemporaneous: the creditor's loan is exchanged for a borrower's assurance to repay. Creditors therefore need to design loan contracts in a manner which induces repayment, and collateral and third party guarantees are often sought. While borrower behavior uncertainty is well understood, there could also be a creditor behavior uncertainty before the loan is repaid: How would the creditor behave if the borrower were to default? Would the creditor be indulgent in providing for rescheduling or would he/she seek to liquidate the collateral? With behavioral uncertainty on either side of a credit transaction, the situation is clearly game-theoretic. The transaction is incentive compatible when it is in the commercial interest of both creditor and borrower to abide by the terms of the contract.[3] However, when judicial redress for the enforcement of contracts is weak, it is well understood that the design of incentive compatible credit contracts becomes problematic. Nevertheless, a bewildering range of informal credit market contracts achieve incentive compatibility without relying on judicial redress.

Responses to Imperfect Information

The intuition behind the theoretical work on credit markets characterized by asymmetric information has four strands of relevance to this chapter. First, in such markets, if interest rates are deregulated, the adverse selection of borrowers ensures that creditors set interest rates so as to maximize

---

[2] The most influential suggestions for reform have emanated from the Narasimham Committee (see Government of India 1991) and from the World Bank (1990).

[3] A noncooperative equilibrium is clearly an incentive compatible outcome.

expected earnings net of defaults, and an efficient market outcome involves credit rationing, even to high quality borrowers (Stiglitz and Weiss 1981). This insight is applicable, however, only when accounting conventions on income recognition and provisioning for bad debts are strong. In India, lending rates by banks and DFIs were deregulated (subject to a floor interest rate) before accounting systems were tightened, and accounting transparency for banks is being phased in over three years. There is considerable evidence that banks have consequently succumbed to adverse selection and sharply raised their lending rates in order to enhance profitability. (Real interest rates for a wide category of borrowers have reached 15 per cent). This can be expected to further worsen the asset portfolio quality of banks and aggravate their recapitalization.

A second strand emanating from imperfect information is that creditors must find ways of acquiring better information on borrowers. One way to do so is for loans to be structured so as to enable borrowers to self-reveal their preferences, and collateral as a signaling device has a strong theoretical rationale.[4] This assumes, however, that collateral can be readily liquidated and, therefore, that contracts enforcement is efficient. A later section of this chapter examines whether, under weaker judicial redress as in India, signaling is feasible, and proposes a loan contract structure for DFIs which would elicit borrower risk revelation. Another option is for creditors to incur expenditures in acquiring borrower information. If creditors are viewed as purchasers of a stream of receivables of uncertain quality, and need to "search" for borrower information, a rationale for price dispersion and collateral is possible.[5] Informal credit institutions typically do charge borrower-specific interest rates and the average interest rate rises as the extent of collateral decreases. When commercial banks are required to lend to borrowers of uncertain quality at fixed and concessional interest rates, crucial issues of bank incentives arise. These are discussed later in this chapter.

A third strand views bank regulation as a means of mitigating adverse selection and moral hazard, and of realigning management incentives so as to be in the interests of depositors. The effectiveness of capital adequacy as a regulatory tool is judged in terms of whether it induces a shift in management priorities, and managerial disincentives range from disallowing the bank from expanding its positive risk assets (the manager becomes

---

[4] Bester (1985; 1987) first proposed the idea of collateral as a signaling device. The importance of market signaling was recognized by Spence (1974).

[5] Even without quality uncertainty, price dispersion is intrinsic to markets where buyers search (see Rothschild 1974). With quality uncertainty, market equilibrium necessarily requires a seller's "warranty" which, in a credit market, amounts to borrower's collateral (see Cooper and Ross 1985). Phlips (1988) and Stiglitz (1987) contain less technical discussions of these issues.

redundant as he/she cannot sanction loans) to the loss of jobs. That managerial incentives may not be aligned to those of bank owners is part of a wider problem[6] and regulatory designs must necessarily encompass the issue of incentive. It is often argued, for instance, that deposit insurance distorts incentives and therefore constitutes a "blank check" resulting in a one-way bet against the government.[7] However, in an under-regulated bank the same incentive distortion will continue, except that its impact will be borne by depositors rather than by the government. Equally, if a bank is well regulated, distortions will be minimized irrespective of deposit insurance. Thus the source of distortion is under-regulation, and deposit insurance merely provides confidence about the safety of deposits.

A fourth strand, which is crucial to the development of well-functioning capital and money markets, is that asset prices are never efficient (in the sense of reflecting the collective information of all agents) and that pricing will be "noisy" so as to compensate investors who acquire costly information (Grossman and Stiglitz 1980). Just how noisy is captured in the bid-offer spreads which are quoted by market-makers and are dependent on the structure of asymmetric information and the inventory costs of holding securities (Glosten and Milgrom 1985). A later section of this chapter examines why these issues are decisive to the setting up of corporate debt markets without which the reform of banks and DFIs may be retarded. If bid-offer spreads rise, markets begin to thin and, beyond a point, trading ceases.

To summarize, information is asymmetrically distributed between banks and their borrowers, as also between banks and their regulators, while incentives are misaligned between banks and their managers. Agents in financial markets consequently incur search costs leading to price dispersion in credit markets, including capital and money markets. A recognition of these issues is vital to the characterization of the financial system and assessments of different reform strategies.

Diversification of Creditor Risks

Banks and DFIs are unable to diversify the risks that face their assisted companies; these companies, in turn, have limited opportunities for hedging their risks. Recession in the Indian economy in the early 1990s has consequently hit bank earnings. The diversification of such risks depends on a deepening of capital markets, money markets, and foreign exchange

---

[6] In widely held firms, manager incentives may not lead to behavior which maximizes shareholder net worth, and short-term bank loans (which put firm management on a "short leash") are seen as the only instrument for curbing manager moral hazard (see Stiglitz 1985). Regulators assume a similar short leash role when the firm is a bank.

[7] In the Indian context see, for instance, Bery (1992).

markets so as to permit banks and DFIs to securitize their loan assets, to access these markets for resources, and to hedge more actively against macroeconomic risks. A broader range of options, futures, and swaps would need to develop, and risk management in banks and DFIs can be successfully handled only if such strongly regulated nonintermediary financial markets develop. The corporate bonds market needs to be specially fostered and existing tax impediments removed.

Changes in India's trade regime are likely to further exacerbate these risks, as several projects earlier sanctioned on the basis of high effective rates of protection find their profitability steeply eroded. Should banks and DFIs share part of the risks to which their assisted companies are subject? As banks and DFIs have a more diversified portfolio than their assisted companies, these credit institutions are arguably systematically less risk averse than their borrowers. The difference in risk aversion could make it beneficial for credit institutions and borrowers to enter into risk sharing implicit contracts, rather than contracts where risks are borne entirely by the borrowers.[8]

Interest Rate Deregulation

The pace and sequencing of interest rate deregulation is perhaps the most contentious issue in financial sector reform. It is, therefore, desirable to explicitly state the theoretical bias of this paper, given that several competing approaches exist in the literature. Intuitive beliefs about the benefits of deregulating interest rates and allowing credit institutions to determine the terms at which they lend are probably best captured in the McKinnon-Shaw models of financial markets, which argue that "financial repression" leads to reduced real rates of economic growth (McKinnon 1973; Shaw 1973). Financial liberalization, underpinned by a market-determined interest rate structure will improve the quality of investment, thereby raising the level of growth and ameliorating the contractionary effects of monetary deceleration which generally accompany an effective stabilization program.

The framework mirrors poorly the current Indian reality, which hardly represents financial repression: nominal deposit rates, which are capped and hence regulated, are substantially higher than the rate of inflation; real lending rates, which are uncapped, are between 10 and 15 per cent; and authentic fears exist of banks and DFIs succumbing to adverse selection in the choice of new borrowers. Recent work by McKinnon (1989) recognizes the importance of moral hazard and adverse selection, and advocates ceilings on deposit and interest rates in "immature bank-based capital markets."

---

[8] This insight was first recognized by Azariades (1975) in an application to the labor market. Villanueva and Mirakhor (1990) discuss its relevance to developing country credit markets.

More significantly, the framework does not provide for informal financial markets, which are a competitive form of intermediation in India. Such parallel credit markets have, however, been incorporated into neo-structuralist credit models of developing economies;[9] this chapter argues for the much stronger relevance of these models. They assume a free funds flow between parallel markets and the banking system, with savers and borrowers having considerable freedom to use either. Household assets are held in one of three forms—parallel market loans, bank deposits, and currency—and allocation is determined using a portfolio framework.[10] The parallel market rate is the relevant interest rate signifying the marginal cost of borrowing, and therefore entering as a variable in the demand function for money.

Raising the real deposit rate under such a framework will deter investment, as lending rates for both bank and parallel market loans rise. This is clearly the very opposite of what the McKinnon-Shaw framework would predict. Further, if reserve stipulations imposed on banks are high and the elasticity of substitution between bank deposits and parallel market deposits is much higher than between bank deposits and currency, increasing bank deposit rates can actually reduce the supply of commercial credit.[11]

Although the enhanced plausibility of neo-structuralist models arises from their inclusion of parallel financial markets, they have two limitations. First, they assume perfect substitutability between bank funds and those of parallel markets and the empirical basis for this is blurred.[12] Second, they lack adequate microfoundations and these must necessarily come from a market characterization in terms of imperfect information. This has implications for the mechanisms of monetary policy. When accounting conventions are tight and banks are at a rationing equilibrium, a contraction of reserve money—achieved for instance by increasing reserve requirements—would make credit rationing more severe. Thus, the level of investment is affected not through an increase in interest rates, but rather through the reduced availability of credit. Thereby, tight money acts directly on real economic

---

[9] Associated particularly with Lance Taylor (1983).

[10] Tobin (1965) introduced the portfolio analysis framework.

[11] Illustratively, this could happen if the two elasticities of substitution were 4 and 1 respectively and if the reserve requirements were 40 per cent. A 1 per cent increase in the bank deposit rate would lead to a net contraction of erstwhile parallel market funds of 4 per cent and an increase in bank credit of 1.6 per cent, while also leading to a decrease in currency of 1 per cent and an increase in bank credit of 0.4 per cent. Overall, commercial credit would decline.

[12] The Das-Gupta et al. (1989) study of four sectors (small-scale industry, wholesale trade, retail trade, and transport operators) finds no econometric evidence of substitutability in credit disbursed. (A weak complementarily hypothesis is more plausible). The deposit market substitutability hypothesis appears not to have been tested.

activity and is not dependent on a high interest rate elasticity of the demand for money (Blinder and Stiglitz 1983).

To summarize: India deregulated its bank lending rates at the start of its financial reform program. With weak accounting conventions, banks could have been expected to raise real lending rates sharply. They did so, thereby risking an adverse selection of borrowers. As accounting conventions tighten, the quality of borrowers will become more transparent, and banks can be expected to get into further distress. The existence of parallel markets implies that investment demand will become depressed as bank deposit rates rise and, if reserve requirements on banks are high, may even reduce the availability of commercial credit. Finally, with information asymmetries characterizing credit markets, monetary policy could act directly on real economic activity by varying the levels of credit rationing, rather than through the interest rate.

This is clearly a stylized view of how monetary policy might work and, as long as the instruments of monetary policy are parametric (varying the reserve requirements, fixing incremental credit/deposit ratios, and imposing selective credit controls), the importance of interest rates in the transmission of monetary policy will become diluted. Interest rates clearly become more decisive when monetary policy is conducted through open market operations, with interest rates on voluntarily held treasury bills and dated government securities providing "benchmarks" to the term structure of interest rates.

MACROECONOMICS: THE CONVENTIONAL WISDOM

The conventional wisdom makes macroeconomic stability a prerequisite to successful financial sector reform, and emphasizes fiscal restraint and monetary stability. A fragile financial system clearly cannot be revived without these; equally, it is vital that one not lose sight of the microeconomic foundations of macroeconomic profligacy. Thus, high reserve pre-emptions do not merely reduce bank profitability, they also alter management incentives by encouraging a risky borrower adverse selection. When the real economy is in recession this can gravely damage a bank's solvency. Excessive fiscal pre-emptions, therefore, introduce an externality into a bank's selection of its other borrowers. The severity of this externality will depend on the legal, regulatory, supervisory, and accounting systems; these are discussed later in this chapter.

The rest of this chapter is organized as follows: several attributes of the functioning of the wide range of informal financial intermediaries are discussed in the next section, followed by a description of the formal financial system, which is dominated by the public sector banks and DFIs. The main issues in the reform of commercial banking are then discussed, and the pre-requisites for restructuring of the country's DFIs are identified. India's capital market, money market, and foreign exchange market, the development of which is crucial to the ability of banks and DFIs to diversify their risks and to raise fresh capital, are addressed in the next section, which is followed by a discussion of several aspects of financial infrastructure, including the legal system, accounting conventions, regulatory and supervisory systems, deposit insurance, and credit guarantees. Selected issues in industrial finance, with decisive microeconomic implications are then presented, followed by some concluding remarks.

INFORMAL FINANCE

Informality as a Banking Style

The Syndicate Bank, one of India's large, nationalized commercial banks, was established in the 1920s with its headquarters in a small town near the country's west coast. Until its nationalization in 1969 the Syndicate Bank specialized in borrowing from and lending small amounts to the poor. In 1968, 50 per cent of deposits and 90 per cent of deposit accounts consisted of deposits of less than Rs 1,500 while 30 per cent of its advances were to the subsequently defined priority sector, as against 8 per cent for the commercial banking system as a whole (Little et al. 1987). Default rates on loans were very low, and administrative costs as well as transaction costs of lending to small-scale industry were probably the lowest in the Indian banking system (Thingalaya 1978). Recruitment and promotion policies favored candidates with a local background, most of whom were out of school but had no university degrees (Bhatt and Roe 1979). For several decades, the bank appointed agents (who were paid a commission) to walk to villages and mobilize deposits. Household thrift was encouraged through the collection of deposits each week, and sometimes every day. Clearly, the bank's operations resembled the informal financial market.

Any such resemblance today, 24 years after the bank was nationalized, would be hard to detect. Government policy has "homogenized" the public sector banking system in several ways, encompassing uniform recruitment and promotional policies, the organizational structure of banks, and employee size and remuneration, and has discouraged special incentive systems which bank management might wish to offer. Banks have also been required to enlarge their territorial reach. The earlier thrust, indeed the capabilities to acquire "information capital" on creditworthy local borrowers has eroded.

Reach of the Informal Sector

Information capital and long-standing debt relationships are integral to informal finance. Clearly, the possibility of replicating informal financial market contracts within the banking system is of interest. First, however, a statistical profile of the informal financial sector is indicated.[13]

Uses of Informal Credit

Almost two thirds of all urban informal credit in the mid-1980s went for trade. Other major users were public limited companies (15 per cent) and small-scale industry (11 per cent). It is estimated that 44 per cent of credit was not tax-accounted (black money). Gross urban bank credit amounted to 77 per cent of the total informal credit deployed. Details are given in Table 6.1.

Table 6.1 Urban Informal Credit

User Sector	Amount (Rs billion)	Per Cent
Trade	517	63
Public Limited Companies	119	15
Small-scale industry	90	11
Others	90	11
Total informal credit	816	100
Estimated black money	358	44
Net recorded informal credit	458	56
Gross Urban Bank Credit	625	77

Source: Das-Gupta (1989).

All sectors use a mix of informal and formal credit, as well as own funds. The dependence on informal credit is particularly strong in the sectors shown in Table 6.2.

There is considerable impressionistic evidence, though little in the way of documentation, that the urban informal finance sector is growing rapidly. Informal rural credit has, however, been decreasing in importance. The All-India Debt and Investment Surveys indicate that 19 per cent of rural households reported outstanding debt in 1981–1982 with the share of informal credit being less than 40 per cent, falling from about 75 per cent in

---

[13] The profile is heavily indebted to Das-Gupta et al. (1989) whose work has substantially enhanced the database on informal urban credit. Informal rural credit has been studied by Panikar, Sen, and Narayana (1988). The results of both studies are summarized in Das-Gupta (1989). Finally, in a strongly interpretative and comparative study, Ghate (1992) has identified similarities across informal financial markets in Asia.

1972.[14] The cooperative and commercial banking sectors have clearly supplanted informal credit, though there are sizeable inter-state variations in these trends.

Table 6.2 Sources of Funds

User Sector	Informal Credit	Formal Credit	Own Funds
Film finance	95	...	5
Traders	53	19	28
Textile trade units	48	10	42
Powerloom units	47	10	43
Garment exports	43	26	31
Small-scale industry	40	32	28

Notes: RBI Surveys cover only units receiving bank assistance. Shroffs do not include Multanis who do receive bank refinance estimated at about 5–10 per cent of their loans outstanding.
Source: Das-Gupta (1989).

Sources of Informal Credit

There is a bewildering diversity of informal finance intermediaries, and deposits constitute the major source of their funds. A sizeable 70 per cent of informal urban credit, however, is not intermediated, but comes from direct sources including trade credit, credit from relatives and friends, and credit from minor, unspecified informal intermediaries. Of the remainder, 18 per cent constitutes mutual finance (chit funds and *nidhis*) while 12 per cent comes from other intermediary lenders. Estimates of advances and deposits in the mid-1980s, which have characterized informal finance, together with a comparison of an estimate of bank finance, are indicated in Table 6.3.

Industrial Structure of Informal Markets

The number of intermediaries in each segment of informal markets as well as the estimated four-firm concentration ratio are preliminary indicators of the industrial organization of these markets (Table 6.4).

---

[14] Reserve Bank of India (1988). However, Panikar et al. (1988) argue that intertemporal comparisons are suspect because of differences in the sampling procedures adopted which underestimate the household incidence of a sampled characteristic, with increasing severity of underestimation in successive rounds.

Table 6.3 Sources of Informal Finance

Sources	Advances		Deposits	
	Amount (Rs billion)	Per Cent	Amount (Rs billion)	Per Cent
Major intermediary lenders of which:	56	12	46	36
(Hire-purchase institutions)	(38)	(8)	(28)	(22)
(Indigenous bankers)	(11)	(2)	(4)	(3)
(Others)	(7)	(2)	(14)	(11)
Mutual benefit intermediaries (Chit Funds and Nidhis)	82	18	82	64
Nonintermediary finance	319	70		
Total advances/Deposits	457	100		100
Gross bank credits/Deposits	625	137	1,084	237

Source: Das-Gupta (1989).

Table 6.4 Industrial Organization of Informal Markets

Market Segment	Number of Firms	Four-Firm Concentration Ratio (per cent)	Market Leadership	Economic Rents	Local Price Dispersal
Finance Corporations	Large	35	No	Small	Small
Hire-Purchase Institutions	Large	88	Yes	No	Yes
Indigenous Bankers	Medium	...	Yes	No	No
Chit Funds	Large	67	No	...	No

... data not available.
Source: Das-Gupta (1989).

Further, there appear to be no barriers to entry and exit, except in the indigenous banking segment which is dominated by the Gujarati Shroffs of Western India. Precise numbers of firms operating in each segment are also not available, though the data base on those incorporated under the Companies Act is more accurate. The number of such nonbanking finance companies (NBFCs) grew from 7,063 in 1981 to 24,009 in 1990.[15] These

---

[15] As reported by the Shah Committee Report on Financial Companies (see Reserve Bank of India, 1992a).

companies are obliged to furnish periodic information about their operations, but in 1990 only 7,772 such companies (about a third of the total) did so. The Reserve Bank of India (RBI) calls them "reporting companies" and regularly releases data about them. The RBI also publishes a more detailed analysis, roughly one per year, on a sample of such companies.

Of the deposits mobilized by reporting companies in 1990, over 46 per cent went to loan companies. Equipment leasing companies mobilized over 16 per cent and investment companies over 15 per cent. Growth of the former has been particularly rapid since 1988, driven by depreciation-related fiscal benefits. The share of deposits mobilized by the latter and housing finance companies also grew during the 1980s. Stagnation and decline in deposit mobilization are, however, evident among loan companies, hire-purchase finance companies, and chit funds.

Comparison with Formal Finance

It is by focusing on the microfoundations of informal credit markets, however, that the strengths of this sector are best appreciated, with implications that spill over into policy for banks and DFIs.

Incentive Compatible Contracts

Informal credit institutions have a far superior portfolio quality than do banks and DFIs;[16] they rarely resort to judicial redress to ensure that borrowers repay, relying instead on varying combinations of the following: investing in information about a borrower's enterprise; building reciprocity with the borrower so that repeated credit contracts can be entered into; lending short term against very liquid assets or against related contracts in product markets; and lending longer term against moveable property as collateral where the right to repossess is relatively straightforward.

Where informal creditors invest heavily in acquiring borrower information, interest charged on loans is generally high. This applies to professional moneylenders as well as to finance corporations. The latter even give one-day loans (where transaction costs would be high) to vegetable and flower vendors and to fishmongers, repayable from the proceeds of the day's sales. In a sample study of 42 such finance corporations it was reported that although 23 had default problems, these were typically temporary, and systematic overdues were small (Das-Gupta et al. 1989). Reciprocity ensured that it was in the interest of borrowers to repay so as

---

[16] There is admittedly an observational caveat to this. Informal intermediaries that pick up substantial bad debts collapse, and therefore cannot be observed, while banks and DFIs are kept alive.

to be eligible for further liquidity whenever they needed it. Similarly, intercorporate funds and trade credit also rely heavily on borrower information, with the latter typically being tied to product purchases. Chit funds, too, depend on reciprocity for their deposit mobilization, and the similarity in economic status or social occupation between borrowers in the more informal funds also provides a mild form of social compulsion not to default. Pawn brokers and the *nidhis* of South India provide consumption credit against gold and jewelry, with the *nidhis* also providing housing finance against real estate collateral. Several studies have documented that *nidhis* have practically no overdues. Finally, the hire-purchase institutions which finance the purchase by relatively affluent borrowers of commercial vehicles which are either new or not more than five years old, use the vehicles as collateral. Although overdues exist, they are eventually collected, and bad debts are small. Auto finance corporations have a similar profile, except that they finance the purchase of much cheaper and older commercial vehicles by less affluent borrowers.

There are two broad implications of this for the formal banking system: first, small, short maturity, repeated loans either based on highly liquid collateral or on accurate borrower profiles, can provide good business ("gold loans" by nationalized banks, largely in Kerala, are testimony to this); and second, informal credit does not provide for industry's investment needs. Start-up capital for manufacturing industry is clearly fraught with risk which informal credit providers are ill-equipped to bear.

Risk Diversification Through Credit

There is considerable evidence of informal credit assisting in diversifying risks associated with demand uncertainty, and this has been elaborately documented for the financing of the textile trade.[17] The number of distribution agents intermediating between textile mills (or powerlooms) and consumers varies between two and six. The shortest chain is mill/powerloom — semi-manufacturer/semi-wholesaler — retailer; and the largest chain is mill/powerloom — trading agent/semi-manufacturer — wholesaler — non-trading agent — retailer.

All intermediaries receive and provide working capital credit of differing maturities. Further, as the distance between the mill and the retailer increases (whether measured by geographical distance, distribution costs or total capital employed), the number of links in the chain increases. The major finance for the distributive trade is tied informal credit given by wholesalers, who structure it in a manner which insulates mills from

---

[17] Much of the empirical work was conducted by Jain, Bhandari, Dholokia, Khurana, and Vora (1982), and supplemented by Das-Gupta et al. (1989).

demand-induced price and quantity risks. Further, informal credit from wholesalers also reduces the costs of disadvantaged retailers (and the reliance on such credit goes down only when demand is stable and markets are easily accessible). For a given distance between wholesaler and retailer, the extent of informal credit used is proportional to retail demand volatility. In this sense, informal credit has a major impact on diversifying risks while insulating the producers. Wholesalers clearly enter into implicit contracts (in the sense discussed earlier in this chapter) with other downstream distribution agents; their incentive for doing so arises from a lowering of average costs consequent to tied credit providing them with economies of scope.

Equity: The Reach of Credit

Government policy has been strongly influenced by the perception that informal credit has been biased against the poor, and is often exploitative and usurious. This has motivated the expansion of formal rural credit, through cooperative credit societies and the expansion of rural bank branches, led to the participation of banks in several government-designed rural developmental programs, and reinforced the rationale for bank lending to the priority sector. Much of the available documentation does, however, cast doubt on whether informal finance is biased against the poor and the small. An analysis of the All India Debt and Investment Survey data of 1982 indicates that, for rural households, informal debt outstanding had a more even distribution across household asset size classes than did formal debt outstanding, with Gini coefficients of 0.2 and 0.6 respectively (Panikar et al. 1988). Further, of the credit outstanding of the lowest asset class, 92 per cent was to the informal sector and 8 per cent to the formal sector. In three crucial credit using sectors, small borrowers are more indebted to the informal sector than are large borrowers, as indicated in Table 6.5.

Table 6.5 Proportion of Borrowing from the Informal Sector (per cent)

	Sector	
	Small	Large
Households	80	41
Traders	72	61
Small-Scale Industries	55	39

Source: Das-Gupta et al. (1989).

Several financial intermediaries, such as chit funds and *nidhis*, do lend sizably to poorer households. Finance corporations do so also although at substantially lower interest rates (Nayar 1982). Auto finance corporations, too, lend to financially weaker transport operators; a study of 90 such corporations in just one town in Tamil Nadu indicated that whereas the annual interest rate charged may be as high as 40 per cent, they have few bad debts, unlike banks, and lend more aggressively (Das-Gupta et al. 1989). Further, during a period when banks in the town had financed the purchase of 30 vehicles, these auto finance corporations had financed as many as 1,082 vehicles.

The perception that informal finance is exploitative is also based on the very high interest rates sometimes charged and the propensity of certain creditors to dispossess borrowers of collateral pledged, particularly land. A recent review of these issues suggests that whereas transactions costs, adjustments for risk, and opportunity cost of funds do push up the lending rates, monopoly profits are visible only in sectors where the returns to investment are high (viz. one-day loans given by finance corporations or hire-purchase loans given by auto finance corporations) (Ghate 1992). Further, collateral is generally sought as an inducement for effecting recoveries, rather than for its forfeiture, and the empirical evidence that land is often forfeited when pledged as collateral is ambiguous.

Efficiency: Impact on the Real Economy

The efficiency of financial intermediation is reflected in its ability to generate value added in the real economy. With a low proportion of debts compared to banks, and with informal credit interest rates being at least as high as bank lending rates, the rate of return generated in activities financed by informal credit is arguably higher than when financed by banks. Further, where commercial banks underfinance an activity either because of supply-side selective credit controls or because of standard norms (such as for working capital) which cannot respond flexibly to demand volatility, informal finance often fills the gap. The marginal rate of return earned by the enterprise from such credit utilization is then very high.

Allocative efficiency is also enhanced by the speed and flexibility with which informal intermediaries respond to borrower needs. Informal intermediaries are superior on five criteria: the speed of business transactions, the coverage of borrowers serviced, the rate of return earned by the enterprise on borrowed funds, the loan recovery rate, and the wider package of services offered to clients (Madhur and Nayar 1987; and Nayar 1982). Further, informal intermediaries are on par with banks on two counts, the average interest rate charged on loans and the intermediation cost. It is only on one count, the speed with which borrowed funds are returned by the intermediary, that banks rate higher.

Regulation of Informal Finance

The diversity of informal intermediaries makes the design of a suitable regulatory environment particularly problematic[18]. The basic thrust of regulation has been on the control of deposits; RBI distinguishes between intermediaries incorporated as companies (NBFCs) and others. For NBFCs, deposits can be mobilized up to a certain multiple of net owned funds (the multiple is 10 in the case of hire-purchase finance and equipment leasing companies, and can be as low as 0.4 for loan and investment companies). Deposit control also involves a cap on deposit rates, a minimum lock-in period and part forfeiture of interest upon premature withdrawal. For intermediaries which are companies, RBI limits the number of depositors to 25 per partner (or individual if it is not an association of individuals) and 250 overall.

Clearly, if an informal intermediary is to grow in size, there will be regulatory pressures for it to incorporate into a company. This will also make it more difficult for it to deal in tax-unaccounted money which is a substantial portion of informal finance. These regulatory controls on deposits are often cited as the reason for the decline in business of the Shroffs in the last twenty years.

Other regulatory mechanisms, which are either less effective or enforceable, have included usury laws targeted at moneylenders, and prevention of the alienation of mortgaged land.

The sheer inventiveness of some of the informal intermediaries, which the formal sector can never hope to emulate, is another reason for caution in regulation. "By pioneering new forms of credit later adopted by the regular banking system, nonbank intermediaries form the cutting edge of banking innovation" (Timberg and Aiyer 1980). Indeed, the myriad ways in which loans are structured make informal credit intensely innovative, benefiting borrower segments which can never hope to be touched by banks.[19] To regulate such informal intermediaries inappropriately may therefore leave

---

[18] The Shah Committee Report on Financial Companies (Reserve Bank of India 1992, Chapter V) contains details of existing regulations imposed on various categories of informal intermediaries.

[19] Bouman (1989, p.125) describes the inventiveness in restructuring the terms of credit as well as in obtaining guarantees and collateral displayed by different categories of informal credit agencies in Sangli District of Maharashtra arising out of the rapid growth of the sugar and dairy industry.

a credit vacuum, while to regulate them more sensitively may require an institution-specific understanding of risks, information which may not be available to the regulator. Certainly, controlling the number of depositors or the value of deposits does not achieve such sensitivity. There is virtue, therefore, in approaching the issue of regulation as a skeptic.

One approach to regulation, therefore, is to confine it to fuller disclosure, as well as to the credit rating of the larger NBFCs whenever deposits are sought, and to recognize that failures of smaller informal intermediaries do not warrant regulatory intervention. There are instances where intervention has actually worsened the situation.[20] For the very large NBFCs, more stringent regulation is certainly desirable; it may therefore be wiser for RBI to concentrate regulatory attention on these, while keeping a vigil on the disclosures of the smaller informal intermediaries. Self-regulation should be encouraged, and there is evidence that where such institutional safeguards exist, they work effectively.[21] Where fraud is detected, the law should clearly be uncompromising.

Such a regulatory "hands off" stance stems from asymmetric institutional information, putting the regulator at a disadvantage. If selection, monitoring, and enforcement are integral to regulation, the informal sector's sheer size and diversity make each of these a regulator's nightmare. Selection of intermediaries for regulation is, therefore, best guided by where failure of the institution is likely to have systemic repercussions, rather than by the desire to protect depositor interests. Deposits taken by informal intermediaries would need to be regulated differently to deposits mobilized by commercial banks.

FORMAL FINANCE

Financial Deepening

India's formal financial system, dominated by commercial banks and DFIs, has expanded rapidly in the last two decades. Bank deposits and other monetary aggregates have risen secularly as a proportion of GDP. So have household financial savings, though their behavior has been erratic since the late 1980s.

---

[20] Fueled by remittances from the Gulf, Kerala saw the emergence of about 12,000 finance companies by March 1987, and several of these took advantage of gullible depositors. RBI's action against several of these companies caused a run on deposits of other well-managed finance companies, which therefore failed. There is, therefore, arguably a case for more discrete regulation (see Das-Gupta et al. 1989).

[21] Associations of traders and several Shroffs' Associations have arbitration committees. Timberg and Aiyar (1980) report that these committees work very effectively.

Table 6.6 Indicators of Financial Deepening
(per cent of GDP)

	1970/71	1980/81	1985/86	1990/91[a]
Total household savings	11.3	16.1	15.7	18.4
Gross household financial savings	4.7	8.9	9.6	8.6
Net household financial savings	3.4	6.3	7.3	...
Currency and demand deposits	13.8	15.4	15.6	17.6
Time deposits	10.1	22.0	26.7	31.1
Broad money	23.9	37.4	42.3	48.7

[a]Provisional, as GDP estimates are quick estimates.
... data are not available.
Notes: Monetary aggregates computed until 1985/86 have adopted the RBI New Series Measures, based on the average of the last Friday of each month. For subsequent years, monetary aggregates represent the sum of the opening year money stock and half the annual increase.
Sources: Government of India (1992); Reserve Bank of India (1991 and 1992); and World Bank (1990).

Data on the flow of funds within the economy also capture structural changes. By compartmentalizing the economy into the household sector, the financial sector (comprising banks and other financial institutions), the nonfinancial public sector, and the nonfinancial private sector, a financial flows comparison between 1986–1990 and 1980–1986 (broadly the second half with the first half of the 1980s) reveals a marginal decline in the importance of banks as financial intermediaries, accompanied by a sizeable increase in the importance of other financial institutions. This is explained by the active mobilization of household savings in the late 1980s by the mutual funds, investment institutions, and deposit taking nonbanking finance companies.

Households acquired a larger proportion of incremental financial assets while the Government's share of incremental financial liabilities rose. The flow of capital funds depicted in Table 6.7 shows that increases in financial assets and liabilities were most dominant in the financial sector.

Two other indicators of financial development are the financial intermediation ratio and the financial deepening ratio, and both have increased during the 1980s.

Changes in the financial structure during the 1980s are also visible in the altered instruments in which financial savings are held. Currency and deposits, as well as nongovernment securities, have gained in popularity. The share of loans and advances has fallen, though they constitute almost a quarter of all financial claims.

Table 6.7 Flow of Capital Funds
(per cent)

Sector	Increase in Financial Liabilities		Increase in Financial Assets	
	1980/81	1989/90	1980/81	1989/90
Financial sector	41	45	44	48
Nonfinancial private sector	20	18	38	39
Nonfinancial public sector	39	38	12	7
Overseas	—	–1	6	6
Total	100	100	100	100

— nil.
Source: Reserve Bank of India (1992a and 1992b).

These indicators of financial deepening and change are clearly based on recorded transactions. Tax-evaded or black transactions are unlikely to be recorded, and financial intermediation is therefore deeper than recorded transactions would reveal.

The rest of this section provides a description of the impact of commercial banks and DFIs, which dominate formal finance. Other deposit-taking institutions include the cooperative banks, the Post Office Savings Bank, and nonbanking companies which accept term deposits. Capital market institutions include Unit Trust of India (UTI) which raises resources through the sale of "units" of small denomination and other mutual funds. The insurance companies, Life Insurance Corporation (LIC) and the four subsidiaries of General Insurance Corporation (GIC), also mobilize household savings and invest them in capital market instruments. Finally, in 1988, RBI established the Discount and Finance House of India (DFHI), a money market institution to trade in treasury and commercial bills and to assist RBI in limited open market operations.

Commercial Banks

The financial system is dominated by commercial banks. With the nationalization of the 14 largest banks in 1969, followed by another 6 in 1980, bank assets belong predominantly to the public sector. Besides these 20 banks, the public sector banking system also includes the State Bank of India (SBI) (whose majority shareholding is with RBI) and 196 regional rural banks (RRBs), each of which is "sponsored" by a commercial bank. The jurisdiction of an RRB is generally one or a few districts within a state, and its mandate extends to rural financing with a particular focus on disadvantaged borrowers. Private Indian banks are generally small (the larger ones having been nationalized). Foreign banks also operate, generally in metropolitan cities, and had 150 branches in March 1991.

The growth of commercial banking since nationalization in 1969 has been particularly striking in the expansion of the branch network (with a thrust on rural branches) and in the rise in deposits, as the indicators in Table 6.8 reveal.

Table 6.8 Growth in Indian Banking

Indicator	June 1969	June 1985	March 1992
No. of commercial banks	89	268	276
of which: No. of RRBs	—	183	196
No. of bank offices in India	8262	51385	60570
Proportion of rural bank offices (%)	22	59	58
Population per bank office	64	15	14
Deposits in India of scheduled commercial banks (Rs billion)	46	771	2376
Outstanding credit of scheduled commercial banks (Rs billion)	36	509	1315

— nil.
Source: Reserve Bank of India (1993).

Banking Policy and Its Outcome

Interest Rates

In April 1992, RBI dispensed with a maturities-specific interest rate structure for deposits, replacing it with a generalized cap. In June 1993, term deposits had an 11 per cent interest cap, the savings account rate was pegged at 6 per cent and current accounts were paid no interest. The inflation rate in June 1993, as measured by the wholesale price index, was around 6 per cent. Term deposit rates, which are generally fixed by banks in the region of 10 per cent for the shortest 46-day maturities, therefore currently earn an unambiguous positive return. Success in the restoration of positive real deposit rates emerged around June 1992, with negative real interest rates being earned earlier on term deposits of the lowest maturity.[22] The recent apprehension, therefore, is not one of "financial repression" but rather of unduly high deposit rates.

---

[22] While in April 1987 a seven per cent interest rate difference existed between the shortest and the longest maturity deposits, it is currently about 1–2 per cent for most banks, with RBI no longer fixing a maturity-specific interest rate.

Lending rates, too, have been rationalized and are deregulated for loans exceeding Rs 200,000; RBI fixed the floor at 17 per cent in June 1993. The rationale for such a floor has never been made explicit and appears to stem from the need for certain key monetary indicators. However, with real lending rates clearly very high, the floor could impede the competitive lowering of the cost of bank credit. For smaller loans, RBI stipulated in June 1993 a lending rate of 12 per cent for loans up to Rs 25,000 and 16.5 per cent for loans between Rs 25,000 and Rs 200,000. The rationalization of lending rates in recent years has been drastic, jettisoning the earlier structure where the interest rate payable often depended on the end use of the loan. There has also been a perceptible reduction in the interest rate subsidy for priority sector loans.

Pre-emptions

Reserve requirements on commercial banks occur through the statutory liquidity ratio (SLR) which constitutes directed investments into the National and State Governments and some of their public sector institutions, and the cash reserve ratio (CRR), which is a pre-emption by RBI. As at June 1993, all scheduled commercial banks (except RRBs) had to maintain an incremental SLR (on domestic net demand and time liabilities) of 30 per cent and an aggregate CRR of 14 per cent. Reserve pre-emptions peaked in 1991 when the SLR was 38.5 per cent and the incremental CRR was 25 per cent. At one stage, therefore, 63.5 per cent of incremental domestic liabilities were being pre-empted.

Directed Credit

For several years, Indian commercial banks have had to direct 40 per cent of their total credit to the priority sector, with direct credit to agriculture being targeted at 18 per cent of total credit, of which half has to go to small and marginal farmers, and 10 per cent of total credit being targeted to weaker sections of society. In addition, one per cent of total credit is targeted at artisans and petty businesses at a highly concessional 4 per cent interest rate. Foreign banks have had lower targets set for them, in recognition of their metropolitan branch location, and were expected to allocate 15 per cent of their outstanding advances to the priority sector by March 1992. They achieved less than 8 per cent. In April 1993, the target was reset for foreign banks through the inclusion of export finance in the priority sector, and foreign banks are expected to reach a priority sector lending ratio of 32 per cent by March 1994, failing which the shortfall will need to be placed with the Small Industries Development Bank of India as a deposit earning interest at 10 per cent. The additional priority for export finance is reflected in a separate 10 per cent advances target set for domestic commercial banks, in addition to the 40 per cent priority sector requirement. Effectively, therefore, half the advances of domestic banks are directed. Lending rates for export are concessional (13 per cent in June 1993) but supported by liberal RBI refinance and the attraction for banks of picking up commissions on foreign exchange transactions.

Expansion of Banks

The licensing of bank branches has, throughout the 1970s and 1980s, been rigidly controlled by RBI, and the five-year branch licensing policy 1985–1990 allotted 5,360 rural and semi-urban centers to banks, in addition to 309 urban centers and 326 metropolitan centers. In the 1990s, branch licensing has eased and the most recent May 1992 guidelines permit banks to close down branches other than in rural areas as well as to swap unremunerative branches in rural and semi-urban areas with other banks without RBI permission.[23] On the opening of new branches, RBI is guided by capital adequacy criteria. The Narasimham Committee has recommended the abolition of bank licensing with autonomy to banks to open branches and to close nonrural branches.

Structure of Bank Credit

In March 1991, 88.5 per cent of the outstanding credit of scheduled commercial banks was owed to the public sector banks, 7.5 per cent to foreign banks and 4.0 per cent to private banks. The credit/deposit ratio was highest at 85.9 per cent for rural areas followed by 62.5 per cent for metropolitan areas, 55.7 per cent for urban areas, and 50.2 per cent for semi-urban areas. Clearly, therefore, after netting for directed investments made by banks, intermediation by banks has benefited rural areas. Further, 22 per cent of credit outstanding involving about 95 per cent of bank accounts are for small loans not exceeding Rs 25,000, indicative of the thrust banks have given to assisting small borrowers, with consequently high transaction costs.

Further, a sizeable 36 per cent of bank credit is provided as cash credit, with another 26 per cent in the form of long-term loans.

The sectoral composition of bank credit indicates that 47 per cent of credit outstanding is to industry, though a sizeable 44 per cent of accounts are in agriculture. Agriculture, trade, personal loans, and professional services account for another 41 per cent of outstanding bank credit.

---

[23] RBI (1992c), pp. 33–38 contains details of existing branch licensing policy.

Quality of Bank Credit

The classification of bank accounts according to their quality of repayment has hitherto been translucent, with the downgrading of assets often being dependent on subjective criteria developed by bank management. The RBI has signaled a change in the accounting convention which will commence from the accounts for 1992/93. Accounting for income recognition will be progressively tightened over three years, while bad debts provisioning will be phased in over two years. Hitherto, banks have classified their accounts into eight categories. As at March 1991, 72.5 per cent of amounts outstanding were classified as satisfactory, another 12.5 per cent as irregular, and only 2.6 per cent as bad and doubtful debts.

Maturity Structure of Deposits

As at March 1991, 15 per cent of bank deposits were current accounts, 28 per cent were savings accounts while the residual 58 per cent were term deposits. About 40 per cent of these term deposits were held for over three years, while term deposits with a maturity of less than a year amounted to 19 per cent. Term deposits in rural areas have higher maturities than elsewhere, as the frequency distribution indicated in Table 6.9 reveals.

Table 6.9 Maturity Pattern of Term Deposits in March 1991

Maturity Period	Percentage of Deposits Held in				
	Rural	Semi-Urban	Urban	Metro	All-India
Up to 6 months	5	6	9	22	13
6–12 months	3	4	5	10	6
1–3 years	38	43	47	38	41
3–5 years	21	22	22	19	21
Over 5 years	33	25	17	11	19
Total	100	100	100	100	100
Per cent of term deposits	15	21	24	40	100

Source: Reserve Bank of India (1993).

Policy for New Private Banks

No private banks have been permitted to be set up since the 1969 nationalization and, in early 1993, RBI announced a policy for encouraging the establishment of new banks. Such banks need to be incorporated under company law with shares necessarily listed on stock exchanges, and have a shareholder capital not less than Rs 1 billion with a minimum promoters' contribution as determined by RBI. Voting rights of individual shareholders cannot exceed 1 per cent of total voting rights (with exceptions for shares held by DFIs).

Development Financial Institutions

DFIs have been established to ensure the availability of adequate investment finance, primarily in the form of long-term project loans. There are currently 59 DFIs in the country. With the exception of the National Bank for Agriculture and Rural Development (NABARD), which lends for agriculture and rural development, the National Housing Bank (NHB), which finances housing activity, and Exim Bank, which provides credit to augment foreign trade, all other DFIs provide finance for industry. Of these, 44 are State Government-controlled institutions: 18 state financial corporations and 26 state industrial development corporations. The asset size of DFIs was about 54 per cent that of all banks (including State Cooperative Banks) in March 1991, as against 35 per cent a decade earlier (Table 6.10).

Table 6.10 Financial Assets of Banks and DFIs
(per cent)

Financial Intermediary	At the end of March		
	1981	1986	1991
Banks of which:	74	70	66
Commercial banks	70	67	63
State cooperative banks	4	3	3
DFIs of which:	26	30	34
All India term-lending DFIs	10	14	15
State level DFIs	3	3	2
Investment institutions	13	13	17
Total	100	100	100

Source: Reserve Bank of India (1992c).

Of the DFIs which lend to industry, about two thirds of annual sanctions and disbursements are provided by the All India Term-Lending Institutions, with the Investment Institutions rapidly increasing their share of assistance.

The sanctions accorded by DFIs include rupee loans, foreign currency loans, direct subscription, and underwriting of equity as well as guarantees to other creditors. The differing focus of the three categories of DFIs which lend to industry is sharply captured in differing proportions of such assistance.

The rapid growth of DFIs, particularly during the 1980s, has been guided by several distinctive policy thrusts.

(i) *Sector-Specific Apex Institutions.* The Industrial Development Bank of India (IDBI) was established as an apex institution for industrial credit, and thereby mandated to coordinate the activities of other DFIs lending to industry. Several other sector-specific DFIs have been set up for similar purposes: Exim Bank for foreign trade, NABARD for agriculture and rural development, NHB for housing, and the Small Industries Development Bank of India for small industry. A major instrument for coordination has been the provision of refinance to other credit agencies within the sector. Refinance dominates the assets of each one of these institutions except for IDBI and Exim Bank.

(ii) *Specialized DFIs.* In recent years, certain specialized DFIs in the private sector have emerged for financing the credit needs of tourism (the Tourism Finance Corporation of India [TFCI]), shipping (the Shipping Credit and Investment Company of India [SCICI]) and technology-based ventures (Technology Development and Information Company of India [TDICI] and Risk Capital and Technology Finance Corporation [RCTFC]). Asset portfolios of such DFIs are subject to potentially larger sector risks and this has led one DFI, SCICI, to broadbase its asset portfolio across all industrial sectors.

(iii) *Resource Mobilization.* Until 1987, debt resources for DFIs were fully provided for through allocations made by the Government or RBI. These allocations, invariably subsidized, were funded either through SLR pre-emptions on banks, or from RBI's profits or, in the case of certain DFIs, out of the government budget. As the Government and RBI cut their resource allocations to DFIs, alternative sources of funds have necessarily been sought. These have included foreign currency borrowings, inter-institutional borrowings, and access to the capital market. Most DFIs have a poor portfolio quality, but have hitherto not been subject to tight accounting conventions which would reveal the extent of their bad debts. Tighter accounting is expected to commence in 1993–1994 and be phased in over two years.

(iv) *Dominance of the Public Sector.* Of the 59 DFIs, only five are outside the public sector. As at March 1992, these five DFIs (ICICI, SCICI, TFCI, TDICI, and RCTC) had provided less than 20 per cent of all DFI cumulative disbursements. ICICI began as a private sector DFI, came within the public sector when its dominant shareholders—the banks, and insurance companies—were nationalized and has subsequently broadened its shareholding in the capital market so as to emerge once again as a private sector DFI. The other four private sector DFIs have been established more recently, the oldest of them (RCTC) in January 1988.

Fragility of Institutions

The statistical profile presented above of banks and DFIs exhibits the rapid growth of assets and an expansion in credit coverage. The territorial span of bank branches has sustained the mobilization of deposits. The maxim that credit institutions can grow only as fast as their resources has diverted attention to the growth of these institutions' liabilities, encouraged by a Government vista of financial deepening in which banks are encouraged to transmute savings into investment. The imperative to raise the savings rate—and more explicitly the financial savings rate—further spurred the growth of these institutions. It is precisely in such a strategy, however, that the seeds of retrogression have been placed.

It is now more widely recognized that the management of a bank's assets is much more crucial to its sturdiness than the growth of its liabilities.[24] This is clearly more than a problem of choosing borrowers, selecting projects, assessing and diversifying risks, and ensuring repayments. It is more centrally the issue of how information is acquired and shared within a financial intermediary, on the incentives for subordinate managers to resist the adverse selection of borrowers, and on the relationship between the available infrastructure (the law, regulation, accounting conventions) and the institution's corporate response.[25] Over the last two decades, fast-growing banks and DFIs have become fragile despite the absence of competition. As financial liberalization enhances competitiveness, the very survival of many of these institutions is at stake.

---

[24] See, for instance, Stiglitz (1989). Capital adequacy in the sense of the Basle norms recognizes this.

[25] De Juan (1991) lists several aspects of the behavior of fragile banks which indulge in "cosmetic management" (hiding past and current losses so as to buy time and remain in control) and "desperate management" (in danger of having to declare a loss or pay no dividends) in a way that mirrors authentically the managerial response of India's weaker banks.

REFORM OF COMMERCIAL BANKING

Commercial banking reform lies at the heart of the restructuring necessary to restore the competitive vitality of the financial system, and this section identifies options for reform which are applicable to banks.

Macroeconomic Policy Support

Fiscal Restraint

While Government pre-emption of bank deposits is targeted to be lowered, the present strategy does not adequately address the pre-emptions of the past. To see why this is necessary one must confront the arithmetic.

SLR preempted 38 per cent of bank deposits in 1992–1993 and is targeted to be reduced to 25 per cent in 1996–1997. These are average SLR targets, implying that incremental SLR must fall more sharply. (Illustratively, if deposits rise by 10 per cent annually, incremental SLR must fall by 11 per cent in order that average SLR is lowered by 1 per cent.) This suggests that a rapid reduction in average SLR within four years necessitates incremental SLR possibly turning negative. The foreclosure of government debt would need to be seriously explored, as it is evident that the 1993–1994 average SLR pre-emption, targeted to fall from 38 per cent to 36.75 per cent, is clearly not tracking the 25 per cent target slated for 1996–1997.

Since 1992, the Government has begun a phased program of disinvesting select public sector equity, and this offers the possibility of funding a government debt foreclosure program. Indeed, it could also be structured so as to create incentives for banks to participate in public sector disinvestment, and thereby raise higher government revenues. Thus, if disinvestment is conducted through auctions and banks are assured that a certain preannounced proportion of their successful bids could be used to immediately retire SLR securities of their choice held by the Government, banks will develop strong incentives to jettison low coupon rate securities in exchange for intermediating in public sector disinvestment.

The foreclosure of government debt will have several other beneficial implications. The large stock of such securities held by banks impedes the creation of an active secondary market for government paper, with coupon rates on securities issued today being appreciably higher than on those issued several years ago. A sizeable clutch of SLR securities held by banks would therefore need to be traded at an appreciable discount, and this would impose unacceptably high losses on banks. As securities depreciation losses have been inflicted upon banks by the Government, debt foreclosure would be an entirely appropriate way for the Government to assist in reviving bank profitability. In the process, fiscal restraint will clearly be strained, particularly as SLR revenues are estimated to have amounted to 17 per cent of the Government's internal financing requirements during 1992–1993. The additional need for larger budgetary allocations to recapitalize banks from 1993–1994 onward will also compound the fiscal difficulty. Options for the Government are limited, however, if its objective of strengthening the profit earning capability of banks is not to be compromised. The privatization of banks can, of course, ease the fiscal burden.

Finally, rapid debt foreclosure also reduces the externality that banks face in the adverse selection of their borrowers, as was discussed earlier in this chapter, and is therefore highly beneficial in correcting distorted management incentives.

Style of Monetary Management

CRR absorbed 15 per cent of deposits in 1992–1993 and RBI intends to reduce this to 10 per cent by 1996–1997. Accordingly, in April 1993, RBI announced a reduction to 14 per cent. As RBI now pays interest on CRR pre-emptions, the efficacy of CRR as a monetary instrument has become blunted. If alternative monetary instruments could substitute, clearly CRR could be brought down to a lower level, perhaps of less than 5 per cent, which would give banks a higher return on their commercial credit.

A change in the style of monetary management towards open market operations would facilitate such a transformation. However, open market operations on treasury bills and dated government securities must necessarily be linked to the development of a primary market for such securities. A later section of this chapter outlines the desirable structure of such a market. Monetary management through open market operations will enable RBI to reduce the need for other parametric regulatory instruments and would facilitate banks earning higher returns on their stock of government securities.

The shift to open market operations also necessitates the development of deeper money markets. This is a contentious issue in the wake of the 1992 securities irregularities and is discussed later. The institutional development of related financial markets is, however, crucial to altering the style of monetary management. These related financial markets, in turn, will help to set key benchmark interest rates which will provide helpful indicators on how RBI and the banks, in turn, could set the term structure of deposit and credit interest rates. Monetary policy through open market operations will thereby integrate investment and credit portfolio decisions in banks, assist banks in arbitrating risk discounted yield differentials between bonds and loans, and sharpen incentives within banks for more efficient asset management. Such efficiency gains will become vital as financial liberalization introduces competitiveness into the system.

Financial Infrastructure Support

Speed in Contracts Enforcement

This is the most crucial prerequisite to the strengthening of banks. India is in need of an urgent revamp of its commercial law, but meanwhile two specialized interventions could assist in the quick enforcement of credit contracts: the first is the establishment of a special tribunal to unclog the regular courts of loan disputes above a minimum size; and the second is to provide legally for private foreclosure of mortgage based loans. These issues are discussed in more depth in a later section of this chapter. The inefficient enforcement of a loan contract lowers the present value of the loan at the point of sanction and will lead increasingly to banks eschewing longer-term lending.

Efficient Bankruptcy and Liquidation Proceedings

There is an emerging consensus that the law on bankruptcy is badly flawed, with sick company revival being unlikely given the late stage at which proceedings commence.[26] Bankruptcy law in India has been enacted under the Directive Principles of State Policy of the Indian Constitution in order to "save the social capital embedded in sick industries" rather than as a means of lowering the transactions costs of secured creditors in liquidating collateral. Unless bankruptcy courts become "courts of convenience" wherein the interests of secured creditors are aggregated and protected, bankruptcy proceedings will be of little interest to creditors. Liquidation, too, moves very tardily, and the liquidation of companies could be accelerated if proceedings are tagged on to bankruptcy courts.

Transparent Accounting Conventions

Transparent accounting conventions reveal how impaired a bank's capital is, and therefore whether depositors' funds are at risk. By prescribing capital adequacy targets to be achieved in future years, RBI sends several signals to banks: that the adverse selection of borrowers is injurious to a banks' ability to expand its asset portfolio; that borrower selection requires greater sophistication, based on investments in information generation and personnel skills; that for a large bank—with operations which are countrywide—a suitable mix of decentralization, incentives, and controls is needed so that subordinate manager behavior is in accordance with bank priorities; and that greater accountability is needed to shareholders of banks with

---

[26] See, for instance, Anent, Gangopadhyaya, and Goswami (1992a and 1992b).

impaired capital. The survival and growth of banks under transparent accounting systems is less likely to put at risk the depositors' interests. Equally, when a bank has reached a position of grave undercapitalization, the options for its revival narrow.

Regulation and Supervision

Clarity in the regulatory rules and supervisory modalities would benefit banks, and these would need to be spelled out by RBI. A classification of banks according to their capital adequacy would also determine the nature of regulatory constraints to be placed on each bank. At one extreme, a severely undercapitalized bank would be discouraged from increasing its commercial credit, while putting fresh deposits into zero-risk, government-guaranteed securities. The price to be paid for lowering depositor risk is the consequential lower interest income earned. Such a strategy could also help to buoy the demand for government securities auctioned henceforth.

Restructuring of Banks

Bank restructuring is a process, not an event, and is usefully sequenced to include the diagnosis of each bank's weaknesses, the short-term profitability support it requires, and measures to provide each bank with longer-term competitive strengths.[27]

Diagnosis

This includes quantitative and qualitative indicators of a bank's weaknesses. Quantitative measures include profitability and capital adequacy under the new RBI accounting convention. As the new accounting norms are being phased in over three years, it would be desirable also to recast the first year accounts to comprehend more precisely the under-provisioning and inadequacy of capital at the initial stage of restructuring. This indicates the extent of the bank's accounting insolvency. In addition, if the value of assets and liabilities of the bank (including SLR securities and the value of collateral mortgaged or hypothecated to it) is "marked to market," one derives a measure of the bank's economic insolvency.

How economically insolvent is the Indian banking system? Opaque balance sheets make estimates of its financial health difficult to compute. One estimate (Varma 1992) notes that a simplified balance sheet of the banking system (normalized to deposits worth Rs 100) could be visualized as shown in Table 6.11.

---

[27] This classification is based on Sheng (1992) with an emphasis on restructuring techniques which are more strongly applicable to public sector banks.

Table 6.11 Simplified Balance Sheet

	Rs		Rs
Deposits	100	Cash	16
Nondeposit resources and net balance of liabilities over other assets	7	Investment	39
Capital and Reserves	2	Advances	54
Total	109		109

Asset depreciation is computed as follows: Roughly 75 per cent of banking sector investments have undergone a depreciation of about 18 per cent. Thus Rs 39 of investments has suffered a loss of Rs 5.26. Further, 30 per cent of advances are in the form of term loans on which concessional interest rates (on priority sector loans) amount to a 11 per cent loss. The loss on Rs 54 of advances is therefore Rs 1.78. Priority sector nonperforming loan losses are estimated conservatively at Rs 1.30, while similar losses on loans to large and medium industries are estimated at Rs 0.97. The total loss is therefore Rs 9.31.

Liabilities depreciation also arises as a result of earlier deposits contracted at lower interest rates. Roughly half the deposits can be argued to have depreciated by 7 per cent, leading to a profit on liabilities of Rs 3.50. Thus the net loss is Rs 5.81. With aggregate deposits as in July 1992 amounting to Rs 2.45 trillion, the aggregate loss would be Rs 140 billion. As capital and reserves amount to 2 per cent of deposits, the banking system has a negative net worth of Rs 90 billion. This is a tentative measure of the economic insolvency of Indian banking.

Other quantitative measures seek a decomposition of a bank's losses and include the following:

— credit losses arising out of poor borrower selection and inadequate collateral;

— inefficient contract enforcement losses preventing the bank from liquidating collateral held by it;

— investment portfolio losses on account of interest rates which are not market related, resulting in depreciation losses as well as lower interest earnings;

— maturity mismatch losses arising from the bank's need to mobilize higher cost resources (often from the interbank money market or by issuing more expensive certificates of deposit);

— inefficiency losses because of overstaffing or overheads, including losses arising from uneconomic branch operation; and

— fraud losses resulting from dishonest practices within the bank.

Disaggregated losses along the above lines would indicate the weaknesses in the bank's business, and the critical areas needing intervention.

This needs to be supported by a qualitative assessment of the bank's solvency. There is a convergence of views among bank supervisors internationally on desirable methods of assessing the strengths of banks, and the CAMELOT rating system could be a useful starting point for bank diagnosis, encompassing capital adequacy, asset quality and concentration, management quality, earnings, liquidity, operations quality, and treasury management.

Short-Term Profitability Support

Such support results in the initial restructuring of banks, and is clearly a damage control intervention which the Government was likely to have initiated in 1993. Where capital is less than 2 per cent of risk weighted assets, experience from other countries suggests that intervention needs to be resolute, as delays invariably increase the cost of restructuring. Techniques of restructuring include flow augmenting and stock transforming solutions.

Flow augmenting solutions include RBI refinance at subsidized interest rates (though this makes a tight monetary policy more difficult to achieve); enabling the bank to earn higher spreads through a lowering of reserve requirements (which enables the bank to expand its credit portfolio) as also through lower priority sector targets; imposing fewer supply-side selective credit controls on the restructured bank, in order that profitable lending can expand; permitting the bank to close down its more uneconomic branches; and harmonizing the tax system so that it is on par with the accounting system. There is some evidence that a purely flow-based solution could be feasible provided capital adequacy is not less than 2 per cent (Sheng 1992).

Stock transforming solutions address capital adequacy and the marketworthiness of banks, and broadly involve liabilities restructuring, capital injection, mergers, and privatization, depending on the extent of bank fragility. The strongest two or three of the public sector banks could probably augment their capital from the market, provided their liabilities are restructured. Thus, the Government may need to realign its share capital and reserves so as to increase share value while retaining the banks' net worth. Earnings per share and dividend paying capacity would need to be adequately attractive, and this may require, in addition, a flow augmenting intervention. Capital injection by the Government should, as far as possible, be subordinated to raising capital from the market, and this needs legislation so as to permit banks to be owned by others besides Government. For the weaker banks, more extensive capital injection by the Government is needed, and these banks are unlikely to be able to draw capital from the market until their balance sheets are spruced up. For the weakest banks, mergers with stronger banks as well as outright sale of these banks' equity by the Government to private sector promoters is needed.

The denationalization of banks is a major issue at present. There are at least four ways in which this could be structured:

— weak denationalization, where the Government retains majority ownership and limits the extent of shareholding which an individual or a company or a group of companies under common management can hold;

— semi-strong denationalization, where the Government retains majority ownership;

— quasi-strong denationalization, where the Government limits the extent of shareholding which an individual or a company or a group of companies under common management can hold; and

— strong denationalization, where the Government imposes no constraints on ownership.

The Government's current intentions are to permit the weak denationalization of banks. This would retain Government control, preclude business houses from gaining even minimal ownership dominance, and retain banks as statutory institutions set up under an Act of Parliament. There is a plausible case for asserting that small investors may be attracted to acquiring share ownership, provided the banks are strong, with adequate returns on investment. Only the most robust of India's nationalized banks can expect to follow this route successfully. For the others, capital injection and bad debt clean-out are prerequisites to following this route, and this would clearly delay denationalization besides requiring the Government to make sizeable budgetary outlays as bank share capital. Tax incentives to new shareholders of such banks could conceivably hasten the process of recapitalizing such banks through private investors.

For the weakest banks, where closure is a serious option, the Government would need to earnestly consider strong denationalization unless it is prepared to inject substantial equity. Private ownership would need to be coupled with the private control of banks. However, the Government appears more inclined to consider quasi-strong denationalization, as recent decisions with respect to DFIs indicate.[28]

Other stock transforming solutions include "carving out" the bad debts of banks at a discount into a separate institution; the Narasimham Committee had recommended the constitution of an Asset Reconstruction Fund (ARF) for this purpose. However, the success in recovery which a centralized ARF would achieve in a geographically large country with recovery suits filed in thousands of civil courts has prompted the Government to be skeptical of the utility of such a carve out.

Longer-Term Competitive Strength

The longer-term restoration of bank profitability is more dependent on macroeconomic support available to the banking system (particularly the legal, regulatory, and supervisory systems discussed earlier) and the internal corporate strengths built up in each institution. Clearly, there can be no definitive blueprint for longer-term success, and the discussion below concentrates on generic issues of relevance to banks.

State of Borrower Net Worth

Unless the real economy performs well, banks are unlikely to acquire longer-term competitive strengths. The flow of capital funds in the economy, discussed earlier in this chapter, provides an indication of sectors which have run revenue surpluses and deficits.

The major deficit increase during the 1980s has occurred in the government sector (inclusive of public sector enterprises) while the largest surpluses have arisen in households as well as overseas. There are banks, such as SBI, which have a large public sector exposure, and public sector restructuring is clearly crucial to strengthening such banks.

---

[28] The issue of successful business groups also controlling banks is a contentious one. Diaz-Alejandro (1985) provides evidence that in Argentina and Chile this has led to severe credit allocation distortions on account of the "creation of oligopolistic power." Equally in several other countries — France, Germany, and Japan among them — links between industry and banking are strong, forged by cross-ownership. What is perceived, however, as a moral hazard problem could be contained through regulation which stipulates strict capital adequacy requirements and full disclosures on credit provided to interconnected business enterprises.

Manager Incentives

The nationalized banking system is unable to reward outstanding managerial performance, monetarily or through accelerated promotions, because of Government restrictions. Recruitment of bank personnel is also made by external agencies. In recognition of banks being independent profit centers, greater flexibility is needed in permitting individual banks to control management incentive structures, and the Government would do well to either delegate these matters to banks or else permit a profit-related incentive structure. Along with management incentives, there should be strong disincentives as well. In particular, it appears desirable that RBI be provided with and decisively exercise powers to replace senior managers of mismanaged banks.

Succession Planning

Government intervention in bank management arises in the appointments of the Chairperson and Boards of banks. Board-level appointments typically bring inadequate professional skills into banks, and the appointments of Chairperson often occur with delays and with no clear plan of succession when a Chairperson retires. Two changes in approach merit consideration: first, Boards should be constituted afresh in a manner which draws considerable private sector talent. Second, the co-option of subsequent members of a Board as well as the appointment of its next Chairperson should draw upon suggestions made by existing Board members in the interests of continuity, based on a more sensitive appreciation of the bank's needs than Government would otherwise have.

Industrial Relations

Bank personnel are generally strongly unionized, and restrictive work practices have strengthened since bank nationalization. Customer service is therefore often very indifferent, automation has been rigidly resisted and, with bank unions confederating into industry-wide unions, wage settlements have typically been uniform across nationalized banks, their terms getting the Government's informal nod. Bank management has, therefore, become increasingly marginalized in the collective bargaining process. Whereas the Government is now moving rapidly to correct this by insisting on work practices being under the exclusive control of bank management and therefore not a determinant of wage negotiations, emerging labor law which has fostered the notion of uniform wages across the banking industry would need to be redesigned so as to underline decisively that emoluments are dependent on bank profitability.

Directed Credit

With the priority sector accounting for 40 per cent and exports for another 10 per cent, half of the credit outstanding is directed. Overdues on priority sector lending do appear proportionately higher than on other credit disbursed, and this has led some reformers to call for jettisoning the priority sector. The Narasimham Committee suggested truncating its size from 40 per cent to 10 per cent and targeting it more precisely at smaller and weaker borrowers.

Since 1992, the extent of concessionality in the priority sector has come down steeply, and directed credit is now largely allocative, rather than a subsidizing mechanism. Such an allocative function could continue to have a rationale,[29] provided bad debts are not proportionately higher. It appears desirable that borrower selection procedures be more rigorously scrutinized, for the proportion of overdues varies appreciably between banks. For several government-sponsored rural development programs, borrowers are often first identified by State Government functionaries, and moral hazard in such identification may account for the dismal recovery rates in such programs. If modalities for borrower selection can be refined, leading to improved recoveries, directed credit may not be as injurious to a bank's health as is commonly believed.

Redefinition of the Borrower Segments

Finally, each bank needs to redefine where its comparative advantage lies, and to accordingly re-identify its profitable borrower segment. Certain loan transactions are incentive compatible when the creditor is an informal intermediary and a larger bank may be unable to mimic such creditor behavior. Equally, as capital and money markets deepen, banks may be able to diversify their risks and quite successfully provide credit to large industrial firms in a manner clearly beyond the capabilities of informal creditors.

It is argued in another section of this chapter that part of the weakness in industrial credit lies in the structuring of project finance, and that if the borrowers were to "market signal" their individual risk perceptions, banks and DFIs may benefit. Agricultural credit markets (which are not analyzed in this chapter) also need to be assessed in terms of whether risks are well shared between borrowers and creditors. The weaknesses of the agricultural credit institutions make such risk sharing particularly vital. Weak institutions are also handicapped in their ability to invest in information on borrower risks.

---

[29] Qualified support for directed credit programs has come most recently from Stiglitz (1993) in the context of export credit targets imposed by governments of several East Asian countries.

One solution, therefore, to the problem of the high information costs of lending to small borrowers may lie in banks channeling finance through informal creditors. Another solution may consist of banks franchising branches in underbanked areas, enabling the franchisee to adopt the style of business of the informal creditor. Innovations of this type need to be more aggressively pursued.

RESTRUCTURING THE DFIS

The DFIs are generally not deposit-taking financial intermediaries and their assets are more specialized than those of banks in their average maturities and their sectoral focus. The DFIs are therefore more vulnerable to financial liberalization and disintermediation, with their earlier resource windows being rapidly shut by the Government, and with a portfolio quality which typically ranges from the indifferent to the very poor. Of the 59 DFIs discussed earlier, all except NABARD and NHB assist industry. It is ironical that the industry financing term-lending institutions, established essentially as a response to market failure (when the capital market was undeveloped in the 1950s and 1960s) now face the prospect of being marginalized as the capital market deepens. Well-rated companies (as well as several not so well-rated) have accessed the capital market directly for an increasing proportion of their resource needs in recent years. The apprehension that companies with an indifferent track record will constitute an increasingly large segment of DFI borrowers is an admission of a possible adverse selection of borrowers. This will further worsen the portfolio quality of DFIs.

The major term-lending institutions are IDBI, ICICI, and IFCI, and these DFIs control key financing decisions for small and medium industry. Projects with costs in excess of Rs 200 million are generally funded in consortium, and where the big investment institutions (such as UTI and LIC) participate in consortium lending they do so without any serious and independent credit appraisal, the promoters' track record being decisive in their decisions to participate. The market structure for industrial finance is thus strongly oligopolistic and therefore disadvantageous to industry, reinforced by IDBI's dominance. As the three large DFIs find themselves under increasing resource mobilization pressure, it is likely that oligopolistic tendencies will be further reinforced with adverse implications for the terms at which industry can borrow.

The DFIs are highly geared institutions with debt/equity ratios typically in excess of 10:1. Their resource needs are therefore largely for debt, and if they are to be competitive in their borrowings they will necessarily need to access the primary capital market. The debt market, however, is at present very thin. A strategy for market-making in debt instruments is needed, reinforced by an active money market which is currently stifled by restrictions imposed by RBI. Unless DFIs can be restructured to make them marketworthy, they will be unable to borrow competitively. Unless they can diversify their asset risks, their profitability will fall sharply.

Several DFIs have been established to provide sectoral focus and coordination, and their "apex" role is generally discharged through refinance. Thus, IDBI provides refinance to SIDCs, SIDBI to the SFCs and commercial banks, and Exim Bank to the commercial banks. (Outside of industry, NABARD provides refinance to agricultural cooperatives and commercial banks, while NHB provides housing refinance to housing finance companies and cooperative housing societies.)

With lending interest rates deregulated, the refinancing function becomes difficult to execute, as resources for the re-financing agency would need to be raised at below market rates of interest. The crisis of a segment of industrial financing is, therefore, partly a crisis of refinance. Further, bank deposit maturities have shortened in the last two years, as deposit rates are no longer regulated by RBI in terms of their maturities. Consequently, the average maturity of DFI borrowings has shortened and, to cope with this, DFIs are lowering their average lending maturities and thereby moving away from their earlier predominance in project finance. This trend is unlikely to be arrested without either a reintroduction of deposit interest rate regulation (which appears undesirable, as it will impede the development of efficient money markets) or else the availability of fiscal concessions to DFIs and other credit institutions in the form of tax-free reserves which are proportionate to the extent of the institutions' assets which are held long term. The nature of business which financial institutions undertake is likely to undergo rapid structural change in the absence of such a fiscal stimulus, with a very real danger of the institution-intermediated long-term debt market facing a collapse.

A restructuring of DFIs would necessitate reform of the legal and regulatory environment, provide for capital market access for DFIs, and facilitate the corporate restructuring needed for each DFI in order to enhance its competitiveness.

Legal and Regulatory Environment

Legal Reforms

Collateral serves little purpose if it cannot be liquidated quickly upon loan default. If comprehensive legal reforms are unlikely, then nonjudicial redress should be permissible through the private foreclosure of loans. An approach to this is sketched in a later section of this chapter.

Transparent Accounting Conventions

The DFIs, too, need to be subject to transparent accounting conventions. The RBI appears set on a course of phasing in greater rigor over two years, commencing 1993–1994. This too is discussed later in this chapter.

Tighter Supervision of DFIs

The DFIs need to be brought under the Board for Financial Supervision, though the style and content of supervision would probably need to be less on-site and intrusive than for banks.

Market Development

Facilitate DFI Access to the Capital Market

The most competitive resource mobilization for DFIs in the absence of subsidized Government and RBI-allocated funds, will necessarily be from the primary capital market. The DFIs must be strengthened to enter the market, with Government restructuring its balance sheets so as to position each DFI on the starting line for capital market access, based on a satisfactory credit rating and approved by the regulatory agency, the Securities and Exchange Board of India (SEBI). Further, as most DFIs have been established under Acts of Parliament, it is vital that they be brought under company law in order that the Government and the private shareholders are invariably treated on par. Whereas parity under a statutory framework could also be realized, private investor perceptions will surely be less favorable, as a statutory DFI would create recurring doubts about whether investors' interests are being compromised. Retaining a statutory cover while opening up market access runs the risk of private investor psychology turning against such DFIs. One statutory DFI, IFCI, has recently been brought under company law and the Government would need to introduce legislation to effect a similar transformation with respect to other DFIs.[30]

Develop the Money Market

This would reinforce a longer-term market for debt, as it would enable primary issues of debt to be first placed in the money market under repurchase offers. The initiatives needed are discussed later in this chapter.

---

[30] The IFCI is being slated for semi-strong denationalization. Although decisions with respect to other DFIs are yet to crystallize, strong denationalization does not appear to be an active policy option.

Provide Fiscal Support for Long-Term Lending

As a transitional measure, and until the long-term debt market is well developed, fiscal benefits to DFIs in the form of tax free reserves out of profits which are proportional to the percentage of their assets held long term would greatly assist the continuance of project finance.

Clarity in Corporate Strategy

Role of Refinance

Refinance is necessarily subsidized, and where Government is unable to provide the subsidy, the DFI itself needs to cross-subsidize. Certain sectors such as agriculture and small industry are very strongly dependent upon refinance for long-term capital needs, with NABARD and SIDBI being the refinancing DFIs. If interest subsidies to DFIs are being phased out, refinance is clearly unprofitable and these institutions would need to instead consider co-financing borrowers along with the primary lender. Thus, currently commercial banks and SFCs obtain about 60 per cent refinance from SIDBI for long-term finance provided to small industry. Instead, SIDBI could co-finance, taking a 60 per cent share in the total loan. Thereby, SIDBI would also take on a borrower credit risk, as against an institutional credit risk as at present. This necessitates SIDBI's credit appraisals being independent of those of existing primary lenders.

Superior Project Appraisal Decisions

Whereas the larger DFIs do appear to invest considerable effort in appraising projects, the bad debts position of DFIs, hitherto not publicly revealed, is widely believed to be uncomfortably high, and rescheduled repayments are customary. Balance sheet transparency will clearly induce greater sophistication in project appraisal, but as the trade regime of the country changes, Indian industry will get nudged towards international competitiveness. India is now committed to a sequenced lowering of its external tariff barriers and, as these fall, project dependency on protection-related profits eases. The DFIs would need to select and structure projects which are globally more competitive. This has implications for the financing of technology upgrading, and DFIs would need increasingly to get into research and development financing.

NONINTERMEDIARY FINANCE

The development of the capital, money, and foreign exchange markets will be critical to the success of banking and DFI reform. Equity and debt, particularly the latter, will need to be raised, and instruments hitherto unexplored in India such as options and futures, will be decisive in the successful hedging of risk.

Capital Markets

There are 23 stock exchanges in India, including a recently established over-the-counter (OTC) Exchange. The dominant exchange is the Bombay Stock Exchange (BSE) which transacts about two thirds of the country's equity trading volume. A varied range of institutions, which participate in and thereby strengthen the securities markets, have also emerged. These include merchant banks, brokerage firms, credit rating agencies, mutual funds, and custodial services. These are supported by about 5,000 brokers and over 100,000 sub-brokers. Equity markets grew rapidly during the 1980s and became an important source of start-up capital for industry. Between 1980 and 1992, the number of shareholders rose from about 2 million to over 15 million, the number of listed companies grew ten times, and market capitalization shot up from Rs 150 billion to Rs 1.5 trillion. The annual volume of trading in 1992 was in excess of Rs 1 trillion. During 1991–1992, nongovernment public limited companies are estimated to have raised over Rs 57 billion from the capital market, mainly through the issue of convertible debentures, but also through equity. About two thirds of this came through rights issues (Table 6.12).

Table 6.12 New Capital Issues by Nongovernment Public Limited Companies (1991–92)

Instrument	Amount (Rs billion)			(Per cent)		
	Public	Rights	Total	Public	Rights	Total
Equity	10	7	17	18	12	30
Convertible debentures	9	26	35	16	45	61
Non-Convertible debenture	—	5	5	—	9	9
Total	19	38	57	34	66	100

— nil.
Source: Reserve Bank of India (1992a).

In addition, the public sector was able to raise, for several years until March 1992, substantial resources from the capital market through the issue of public sector undertaking (PSU) bonds. Finally, the Government also issues treasury bills and dated securities of varying maturities, but secondary markets for these are either weak or nonexistent as most government securities have traditionally been held by banks in fulfillment of reserve requirements.

Growth of the capital market has taken place without adequate regulation and supervision. Brokers have indulged in excessively dangerous speculation and, as major securities trading irregularities which surfaced during 1992 revealed, also financed such speculation with bank funds. Investors have had few enforceable remedies against delays in securities transfers or in receiving payments, and several undesirable work practices in capital market institutions have consequently developed. The Government has therefore established a regulatory agency, SEBI, to which certain powers of regulation and supervision were delegated in 1992. In late 1992, the Government repealed the Capital Issues (Control) Act, providing freedom to companies to access capital markets without Government approval. Government control over the pricing of securities at the stage of issue has also been disbanded. Nevertheless, several enduring weaknesses need to be addressed, both to improve the efficiency of capital markets, as well as to create markets for debt. In the absence of such reforms, DFIs will be particularly vulnerable.

Capital Market Improvements

Primary Markets

Although equity offers to the public through a prospectus disclosure have enhanced information revelation by issuing companies, hitherto SEBI has only permitted fixed price offerings, and there continues to be a pricing restriction on initial public offers (IPOs), except under certain clearly specified conditions. There is considerable evidence that IPO prices typically rise rapidly as between issue and listing (which can take up to three months) and thereafter for several months. There is, therefore, clearly an investor price discovery at work, indicative of inadequately researched information on the initial offers being publicly available. The price discovery could be aided by more flexible forms of public offers being permitted, with "bought out deals" and other forms of price flexibility being permitted. India has, as yet, no concept of a qualified institutional buyer (QIB) and therefore lacks mechanisms which permit equity offers to QIBs through auctions or syndicated offers. The success of such offers to agents whose equities research is typically superior would represent market signaling which would aid price discovery by other investors. In the absence of such mechanisms, excess demand for certain scrips is periodically observed, spawning pre-issuance "grey markets" which operate illegally. The regulatory design and price discovery by investors in the market for fresh equity issuance in India remains largely under-researched.

Malpractices in fresh equity issuance have also developed through the preferential allocation of equity to select shareholders, generally representing the management in control of the company, at prices which are at drastic discounts to market prices. With the repeal of the Capital Issues (Control) Act in 1992, several companies—generally owned from overseas—have put through such preferentially priced equity offers, occasionally at prices as low as one tenth the market price.

Secondary Markets

Reforms in stock exchange trading have come rapidly under SEBI's regulatory vigil, with several malpractices being curbed. Settlements in stock exchange transactions are typically slow, and uncertain. For specified shares (effectively a forward list), settlement can be carried over to the next period of settlement—which typically lasts a fortnight—through "badla" or contango. Although contango-based systems of stock exchange trading do enhance liquidity, they also endanger the safety of exchange trading and permit market manipulation through futures and options transactions in certain shares, dominating transactions for immediate settlement. There have been certain settlement periods, particularly in 1992 when major securities irregularities were detected, when short sales to long buyers have swung certain share prices without underlying sale transactions for immediate settlement being visible. The "badla" system clearly lends itself to market manipulation and SEBI has been relentless in seeking to curb this.

Poor market practices are also manifested in terms of the possibility of a trade executed on the floor of the BSE being cancelled by either the selling or buying broker by the end of the day, the inadequacy of audit trails being recorded by the BSE, capital adequacy norms for brokers not being established, and a sizeable proportion of "bad deliveries" arising with companies not immediately registering sales executed on stock exchanges. Enforcement against malpractice is widely believed to be poor, and the scope for improvement in market practices will require radical changes in the legal and regulatory environment. An electronic depositories-based system of stock exchange trading, clearance, and settlement with securities ownership recorded through book entries, provides the only medium-term assurance that several of these practices will be eliminated.

Creation of Debt Markets

The development of robust primary and secondary markets for Government and corporate debt will determine whether long-term financial assets will be held voluntarily by investors. In 1992, the Government began auctioning treasury bills and dated securities. The RBI, which conducts the auctions, fixes an unannounced reserve coupon rate above which bids will not be accepted, and unallocated bonds are purchased by RBI. Hitherto, a relatively small proportion of Government bonds (though no State Government bonds) have accordingly been auctioned. Several institutional innovations will be needed if markets for government and corporate long-term debt are to develop.

Government Bonds

The creation of a primary market for Government securities will necessarily lead to more active trading by RBI in the secondary market and will accordingly tilt the conduct of monetary policy in favor of more active open market operations. This is certainly the historical experience elsewhere. Countries differ, however, in the manner in which auctions are structured and in the participants, instruments, regulations, and market development needed to create a continual demand for such securities. A blueprint for a government securities market for India has recently been proposed (IMF 1992) which argues that flexiprice, quantity-constrained, repeated auctions constitute the most appropriate mechanism for selling such securities. The blueprint further suggests the following three categories of participants:

— Government securities dealers (GSDs), who would have the exclusive right to participate in the primary auctions, to trade in government securities, and to access the Inter-Dealer Broker network;

— Primary dealers, a subset of GSDs, who would also be expected to underwrite securities auctions and to act as market makers by providing continual two-way quotes and by creating investor demand for such securities. In exchange, RBI would conduct its open market operations exclusively through them, and also possibly provide "last resort financing" for their inventory positions; and

— Inter-dealer brokers (IDBs), who would collect bids and offers from GSDs and thereby enable GSDs to trade anonymously with each other, with IDBs picking up commissions.

The incentives and the risks implicit in such a framework need to be assessed in the context of institutional development in India. While repeated flexiprice quantity-determined auctions will enable the Government to raise a predetermined quantum of debt each year, the same objective could, however, be realized through repeated auctions where Government fixes a reserve price which is periodically altered in response to the quantity of securities sold in earlier auctions.[31] The crucial issue, therefore, is the

---

[31] The success of this is admittedly dependent on whether dealers who bid for securities fail to recognize a "pattern" in RBI altering its reserve price. The situation is game-theoretic: more precisely, there is an informationally dynamic multi-period game. See, for instance, Fershtman (1987) and Friedman (1980).

acceptable trade-off between price volatility and quantity uncertainty and, more fundamentally, on the relative independence of monetary and fiscal policy (price volatility arguably weakens RBI's interest rate stability; quantity uncertainty complicates the Government reaching its fiscal target). The acceptable trade-off must depend on a prior assessment of how price and quantity uncertainties in securities auctions should operate.[32]

The sizeable stock of SLR securities held by banks will clearly also retard the development of the securities market because of the depreciation losses and lower yields earned by banks. Debt foreclosure by the Government would greatly facilitate the creation of the market. Further, as banks will be dominant investors in Government securities, there would be a conflict of interest if they were to also participate in primary auctions. A clear institutional specification of dealers will therefore be needed. Finally, the incentives available to dealers to participate in auctions need careful assessment. Primary dealers would be obliged to actively underwrite auction sales in return for exclusive participation in RBI's open market operations. If the latter are slender, then these dealers could form a cartel to keep interest rates higher than they need be. Further, the business of IDBs is contingent on GSDs trading anonymously with each other. The IDBs, therefore, provide details of "inside" two-way quotes on offers by GSDs. Anonymous trading introduces a moral hazard problem, and violations may be hard to detect.

Corporate Debentures and PSU Bonds

It is often argued that one of the reasons for the unpopularity of debt instruments in the primary market is that, for much of the decade leading to 1992, returns on equity were substantially higher, with issue prices being fixed by the Government. However, PSU bonds were readily marketable, though the securities irregularities of 1992 indicated that it was their intrinsic value as collateral, to facilitate ready-forward money market transactions, which accounted for their easy marketability. Despite returns on equity having fallen in 1993, debt markets continue to be unattractive.

As with government bonds, well-regulated dealer markets are also crucial for developing secondary markets for corporate debentures and PSU bonds, and certain dealers would need to act as market-makers, quoting two-way prices. The success of market-making, however, would depend on the narrowness of the spreads quoted by these dealers, as well as the absence of undue volatility in their bid-offer prices. Market-makers also face competition from "informed traders" (including investors) who under conditions of asymmetric information and being under no obligation to quote

---

[32] Goodhart (1989) discusses these trade-offs and emphasizes that flexiprice auctions demand considerable "market preparation" by the Central Bank involving a continual dialogue with the main bidders.

both offer and bid prices, may quote a finer rate on any one price than the market-maker would be prepared to quote. This would depend on the costs of bond inventory accumulation and bond value depreciation which the market maker faces.[33]

The Government initiated in 1991 the proposal to establish a National Stock Exchange (NSE) which would permit institutional membership and eventually unify trading within the country through electronic access and screen-based transactions. The NSE anticipated commencing operations in late 1993 with an initial focus on debt instruments. Issues pertaining to the successful diversification of risks by market-makers would need to be carefully assessed by the NSE.

Money Markets

Money market liberalization has followed in the wake of the Vaghul Committee Report of the Working Group on the Money Market (Reserve Bank of India 1987). The interbank market is largely deregulated, with no interest rate curbs on either overnight call money or on notice money which can be up to 14 days. The RBI has also sought to develop a market for commercial bills, by nudging banks into providing up to a quarter of their working capital limits to large and medium companies against bills. Other money market instruments are commercial paper (which companies having a minimum of Rs 50 million as working capital limit and a good credit rating can float), certificates of deposit (which banks can issue up to 10 per cent of their outstanding deposits), and treasury bills (earlier primarily the 162 day bill, but from 1992 also the longer dated 364 day bill). The RBI has also introduced guidelines for banks and DFIs setting up Money Market Mutual Funds, though no such Fund had been approved as of June 1993.

There are elaborate guidelines on the nature of money market transactions which are permissible, and very serious securities irregularities which surfaced in 1992 revealed that several banks and DFIs (and particularly the foreign banks) had violated these guidelines. The mechanisms for violations were the repo (repurchase offer or ready-forward) deal and the short-term portfolio management scheme (PMS), both of which enabled banks to overcome the severe reserve requirements imposed on their deposits. The repo was managed through the purchase and sale of selected

---

[33] Glosten and Milgrom (1985), which contains a discussion of these issues, emphasizes that the success of market-making depends on the ratio of uninformed to informed traders; informed traders' accurate and asymmetric information on expected "finer" prices; the costs and risks of absorbing inventory fluctuations; and relative risk-aversion between informed traders and market-makers.

securities which the banks held (government securities, PSU bonds, and units of UTI), while the PMS transactions were off the balance sheets of banks which offered the PMS.

The RBI has consequently further tightened its restrictions on legitimate money market transactions; the current position is summarized below:

Legitimate Transactions	Illegitimate Transactions
Bank to Bank Lending	
Interbank call and notice money	Repos in dated government securities or PSU bonds or UTI units
Rediscounting of bill discounted by other banks	
Repos in treasury bills	
Bank to Nonbank Lending	
Commercial paper	Repos with nonbank agents
Discounting of corporate bills representing real transactions	Discounting of accommodation bills
Rediscounting, within sanctioned working capital limits, of bills discounted by nonbank discount houses	Rediscounting outside sanctioned working capital limits
Nonbank to Nonbank lending	
Intercorporate transfers maturities	Portfolio management schemes for less than a year
Discounting of corporate bills by nonbank discount houses	
Nonbank to Bank Lending	
Certificates of Deposit	

As an active and deep money market is also crucial to the creation of robust debt markets (as new debt instruments are often conveniently first placed in money markets as repos), the Government and RBI would need to consider how several kinds of transactions currently forbidden can be made legitimate. As incremental reserve requirements on banks go down rapidly, RBI would be less concerned about money market transfers arbitrating bank deposits. As the capital market becomes more prudentially regulated, fund transfers from banks to the stock market for overtly speculative purposes would also become unprofitable. Both these are prerequisites to the development of an active money market. The establishment of electronic clearance, settlement, and depositories for major securities would also ensure that repo transactions are invariably backed by these securities as collateral.

Foreign Exchange Markets

The management of the country's balance of payments since mid-1991 has transformed an unprecedented payments crisis into a comfortable exchange reserves position. Whereas stabilization in the short term was achieved through external exceptional financing from the IMF and other multilateral institutions; in the longer run, changes in trade policy and its financing are expected to reduce the trade deficit. India now has a unified exchange rate which is determined by the forex market, a large measure of convertibility on the current account (aside from a compact negative list requiring licensing) and convertibility, too, for the import of capital goods. The capital account is otherwise not convertible. Key indicators of balance of payments in recent years are shown in Table 6.13, and reveal that the crisis of 1990–1991 was contained in the following year.

Table 6.13 Balance of Payments
(US$ million)

	1988–1989	1989–1990	1990–1991	1991–1992
Trade balance	−9361	−7456	−7750	−3078
Current account invisibles	1364	615	23	243
Capital account balance	8064	6974	5235	5628
Overall balance	67	137	−2492	2793
Net IMF borrowings	−1068	−877	1214	781
Increase in official reserves	−1001	−740	−1278	3574

Source: Government of India (1993).

The unification of the exchange rate occurred in 1993. A year earlier a dual exchange rate mechanism governed foreign exchange transactions whereby 40 per cent of export proceeds and inward remittances had to be converted at an RBI determined official exchange rate while the rest was converted at the market rate of exchange. Imports of specified bulk commodities and government purchases would be at the official rate while almost all other imports were at the market rate. A weighted exchange rate was adopted for certain categories of imports under licenses for exportable products, such as gems and jewelry.

As the foreign exchange markets develop it will be necessary to permit corporate enterprises to hedge foreign currency risks more actively. At present there are strict limits imposed upon members of the Foreign Exchange Dealers Association of India (FEDAI) on the extent to which they can commit themselves to forward foreign currency contracts, and more active and longer-term intra-currency risk management is not permitted. The RBI has hitherto not allowed banks to offer their corporate clients options and cross-currency swaps which would enable companies to hedge their risks clearly; existing foreign exchange policy is being guided by the need to minimize bankers' risks in forex management, rather than to hedge more fully the risks of corporate enterprises.

FINANCIAL INFRASTRUCTURE

A recurrent theme of this chapter is that incentives for rigorous borrower appraisal and for assessments of risk are closely dependent on the legal and accounting systems, the nature of regulation and supervision that financial intermediaries are subject to, and the presence of external deposit and credit guarantees. This section assesses the financial infrastructure that supports India's banks and DFIs.

Legal System

The process of commercial law in India has become increasingly emasculated, and judicial redress is painfully slow. This has disturbing implications for much of the economic reform packages unfolded by the Government since 1991, and unless reforms in commercial law are introduced, diminishing marginal returns to several other initiatives may be the consequence. Without a speedy enforcement of contracts, investments may not be attracted to several sectors; enforcement is currently impeded by at least three deficiencies. First, the procedural law is cumbersome with a separation between obtaining a decree and getting it enforced. (Common law imperatives appear to preclude the two being fused into a single judgment.) Second, the courts are clogged, litigation is time-consuming, and court practice is biased against quick disposal. Third, the appellate process is often liberally multi-layered, leading to further delays. Arbitration, for instance, is only the first step in a meandering legal process.

In the financial sector, the Government's response has been to introduce legislation to provide for a special tribunal to adjudicate credit disputes, where the outstanding amounts are above a specified limit. The legislation is expected to be introduced in 1993 and its intent is to hasten the adjudication of such credit defaults. This should certainly assist in speeding up litigation, though there is the lurking apprehension that, in the course of time, the tribunal too will become clogged (as has happened to tribunals in other sectors), particularly as the process of law will not alter radically.

Private Foreclosure

If the institutions of judicial redress cannot be restructured in a timely manner, one solution may be to provide for nonjudicial redress; the private foreclosure of loans illustrates the advantages of this. When a credit institution provides a mortgage-based loan, the conditions of the contract will stipulate that in the eventuality of default as signaled by the credit agency, matters would be put in the hands of a private receiver who would be entitled to take possession of secured assets, put them for sale, and distribute the proceeds to creditors.[34] The process will not be appealable in a civil court of law. Besides assisting creditors in speedily recovering loans given for unsuccessful commercial ventures, private foreclosures make possible the emergence of institutions in the private sector which perform without court intervention the role of receiver and liquidator.

Such a procedure for redress would need amendments to several enactments including the Transfer of Property Act, 1882,[35] the Companies Act, 1956, the Presidency Towns Insolvency Act, 1909, the Provincial Insolvency Act, 1920, the Indian Contract Act, 1872, and the Code of Civil Procedures, 1908. In the case of the last two enactments, receivers would need to be given the rights to seize properties in pursuance of private foreclosure without court intervention.

Bankruptcy Law

Bankruptcy proceedings in respect of medium and large industrial companies are currently heard by a Board for Industrial and Financial Reconstruction (BIFR) under the Sick Industrial Companies (Special Provisions) Act, 1985. A company is declared "sick" when its net worth has been fully eroded, two successive years of cash losses have occurred, and the company has been incorporated for at least seven years. The BIFR attempts, in the first instance, to rehabilitate a sick company by persuading its major financial stakeholders (creditors and promoters) to agree to a financial

---

[34] It would clearly be desirable to exclude from private foreclosure loans where agricultural land is mortgaged, as the desire to dispossess farm owners may lead to fraudulent documentation.

[35] The anachronistic provisions of this Act are evident in foreclosure being available only through a court (Section 67) unless the mortgage is an "English Mortgage and neither the mortgager nor the mortgagee is a Hindu, Muhammadan, Buddhist etc., or where the mortgagee is the Government or where the mortgaged property is situated in Calcutta, Madras, Bombay and other specified towns." The rationale for this piece of legislation is rooted in India's colonial history.

restructuring of the company's liabilities, together with possible further investment for modernization. The BIFR has no authority to compel the acceptance of a restructuring package, and the package must therefore win the stakeholder's acceptance. If it does not, BIFR is obliged to either look for a new promoter to take over the company on terms acceptable to the creditors or to order the liquidation of the company.

A major critique of India's bankruptcy law is that it is not structured in the interests of its secured creditors. The RBI guidelines for creditors and promoters giving further credit to sick companies do not necessarily ensure that creditor interests are protected. Companies come within BIFR's jurisdiction at a stage where rehabilitation is difficult, and such credit therefore leads to the evergreening of companies. Where companies are ordered to be liquidated, proceedings for liquidation need to be conducted in High Courts, and typically take several years, by which time the value of realizable collateral becomes paltry.[36] One option would be to invest BIFR with the authority to liquidate companies wherever it has recommended winding up.

Accounting Conventions

The accounts of banks for 1992–1993 will, for the first time, bring greater transparency to balance sheets. The transparency will be progressively enhanced over the following two years and relate to income recognition, provisioning, and capital adequacy.

Income Recognition

An amount due to a bank is classified as past due if it has not been paid within 30 days of the due date. An asset is regarded as nonperforming if interest on it is past due for a year as at the end of March 1993, for nine months as at the end of March 1994 and for six months as at the end of March 1995 and subsequent years. The new norms stipulate that banks cannot book any income on nonperforming assets.

Provisions for Bad Debts

All assets are to be classified into the following four groups: standard assets, which have hitherto not become nonperforming; substandard assets, which have been nonperforming for less than a year; doubtful assets, which have become nonperforming for over a year; and loss assets, where the prospects of recovery are very slender. No provisions need to be made for

---

[36] A recent media report observed that it takes decades to wind up companies and that three companies have been under liquidation in the Calcutta High Court for over 50 years.

standard assets, 10 per cent needs to be provisioned for substandard assets, 100 per cent of the security shortfall together with 20–50 per cent of secured default needs to be provisioned for doubtful assets (commencing at 20 per cent and rising by 10 per cent each year) and 100 per cent needs to be provisioned for loss assets. Investments of banks need to be classified as permanent and current investments, and only the latter needs to be provisioned for upon depreciation. During 1992–1993, not less than 30 per cent needed to be classified as current investments and this ratio is being progressively increased to 50 per cent. At least 30 per cent of total provisioning needed to be made in 1992–1993 and the rest the following year.

Capital Adequacy

Guided by the Basle Committee framework, a risk weighted assets ratio will constitute the measure for capital adequacy, with balance sheet assets, nonfunded items, and other off-balance sheet exposures being assigned prescribed risk weights. Balance sheet assets typically have a 100 per cent weight. If, however, they are backed by a Central or State Government guarantee or a guarantee of a Government credit guarantee institution, they acquire a zero weight. Banks need to reach specified unimpaired capital levels, which will be a percentage of aggregate assets weighted by their risk weights. Foreign banks operating in India were expected to reach an 8 per cent capital adequacy by March 1993. Indian banks with overseas branches are expected to attain this level by March 1994, while other Indian banks have time until March 1996. Further, capital is bifurcated into Tier I (equity and unrevalued reserves) and Tier II (revaluation reserves and other forms of shareholding, as well as subordinate debt). Tier II capital cannot exceed Tier I capital for purposes of capital adequacy computations.

The Government has also signaled its intention to introduce legislation to permit nationalized banks to access capital from the market, and it is likely that the stronger banks will be encouraged to do so. If disclosure of the balance 70 per cent is to be made, existing equity must be effectively written down by that amount, diluting starkly the book value of the share. The price at which new equity can be placed comes down sharply, affecting banks' ability to make the balance provisions in 1994–1995 without further capital injection from the Government.

For DFIs, similar transparent accounting conventions are expected to have been phased in over two years commencing 1993–1994. The adequacy of existing provisioning has varied considerably between the DFIs.

Regulation and Supervision

The RBI regulates and supervises the functioning of commercial and cooperative banks. The Government, as the owner of the nationalized banks, also issues periodic guidelines to these banks in matters of organizational structure, personnel management, internal bank vigilance, and handling of sensitive issues in industrial relations, besides laying down policy parameters for the flow of credit to selected sectors. Not surprisingly, a regulatory overlap does lead to fragmentation in control, particularly when the regulatory regime is in flux.

The regulatory stance would need to shift gradually from the parametric to the prudential. Parametric regulation fixes prices (viz. interest rates) and quantities (viz. sectoral credit allocations) and imposes microeconomic constraints (viz. not to share in term loans financing for projects above a certain size). Prudential regulation looks at aggregate risks to which the portfolios of banks are exposed and seeks to contain these risks. Illustratively, selective credit controls (being supply driven) are parametric, while capital adequacy requirements are prudential.

The RBI has recently announced its intention to constitute a Board for Financial Supervision. At present, supervision is being exercised primarily through bank inspections, which are of two kinds: a Financial Inspection of each bank every two to four years which involves visits to the head office, controlling offices and selected branches; and an Annual Financial Review, hitherto confined to public sector banks, which earlier relied on the banks' Management Information Systems, but has in the last two years also covered the inspection of large branches having deposits in excess of Rs 50 million.

The Government's intention appears to be to also bring the DFIs within the ambit of the Board for Financial Supervision, and it would be desirable that the manner of supervision (with the most appropriate combination of on-site and off-site supervision) be clearly defined for both banks and DFIs. There is certainly a case for DFI supervision being less intrusive and more prudential than for banks. Bank failures would generally lead to stronger "ripple effects" within the financial system than would DFI failures, with implications for deposit security and the payments system.

Elsewhere, the regulation and supervision of banks are among the more intractable issues confronting central bankers, particularly as the risks in owning and trading in several new financial products is often unclear. The debate in India, on the manner of bank and DFI supervision, is yet to begin.

Deposit Insurance and Credit Guarantees

Deposit insurance and credit guarantees raise some of the more difficult issues on the safety of and confidence in the banking system. Deposit insurance reassures depositors about the safety of their funds, and in the absence of effective regulation can lead to heavy payouts to depositors. Credit guarantees can also similarly distort bank incentives, encouraging the adverse selection of borrowers.

In India, both credit guarantee and deposit insurance schemes are operated by the Deposit Insurance and Credit Guarantee Corporation (DICGC), a subsidiary of RBI. The DICGC has operated six guarantee schemes for two decades, of which three were discontinued in 1992 because of negligible guaranteed advances. Two schemes for small borrowers and one for small-scale industrial borrowers remain. However, guarantee fee receipts have fallen short of claims for some years, leading to withdrawals from the deposit fund for settlement. Credit guarantees have thereby weakened the safety of deposits, and led DICGC to raise sharply the guarantee fee. DICGC-guaranteed loans are treated as zero risk assets in the accounting convention and, in that sense, guarantees become a substitute for additional capital. In the process, DICGC has become under-capitalized.

The deposit insurance fund has hitherto been self-financing, though the insurance cover is available only to deposit accounts in commercial and cooperative banks. In 1992–1993, DICGC paid out Rs 76.40 million towards settlement of claims in respect of one commercial and four co-operative banks, and made a provision of Rs 509.70 million for another commercial bank under liquidation on account of severe irregularities in recent securities transactions. The deposit fund's solvency is likely to be more severely tested in the years ahead.

ISSUES IN INDUSTRIAL FINANCE

This section examines four aspects of the market for industrial finance, and emphasizes certain microeconomic issues of relevance to credit institutions.

Market Signaling under Weak Contracts Enforcement

Cynical bankers in India relate two types of anecdotes about the way some entrepreneurs have been able to manipulate financial markets to their advantage. The first refers to the ability to use a promoted project to extract rather than to inject capital, by overcapitalizing the project, keeping the promoters contribution to the bare minimum, relying more strongly on debt as a means of financing, and overinvoicing project purchases. If the project fails, the promoters apply immediately to their creditors for various concessions invoking indicative guidelines of RBI which encourage the sanction of such concessions. The second anecdote relates to the ease with which personal guarantees of promoters can be obtained in the full knowledge that its enforcement will be obstructed in courts of law. It is often observed that whereas India has a large number of sick industries, it has very few sick industrialists. Both issues focus upon the pervasive influence of moral hazard in industrial finance under a weak contracts enforcement regime.

How should DFIs react to this? Historically, it is indeed correct that until the early 1990s projects in India were highly leveraged and, more dangerously, the extent of leveraging increased with project size. Further, the promoters equity could be funded through mutual funds, implying that the promoters often brought in little in the way of "owned funds." Mutual funds found it beneficial to contribute to promoters' equity because Government pricing of new public issues or of rights issues was often appreciably lower than the subsequent stock market price. The scenario has altered radically since 1992. The Government no longer prices equity offered by companies; SEBI has banned mutual fund contributions to promoters equity; and IDBI has tightened the structure of project financing. For new projects costing up to Rs 2 billion, debt/equity ratios are not to exceed 1.5, the promoter's contribution will be at least 25 per cent (and preferably 30 per cent), and the core promoter's contribution is to be 15 per cent, wherein promoters are forbidden from disposing of their equity without approval of the DFIs. For projects costing over Rs 2 billion, the norms are more liberal. Debt/equity ratios can go up to 2.0, promoter's contribution comes down to 20 per cent and the core promoter's contribution needs to be as low as 12.5 per cent.

The new norms are clearly an improvement but are they good enough in discriminating between promoters of projects when the enforcement of contracts is weak? The DFIs would do well to consider offering a menu of project finance conditions in a manner that enables project promoters to reveal their preferences about project risks.

The recent theoretical literature on signaling in credit markets assumes an efficient contracts enforcement system wherein collateral is the incentive device to separate the high risk from the low risk borrowers.[37] Where the legal system is weak, clearly collateral cannot perform a signaling role, and the incentive mechanism for borrower self-selection would need to be structured differently.

If the core promoters equity is endorsed in favor of the DFI with permission accorded for the DFI to sell such equity and appropriate the proceeds upon loan default, then the DFI could use the extent of the core contribution, the size of the loan sanctioned, and the interest rate charged to design a self selection mechanism. Where a borrower believes he/she is in a low-risk category, the borrower will provide higher core equity and take a small loan at a lower interest rate. A high risk borrower will minimize on core equity, push up the loan sought, and agree to pay a higher interest rate. Borrowers will accordingly signal their risk by accepting a project financial structure in the knowledge that they forfeit their core equity upon default.

---

[37] Bester (1985 and 1987) was the first to introduce collateral as a signaling device.

Existing project finance norms of DFIs favor standardized guidelines across projects of the same size to avoid allegations of discrimination. However, discrimination of borrower risk is the very essence of project selection and DFI response to asymmetric information must encourage the self-revelation of borrower risk through market signaling. This is particularly vital under a weak contracts enforcement regime.

Risk Diversification for Working Capital

The adequacy of working capital sanctioned to large companies is becoming an increasingly contentious issue between bankers and industry. Existing RBI guidelines favor the establishment of bank consortia for such companies, and discourage working capital financing outside of these consortia. Working capital limits are to be computed in terms of maximum permissible bank finance (MPBF) and are determined by norms specified for each industry, with banks conducting an annual review of each company's accounts to appraise performance and assess the next year's working capital needs. The norms are continually reviewed in consultation with industry by a "Committee of Direction" in RBI. Banks are permitted to provide temporary deviations from the norms and RBI consequently argues that existing arrangements combine discipline with flexibility.

Industry perceptions are often markedly different, and emphasize delays and rigidities in banks responding to the need for revising upwards an MPBF limit. Existing arrangements appear obsessive about ensuring that companies do not draw funds in excess of MPBF, but do not necessarily provide a solution when companies are underfinanced. Such underfinancing can occur because certain expenses incurred by a company cannot be included for the computation of MPBF or because there has been a sudden step up in business procured by the company. Equally, however, several companies have drawn funds in the past far in excess of reasonable working capital requirements by getting nonconsortia banks to discount "accommodation bills" or else to rediscount such bills which have earlier been discounted by financial services companies. Working capital financing of large companies by banks is therefore fraught with moral hazard.

There is inadequate recognition that spanning these issues is the more fundamental difficulty that banks face of being unable to successfully diversify their risks on working capital loans through markets for short-term corporate debt. Unless such markets deepen in India, bankers' risk-aversion will continue to manifest itself in terms of inflexible and unyielding lending procedures. Existing lending modalities are therefore characterizeable as an institutional response towards risk.

Risk in providing working capital to large companies has three facets: risk of size; risk of high capital loss as some companies move into a position of distress and then into sickness; and systemic risk associated with a down-turn in the business cycle or recession in the economy. With banks providing working capital whose asset value is not market determined, the risk of being locked into a high capital loss situation is appreciable and nondiversifiable.

The desirability of developing a more widespread primary market for short-term corporate debt was discussed earlier. Until such markets are established banks may find it advantageous to move rapidly to sanctioning working capital to large companies predominantly in the form of commercial paper or other securitizable debt, and existing companies with cash credit facilities should be subject to such a phased programmed of securitization. Banks would benefit by creating, in tandem, secondary markets for such corporate debt. Money market mutual funds would generate a demand for such debt, and RBI's speedy approval of applications to set up such funds would therefore be beneficial.

Consortia arrangements for companies would not then be subject to an RBI straitjacketing, but would constitute voluntary and flexible arrangements between participating banks and companies. Rollovers of facilities offered by a consortium would be repriced with reference to the company's financial health. The MPBF concept would thereby gradually give way to a bankers' assessment based on indicators of company viability and solvency. As capital markets deepen, merchant bankers will coordinate the working capital requirements of large companies.

The RBI's guidelines on MPBF amount to parametric regulation, as defined earlier in this chapter. A risk diversification approach to working capital is akin to prudential regulation. A change in RBI's regulatory stance to working capital financing could benefit banks and well-managed large companies.

Financing Sick Companies: Mortality or Morbidity

India has a long tradition of keeping alive very sick industrial companies through a combination of barriers to exit, government taking over the management or nationalizing such companies, and through various fiscal and credit reliefs being provided. Existing bankruptcy law, discussed earlier, further reinforces this.

A company is declared sick and comes before BIFR once its net worth is fully eroded. Recent econometric studies of BIFR-sanctioned packages reveal that it is possible to predict such impending sickness at least two years prior to its occurrence (Anant et al. 1992a, 1992b). The ability to turn around a sick company at this late stage must be doubtful and yet about two thirds of BIFR cases end in the sanction of rehabilitation packages. The reliefs involve the partial write off of loans, rescheduling of principal repayments, the capitalization of interest and fresh finance (generally for modernization) being brought in by DFIs and banks.

A study of BIFR packages indicates that they envisage high debt/equity and total liability/equity ratios which are typically higher than the average for non-BIFR firms, implying a worse financial structure with weaker insurance against years of poor business activity. Further, as fresh funds are brought in at lower than market interest rates, DFIs and banks forfeit the higher interest income they could otherwise have earned. Why, then, have they been doing so?

One explanation is to be found in the joint impact of an inefficient legal system and translucent accounting norms in structuring management incentives. Prolonging the morbidity of companies without a future, even injecting further funds into them, has hitherto not necessitated large provisioning by DFIs and banks. The provisions would have been higher if BIFR had ordered liquidation, without the prospect of the DFI or bank making appreciable recoveries, given the several years that liquidation typically takes. By prolonging the morbidity of companies, banks have retained a balance sheet advantage. With enhanced balance sheet transparency, however, this advantage will disappear, and while the credit institutions may continue to be agreeable to restructuring past liabilities, they will be cautious about infusing fresh finance.

The financing of sick companies indicates that DFI and bank motivation is decisively shaped by the efficacy of the prevailing legal and accounting systems. Short-term management horizons induce DFIs and banks to prolong the life of a sick company, although this is injurious to the credit institutions over a longer horizon. Incentives in the long term will be consistent with those in the short term only if contracts enforcement is speedy and accounting conventions are tight.

Financing Small-Scale Industries

Small industrial undertakings cover a diverse range from enterprises using modern technology with high capital productivity, to artisans and very tiny village and cottage industries, whose products are generally sold locally.[38] For enterprises based on modern technology, credit appraisal and financing decisions by bankers are not dissimilar to those which need to be taken for larger industries. Information asymmetries become more acute, however, when low technology based enterprises, or those in the village and cottage sector, have to be appraised for credit sanction.

---

[38] Investment in plant and machinery should not exceed Rs 6 million (or Rs 7.5 million for ancillary and export-oriented units) for the enterprise to qualify as small. If the investment does not exceed Rs 500,000, the enterprise is labeled as tiny.

Whereas informal creditors would incur search costs, banks and DFIs have financed on the basis of borrower eligibility (rather than creditworthiness) at concessional interest rates. Such loans risk higher default, and provide lower returns to the bank even when repayments are regular. With the logic of borrower selection being so adverse to the bank's interests, the bank begins cutting down on its search transaction costs. For several rural development programs this has been institutionalized through state government functionaries selecting borrowers and asking banks to finance them.[39]

The rigorous appraisal of small industrial projects also gets discouraged on account of the availability of refinance. Although the portfolio quality of SFCs ranges generally from the indifferent to the poor, their defaults to the refinancing agency (earlier IDBI and now SIDBI) have been kept low through the refinancier evergreening the SFCs. Increasingly, therefore, fresh refinance from SIDBI is being used to enable some SFCs to repay their dues to SIDBI. This delays SIDBI's own behavioral response to the deterioration in the portfolio quality of small industry lending, retards a recognition of the crisis of SFCs, and encourages the continuance of earlier techniques of small industry project appraisal.

The belief that small projects should have a higher proportion of debt has also financially debilitated small industry. Until 1991, debt/equity ratios were as high as 4:1, and promoters could also draw up to 15 per cent of the project cost as subordinated loans in lieu of equity (in the form of "seed money"). Although the debt/equity ratio has been lowered to 2:1 recently, the damage caused to small industries financing through liberal debt for several decades is visible in the large proportion of unsuccessful enterprises. The structure of project finance has contributed to the financial fragility of small industry.

These issues must be addressed and correctives incorporated, before a judgment can be formed on whether small industry financing is intrinsically more risky then the financing of larger enterprises. As Indian industry further matures, the small industrial base will necessarily grow rapidly, as has happened in several East Asian countries, making this sector a profitable one to finance.

CONCLUDING OBSERVATIONS

The reform of the financial sector is a process and not an event, necessitating a restoration of the competitive efficiency of existing banking and financial

---

[39] This is so, for instance, in the Integrated Rural Development Program (IRDP) with banks either sharing borrower selection responsibility with the State Governments, or sometimes delegating this function to the State Governments. The adverse selection of borrowers is the natural outcome. The IRDP recovery rates are less that 40 per cent.

institutions as well as the emergence of new institutions characterized by innovative management styles. One concept (Tobin 1984) of the efficiency of a financial system comprises the following:

— Information arbitrage efficiency, which is inversely related to the gain available through the use of common and public information;

— Fundamental valuation efficiency, which is reflected by the extent to which the discounted present value of benefits from a financial asset are reflected in its price;

— Full insurance efficiency, which depends on the extent of hedging possible against future contingencies; and

— Functional efficiency, which is inversely proportional to the combined transactions costs of borrowers and lenders.

Policy initiatives which encourage the emergence of new private sector banks and financial institutions and create incentives for stronger management efficiency in existing institutions through their privatization would probably achieve gains in information arbitrage efficiency and functional efficiency. The other two measures of efficiency would show improvements only if appropriate financial markets for the proper valuation and risk bearing of financial assets deepen. The creation of more genuine money markets as well as a broadening of capital markets to incorporate the flotation of longer-term debt securities would therefore be crucial to realizing efficiency gains in the financial sector in the composite sense identified by Tobin. The precise style of privatization adopted by the Government of India, choosing between weak, semi-strong, quasi-strong, and strong denationalization, as discussed in an earlier section of this chapter, will also prove decisive to the pace and vigor with which information arbitrage and functional efficiencies can be restored within the financial system.

In sequencing the reform it may be helpful to borrow an investigative framework from physics by distinguishing between boundary conditions, initial conditions, and the dynamics of bank restructuring. Boundary conditions, which are outside the control of banks and DFIs, include the legal system (both for efficient contracts enforcement and for bankruptcy and liquidation), fiscal restraint, monetary and exchange rate stability, effective supervision, transparent accounting conventions, and well-regulated capital and money markets. Those boundary conditions which are not conducive to financial efficiency at the commencement of reform need to be strengthened very quickly if reform is to succeed. The reform of the legal system, in particular, though decisive to the success of reform, has hitherto not been seriously addressed. Initial conditions determine the path taken by reform and the CAMELOT indicators (discussed earlier) at the start of reform will determine the "trajectory of reform" in its impact on individual banks and DFIs. Finally, the dynamics of individual bank and DFI restructuring involves measures taken by their managements to show improvements in these CAMELOT indicators. Such an investigative framework helps to bring out the complex interrelationships in the sequencing of reform.

Acknowledgments

I should like to thank Ashok Khanna, Larry White, and, especially, Rahul Khullar for their helpful discussions on an earlier draft of this chapter.

References

Anent, T.C.A., S.Gangopadhyay and O.Goswami. 1992a. "Industrial Sickness in India: Initial Findings, Report 1". *Studies in Industrial Development, Paper No. 2*. Ministry of Industry, Government of India.

──────. 1992b. "Industrial Sickness in India: Characteristics, Determinants and History, 1970–1990, Report 2." *Studies in Industrial Development, Paper No. 6*. Ministry of Industry, Government of India.

Azariades, C. 1975. "Implicit Contracts and Underemployment Equilibria." *Journal of Political Economy*, 83, 1183–202.

Bery, S.K. 1992. "India: Reform of the Financial Sector." Paper presented at the FICCI Conference on "Banking and Finance: Monetary and Fiscal Policies." Bombay.

Bester, H. 1985. "Screening vs Rationing in Credit Markets with Imperfect Information." *American Economic Review*, 75, 850–5.

──────. 1987. "The Role of Collateral In Credit Markets with Imperfect Information." *European Economic Review*, 31, 887–99.

Bhatt, V.V., and A.R. Roe. 1979. "Capital Market Imperfections and Economic Development." World Bank Staff Working Paper 338.

Blinder, A., and J.E. Stiglitz. 1983. "Money, Credit Constraints and Economic Activity." *American Economic Review*, 73, 297–302.

Bouman, F.J.A. 1989. *Small, Short and Unsecured: Informal Rural Finance in India*. Oxford University Press.

Cooper, R., and T.W. Ross. 1985. "Product Warranties and Double Moral Hazard." *Rand Journal of Economics*, 16, 103–13.

Das-Gupta, A., C.P.S. Nayar and Associates. 1989. "Urban Informal Credit Markets in India." Study prepared for the Asian Development Bank. New Delhi.

Das-Gupta, A. 1989 ."Reports on the Informal Credit Markets in India: Summary." Report prepared for the Asian Development Bank. New Delhi.

Diaz-Alejandro, C. 1985. "Good-Bye Financial Repression, Hello Financial Crash." *Journal of Development Economics, 19*, 1–24.

De Juan, A. 1991. "From Good Bankers to Bad Bankers: Ineffective Supervision and Management Deterioration as Major Elements in Banking Crises." World Bank.

Fershtman, C. 1987. "Alternative Approaches to Dynamic Games," in Bryant R.C. and R. Portes, eds. *Global Macroeconomics: Policy Conflict and Cooperation.* IEA Conference Proceedings.

Friedman, J. 1980. *Oligopoly and the Theory of Games.* North-Holland, Amsterdam.

Ghate, P. 1992. *Informal Finance: Some Findings from Asia.* Hong Kong: Oxford University Press for Asian Development Bank.

Glosten, L.R., and P.R. Milgrom. 1985. "Bid, Ask and Transaction Prices in a Specialist Market with Heterogeneously Informed Traders." *Journal of Financial Economics, 14*, 71–100.

Goodhart, C.A.E. 1989. *Money, Information and Uncertainty (2nd Edition).* London: Macmillan.

Government of India. 1991. "Report of the Committee on the Financial System." New Delhi.

_____. 1992. "Economic Survey, 1992–93." New Delhi.

Grossman, S., and J.E. Stiglitz. 1980. "On the Impossibility of Informationally Efficient Markets." *American Economic Review, 70*, 393–408.

International Monetary Fund. 1992. "India: Development of the Government Securities Market." Mimeo.

Jain, A.K., L. Bhandari, N. Dholokia, R. Khurana, and M.N. Vora. 1982."Distribution of Mill Made Cotton Textiles in India: Summary and Conclusions." Ahmedabad: Indian Institute of Management.

Little, I.M.D., D. Mazumdar, and J.M. Page. 1987. *Small Manufacturing Enterprises: A Comparative Analysis of India and Other Economies.* Oxford University Press.

McKinnon, R.I. 1973. *Money and Capital in Economic Development.* Washington, DC: Brookings Institution.

_____. 1989. "Financial Liberalization and Economic Development: A Reassessment of Interest-Rate Policies in Asia and Latin America." *Oxford Review of Economic Policy.* 5, 29–54.

Nayar, C.P.S. 1982. "Financial Corporations: A Study of Unregulated Banks." Madras: Institute for Financial Management and Research.

_____. 1984. "A Study of Non-Banking Financial Intermediaries." Madras: Institute for Financial Management and Research.

Panikar, P.G.K., C. Sen, and D. Narayana. 1988. "The Informal Credit Markets in Rural India." Study prepared for the Asian Development Bank. Trivandrum.

Phlips, L. 1988. *The Economics of Imperfect Information.* Cambridge: Cambridge University press.

Ramani, S. 1986. "A Case Study of Nidhis in the City of Madras." Unpublished Masters Thesis. University of Madras, Madras.

Rothschild, M. 1974. "Searching for the Lowest Price When the Distribution of Prices is Unknown." *Journal of Political Economy,* 82, 689–711.

Reserve Bank of India. 1987. "Report of the Working Group on the Money Market." Bombay.

_____. 1988. "All India Debt and Investment Survey, 1981–82." Bombay.

_____. 1991. "Quick Estimates of National Income 1989–90." *RBI Bulletin.* March.

_____. 1992a. "Flow of Funds Accounts of the Indian Economy: 1980–81 to 1985–86." *RBI Bulletin.* January.

_____. 1992b. "Flow of Funds Accounts of the Indian Economy: 1986–87 to 1989–90." *RBI Bulletin.* January.

_____. 1992c. "Quick Estimates of National Income 1990–91 - A Review." *RBI Bulletin.* July.

_____. 1992d. "Report of the Working Group on Financial Companies." Bombay.

_____. 1992e. "Performance of Financial and Investment Companies, 1988–89." RBI Bulletin. May. 849–73.

_____. 1992f. "Report on Trend and Progress of Banking in India (July-June)." Bombay.

_____. 1993. Banking Statistics: Basic Statistical Returns. Vol 20.

Shaw, E.S. 1973. *Financial Deepening in Economic Development.* New York: Oxford University Press.

Sheng, A. 1992. "Bank Restructuring: Techniques and Experience." Mimeo. World Bank, Washington, DC.

Spence, A.M. 1974. *Market Signaling: Information Transfer in Hiring and Related Processes.* Cambridge, Mass.

Stiglitz, J.E., and A. Weiss. 1981. "Credit Rationing in Markets with Imperfect Information." *American Economic Review,* 71, 393–410.

Stiglitz, J.E. 1985. "Credit Markets and the Control of Capital." *Journal of Money, Banking and Credit,* 17, 133–52.

_____. 1987. "The Causes and Consequences of the Dependence of Quality on Price." *Journal of Economic Literature,* 25, 1–48.

_____. 1989. "Financial Markets and Development." *Oxford Review of Economic Policy,* 5, 55–68.

_____. 1993. "The Role of the State in Financial Markets." Paper prepared for the World Bank's Annual Conference on Development Economics. World Bank. Washington, DC.

Taylor, L. 1983. *Structuralist Macroeconomics: Applicable Models for the Third World.* New York: Basic Books.

Thingalaya, N.K. 1978. "Innovations in Banking: The Syndicate's Experience." World Bank Development Economics Department Domestic Finance Study 46.

Timberg, T., and C.V. Aiyar. 1980. "Informal Credit Markets in India." *Economic and Political Weekly*. Annual Number.

Tobin, J. 1965. "Money and Economic Growth." *Econometrica*, 33, 671–84.

———. 1984. "On the Efficiency of the Financial System." *Lloyds Bank Review*, 153, 1–15.

Varma, J.R. 1992. "Commercial Banking: New Vistas, New Priorities." Mimeo. Ahmedabad: Indian Institute of Management.

Villanueva, D., and A. Mirakhor. 1990. "Strategies for Financial Reforms: Interest Rate Policies, Stabilization and Bank Supervision in Developing Countries." *IMF Staff Papers*, 37, 509–534.

World Bank. 1990. "India: Financial Sector Report. Consolidation of the Financial System." Washington, DC.

CHAPTER SEVEN

DEVELOPMENT OF THE FINANCIAL SECTOR IN PAKISTAN

NADEEM UL HAQUE AND SHAHID KARDAR

INTRODUCTION

In recent years, as external credit flows to developing countries have declined, the need for an efficient and viable financial sector has grown. An efficient financial system collects domestic savings and allocates the collected resources to the best possible investment opportunities, allowing for better domestic resource mobilization and utilization, and reducing the reliance on external financing. This chapter acknowledges this role of the financial sector in Pakistan and attempts to understand the nature of the country's financial markets, to determine factors that inhibit financial intermediation, and to identify policy measures that might be taken to facilitate the development of the financial markets in the country.[1]

Throughout the 1980s, Pakistan has been able to achieve strong growth with low inflation while meeting much of its external payment obligations even though recorded domestic investment and saving rates remained low and fiscal deficits were quite large. Financial intermediation has been geared toward the financing needs of the Government and those of directed credit. During this period, the degree of integration of the Pakistani economy and especially the financial markets with the rest of the world has increased sharply as out-migration of workers from Pakistan created large remittance inflows and developed a well-functioning informal foreign exchange market. The economy has been opened and *de facto* convertibility almost achieved. The financial sector was unable to respond to these developments, mainly because the Government owned many of the institutions and retained a tight hold on entry to financial markets. Thus, financial intermediation took place mainly in the informal sector and outside the official financial system.

Pakistan has reached an interesting point in its history and development. Politically, the country has experimented with differing forms of government and is now committed to democracy. Economically, the free

---

[1] In recent years, only one comprehensive study has been conducted on the financial sector (World Bank 1990). Two additional studies on specific segments of the market have been done: one on housing finance by Australia-New Zealand Grindlays Bank (1992) and the other on government debt auctioning system.

market economic approach to development is now accepted by all factions, and economic reform and development are now important elements of the political debate. In line with these developments, economic liberalization was initiated in 1990 and is currently being carried forward. In 1991, financial reform in the country was initiated with a strong commitment to market-oriented development of the financial sector. This reform is still underway and will need to be strengthened substantially if modern, competitive financial markets are to be developed.

FINANCIAL SECTOR: SIZE, DEPTH, AND SOURCES OF FUNDS

In the last few years—during which financial sector liberalization has taken place—financial assets as a percentage of GNP have increased from 85.3 per cent in FY1989/90 to 90.6 per cent in FY1991/92[2] (Table 7.1). The largest element in private sector financial wealth is the government debt instrument, which constitutes slightly less than half of financial wealth. Over three years, the share of government debt instruments has decreased by about 3 percentage points—from 49.7 per cent to 46.9 per cent of total financial assets.[3]

During the period FY1989/90 to FY1991/92, currency in circulation as a percentage of total assets declined while demand deposits rose; time deposits have remained more or less unchanged. The degree of monetization does not appear to be increasing rapidly in Pakistan. Both M1 and M2 as a percentage of GDP appear to have evolved on a flat trend over a 10-year period (Figure 7.1). For the limited data available for M3, the same trend is evident.

Figure 7.2 presents comparative evidence on the degree of monetization. As shown, between 1950 and 1992 Pakistan exhibited the same behavior as other South Asian countries. As growth accelerates, the ratio of M2 to GDP appears to increase at a fast pace.

Sources and Uses of Funds

Savings originate mainly in the noncorporate and household sector, which has supplied between 50 and 80 per cent of all savings during the five-year period, FY1985/86 to FY1989/90 (Table 7.2). Savings in public

---

[2] Fiscal year (FY) in Pakistan ends 30 June.

[3] However, if instruments issued by public sector corporations such as the Water and Power Development Authority (WAPDA) and Civil Aviation Authority (CAA) are included under government debt, the decline in the share of government debt instruments is only marginal. There is essentially a switching from one instrument to another, both guaranteed by the Government.

Table 7.1 Financial Assets in Pakistan

	FY1989/90		FY1990/91		FY1991/92	
	Rs million	Per cent share	Rs million	Per cent share	Rs million	Per cent share
Currency in circulation	115,067	15.11	136,967	14.86	151,819	13.71
Demand deposits	119,704	15.72	144,845	15.72	179,361	16.19
Time deposits	80,241	10.54	89,223	9.68	116,510	10.52
Other deposits	2,209	0.29	3,113	0.34	5,322	0.48
Government debt instruments (excluding DBC and FC repayment bonds)	378,345	49.68	444,514	48.23	519,621	46.91
Paid-up capital	25,100	3.30	30,403	3.30	50,534	4.56
Life Insurance Fund	16,193	2.13	19,382	2.10	25,383	2.29
Deposits with scheduled DFIs	24,740	3.25	37,596	4.08	42,045	3.80
WAPDA bonds	–	–	15,577	1.69	17,008	1.54
Total	761,599	100	921,620	100	1,107,603	100
Growth in financial assets				21.01		20.18
Per cent share of total financial assets in GNP		85.3		88.2		90.6

DBC: Dollar Bearer Certificate.
FC: Foreign Currency
WAPDA: Water and Power Development Authority.
Source: State Bank of Pakistan.

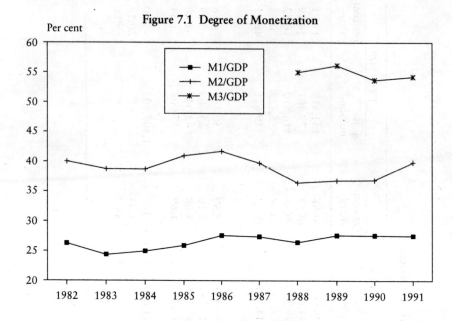

Figure 7.1 Degree of Monetization

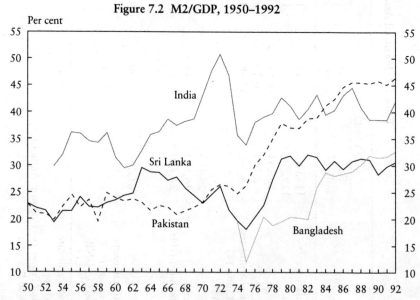

Figure 7.2 M2/GDP, 1950–1992

sector enterprises, though erratic, are sizeable and rank next only to those from the noncorporate and household sector. The share of savings in the private corporate sector, though much smaller than public sector enterprise savings in total savings, has been on the increase, going from

Table 7.2 Flow of Funds

	Amount (Rs million)	Federal Government	Provincial and Local Government	Public Sector Enterprise	Other Public Inst. (per cent)	SBP	Scheduled and Coop. banks	NBFIs	Insurance	Private Corporate business	Non-corporate and Household	Rest of the World	Total
FY1985/86													
Sources of Savings	96,545	−0.10	5.56	19.08	1.86	0.65	3.00	0.73	0.23	2.28	62.76	12.96	100
Investment	96,545	0.30	9.42	33.38	0.67	0.08	0.50	0.15	0.15	10.83	38.43	–	100
FY1986/87													
Sources of Savings	109,540	−16.29	3.08	8.10	1.69	0.55	1.34	0.63	0.20	6.50	88.94	5.17	100
Investment	109,540	0.34	13.37	25.74	0.67	0.08	0.48	0.13	0.11	8.57	44.53	–	100
FY1987/88													
Sources of Savings	121,666	−17.34	11.58	19.45	1.77	0.19	0.95	0.81	0.32	6.32	57.12	16.83	100
Investment	121,666	6.82	16.25	31.98	0.60	0.05	0.53	−0.01	0.10	22.86	20.82	–	100
FY1988/89													
Sources of Savings	145,570	−17.33	8.85	14.24	1.62	−0.15	1.28	0.41	0.24	7.35	65.92	17.67	100
Investment	145,570	6.03	12.14	24.76	0.59	0.04	0.34	0.08	0.09	22.05	33.28	–	100
FY1989/90													
Sources of Savings	162,076	−2.86	1.75	19.73	1.78	0.12	1.86	0.58	0.17	10.87	48.05	17.95	100
Investment	162,076	6.37	14.54	29.93	0.35	0.05	0.39	0.13	0.08	36.79	11.39	–	100

– not applicable.
Source: Flow of Funds Account.

nearly 2.3 per cent in FY1985/86 to nearly 10.9 per cent in FY1989/90. Savings from the rest of the world are understandably large, rising consistently as a percentage of total savings, except in FY1986/87. Owing to the sizeable budget deficit, the Federal Government's savings are consistently negative. Provincial Governments' savings are positive largely because of Federal Government grants.

The noncorporate and household sector, with somewhat erratic trends, and public sector enterprises are the principal users of savings. Together they absorbed 71.8 per cent of the resources in FY1985/86 and 70.3 per cent in FY1986/87. However, over the years their share has been declining. Correspondingly, the share of private corporate sector has increased, from 10.8 per cent in the total investment in FY1985–86 to 36.8 per cent in FY1989/90. The Federal and Provincial Governments are large users of resources, varying between 15.8 per cent and nearly 23.1 per cent.

FINANCIAL INSTITUTIONS: TYPE AND PERFORMANCE

The financial sector in Pakistan has evolved over the years in response to the Government's plans for the growth and development of the country. The sector comprises the State Bank of Pakistan, four state-owned banks, two newly privatized banks, about 10 newly licensed commercial banks, two provincial banks, eight development financial institutions, about 30 leasing companies, 45 modarabas, 10 investment banks, three stock exchanges, two large publicly floated mutual funds, 58 insurance companies, and some nascent discount houses, housing finance companies, and government bonds.

In an effort to make the financial sector subservient to the dictates of economic planning, the Government has stifled growth and innovation, eliminated price, product, and institutional competition, reduced policing and supervision, and increased its ownership of the institutions within the sector. This policy has led to increased losses in the sector which have had to be borne by the taxpayer and the depositor.

The Government is the dominant agent in the financial markets, both as a borrower and a provider of financial services (Table 7.3). Government-controlled institutions such as the nationalized commercial banks and development financial institutions (DFIs) account for over 80 per cent of all loans and a little less than half of the sum of all deposits and government borrowing. In addition, government bonds, which are used primarily for financing the deficit of the Government, constitute more than 40 per cent of financial wealth. In the securities market, the largest players are the two public sector investment promotion agencies that are operating large mutual funds.

The financial institutions in Pakistan can be classified into two broad categories—banking companies and nonbank financial institutions—with both being controlled primarily by the central bank, the State Bank of Pakistan. They are also regulated, depending upon the nature of the institution, by the Corporate Law Authority, the Ministry of Finance, and the Religious Board. A brief description of these institutions is presented below:

Table 7.3 Financial Sector Overview, 1991

	Numbers	Assets (Rs billion)	Per cent of total	Deposits (Rs billion)	Per cent of total
Commercial banks	43	525.3	67.0	443.5	42.7
Nationalized commercial banks		439.2	56.1	378.0	36.4
Privatized	2	73.6	9.4	60.1	5.8
State-owned	4	366.5	46.8	318.7	30.7
Foreign banks	21	81.4	10.4	64.0	6.2
New banks	10	...	...	...	...
Specialized banks	4	...	...	...	...
Provincial banks	2	4.7	0.6	1.5	0.1
Nonbank financial institutions	128	121.1	15.5	38.3	3.7
DFIs	10	102.8	13.1	26.9	2.6
Industrial	7	60.4	7.7	24.7	2.4
Other	3	42.4	5.4	2.2	0.2
Leasing companies	50	6.3	0.8	1.3	0.1
Modarabas	45	6.7	0.9[b]	6.7	0.6
Investment banks	11	5.2	0.7	3.4	0.3
Housing finance companies	2	0.1	0.0	...	...
Securities markets	3	114.7	14.6	114.7	11.1
Publicly-owned mutual funds	2	17.7	2.3[b]	17.7	1.7
Investment Corporation of Pakistan	1	2.5	0.3[b]	2.5	0.2
National Investment Trust	1	15.2	1.9[b]	15.2	1.5
Insurance companies	58	22.4	2.9	...	0.0
Life	4	19.2	2.5	...	0.0
General	54	3.2	0.4	...	0.0
Domestic	44	2.6	0.3	...	0.0
Foreign	10	0.7	0.1	...	0.0
Government bonds		...	...	441.1	42.5
Domestic	28[a]	...	...	431.2	41.6
Foreign	6[a]	...	...	9.9	1.0
Total	266	783.48	100	1,037.6	100

[a] Number of instruments.
[b] Mutual fund deposits are being taken as equivalent to asset holdings.
... data not available.

Commercial Banks

Commercial banks were nationalized in 1974 and are now in the process of being privatized. The private sector has been allowed entry into commercial banking since 1992.[4] Two nationalized commercial banks have been privatized since 1990: Muslim Commercial Bank was sold by auction/negotiation, while Allied Bank was sold to its employees. The market share of the nationalized commercial banks has been declining over the last five years; this rate of decrease in market share is expected to accelerate with the induction of the new private banks.

The three nationalized commercial banks in Pakistan—Habib Bank Limited, United Bank Limited, and National Bank of Pakistan—have a large branch network that facilitates deposit mobilization (Table 7.4). The newly privatized banks have acquired the branch network that will allow them to expand and compete with the state-owned banks. While these institutions play an important role in financing short-term credit requirements, their success in raising deposits ensures that they have a significant surplus of funds that can be lent or invested in government securities.

The foreign banks in Pakistan are branch operations; they are not separate legal entities. Much of their success can be attributed to their superior management skills and better access to international financial markets. A large fraction of foreign currency deposits (reportedly as much as 80 per cent) are with foreign banks, partly because of their marketing efforts and partly because of their credibility as international banks. Foreign banks are only allowed four branches within the country.

Performance of Commercial Banks

As can be expected, the banking sector has operated inefficiently because of nationalization, though with the recent entry of private banks in the country and the subsequent competition, this is changing. Foreign banks have shown the highest growth rate in terms of deposits in the last three years. State-owned banks averaged an annual growth of 12.7 per cent over the three-year period 1989–1992, while the recently privatized nationalized commercial banks averaged 18.2 per cent and foreign banks, 31.4 per cent (Tables 7.5 and 7.6). Foreign banks are also more efficient in terms of controlling administrative costs. For example, growth of administrative costs averaged 19.7 per cent for state-owned banks, 18.3 per cent for the recently privatized banks, and 17.6 per cent for foreign banks. As a percentage of assets, administrative costs are the lowest for foreign banks.

---

[4] To induce competition, several financially sound groups were allowed to set up new commercial banks; ten such institutions have been established. Each of these new banks began with an identical initial business strategy, i.e., Rs 300 million in capital distributed widely through the stock market and the establishment of five branches within 18 months. Unfortunately, one of these banks has already shown signs of trouble.

Table 7.4 Size of Commercial Banks, 1991

Institution	Assets		Deposits		Branches	
	Rs million	Per cent	Rs million	Per cent	Rs million	Per cent
Nationalized commercial banks	439.9	83.6	378.8	85.4	7,023.0	97.3
Privatized	73.6	14.0	60.1	13.5	2,035.0	28.2
State-owned	366.3	69.7	318.7	71.8	4,988.0	69.1
Foreign banks	81.3	15.5	64.0	14.4	67.0	0.9
New banks[a]	...	...	...	...	...	...
Provincial banks	4.7	0.9	0.9	0.2	125.0	1.7
Total	525.9	100.0	443.7	100.0	7,215.0	100.0

[a]Financial data not yet available.
... data not available.
Source: State Bank of Pakistan.

Table 7.5 Profitability and Costs of Administration

	Total administration cost as per cent of total assets[a] (3 yr avg)[b]	Pre-tax profit as per cent of deposits (3 yr avg)	Average annual growth of deposits over 3 years	Average annual growth of administration costs over 3 years
Commercial banks[c]	1.94	1.13	15.49	19.19
Nationalized commercial banks	2.11	0.57	13.53	19.39
Privatized	2.43	0.57	18.24	18.31
State-owned	2.06	0.57	12.74	19.67
Foreign	0.85	5.03	31.43	17.65
New private[d]	0.62	2.47	...	...
Provincial[d e]	0.84	5.86	58.61	67.50
Nonbank Financial Institutions				
DFIs	1.23	3.87	0.36[f]	26.61[f]
Investment banks	1.02	2.39	−4.27[f]	105.48[f]

[a]End-June basis. Data for RDFC not included due to nonavailability of Annual Reports while the data for ICP and NIT have been shown under the heading of Mutual Funds in another table.
[b]Average of three years ended December 1991.
[c]In case of commercial banks, administration cost includes salaries, allowances and other benefits.
[d]The data pertain to the year 1992 only.
[e]Bank of Punjab only.
[f]Average annual growth for two years.
... data not available.
Notes: Data regarding deposit mobilization by Saydi-Pak, Pak-Kuwait and Pak-Libya are not available. Data in respect of investment banks have been compiled from the operating balance sheets of these institutions.

The gross revenues of foreign banks have grown at a much faster pace than the rate of growth of their costs. For Pakistani banks, the picture is much more bleak (Table 7.7). For Pakistani banks, the difference in the growth rates is extremely narrow, if not negative. This conclusion even holds for FY1990/91, when post-financial reforms profits of banks rose significantly.

Bank deposits have grown significantly in recent years primarily because of the rapid growth of money supply which increased by around 18 per cent between 1990 and 1991 and 20.6 per cent in FY1991/92. Money has also flowed into the banking system from the informal economy with the change in incentives following the recent slide of the stock market and the significant slump in the real estate market.

Table 7.6 Comparison of Privatized and State-Owned NCBs
(per cent)

	1990	1991
Spread		
Privatized banks	8.5	4.1
State-owned nationalized commercial banks	6.9	4.7
Deposit Growth (1985–1990)		
Privatized banks	4.0	7.2
State-owned nationalized commercial banks	2.1	0.5
Administrative expense/Total assets		
Privatized banks	4.1	3.9
State-owned nationalized commercial banks	3.1	3.3
Growth in administrative expense (1985–1990)		
Privatized banks	16.8	14.2
State-owned nationalized commercial banks	2.0	25.6

Table 7.7 Growth Rate of Gross Revenues and Total Costs of Banks
(per cent)

	FY1986/87		FY1988/89		FY1990/91	
	Gross Revenue	Total Costs	Gross Revenue	Total Costs	Gross Revenue	Total Costs
Foreign:						
Grindlays	35.4	24.3	26.8	18.7	63.3	20.1
Standard Chartered	49.0	41.9	52.5	25.0	49.6	26.7
Bank of America	23.8	6.9	18.4	43.5	57.8	19.7
American Express	49.6	15.8	20.6	24.3	46.0	48.5
Citibank	76.2	18.4	(13.8)	33.1	121.1	88.9
Pakistani:						
National	23.9	26.2	7.7	7.4	42.7	28.9
Habib	14.8	20.4	9.7	9.6	22.9	29.0
United	16.9	19.0	10.4	11.0	12.9	15.2
Muslim Commercial	28.3	26.3	10.2	10.1	16.5	14.4
Allied	21.6	21.6	12.6	13.1	17.1	13.8

Source: Estimates from financial statements of banks.

Demand deposits constitute about 47–50 per cent of total bank deposits which is much higher than other countries (Table 7.8).[5] Around 57 per cent of deposits with scheduled banks are with the nationalized commercial banks with their large network of branches. Their share has declined from 65 per cent in 1991, partly because of establishment of new banks and partly because of the concerted efforts made by private and foreign commercial banks to mobilize deposits, particularly in the form of foreign currency accounts. As a result, foreign currency deposits increased by $1.4 billion between May 1991 and December 1992, because of the Government's exchange reforms and the diversion of home remittances into foreign currency accounts. Today foreign currency deposits are estimated to be $2.7 billion.

Table 7.8 Performance of Commercial Banks' Scheduled Deposits
(Rs billion)

	30 June 1991	5 April 1993		
		Demand	Time	Total
Commercial banks:				
Nationalized commercial banks	194.0	131.0	119.5	251.1
New Pakistan private commercial banks[a]	44.5	53.4	47.9	101.3
Foreign banks	59.8	21.0	62.7	83.7
Specialized banks	5.9	1.4	4.3	6.2
Total	304.2	207.4	423.9	442.3

[a] Including the two recently privatized, nationalized commercial banks.
Source: Banking Regulation Department, State Bank of Pakistan.

Profitability and Productivity of Commercial Banks

In nominal terms, the spreads (Table 7.9) have risen steadily from 5 per cent in FY1986/87 to 6.2 per cent in December 1992. In real terms, the weighted average yield on deposits fell from 3.9 per cent in FY1986/87 to −4.6 per cent in FY1990/91, and −1.1 per cent in December 1992. The weighted average yield on loans also declined; after declining from 8.9 per cent in FY1986/87, it went to 1.1 per cent in FY1990/91. Thereafter it climbed rapidly to 5.14 per cent by December 1992.

---

[5] Perhaps this explains why the costs of funds have been low for commercial banks. See Dattels (1992).

Table 7.9 Weighted Average Interest Rates on Deposits and Advances
(per cent)

	Weighted average rate of return on deposits		Weighted average rate of return on advances		Differential
	Nominal	Real	Nominal	Real	Nominal
FY1986/87	7.49	3.89	12.51	8.91	5.02
FY1987/88	7.37	1.07	12.59	6.29	5.22
FY1988/89	7.52	−2.88	12.59	2.19	5.07
FY1989/90	7.70	1.70	12.93	6.93	5.23
FY1990/91	8.06	−4.64	13.82	1.12	5.76
FY1991/92	8.50	−1.10	14.55	4.95	6.05
December 1992	8.74	−1.06	14.94	5.14	6.20

	Weighted average rate of return on deposits		Weighted average rate of return on advances		Spread	
	Islamic Modes	Interest Based	Islamic Modes	Interest Based	Islamic Modes	Interest Based
FY1987/88	7.31	7.67	13.38	10.70	6.07	3.03
FY1988/89	7.44	7.95	13.14	10.89	5.70	2.94
FY1989/90	7.60	8.23	13.59	10.59	5.99	2.36
FY1990/91	7.67	9.82	14.38	10.73	6.71	0.91
FY1991/92	8.27	9.35	15.06	10.72	6.79	1.37
December 1992	8.53	9.53	15.41	11.11	6.88	1.58
Average spread (FY1987/88 / FY1991/92)					6.36	2.03

Source: State Bank of Pakistan.

Foreign banks seem to be the main beneficiaries of the relatively large spreads, especially as the nationalized banks have to deal with problems of large overheads, increasing inefficiencies and nonperforming loans.[6] The share of foreign banks in total after-tax profits of all scheduled banks increased from 17.5 per cent in 1986 to 53.7 per cent in 1991. Similarly, the return on total assets of foreign banks ranged between 1 per cent and

---

[6] Comparisons between foreign banks and local banks should be made bearing in mind that these banks are not engaged in the type of activities carried out by nationalized commercial banks. They are neither competing in the same markets nor subjected to the same degree of political pressure to hire staff and advance/reschedule loans for them to have similar cost profiles as nationalized commercial banks.

2.4 per cent, while that of Pakistani banks ranged between 0.1 per cent and 0.3 per cent.[7] Customer advances of foreign banks grew at around 14.6 per cent per annum between 1989 and 1991, compared with the annual growth of just under 9 per cent in advances made by the nationalized commercial banks.

The three-year averages of the profitability indicators (Table 7.5) for nationalized commercial banks, private commercial banks, DFIs, and nonbank financial institutions are illuminating. In the case of nationalized commercial banks, total administrative costs were 2 per cent of total assets compared with 0.85 per cent in the case of foreign, 0.6 per cent in the case of private (data for the group is only available for 1992), and 0.8 per cent for provincial banks. Pre-tax profits as a percentage of deposits are 0.6 per cent for nationalized commercial banks, 5 per cent for foreign, 2.5 per cent for new private, and 5.9 per cent for provincial banks compared with 3.9 per cent for DFIs, and 2.4 per cent for investment banks (Table 7.5). Administrative expenditures of banks have also increased, which has led to reduced profitability. Banks now maintain a large staff—number of persons per branch has gone from 9.3 in 1974 to 13.7 in 1993—at an increased cost per employee. Personnel cost per employee has risen from Rs 10,984 in 1974 to Rs 109,413 in 1992.[8] Data on productivity ratios, i.e., the cost of earning Rs 100, illustrates the inefficiency of Pakistani banks. For foreign banks, staff costs have increased at 12 per cent per annum; for Pakistani banks, the annual growth is 19 per cent (Table 7.10).

Nonperforming Loans

Over the years, the capital base of nationalized commercial banks has been severely affected by the poor quality of bank loans made primarily on political and uneconomic grounds. As a result, the single most formidable problem facing the banks is the heavy burden of nonperforming loans. Though rescheduling of loans is common, the total advances of nationalized commercial banks categorized as bad and doubtful debts are Rs 56 billion (Table 7.11), of which Rs 46 billion are classified as advances related to the private sector. Just under 23 per cent of the private sector's classified debt pertains to advances under mandatory targets and concessional credit schemes.[9] In 1989, the State Bank of Pakistan estimated that

---

[7] These data are estimates derived from financial statements of banks.

[8] Though salaries remain low, expensive and inefficiently provided perquisites are the principal means of payment to employees. Political pressures have also resulted in top heavy management at most nationalized commercial banks.

[9] An analysis of the classified loans by number and size of loan shows that 98.4 per cent of the loans are below Rs 1 million each, making up 17.3 per cent of the total classified amount, compared with 0.04 per cent of the number of loans above Rs 50 million each that made up 36.2 per cent of the classified amount.

around 14.2 per cent of the lending portfolio of nationalized commercial banks were made up of nonperforming loans.

Table 7.10 Productivity Ratios of Banks
(per cent)

	1986	1988	1991
Foreign:			
ANZ Grindlays	58.8	45.6	25.2
Standard Chartered	36.2	30.9	15.9
Bank of America	45.1	33.3	17.9
American Express	50.0	30.5	28.5
Citibank	42.1	37.7	39.0
Pakistani:			
National	85.1	87.8	80.4
Habib	70.9	76.4	88.6
United	86.9	89.1	93.2
Muslim Commercial	84.0	81.7	88.2
Allied	91.6	91.9	90.0

Source: Estimates from financial statements of banks.

Table 7.11 Classified Advances of Nationalized Commercial Banks
(31 December 1991)

		Rs billion
Classified loans disaggregated by sector:		
Total loans		56.0
Foreign constituents		4.8
Public sector		5.2
Private sector		46.0
Classified loans disaggregated by number and size:	Number	Rs billion
Up to Rs 1 million	381,327	9.8
Rs 1 million to Rs 5 million	4,341	6.8
Rs 5 million to Rs 10 million	766	5.0
Rs 10 million to Rs 50 million	685	14.2
Over Rs 50 million	151	20.2
Total	387,270	56.0

The position of the DFIs is even more delicate than that of commercial banks. Provisions against bad loans have increased from 0.1 per cent of assets in 1980 to 0.9 per cent in 1991. It is estimated that about Rs 13 billion was held in nonperforming loans, representing 30 per cent of the portfolio, Rs 6 billion of which was in respect of cases that had been in litigation for more than one year.

The problem of debt recovery is not simply a technical issue. Not only have political pressures affected the quality of the loan portfolio of banks, they have also been instrumental in preventing banks to proceed against persistent defaulters and have resisted attempts to improve the enforcement mechanisms.

Development Finance Institutions

DFIs primarily focus on providing long-term debt and constitute a small percentage of the total financial assets. Until recently, they relied almost entirely on government funding or lines of credit from multilateral agencies, guaranteed by the Government. Now, locally mobilized deposits make up one third of the assets of the DFIs, making supervision of these institutions important for their own development as well as for the survival of the financial system. Consequently, in the recent reforms, DFIs were brought into the regulatory fold of the State Bank of Pakistan.

Industrial Development Financial Institutions

There are seven DFIs that have been chartered to target lending to industrial projects. Of these, the four largest—National Development Finance Corporation, Banker's Equity Limited, Pakistan Industrial Credit and Investment Corporation, and the Industrial Development Bank of Pakistan—account for more than 90 per cent of long-term loan disbursements. All of these DFIs pay taxes at the regular corporate tax rates. Pakistan has also entered into joint-venture DFIs with Kuwait (Pakistan Kuwait Investment Company), with Libya (Pak-Libya Holding Company Limited), and with Saudi Arabia (Saudi-Pak Industrial and Agricultural Investment Company Limited). These companies are owned jointly by the Governments of the two countries, and work to provide project finance to various ventures. Currently, however, their level of activity is fairly limited; each of these companies is less than one tenth the size of the public sector DFIs.

The DFI sector remains almost totally monopolized by the state-owned agencies. The only competition that can be expected for the DFIs may occur once the investment banks have had a chance to develop. It is also not quite clear what the difference between a DFI and a bank is. Just as there is concern over the portfolios of nationalized commercial banks, because of unsound loans that were politically motivated, there is every reason to be concerned about the financial health of the DFIs. Unfortunately, no data are available to refute these concerns.

Financial Development Financial Institutions

There are two large government organizations that provide mutual fund services in Pakistan. These organizations place funds in publicly traded stocks, as long-term investments or for short-term trading.[10] National Investment (Unit) Trust is an open-ended mutual fund which was established to support new entrants to the equity market. It retains the option to purchase 15 per cent of all new public share issues from the issuers.[11] The Trust has a large portfolio of shares and has appointed at least one director on the boards of many companies.

The Investment Corporation of Pakistan manages 21 close-ended mutual funds. In addition, it underwrites new share issues and promotes awareness about the stock market. It has, since its inception, sponsored the listing on the Karachi Stock Exchange (KSE) of 21 close-ended mutual funds. The pre-emptive rights of the National Investment (Unit) Trust over new issues are not available to the Investment Corporation. As a financial institution, the Corporation has played a significant role in the underwriting of public flotations and in establishing underwriting syndicates.[12]

The financial development financial institutions are the main institutional investors on the equities market. Their role as a disciplining force on company management is critically important. It is generally believed that political pressure is applied on them to support the market when necessary. It is reported that, during the market slide in the latter half of 1992, roughly Rs 10 million per day was invested in the market during the period August–October 1992.[13]

---

[10] Recently, some modarabas and private sector companies have begun to participate in the market for equities. Thus, there may be some competition developing in the mutual fund market. The international markets are also joining in this competitive scenario as the first international placement of a Pakistan Fund was arranged by Citibank in 1992.

[11] A pre-condition for the Controller of Capital Issues approval for a new issue is the right of first refusal to NIT to 15 per cent (until recently this was at 20 per cent) of the capital to be issued to the general public. It is not, however, bound to accept this preferential right. It has to hold the shares it purchases for two years and, hence, cannot disinvest in a poorly performing company.

[12] The dividend rate on NIT units has fluctuated between 16 per cent to 18 per cent in recent years. The rate of return on an investment on ICP mutual fund averaged 27 per cent between 1988 and 1991 and jumped to 41 per cent in 1992.

[13] Thus the National Investment (Unit) Trust is thought to have bailed out well-connected speculators and brokers who had made poor investment decisions.

Nonbank Financial Intermediaries

Following recent financial liberalization, entry into nonbank financial intermediaries has increased significantly, concentrating much of the competition in the financial markets in this area. However, these intermediaries have a very small share of the market as yet and have not quite gained the acceptance of the market. The lack of an adequate regulatory and supervisory system works against these companies as they continue to be looked upon with suspicion. Licensing and allocation of credit for political purposes are both examples of the failure of supervisors. Moreover, earlier experiences of the nonbank financial intermediaries, the finance company crash of 1978 and again in 1987, and then the crash of the cooperatives, each resulted in a complete loss of deposits without any indictments or supervisory action on behalf of the interests of depositors.[14] The new wave of nonbank financial intermediaries will have to deal with this legacy and build up a reputation that can be trusted.

Leasing companies are a relatively new phenomenon in Pakistan, dating from the late 1980s. Leasing provides a means to finance growth without using normal overdraft facilities and has the advantage of enabling lease expenses to be claimed on taxes. Leasing has grown quite rapidly over the past few years at a rate of about 70 per cent. Competition has increased; there are now more than 16 leasing companies as well as several modarabas that are engaged in leasing. However, though the industry has become quite competitive, it is still dominated by four large players: the National Development Leasing Company, a government-owned subsidiary, has about 30 per cent of the leasing business; First Grindlays Modarabas has about 17 per cent; Orix leasing has about 14 per cent; and First B.R.R. Capital Modarabas has about 11 per cent. Leasing practices are varied with some companies requiring additional collateral to be pledged, while others prefer to lease only to better-known companies. In general, leasing is restricted to equipment with a ready resale value. Leases have begun to finance longer maturities and the proportion going to finance equipment has increased, owing to tax incentives. Despite the increase in the size of leasing in capital markets, leasing only provides about 3 per cent of total capital funds which is quite low compared to about 7 per cent for other Asian countries.

A *modarabas*, an Islamic mode of finance revived in the 1980s, is conceptually similar to a close-ended limited partnership where a management company provides expertise while investors provide capital. Modarabas

---

[14] In fact, the State Bank retains no data on those crashes and, as noted by Said (1993), probably exacerbated the crash in 1987. See also Samad (1993).

were especially popular with sponsors because they had the advantage of being exempt from income tax. This exemption has now been withdrawn for modarabas that became operational more than three years ago. Modarabas can only engage in activities that are approved to be Islamic by the Religious Board which has been constituted for this purpose.

At present there are 45 modarabas operating in Pakistan. The better known have been established by banks and by business houses as an alternative form of raising finance. Modarabas' shares are traded on the stock market and their investment field is unrestricted so long as it is sanctioned to be Islamic. Nearly 75 per cent of their income comes from leasing, with bank and stock market investment accounting for a very small proportion. The number of modarabas has grown rapidly, particularly in 1992. The total value of the paid-up capital of modarabas is over Rs 5 billion. In the portfolio management business, modarabas are competing with the Investment Corporation of Pakistan and the National Investment (Unit) Trust as they have also expanded the portfolio choice available to small investors.

Investment banks were introduced in 1989 as a gradual approach to introducing competition in the banking sector. At present there are nine investment banks in Pakistan. However, the contribution of these banks in the financial system has remained marginal. The bulk of these banks' business is in the area of quasi deposit taking and short-term finance to selected Pakistani companies in the stock market. These banks have not yet made much impact in the market and constitute less than 1 per cent of financial assets.

Housing finance companies are a recent entrant into the financial markets. Historically, there has only been one housing finance institution in Pakistan; that one is government-owned. Approval has now been given to eight new companies to enter the field. One is already in operation while three were expected to commence operations during 1993.

FINANCIAL MARKETS AND INSTRUMENTS

The following brief description of the financial markets provides some indication of the extent to which these markets meet the investment, liquidity, and risk-sharing needs of the economic agents.

Money Markets

Interbank Overnight Funds Market. The call market is basically an overnight market for unsecured, i.e., noncollateralized loans between banks and some DFIs. The interbank call money market plays a significant role in monetary management as it is essentially a market for trading cash reserves among banks. The interest rate is determined purely on supply and demand for loanable funds (excess cash reserves of banks).

Ceiling Trading. Bank-by-bank credit ceilings, which were in force in Pakistan between 1972 and 1992, and credit deposit ratios, which have been in force since 1 July 1992, place credit limits on banks. Banks are allowed to trade this credit limit at mutually agreed terms. The ceiling rate generally remains higher than the interbank call money market rate, sometimes as much as 10 per cent higher. Generally, larger Pakistani banks (nationalized commercial banks) are buyers in the ceiling market while the smaller banks, including foreign banks, are sellers.

GOP Short-Term Treasury Bills. Six-month treasury bills have been auctioned at a discount twice a month since March 1991. Commercial banks have been the largest investor in these bills, which they use mainly for meeting liquidity requirements.[15] Contrary to expectations, a secondary market has not developed in treasury bills, mainly because the high liquidity requirements imposed on the banks and the lack of suitable alternatives force them to hold the bills to maturity rather than trade them.[16]

Repurchase Agreement Transactions. Repurchase agreements or repos have been gaining in popularity as an alternative to the overnight funds money market instrument. This is because they are collateralized and can be traded in a wider market comprising nonbank institutions, DFIs, corporations, and even individual investors. The repos market has grown quite strongly since its introduction in March 1991, from about Rs 2.5 billion in March 1991 to about Rs 14 billion in December 1992.

Debt Markets

The debt market is based primarily on government debt since commercial paper has not been granted official sanction.

Federal Investment Bonds. As of 20 May 1993, the total investment in these bonds was about Rs 131,566 million. The share of banks' investment

---

[15] Treasury bills are held mainly by the banks. Total sales in auction was Rs 206,784 million by 31 May 1993, of which banks had purchased about Rs 196,434 million while the nonbank sector had purchased about Rs 10,350 million (94.99 per cent and 5.01 per cent, respectively).

[16] The rapid growth in repurchase agreements has perhaps also impeded secondary market growth in government paper.

was 77.72 per cent. Ten-year Federal Investment Bonds are popular; they constitute 61.3 per cent of such bond holding. Interestingly enough, next to the 10-year bond, investors prefer the three-year bond which pays out 13 per cent per annum compared with the 14 per cent return on a five-year bond.

Auctions of the Federal Investment Bonds are not completely market determined in the sense that the Government reserves the right to accept the number and amount of bids tendered in each auction. The Government also indicates the cut-off price while accepting bids.

Other Government Papers. Several savings certificates issued by the Government are available through National Savings Centers and post offices. Only individuals are allowed to invest in these schemes and the return on these certificates is tax-exempt. Defence savings certificates are of a similar nature. Together these constitute about 22 per cent of government debt.

DFI Papers. A number of savings instruments and investment certificates are issued by the state-owned DFIs. Most of the DFI and public sector papers require no identification of the owner. The Government has now decided to discourage issue of bearer bonds and encourage floating of registered bonds instead.

Commercial Papers. There is no market for corporate debt in Pakistan. Corporate debt instruments, such as Term Finance Certificates, are sold to a syndicate of banks and other financial institutions for their own accounts. To lessen the burden on budgetary financing, the Government has allowed some public sector agencies such as the Water and Power Development Authority and Civil Aviation Authority to raise funds from the market by issuing bonds.

Equities Markets

There are three stock exchanges in Pakistan: one each in Karachi, Lahore, and Islamabad. The KSE, which was established in 1948, and where 90 per cent of the current trading takes place, is the oldest. As of 31 May 1993 there were 630 companies listed on the KSE, having increased from 362 in 1985.[17]

---

[17] Current membership of the KSE is around 200 of which around 100 are active; trading is dominated by 15 to 20 members. New members are required to purchase an existing seat. In June 1990, the constitution of the KSE was amended to permit corporate membership. Corporate members are required to have a minimum paid-up capital of Rs 20 million. Three such members have recently incorporated.

Stock market activity has increased sharply in recent years. The State Bank of Pakistan index of share prices almost doubled between June 1991 and June 1992, having risen steadily from 260.6 in FY1987/88 to 393.5 in FY1990/91 and thereafter to 760.2 in FY1991/92. The aggregate market capitalization[18] at the KSE is approximately Rs 200 billion, having grown from just Rs 38 billion in FY1987/88.[19] A strong demand for equities has accompanied market growth: new offerings increased from Rs 449 million in FY1987/88 to Rs 6,415 million by FY1991/92, at a rate of 9.5 per cent per annum. Strong demand for stock has been evident in recent years with considerable oversubscription for new issues. The amount subscribed as a ratio of amount offered was 8.0 in 1989 and thereafter declined steadily to 6.9 in FY1991/92.

The average daily turnover in FY1987/88 was around 500,000 shares, growing to over 2 million shares in FY1991/92 and during the first half of 1993 it has occasionally touched the 7.5 million mark.[20] Approximately 300 equities are now traded on an average day. Among these, trading in most shares is in insignificant volumes. Trading in shares of 10 companies is estimated to account for over 30 per cent of turnover in terms of value. However, the market lacks depth and remains dominated by informed individuals.

Trading consists of long positions in equities as short-selling and derivatives are not allowed. The only mutual funds that have been allowed until recently have been those run by public sector financial DFIs, such as the National Investment (Unit) Trust and the Investment Corporation of Pakistan. International capital has only very recently been allowed access to the stock market.

Foreign Exchange Markets

The State Bank of Pakistan fixes both the spot and forward exchange rates. In the spot market, the State Bank maintains a managed float system, pegging to an undisclosed basket of foreign currencies. The rate is managed so as to maintain the country's competitiveness. In the forward market, the

---

[18] Market capitalization is calculated by multiplying the market price with the issue share capital, even though less than 15 per cent is actually traded in the market.

[19] Although it ranks among the top Asian countries in terms of numbers of listed companies, Pakistan has one of the smallest markets in terms of market capitalization and trading volumes. The share of Pakistani equities in emerging markets is a mere 1.6 per cent.

[20] The stock exchange has set fees for institutions and individual clients that are linked to market prices and range from Rs 0.09 for institutional clients and Rs 0.10 for other clients per share for a share with a price less than Rs 20 and Rs 0.64 and Rs 1.50, respectively, if the share price exceeds Rs 100.

State Bank has arbitrarily fixed a premium of 3.5 per cent for six-month cover and about 7.5 per cent for 12-month cover. These premia bear little or no relationship to market considerations or to past or expected exchange rate behavior.[21] These are the only types of forward cover on exchange risk that are available. Only traders and importers of capital equipment can avail of this facility.

The large availability of workers' remittances as well as the growth of the informal economy has made the currency virtually fully convertible in recent years. Since 1982, when the country adopted the managed float with a peg to the dollar, no serious foreign exchange shortages have been experienced. In fact, the black market rate, i.e., the *hundi* rate and the foreign exchange bearer certificate (FEBC) rate, the rate derived from the secondary market trading in FEBCs, have both been virtually identical and have differed from the official rate by only a small margin of 6–10 per cent.[22] The black market premium shows little sensitivity to changes in domestic policy or the level of reserves, which suggests the possibility of large inflows in the informal sector (Haque and Montiel 1991a; Montiel et al. 1993).

Insurance

The State Life Insurance Corporation (SLIC)—which until recently had a monopoly on life insurance—continues to hold the bulk of the life insurance business and, consequently, is a dominant player in the insurance market. Along with the financial development financial institutions and the modarabas, SLIC is also a major institutional investor holding a portfolio consisting of shares of 206 companies.

Unlike the SLIC, the general insurance companies are subject to fewer restrictions. For example, they are free to choose their own portfolio of investment and can invest in any paper that can be transacted in the stock market. However, a new condition that might be enforced upon them (presently under discussion) would be a restriction on the maximum permissible holding to the lower of 2.5 per cent of the equity of the new company or 2.5 per cent of the insurance company's portfolio. General insurance companies prefer investments in equities because the solvency requirement test values their investments at market prices.

---

[21] This increases the exposure of the State Bank and the Government. The country's external debt—which is of a long-term concessional nature—is now being serviced through short-term resources generated by multi-currency short-term loans/commercial borrowing, mobilization of foreign currency accounts, and interbank borrowing. This results in a mismatching in the assets and liabilities of the foreign currency portfolio of the country and could lead to increased exposure to interest and exchange rate risks.

[22] See Haque and Montiel (1992a) for a more detailed discussion of exchange rate policy in Pakistan.

Informal Financial Markets

Pakistan has a dynamic, parallel/nonformal sector which has seldom been studied. Essentially three types of financial activities take place in the informal sector:

(i) a market for trading foreign currency. The market for foreign currency is semi-legal because of the liberalization introduced by the Government on the capital account (a market in which, interestingly, banks are not allowed to participate). The transactions on the trading account are still controlled and have to be routed through the formal banking system, though imports not paid for through a letter of credit are allowed entry on payment of a penalty.

(ii) an informal credit system, a component of which is the system of granting credit within the stock market.[23]

(iii) a futures market. This is a recent phenomenon where information technology allows trading on the international futures markets. It is classified here as informal only because it is an unregulated activity.

ADEQUACY OF THE FINANCIAL INFRASTRUCTURE

As discussed earlier, an important function of the financial system is to allow the arbitrage of asymmetric information. This requires that a capable regulatory system be in place. In Pakistan, the financial markets are currently regulated by various government agencies, including the Ministry of Finance, the State Bank of Pakistan, the Corporate Law Authority, and the Controller of Capital Issues. The existence of multiple regulators hinders the smooth functioning of the country's capital markets and leads to fragmented monitoring.

The Ministry of Finance is at the apex of the control of financial markets. Regulation and supervision of financial markets are primarily in the hands of two government agencies: (i) the State Bank of Pakistan is the main regulatory agency for the banking and, more recently, for the non-bank financial institutions (the latter are controlled administratively by the Ministry of Finance); and (ii) the Corporate Law Authority is mainly

---

[23] Another common type of financial transaction in the informal financial markets—rotating mutual credits—has not, however, been examined. Rotating mutual credit involves a group of members, each of whom contributes a fixed amount. The size and practices of these groups can vary significantly from one group to the next.

responsible for the regulation of capital markets. In addition, the Pakistan Banking Council, which was created as an umbrella organization for running the nationalized commercial banks, remains in existence despite privatization and increasing nationalized commercial bank autonomy. Two provincial banks have been established but because of constitutional protection, they fall outside State Bank supervision.

The Corporate Law Authority was initially established to handle legal matters pertaining to companies. However, many such issues are still being handled by other regulatory agencies, which complicates the process of streamlining the regulatory system. In essence, the financial sector's regulatory and monitoring systems are too fragmented: the country needs a regulatory body which is independent and professional. Ideally, the State Bank of Pakistan should be in charge of the money market and the banking and deposit-taking institutions that are related to the operation of monetary policy, and the Securities and Exchange (SEC)—perhaps a strengthened and improved Corporate Law Authority—should be in charge of all issues pertaining to the functioning of the securities market.

Institutional Capability of the Country's Regulatory Agencies

The regulatory bodies do not have the capacity—in terms of financial resources and staff members and skills—or the autonomy to perform their supervisory functions and enforce regulations. Until the introduction of the new prudential regulations, the inspection department of the State Bank was essentially engaged in checking that banks were meeting their credit-related targets. The State Bank has tended to focus on monitoring compliance with ratios rather than on analysis of the implications of the prudential regulations.

Current skill levels are weak in the areas of risk assessment of new instruments, computerization of accounts, audit of automated accounting systems, lending in the form of overdrafts, assessment of credit exposure to check credit concentration, and capital adequacy based on risk-weighted assets and complex lending structures which make assessment of credit quality difficult. For assessing capital adequacy, the ability to analyze credit or to conduct a risk analysis of interest rates will have to be strengthened significantly.

In part, the weakness in the supervisory capacity of the State Bank is because officials are overburdened by routine work (and there is no incentive structure) and in part because of the constraints imposed by personnel policies. Restrictions on pay scales make it difficult for the State Bank to recruit or retain good professionals. Its freedom in this area is further hampered by the contractual agreement with the trade union whereby the State Bank of Pakistan can only hire for entry level positions and cannot easily recruit experienced professionals directly at the officer level.

Weak Enforcement of Regulatory Mechanisms

Despite the elaborate legal system governing such activities as accounting and auditing standards, reporting mechanisms, and shareholders' meetings, its application is weak, largely because of poor enforcement.

Information Disclosure

An important prerequisite of smooth transition from a regulated to a deregulated system is the institutional capability and integrity of the domestic financial sector. Fair trading, diligence, and information disclosure are the basic pillars of a dependable financial structure. Flow of reliable and detailed information to investors is critical for the development of financial markets.

The State Bank and the Capital Law Authority have begun to enforce the various regulations governing the operations of the financial institutions. These institutions, mainly because of lack of institutional capacity, face difficulties in enforcing these regulations pertaining to disclosures, ensuring compliance with accounting standards and legal requirements, and levying penalties in case of noncompliance.

A Credit Information Bureau was established in the State Bank during FY1990/91 to maintain data pertaining to borrowers of all lending institutions. A record of all borrowers having liabilities of Rs 1 million or above is kept on computer and is updated quarterly. To assess distribution of credit and determine groups/industries having potentially large overdue payments, borrowers are also classified according to type of industry and geographical location. The Bureau makes available these data to any lending institution seeking to assess the creditworthiness of a potential borrower. As the Bureau became fully operative only from 1 January 1993, it is still too early to assess its effectiveness.

Auditing and Accounting Standards

Information disclosure on capital adequacy requirements and the limits placed on the scale of activities are only satisfactory if there is an appropriate legal framework and if that framework enforces proper accounting practices. Without the consistent application of generally accepted accounting and auditing conventions, there is a tendency to use collateral as a signal for project viability.[24]

Though the auditing standards and legal disclosure requirements are comparatively high, at least for public quoted companies, and in line with

---

[24] However, as will be discussed later, collateral is not easy to collect so the efficacy of this signaling device is also in question.

international practices, there is poor regulation of auditing firms which underlies the poor compliance with standards.

While the Institute of Chartered Accountants has on average proceeded against 10 firms per annum, it has only penalized six firms during the previous five years. Of the 26 complaints received by the Institute between 1978 and 1986, only four dealt with noncompliance with disclosure requirements or failure to qualify an audit report. In none of the 26 cases were any penalties levied. In the four cases, the Council of the Institute of Chartered Accountants expressed displeasure but decided to waive action because of the confession of the offending firm.

Prudential Regulations

Banks were nationalized in 1974 because of the suspicion of interlocking ownership of financial, commercial, and industrial interests and increasing concentration of ownership. As financial markets are opening up, there is a greater need for regulations that deal with issues relating to concentration in the financial markets and matters concerning conflict of interest, and the security and safety of deposits. Therefore, in 1992 the State Bank of Pakistan issued a set of prudential regulations for nonbank financial institutions and scheduled banks (Box 7.1).

The importance of developing and enforcing adequate prudential regulations for the development of competition in the financial markets has been clearly outlined by research as well as experience of financial reform in developing countries (Vittas 1992; Faruqi 1992). Issues of moral hazard and adverse selection are largely dealt with through prudential regulations. However, for prudential regulations to succeed, economic policy must be clear on what can be expected of the financial sector in general, and the banking system in particular.

The need to strengthen bank supervision with adequate risk capital requirements, and regulations that limit the exposure of the banking system to volatility in equity and real estate markets, is critical in a period of financial and economic reform. Success with such reforms is likely to induce capital inflows and threaten domestic demand management.[25] Domestic policy should adjust to this possibility if the reform is not to unravel.

Debt Recovery System

The framework for recovery of debt in Pakistan is not effective. The legal framework for loan recovery is cumbersome and time-consuming, requiring several years in court. The court system and the overloaded law enforcement agencies are not equipped to handle the area of "law and economics."

---

[25] See Calvo et al. (1993) and references therein for experiences of countries with capital inflow problems.

Box 7.1 State Bank of Pakistan's Prudential Regulations

The key capital adequacy requirements established by the State Bank of Pakistan during the latter half of calendar year 1992 are as follows:[1]

- To reduce the exposure of a bank or nonbank financial institution, contingent liabilities of a financial institution cannot exceed 10 times its paid-up capital and general reserves.

- The total credit-related facilities that a borrower can avail cannot exceed 10 times the borrower's capital and reserves—based on certified financial statements. In the case of nonbank financial institutions, the total exposure of the institution to a single borrowing entity cannot exceed 20 per cent of its equity. In the case of listed companies, exposure cannot exceed 20 per cent of the total assets of a nonbank financial institution.

- Shareholders and directors of banks with holdings of 5 per cent or more of the share capital of the concern cannot borrow from banks in which they have an interest.

- New lending in the form of a long-term loan is to be made available on the basis of a debt/equity ratio not greater than 60:40. The existing position of each borrower must be regularized by 30 June 1996.

- By 1 January 1994 the current asset/current liability ratio of a borrower should be 1:1.

- If interest or principal is overdue (past due) by 90 days from the due date the income from such nonperforming assets cannot be recognized. Further, a provision has to be made on the difference between the outstanding balance and the liquid assets realizable without recourse to a court of law. The provision should be 2 per cent for a past due by 90 days; 25 per cent if overdue for 180 days; 30 per cent if outstanding for one year or more; and 100 per cent overdue beyond two years from the due date. Compliance with this provision became mandatory from 30 December 1992.

- A nonbank financial institution is required to hold at least 70 per cent of its assets in the form of its principal line(s) of business.

- At least 3 per cent of the credit facilities are to be earmarked for small entrepreneurs.

---

[1] The target set by the Government was that by 31 March 1993, the capital of nationalized commercial banks should be a minimum of 3 per cent of their callable liabilities. The targeted adequacy was achieved in two banks whose management has been passed onto the private sector and by one of the remaining nationalized commercial banks. The remaining two nationalized commercial banks need over Rs 15 billion to achieve this ratio. The recent liberalization of interest rates and the availability of new government debt instruments at virtually market rates is expected to facilitate this recapitalization.

To strengthen the debt recovery framework and expedite cases of loan recovery, Special Courts were created to handle the loan cases of pre-1985 interest-based lending. In addition, Banking Tribunals were set up to expedite loan recovery cases of noninterest financing. Both seem to have improved recovery performance somewhat, creating the impression that the formidable challenge of foreclosure can be addressed through an appropriate strengthening of the legal framework and the judicial machinery.

The Special Courts and Banking Tribunals suffer from the same problems as the rest of the public sector: inadequate human capital, a rigid salary structure that is incompatible with human capital requirements for productivity, poor quality infrastructure, inadequate incentives for improved productivity, and subservience to political priorities. The tribunals remain inadequately staffed and lacking in space and facilities. The number of tribunals, especially in larger cities, remains inadequate to handle the backlog.

To make these tribunals more effective, the number of tribunals should be increased; an incentive structure should be put in place to attract quality staff; productivity should be emphasized; and funds for necessary infrastructure should be made available. Rather than differentiate between various kinds of loans, banking tribunals and special courts should be merged into one organization.

Finally, besides improving the efficacy of the judicial machinery, there must be a political commitment to implement any loan recovery program put into effect.

Weak Payment and Information Systems

Service standards in the banking system are poor. In particular, the payment system is weak, inefficient, and undeveloped. Real time quotes are not available on the stock market. Intercity check cashing can take more than a week and bank charges can be as high as Rs 100.

In Pakistan availability and, more importantly, reliability of information is a major hindrance to the investment of savings in financial instruments.

The liquidity aspect of an instrument is affected by a variety of factors including restrictions on encashability at the issuing office, legal constraints to transferability of instruments, and fiscal impediments (particularly in the form of stamp duty) to transferability. Only prize bonds, Foreign Exchange Bearer Certificates and Dollar Bearer Certificates can be encashed anywhere in Pakistan; all other instruments can only be encashed at the issuing office/branch. Similarly, interest can only be collected from the office that issued the instrument.

The legal framework only allows transfers of shares. Instruments that qualify as receipts of deposit, being non-negotiable instruments, are non-transferable. A standard Fixed Deposit Receipt, for example, can only be transferred through assignment. Also, a transferable instrument attracts stamp duty unless the transacting party is a financial institution.

In addition, registration procedures can take as long as 45 days. This could be avoided if institutions were to establish a facility for warehousing of shares by maintaining subledger accounts. However, this could only occur if financial institutions were given certain exemptions, for instance in respect of ceilings.

ECONOMIC AND FINANCIAL POLICIES

Constraints of Financing a Large Fiscal Deficit

Over the last decade-and-a-half, the Government's deficit has generally been quite large, varying from between 5.3 per cent and 8.9 per cent of the GDP. Government borrowing at a highly concessional rate from the banking system—particularly from the State Bank of Pakistan—was considered a less painful option than either revenue mobilization or containment of expenditure.[26] The low yield on government securities had obvious implications for profitability of commercial banks and their efforts at deposit mobilization.

To avoid inflationary finance, the Government financed its deficit from nonbank sources where it operated a number of small saving schemes, such as the Defense Saving Certificates, National Deposit Certificates, *Khas* Deposit Certificates, Premium Saving Certificates, and Saving Bank Accounts. These instruments had attractive rates. The effective yield on a number of them was considerably higher because of income tax exemptions and the eligibility of some to qualify for investment allowances for income tax purposes. To finance its deficit, the Government discourages deposit mobilization by the banks, requiring them to maintain high liquidity ratios while also offering higher rates on government bonds than those allowed on deposits. As a result, there has been an accumulation of public debt which has become a major constraint on fiscal policies.[27]

Financial Repression and the Quasi-Fiscal Deficit

Through much of the country's history, financial markets have been repressed by policy. Direct controls on deposit and loan rates have been

---

[26] Until recently, the Ministry of Finance could borrow directly from the State Bank of Pakistan, without limit, at a rate of 0.5 per cent on 90-day ad hoc treasury bills, while commercial banks held government-approved securities at 6 per cent to meet a liquidity ratio of 35 per cent. (See also Haque and Montiel 1989 for an analysis of fiscal policy.)

[27] See Haque and Montiel (1993) for an analysis of the effect of the accumulated debt stocks on current fiscal policy choices.

maintained. The principal instrument for monetary control has been credit controls involving both bank-by-bank credit ceilings and directed and mandatory credit by way of subceilings for selected sectors. Though recently reduced, concessionary finance to various sectors has been quite significant.[28] Nearly 35 per cent of credit to private sector and public sector enterprises were at subsidized rates, amounting to a total subsidy of Rs 9 billion (Table 7.12). The schemes and the interest rate charges on these subsidized schemes are shown in Table 7.13.

These policies have resulted in a distorted and regulated interest rate structure.[29] While credit ceilings served as an effective instrument of credit control, they adversely affected commercial banks' incentive to mobilize deposits. Competition among banks was limited as was their ability to respond flexibly to demands of the economy.[30] The differential subsidization of interest rates on advances for priority sectors severely distorted the incentive structure, weakening the motivation for both the mobilization of resources and the screening and collection of loans. The viability of banks as financial intermediaries was threatened as a result, while the quasi-fiscal deficit of the Government was substantial (Box 7.2).

Interest rates on deposits remained negative in real terms from the beginning of the decade (Table 7.14); deposit rates have remained below international rates as implied by the uncovered interest parity for much of the period. The difference was as high as 14 percentage points. In fact, the two tended to come closer together only when international interest rates declined substantially in 1991–1993.

Prior to June 1985, the State Bank of Pakistan determined the domestic interest rate structure. From 1 July 1985, interest rates were abolished and profit and loss sharing put into effect, as required by the dictates of Islam. Commercial banks now obtain clearance from the State Bank for the announcement of their rates of profit sharing which are determined on an *ex post* basis. The profit-and-loss-sharing system, however, does not apply to government securities and foreign currency deposits—both continue to offer pre-announced fixed interest.

---

[28] Concessional credit is available for the purchase of locally manufactured machinery, refinancing of exports, production loans for small farmers, and credit lines to the Agricultural Development Bank of Pakistan (ADBP) and the Federal Bank for Cooperatives (FBC) at considerably lower than market rates of return. The banks are attracted to credit for exports because of the commissions they earn on foreign exchange transactions.

[29] The direct ownership of much of the financial sector institutions by the Government translated into reduced competition in the financial sector. The nationalization of banks in 1974 also served to strengthen the allocation of credit by nonmarket means.

[30] Banks tried to circumvent credit ceilings in a number of ways: until the penalty on banking credit ceilings was raised to a prohibitively high level, banks would accommodate their clients and pay a penalty which they were able to retrieve through other business received from such clients. Also, banks made liberal use of their off-balance sheet items.

Table 7.12 Estimates of the Quasi-Fiscal Deficit
(Rs million)

	Subsidized lending rates			Amount of outstanding credit			Amount of subsidy[a]		
	1990–91	1991–92	1992–93	1990–91	1991–92	1992–93	1990–91	1991–92	1992–93
A. Private sector									
Concessional credit									
LMM	7.00	8.00	8.00	9,800	12,864	12,864	882	1,029	1,415
Export finance	7.00	8.00	8.00	15,708	20,381	24,151	1,414	1,630	2,657
YIPS	8.00	9.00	9.00	516	862	862	41	60	86
Government-Sponsored									
Schemes (SBFC)	11.00	11.00	11.00	298	250	250	15	13	20
Self Employment (SBFC)	10.50	10.50		20	20		1	3	
Power Looms (SBFC)	8.00	8.00	8.00	23	23	23	2	2	3
Small-scale Industry									
(SBFC)	11.00	11.00	11.00	194	236	236	10	12	19
Loans to small									
farmers (Banks) i)	7.00	10.50	10.50	765	443	443	69	24	38
ii)	0.00	0.00	0.00	835	944	944	134	151	179
Advances by ADBP	12.50	12.50	13.50	42,016	43,908	44,823	1,471	1,537	2,465
Advances by FBC	9.33	9.33	9.33	1,836	1,902	1,700	122	127	164
Advances by HFBC	12.34	12.34	12.34	15,571	15,996	16,296	570	585	1,085
Self employment (Banks)		15.00	15.00			11,540			462
Total				87,562	97,829	114,152	4,730	5,171	8,595

	1990-91	1991-92	1992-93	1990-91	1991-92	1992-93	1990-91	1991-92	1992-93
B. Government Sector (Commodity operations)									
i) Central	10.8	10.8	12.5	11,151	15,131	17,577	580	787	1,143
ii) Provincial	10.8	10.8	12.5	7,523	7,738	7,738	5,099	391	402
Total				18,674	22,869	22,676	971	1,189	1,474
C. NBFIs							3,958	3,000	3,000
D. Loans on foreign exchange guarantees							5,000	4,000	4,000
Grand Total-Quasi Fiscal Deficit (QFD)							14,659	13,360	17,069
Per cent share of QFD in GDP							1.43	1.11	1.26
Per cent share of QFD in government revenues							8.68	5.74	...
Per cent share of QFD in total fiscal deficit							16.43	14.83	...
Per cent share of QFD in total fiscal expenditure							5.68	4.14	...
Memorandum items									
Per cent share of subsidized credit in total credit to private sector				39.61	38.93	37.66			
Per cent share of subsidized credit in total credit to private sector and PSEs				34.80	34.73	34.23			
Per cent share of (B) in total credit to government sector				9.76	8.54	6.72			
Per cent share of QFD in total domestic credit				26.05	23.70	22.15			

[a] Market rates for calculating opportunity cost for the three years were taken to be 16, 16 and 19 respectively.

Notes: (i) Data for 1992-93 are as of 20 May 1993; (ii) For LMM and SBFC credit schemes, data have been incorporated to that of 1991-92 for 1992-93.

Table 7.13 Existing Lending Rates on Concessionary Financing

Name of the Scheme for Investment Credit	Lending Rates[a] (per cent)	Date of Effectivity	Subsidy Provided Through Budget
Locally manufactured machinery			
Local sales	8	1 Aug 1991	No
Export sales			
Pre-shipment	8	1 Aug 1991	No
Post-shipment	8	1 Aug 1991	No
Loans to Power Looms Sector	8	1 Aug 1991	No
Small Business Finance (SBFC) Scheme for doctors, engineers and lawyers	11	1983/84 Doctors 1984/85 Engineers 1985/86 Lawyers	
Projects under Youth Investment Promotion Scheme (YIPS):			
For rural area	8	26 Nov 1991	No
For urban area	10	26 Nov 1991	No
Tourism Projects of Hazara Division, Coastal Areas Sehwan Sharif and Multan	8	26 Nov 1991	Difference determined by the Government
Concessional Financing Scheme for Northern Areas and Azad Kashmir through SBFC	8	1 March 1992	Difference determined by the Government
Industrialization and employment in Northwest Frontier Province	3 less than normal rate	1 July 1991	No
Self-Employment Scheme	13	April 1992	No
Export Finance Scheme	8	1 March 1992	No
Agricultural Production Loans to small farmers by Agricultural Development Bank of Pakistan (ADBP)	12	1 July 1990	No
Agricultural Production Loans to small farmers for months by commercial banks	7 (8 mos) (10.5/mo)	11 July 1990	No

[a]Market lending rates vary between 18 and 22 per cent per annum in the formal sector.

> **Box 7.2 Quasi Fiscal Deficit**
>
> The quasi fiscal deficit arising from concessional credit, exchange loss on foreign exchange guarantees, and losses on nonperforming loans (default on central bank finance) is calculated at about Rs 17 billion or about 1.3 per cent of GDP in FY1992/93. During FY1992/93, the subsidy on credit was estimated to be about Rs 10 billion—Rs 8.6 billion to the private sector and Rs 1.5 billion to the government sector. Losses on foreign exchange operations arising from the foreign exchange cover provided to foreign currency deposits was estimated to be around Rs 4 billion in FY1992/93. Losses on defaults by the financial institutions on the State Bank of Pakistan concessional lending schemes was about Rs 3 billion per annum for FY1990/93.

Table 7.14 Interest Rates on Deposits

	Weighted Average Rate of Return on Deposits Nominal	Weighted Average rate of Return on Advances		Differential		Interest Parity Rate[a]
		Real	Nominal	Real	Nominal	
FY1986/87	7.49	3.89	12.51	8.91	5.02	
FY1987/88	7.37	1.07	12.59	6.29	5.22	
FY1988/89	7.52	−2.88	12.59	2.19	5.07	11.33
FY1989/90	7.70	1.70	12.93	6.93	5.23	21.64
FY1990/91	8.06	−4.64	13.82	1.12	5.76	13.68
FY1991/92	8.50	−1.10	14.55	4.95	6.05	14.79
Dec. 1992	8.74	−1.06	14.94	5.14	6.20	8.92

[a] The uncovered interest parity rate is measured as the 3-month LIBOR plus actual changes in the exchange rate against the US dollar.

The State Bank retains a fair amount of control over the determination of interest rates and the profit-sharing rate: it prescribes methods of calculating service charges in the case of interest-free loans and prescribes minimum and maximum rates for trade-related and certain types of investment financing. In recent years, some efforts were made to allow the interest rate structure to move upward: rates on national savings instruments have been rationalized; rates on subsidized credit have been revised upward though some still remain negative in real terms. General purpose credit is now available at market prices subject to a maximum of 20 per cent.

Low deposit rates have prevented savings from flowing into the financial sector at a rapid rate, and may have contributed to the financial disintermediation in the latter part of the 1980s when banks were by-passed in a large number of financial transactions, payments, and debt settlements. For five years ending in FY1990/91, time deposits remained more or less unchanged and demand deposits increased only slightly, while the bulk of the increase in money supply was in the form of currency.

Depositors also bear the cost of the nonperforming loans that were accumulated in the past, and the high administrative expenditures of banks that arise out of overstaffing and enhanced union activity. Recently, increased competition in the banking sector, especially among the new private banks, has led to more competitive rates being offered.

In January 1992, the concept of a 65 per cent credit/deposit ratio was introduced. This was later changed to 30 per cent on local currency deposits and 40 per cent on foreign currency deposits, and then changed again to a uniform ratio of 30 per cent. In principle, the credit/deposit ratio operates as a reserve requirement. Within the allowable credit, banks are required to finance priority sectors at subsidized rates. Though there is also a reserve requirement in the shape of the liquidity ratio (45 per cent of deposits, of which 5 per cent has to be held with the State Bank of Pakistan), the credit deposit ratio renders the liquidity requirement redundant. This fact is important because the Government's cost of servicing domestic debt is at a lower rate than it would have been without a credit deposit ratio requirement. The credit deposit also leads to inefficient allocation of credit as banks discriminate on a non-price basis: it is estimated that about 65 per cent to 70 per cent available credit in this regime is made available to prime borrowers.

Recent Financial Sector Reforms

In 1989, the Government began to implement financial sector reforms to address the factors that inhibited the efficient functioning of the financial sector. The basic thrust of these reforms is privatization, deregulation, liberalization of the environment for savings and investment, and the dismantling of a regime of controls established by the Government to regulate economic activity. It is expected that the new system of financial arrangements will allow efficiency to be rewarded while making economic management more transparent and less discretionary.

The financial sector reforms that have been implemented, thus far, basically fall into the broad categories of improved government debt management, monetary management policies, and organization and supervision of the banking system.[31] These reforms are based on the following key measures:

---

[31] This financial sector reform was supported by a World Bank Financial Sector Adjustment Loan (FSAL).

(i) improvement of the competitive environment in which banks operate through privatization and greater reliance on market forces;

(ii) liberalization of exchange controls;

(iii) deregulation of the structure of interest rates and a gradual dismantling of the system of directed credit;

(iv) permission for foreign ownership of equity; and

(v) improved bank supervision.

Public Debt Auctions

As a step in the direction of interest rate liberalization, an auction system has been introduced for government securities, with commercial banks as the primary dealers.

Two debt instruments—six-month treasury bills and 3–10-year Federal Investment Bonds—have been introduced. However, auctions of government securities are not conducted free of government intervention: a cut-off price, which may or may not be known, is determined by the Ministry of Finance; the amount of the auction is not pre-announced; and the Government reserves the right to reject all bids or even to sell substantially larger amounts than expected. By 27 May 1993, 59 auctions had been held with total investments in treasury bills of nearly Rs 207 billion, of which nearly 95 per cent is held by banks. The share of the nonbank sector is rather small.

Competitiveness in the Financial Sector

Privatization of nationalized commercial banks and permission for new banks to enter in the market was an important element of the recent reform package. Though only two of the six nationalized commercial banks have been privatized so far, it is intended that at least two more do so. In addition, 10 commercial banks, 10 investment banks, 30 leasing companies, and 45 modarabas have been floated to increase the number of financial agencies in the free market and to promote resource mobilization and financial intermediation.

Foreign Exchange Reform

Prior to February 1991, foreign exchange markets consisted of: (i) trade in Foreign Exchange Bearer Certificates (introduced in 1985; rupee-denominated assets bought with dollars, convertible into rupee or foreign exchange at the prevailing exchange rate)[32]; (ii) operation of Foreign Currency Accounts allowable only to Pakistani nationals abroad and for a limited period after their return to the country; (iii) use of informal *hundi* markets involving the sale and purchase of foreign currencies in the black market;[33] and (iv) the official market where foreign exchange was rationed according to official dictates.

In the spring of 1991, foreign exchange controls were relaxed (short of convertibility), perhaps in recognition of the large informal foreign currency trading. Several new debt instruments, denominated in foreign currencies, were introduced. The market now encompasses: operation of Foreign Currency Accounts, Foreign Exchange Bearer Certificates, Dollar Bearer Certificates, and money changers authorized by the State Bank of Pakistan.

DEVELOPING THE MARKET FOR EQUITIES

Investor Perceptions and Returns to Stockholders

For the equities market to develop, investors must perceive it to operate in a transparent manner. Unfortunately, the stock market is not perceived by the average investor to be either efficient, fair, or judicious in terms of providing the investor a fair return. The rules regarding conflict of interest are vague, and enforcement of the rules that do exist is limited.

The bulk of earnings from stock holdings are derived from capital gains. During the five-year period, 1987–1992, returns in the form of cash and stock dividends were approximately 3–5 per cent on well-constructed portfolios. Including rights issues and capital gains, the average rate of return is around 17.3 per cent. Thus, even disregarding the exceptional increase in the index after the recent reforms, the yields from judicious investments in the stock market were quite attractive.

In view of the weak regulation and supervision of companies and the inability of the small investor to obtain reliable information, the typical small investor prefers investment in multinationals. As a result, the multinationals have high P/E ratios, at times exceeding 50.

A large fraction of the firms listed on the KSE do not pay out dividends. In FY1989/90, about 45 per cent of the listed companies paid dividends; this proportion declined to about 26.1 per cent in FY1991/92, perhaps as a response to the slowdown in manufacturing activity.

---

[32] The exchange risk on Foreign Exchange Bearer Certificates is borne by the bearer of the bond since at the time of honoring the claim the Government pays out the dollar equivalent of the face value of the bond. In recognition of the rupee denomination of the bond, the interest rate on the Certificate is fixed at 15 per cent.

[33] The *hundi* market, though illegal, was very popular with workers abroad because of its easy access, better exchange rate, and the speed with which the money was delivered to the beneficiary.

Limited Availability of Stock and the Institutional Investor

The stock market in Pakistan remains very thin: only limited quantities of a firm's stock are traded and price jumps are quite common. As a result, share prices exhibit a great deal of volatility because of fluctuations in the market. These prices cannot, therefore, be regarded as true indicators of the fair value of shares; it is difficult to judge whether these prices would prevail if large blocks of shares were to be unloaded in the market.

The credit regime in the country allows for considerable leverage in the ownership of companies. Cheap credit, resulting from financially repressive policies, allows for the sponsor/owners of companies to lock away a large share of equity. Nonmarket access to credit as well as inadequate enforcement of credit terms makes equity a less desirable form of financing. According to estimates from the State Bank of Pakistan, the typical share-holding pattern of a company (even a multinational) listed on the stock exchange is about 60–65 per cent with sponsors, about 20–25 per cent with institutional investors, and about 10–15 per cent with individual investors. In FY1990/91, less than 8 per cent of the equity of the largest 20 domestic firms were held by small shareholders, about 33 per cent by the institutional investors and other public sector agencies, while about 54 per cent were held by the large shareholder (e.g., board of directors and their proxies).

Prior to the 1991 reforms, about 10 per cent of equity of private domestic companies on average were available in the market. After the reforms, foreign investors were allowed into the market and there was some divestment by some institutional investors. As a result, the available float have been increased to around 20–24 per cent (including the stock held by foreign buyers) for the companies already listed on the stock exchange.[34] For new companies, regulations require about 30 per cent to be reserved for individual investors. It is not clear, however, whether sponsors manage to obtain some percentage of equity in the names of friends and relatives since there is no policing of such matters.

Liberal availability of credit from public sector banks and DFIs to blue-chip companies and well-connected sponsors and the weak enforcement of recovery procedures has made debt the preferred mechanisms for raising capital. The availability of credit at concessional rates has also constrained the development of capital markets. More recently, however, growing costs of plant and equipment have forced entrepreneurs to increasingly rely on the capital market to raise capital for investment. The analysis of the debt/equity ratio shows that around 45 per cent of the private sector's capital employed in the enterprise comprised debt compared with around 50 per cent for the

---

[34] The level of foreign investment in the stock market is currently $300 million, which represents 3.5 per cent of market capitalization and 12–15 per cent of market float.

public sector. The debt/equity ratio for the domestic private sector would have been slightly higher had multinationals, with an average debt/equity ratio of 25:75, been excluded from the overall figures for the private sector. The debt/equity ratio has been estimated on the basis of long-term debt. If the "permanent element" of the overdraft facilities extended by banks to these enterprises were to be included in the debt component, the debt equity ratio would be substantially higher.

Changes in regulations will help improve the relative attractiveness of equity. Banks have begun to respond to the State Bank's prudential requirement of a debt/equity ratio of 60:40. In addition, the Corporate Law Authority recently relaxed the condition that all new issues had to be priced at Rs 10 and distributed by means of a lottery. By allowing more market-determined pricing of new issues, equity financing has been made somewhat more attractive.

Privatization

The stock value of firms reflects market value only in respect of the relatively small percentage of shares traded in the market. Between 80 and 85 per cent of the shares are tightly held by either the sponsors or public sector institutions, such as the National Investment (Unit) Trust, Investment Corporation of Pakistan, the State Life Insurance Corporation, and the DFIs. The first three are together estimated to hold about 21 per cent of the estimated value of market equity.[35] Whether these public sector institutions perform the role of the institutional investors in a well-functioning market remains an open question. Certainly, very few of these institutions appear to be acting as a disciplinary force in the management of companies.

Cumbersome Procedures for Trading Stocks

Settlement involves physical delivery of securities and checks rather than electronic transfers. The process of share transfers and registration is cumbersome and time-consuming, requiring about 45 days of fairly detailed paperwork, all of which negatively impacts on trading volumes. Securities have to be counted, transfer deeds have to be verified against original certificates, and identification numbers of certificates have to be noted. There is also the risk of holding large numbers of negotiable instruments. The

---

[35] The State Life Insurance Corporation (SLIC) has a portfolio consisting of shares of 206 companies. It currently holds 3.1 per cent of the total market capitalization of KSE; the average holding period is five years. In 1992, it traded 3.7 per cent of its portfolio. SLIC's maximum exposure in the case of new companies is limited to 2.5 per cent or lower of the equity of the new company or its own portfolio. The yield that it is earning on its assets is 15 per cent.

settlement period is long, from 8 to 14 days. Because of lack of transparency and nonformalized reporting requirements, almost 50 per cent of the transactions are not reported. Only institutional and foreign investors require that brokers report their transactions through the recently established electronic market information system, FASCOM. Brokers do not wish to invest in setting up the same procedure for smaller clients.

Several areas remain where policy tends to be non-uniform in its approach. For example, there is a withholding tax of 10 per cent on dividend income, which makes equity investment relatively less attractive than the tax-free government bonds or foreign exchange accounts. There are also restrictions that limit the participation of foreign competition. All foreign buyers are required to register their purchases, and sales of shares by foreign buyers are not allowed until the shares have been registered, unless the buyer operates a special convertible rupee account.

Limited Use of Current Information Technology

If the securities market is to realize its potential as a meaningful financial market, the use of modern information technology must be relied upon for all aspects of the transaction. Though FASCOM, a system for the provision of market information via electronic means, has been made available, its use is quite limited. Until the summer of 1993, the system was available only in Karachi.

The automation of the settlement process and the clearing system must involve the introduction of trade reporting tickets to facilitate the introduction of real-time trade matching and reporting.

CONSTRAINTS ON THE DEVELOPMENT OF SECONDARY MARKETS FOR GOVERNMENT DEBT

As treasury bills and Federal Investment Bonds are non-redeemable before maturity, the secondary market is supposed to provide the liquidity to those holding these securities. However, for the reasons discussed below, the secondary market for government debt has not developed. The primary factor affecting the smooth functioning of the secondary market appears to be the lack of trust among the key players in the market.

(i) *Inconsistency in Government Actions*. The Government has not been consistent in accepting or rejecting bids. At times it has accepted offers for treasury bills that were several times more than the amounts for which bids were invited. At other times it has rejected bids without either assigning or indicating through its actions the reasons for rejection. This inconsistency is the result of a debt-auctioning process driven by fiscal considerations rather than a debt management/monetary policy strategy.[36]

(ii) *Lack of Competition Among Primary Dealers.* As the Government does not pre-announce the amount to be tendered, it has not generated competition among the primary dealers for a specified number of securities. As a result, the primary dealers compete with each other for what they expect will be the cut-off price that will be accepted by the Ministry of Finance. Some banks take advantage of the price auctioning system by offering a large bid at a specified price, which puts pressure on the Ministry of Finance to accept large bids irrespective of the cost. Banks are also not informed about dates of future auctions. The price in the secondary market tends to fall near auction time because the expectation of primary dealers is that they will obtain a better rate in the primary auction than in the secondary market.

(iii) *Length of Time Taken for Processing Bids.* The gap of two to four days between the submission and acceptance of offers is too long; the interbank market effectively stops functioning for about five working days every month.

(iv) *Other Attractive Government Securities.* Though it issues Federal Investment Bonds, the Government impacts on their marketability by issuing other paper which has vastly superior qualities in that it is bearer and exempt from *Zakat*. These competing instruments reduce the attractiveness of Federal Investment Bonds and with it that of the secondary markets.

(v) *Easy Access to the Repo Facility.* The current policy of the State Bank is to conduct repo at 15 per cent; instead of the market, the State Bank's repo window is providing the liquidity. Ostensibly, the objective is to force banks to focus their efforts on cash management. Recently, the State Bank tightened the repo facility by denying access to the window if the party concerned was a net lender during the day. The withholding tax treatment also encourages financial institutions to opt for a repo; the return on a repo transaction is treated as a capital gain and hence not subject to a withholding tax at source.

---

[36] In the case of federal investment bonds in particular the auctions have been held too frequently—making it difficult for a secondary market to develop.

(vi) *Withholding Tax Structure.* A major deterrent to the development of a market in which treasury bills and Federal Investment Bonds can trade freely is withholding tax. Treasury bills and Federal Investment Bonds are sold cum-interest and, on maturity, the holder of the instrument is liable to withholding tax on the entire profit.

The withholding tax structure also segments the market. Foreign dealers are subject to a tax rate of 60 per cent, while domestic institutions and individuals pay rates of 30 per cent and 10 per cent, respectively, resulting in an after-tax return of 4.5 per cent on a three-year Federal Investment Bond for a foreign participant as against 9.1 per cent in the case of a domestic institution. Foreign participants avoid this liability by a repo of the security to a domestic participant, withholding tax being deducted only on maturity or payment of interest. In anticipation of problems, dealers advise their customers to physically present their coupon on the profit payment date. This defeats the purpose of having a book-based system of government securities.

The tax structure also makes it difficult, and time-consuming, for the State Bank of Pakistan to keep track of the ownership of the document to ensure that the correct tax amount is deducted. Trading is not allowed seven days before a coupon date to enable the State Bank staff to calculate the withholding tax deductions. The problem is compounded by the lack of automation in the clearing and settlement systems, leaving the transactions subject to manual checking and monitoring of different securities and maturity dates. These problems will become more acute over time as the number of issues increases and there is wider distribution of this paper.

(vii) *Cumbersome Reporting Requirements.* The time involved to handle the reporting requirements paper work acts as a disincentive to transactions in Federal Investment Bonds.

The development of secondary markets will take time as primary markets themselves are in the early phases of their development.

INCREASED COMPETITION IN INSTRUMENTS

Differential Fiscal Treatment of Financial Instruments

A major impediment to the development of financial markets is the differential tax treatment of financial instruments. The tax structure favors some financial instruments over others and some investors over others, resulting in the segmentation of the market. For example, the tax rate for individuals holding treasury bills and Federal Investment Bonds is 10 per cent, while for domestic institutions it is 30 per cent, and for foreign financial institutions, 60 per cent.

Tax shelters are directed toward contributions to public sector saving vehicles and instruments, thereby creating a bias against private borrowers. For example, bank deposits, treasury bills and Federal Investment Bonds are subject to withholding taxes (at rates higher for banks than individuals), while income from Bearer National Fund Bonds, Special National Fund Bonds, instruments of National Savings Schemes (such as Mahana Amdani Accounts), the income from the first three issues of Water and Power Development Authority's bearer bonds and first two issues of its registered bonds, foreign exchange bearer certificates, dollar bearer certificates, and foreign currency accounts are exempt from income taxes. In fact, the investment in dollar bearer certificates, the Bearer National Fund Bonds, the Special National Fund Bonds and balances in foreign currency accounts are also exempt from wealth tax. In addition, withholding tax is not levied on the income earned from these bonds. On prize bonds, however, there is a withholding tax of 7.5 per cent if the prize money exceeds Rs 24,999. For the private sector, the most formidable constraint to resource mobilization is the fiscal concession granted to incomes earned from instruments issued by the Government.

The administrative division of the financial institutions is made more complex by the different fiscal treatment and regulatory mechanisms, which themselves tend to change frequently over time. A market niche is being created for different types of financial institutions—investment banks, leasing companies, and modarabas—because of fiscal concessions for some or because of discrimination against others through the regulatory system. This has led to the proliferation of new types of financial institutions and has distorted the relationship between financial activities and institutions. Examples of regulation which adversely affect operational activity include a relaxed set of rules on bad debt provision in the case of DFIs, a higher rate of taxation of banks compared with nonbank financial institutions, favorable rules on depreciation of fixed assets in the case of leasing companies, the imposition of an excise duty on loans from banks while other loans are exempt, a lower rate of tax for modarabas, and credit/deposit ratio requirements that discriminate against banks.

Limited Availability of Financial Instruments

As discussed above, there are basically only two types of instruments available in the market—government debt and equity. In the debt markets, secondary markets are not developed while in the equity markets, cheap credit and weak regulation fosters thin trading on limited information.

In Pakistan, financial arbitrage is limited by the fact that the formal markets limit trading only in long positions in equity only. No derivative instruments underpin such trading, and short-selling and futures trading is not allowed. Regulatory agencies actively discourage the development of these markets and instruments.

Potential New Instruments

Several new instruments could be introduced to make the financial markets more efficient, though such development is best left to the markets. The following is therefore for purposes of illustration only.

Commercial Paper

Large corporations with a good performance record (e.g., banks and multinationals) could, theoretically, issue their own commercial paper directly, i.e., without the involvement of banks in the capacity of intermediaries. A larger number of companies have the potential to issue commercial paper. In particular, the subsectors of textile and sugar have a large base and could use the market to satisfy their seasonal demand for credit by floating commercial paper. The Government has liberalized many of the rules pertaining to the public issue of commercial paper, though corporate bonds have only been issued by public sector corporations. For several reasons, the private sector has not been able to issue these instruments.

(i) Bond-issuing costs (including interest and administrative costs) are usually higher than bank-financing costs and issuing procedures are time-consuming. Therefore, bank credit is preferred.

(ii) There is little incentive for "blue chip" companies to float commercial paper because they have good access to bank credit.

(iii) High stamp duty (3.5 per cent in Sindh) and registration fees are major deterrents to the development of corporate commercial paper. In any case, the law does not define commercial paper. What level of stamp duty would be levied on which instrument is not clear.

(iv) One of the prerequisites of a viable private debt securities market is the ready availability of credit ratings. In Pakistan, however, there is inadequate information available within the system, even on "blue chip" borrowers. There is no credit rating system for assessing the creditworthiness of issuers of instruments. The State Bank has only recently established a Credit Information Bureau.

(v) The development of a convertible bond market is also constrained by legislation which does not permit companies to issue shares without giving existing shareholders the right of first refusal. Companies should be allowed to issue a percentage of the issued capital, perhaps every three years, without having to uphold pre-emption rights of existing shareholders.

(vi) The charters or rules and regulations governing provident and pension funds do not permit them to invest in corporate debt.

(vii) The State Bank has also imposed several restrictions that make public issues of commercial paper difficult. Specifically, private sector companies cannot issue securities in bearer form, placing them at a competitive disadvantage vis-á-vis public sector institutions, which have been allowed to issue bearer securities for the past several years. The Ministry of Finance recently decided that government-owned enterprises will not be allowed to issue bearer certificates.

(viii) Banks are not permitted to provide bank guarantees on any security issued by a private company. This makes it particularly difficult for companies, other than blue-chip entities, to raise money by issuing securities. The imposition of these restrictions has made the issue of redeemable capital securities by the private sector difficult. Consequently, there has been little operational activity in this segment of the financial market.

Convertible Bonds

A consequence of inadequate supervision is that accounts presented by corporations are generally not considered accurate: Share values and dividend declarations are not considered to reflect the true market values. In such an environment, convertible bonds may be an important source of finance as well as a means of management discipline, creating an incentive for management to perform or risk losing capital.

Factors Hindering Development of Housing Finance Markets

Several difficulties in the interpretation of the tax code have hindered the development of housing finance companies. First, in the case of a self-occupied property, the Government does not permit the taxpayer to set off the expenditure on debt servicing (the financial charges component) against his other income. Second, housing finance companies can be deemed to be indulging in trade on property, even though their charter specifically prohibits them from such activities. They could be held liable to a 2.5 per cent turnover tax. Third, under a mortgage agreement that registers the property in the name of a company, the occupant of the house could be treated as a tenant for property tax purposes and, therefore, makes the housing finance company liable for property taxes at rates applicable to rented properties, or 25 per cent of the "annual rent."

Another important factor hindering the growth of housing finance is the high cost of registering property-related transactions. The costs in urban areas in Sindh include a 1 per cent registration fee, an 8.5 per cent provincial stamp duty, a 2.5 per cent property transfer fee charged by local government, a 5 per cent capital value tax on residential plots exceeding 240 square yards, and other charges (professional charges, brokerage fees and court fees). In the case of a loan, there is a 1 per cent charge for registration of the mortgage documents, a 3 per cent stamp duty for a nonbank financial institution (the stamp duty is 1 per cent in the case of a banking company) plus charges for fire insurance and professional input.

POLICIES FOR THE DEVELOPMENT OF FINANCIAL MARKETS

Fiscal Prudence and Stable Policy Environment

It is generally agreed that a stable macroeconomic environment is critical to the success of any reform effort (see Khan and Sundrarajan 1992). A stable price regime supported by sustainable fiscal and current account deficits which are capable of being financed are key components of any reform program. Unfortunately, inflationary expectations associated with large fiscal deficits have geared economic and financial decisions toward short-term views. Commercial banks are focusing their activities on short-term lending, most of which are in the form of overdraft facilities. Their long-term lending activity is essentially in the capacity of an intermediary for the State Bank's line of credit for locally manufactured machinery. Similarly, leasing companies are primarily engaged in entering into two to three-year leases; five-year leases have become less common. The lack of relevant and reliable economic indicators also makes it difficult for players to formulate their expectations regarding the long term. The current level of the fiscal deficit will, therefore, need to be reduced if inflation is to be reduced to less than the current level of close to 10 per cent.

The Government has historically financed its budget by forced levies on the financial system through a system of credit rationing and obligatory portfolio requirements, requiring an investment in government bonds at below market rates. The financial sector has been providing "disguised" revenue to the Government as it captures a large fraction of bank deposits at artificial interest rates (Haque and Montiel 1991 and 1992). The budget deficit is about 8 per cent of GDP. As the demand for finance by the Government increases, the level of interest rates rises and the private sector is denied credit through credit rationing rather than on the basis of price.

An uncertain political and economic environment discourages long-term investment and institution-building. In the past, policy has frequently been changed because of the revenue needs of the Government and the political pressure of lobby groups; that has always occurred at the expense of efficiency. For example, in January 1992, the Government ostensibly dismantled the structure of credit ceilings and introduced the concept of a 65 per cent credit/deposit ratio. The credit/deposit ratio was then changed to 30 per cent on local currency deposits and 40 per cent on foreign currency deposits, and then changed again to a uniform ratio of 30 per cent.

Perhaps the factor that will inhibit the development of financial markets most is the lack of political will and the determination to proceed against all those responsible for perpetrating business and financial frauds.

Market-Oriented Monetary and Credit Policy

Market-based credit allocation, as a part of a coherent monetary policy that relies on indirect control of monetary aggregates in line with domestic inflationary developments, has been shown to be important to the development of financial markets (Khan and Sundrarajan 1992). Financial intermediation will improve if monetary policy allows market forces to determine credit allocation. Directed credit, with nearly 14 different concessional credit schemes, is causing considerable misallocation of resources, as is evidenced in the reserve position of commercial banks.[37] Rationing of concessional credit has led to distortions in credit allocations and has often facilitated the establishment of industrial units of doubtful viability.

Villanueva and Mirakhor (1990) note that interest rate deregulation must keep pace with the development of market supervision in order to

---

[37] For FY1991/92, the mandatory credit targets for commercial banks comprised Rs 6.7 billion for agriculture and Rs 2.6 billion for small-scale industry. The share of small-scale industries in credit extended to the manufacturing sector (both private and public sector enterprises) averaged just over 25 per cent over the previous five years. For self-employment (including the duty free import of around 80,000 yellow cabs), the amount advanced during FY1992/93 was approximately Rs 11.5 billion.

prevent adverse selection in the credit markets from exposing bank portfolios to excessive risk. In Pakistan, there is an urgent need to improve supervision and to harden the budget constraint for financial institutions so that expectations, implicit or otherwise, of a bailout are removed.

Reform of the Financial Sector

While numerous ideas relating to the reform of financial markets and institutions have been discussed in earlier sections, the broad strategy of reform remains to be developed. Currently, the Government maintains a dominant role in the sector, not as the guardian of the system but as the largest provider of financial services in the system. The system is dominated by inefficiently run, overstaffed, government-owned financial institutions that are subject to considerable political interference and have, as a result, developed a poor quality portfolio of assets. Competition from foreign banks and the newly created banks provides only limited competition. The country's money and capital markets are small and trading is generally carried out in government securities and in the equities of a relatively small number of companies. Cumbersome legal processes for contract enforcement and the indifferent application of accounting, auditing, and supervision standards have contributed significantly to the poor performance of the country's financial institutions.

The financial sector as it stands in Pakistan today is, therefore, unable to perform many of the functions expected of it. Risk-sharing arrangements, hedging, and portfolio diversification are all limited by the lack of availability of instruments. Market solutions to dealing with asymmetric information remain limited in view of the difficulties of contract enforcement, and inadequate regulation and supervision of financial institutions. The inordinately large role of the Government in this sector as a supervisor, regulator, and owner of institutions and instruments has resulted in limiting arbitrage so necessary to market efficiency. Moreover, the moral hazard associated with such an arrangement has led to the nationalized banks holding a large portfolio of bad debts, the cost of which will be borne by the taxpayer.

Recommendations on reform measures include the following:

(i) *instituting deregulation, liberalization, and privatization with an emphasis on market and instrument development.* This would include facilitating entry by eliminating licensing and/or cumbersome regulatory procedures; encouraging price competition, including market determination of interest rates; developing a secondary market trading in areas such as government debt instruments and commercial paper; allowing shortselling and development of derivatives if market participants so choose; and developing a uniform system of regulation and fiscal treatments of financial instruments and institutions.[38]

(ii) *improving regulation and supervision of financial markets and institutions.* The regulatory agencies—such as the State Bank of Pakistan and the Corporate Law Authority—need internal reforms, including the installation of qualified and suitably trained staff to ensure compliance with sound business practices. The departments in each need to be streamlined. Both agencies need to be made autonomous in terms of their roles and responsibilities. Supervision must be vigilant and punitive in the interests of its own credibility.

(iii) *improving the information and payments systems and technology.* The lack of transparency and credibility of information affects the development of a modern financial system. The integrity of trading in financial markets is preserved by the timely flow of information relating to trades and fundamentals. If financial markets are to handle large transactions in an expeditious manner, clearing house and depository functions are necessary elements of the enabling environment. Long delays in intercity check cashing and lengthy requirements for stock registrations are serious impediments to market development.

(iv) *improving the legal system.* The inability to enforce contracts through legal channels is a root cause of inefficiency in the Pakistani economy. The legal system must be made more responsive to the needs of financial liberalization. This will require a supportive legal, administrative, and regulatory structure for protecting bank depositors and an adequate bankruptcy law for enforcing foreclosures.

As there are fairly large complementarities among these four elements of the reform strategy, a natural sequencing of measures suggests itself. Efficient financial intermediation would require that movement be made on all these fronts. However, experience with poor regulation and supervision, which has stifled market development on the one hand and created increased financial losses and tax burdens on the other, suggests that liberalization and deregulation should be immediate. Meanwhile, the efforts of the authorities should be directed toward the development of an adequate regulatory and supervisory system which includes relatively conservative prudential

---

[38] Regulatory mechanisms and fiscal incentives should be linked to activities (e.g., lending, accepting deposits, leasing, underwriting) rather than to institutions. Similarly, all securities should be subject to a uniform policy in respect of Zakat, wealth tax and withholding tax.

regulations and well-designed audit mechanisms for early detection of problems. The possibility of a deregulated market seeking market solutions to developing the information and payments system should be explored. Improved information flows and the entry of the private sector in generating information will generate self-regulation and enforce a reputational bonding (Holmstrom 1979; Hart 1981). The design of the reform should attempt to enlarge the role of the private sector not only in trading but in delivering as much of the infrastructure for trading as is possible. State interference should be limited only to where it is entirely necessary such as the development of the legal infrastructure to enable more efficient contracting. However, this is part of a larger structural change as it is tied to the issue of legal and judicial reform as a whole.

CONCLUSION

Financial markets in Pakistan remain under considerable government control not because of tight regulation and supervision but primarily because the Government owns much of the financial sector and is the main user of credit in the system.

The regulatory and supervisory system remains weak and inefficient, relying on direct controls and lacking the flexibility to deal with the dictates of global financial markets. The regulatory agencies need internal reforms, including qualified and suitably trained staff to ensure compliance with sound business practices and to enable streamlining of the various departments to make them more efficient and productive, and to make staff aware of modern commercial banking practices. Their autonomy with regard to their own markets would be an important step in defining responsibility.

The accounting and auditing conventions and practices need to be strengthened so that reliable information is more readily available. Negligence, misreporting, concealing information, and false certification of information and auditing should be swiftly penalized and made public. Credit ratings should be made available at all levels to allow reputation effects to work in the market for credit. Considerable private sector participation could be initiated in this area with appropriate incentives from the regulatory agencies.

The legal process is slow in enforcing contractual obligations. Delays are caused by court practices, a clogged judicial system, and defects in regulations on procedural matters which render the process ineffective in dispute resolution. The inability to enforce contracts through legal channels is one of the root causes of the inefficiency in the Pakistani economy.

A fiscal policy based on large deficits has been soaking up liquidity in the financial system and has inhibited the development of the financial markets by crowding out credit to the private sector. At the same time, policies based on directed credit for planners have distorted interest rates and misallocated credit. While establishing macroeconomic policies consistent with sustained noninflationary growth, however, efforts must be made to ensure that the fiscal treatment of financial instruments and institutions is uniform. Regulatory mechanisms and fiscal incentives should be linked to activities (lending, accepting deposits, leasing, underwriting) rather than to institutions. Similarly, all securities should be subject to a uniform policy in respect of Zakat, wealth tax and withholding tax.

Over the longer term, increased private sector participation in financial markets will be required. It is only by allowing private sector entry into setting up financial instruments as well as developing instruments, that adequate competition and arbitrage will take place. Increased competition will also lead to the development of more formal participation of the private sector in developing and implementing financial sector initiatives and improving self-regulation.

Legislative and fiscal changes may be required to stimulate the development of secondary markets in government debt and the stock market. The Government must initiate a system of regular auctions reasonably spaced to give the secondary market time to develop.

A four-pronged strategy for reform has been suggested: deregulation, liberalization and privatization with an emphasis on market and instrument development; improvement in regulation and supervision of financial institutions and markets; improvements in the payment system and the information technology; and improvements in the legal system for better contract enforcement including debt recovery or collection of collateral. The design of the reform should attempt to enlarge the role of the private sector not only in trading but in delivering as much of the infrastructure for trading as is possible. Government interference should be limited only to where it is entirely necessary such as the development of the legal infrastructure to enable more efficient contracting. This, of course, would be part of a larger structural change of the legal and judicial system.

Bibliography

Australia-New Zealand Grindlays Bank. 1992. "Capital Markets in Pakistan: Resource Mobilization for Housing Finance Companies." December.

Black, F., and M. Scholes. 1972. "The Valuation of Option Contracts and a Test for Market Efficiency." *Journal of Finance.* 27(2) May, pp. 399–417.

Calvo, Guillermo A., Leonardo Leiderman, and Carmen Reinhart. 1993. "The Capital Inflows Problem: Concepts and Issues." IMF Paper on Policy Analysis and Assessment. PPAA/93/10 July.

Cho, Yoon Je. 1986. "Inefficiencies from Financial Liberalization in the Absence of Well-Functioning Equity Markets." *Journal of Money, Credit and Banking.* 18(2) May.

Dattels, Peter. 1992. "Development of Secondary Financial Markets in Pakistan and the Emerging Market for Government Securities." Address to the Management Association of Pakistan, 23 January.

Eatwell, John, Murray Millgate, and Peter Newman. 1989. *The New Palgrave: Finance.* New York: McMillan.

Faruqi, Shakil. 1992. "Financial Sector Reforms in Asian and Latin American Countries." Washington, DC: Economic Development Institute, World Bank.

Grossman, Sanford, and Oliver Hart. 1980. "Takeover Bids, Free Rider Problem, and the Theory of the Corporation." *Bell Journal of Economics.* 11(1):, pp. 42–64.

Grossman, Sanford, and J. E. Stiglitz. 1980. "On the Impossibility of Informationally Efficient Markets." *American Economic Review.* 70(3) June, pp. 393–408.

Hakansson, N. "Changes in the Financial Market: Welfare and Price Effects and the Basic Theorems of Value Conservation." *Journal of Finance.* 37(4) September, pp. 977–1004.

Haque, Nadeem Ul. 1988. "Fiscal Policy and Private Saving Behavior in Developing Countries." IMF *Staff Papers.* June. Vol. 35, No. 2. pp. 316–335.

Haque, Nadeem Ul, and Peter J. Montiel. 1989. "Consumption in Developing Countries: Tests for Liquidity Constraints and Finite Horizons." *Review of Economics & Statistics*. August.

_____. 1991a. "Capital Mobility in Developing Countries—Some Empirical Tests." *World Development*. October. Vol. 19 No. 10. pp. 1391–1398.

_____. 1991b. "Macroeconomics of Public Sector Deficits: The Case of Pakistan." Paper prepared for World Bank Project on "The Macroeconomics of Public Sector Deficits." Working Paper No. WPS 673. Country Economics Department, World Bank. May.

_____. 1992a. "Policy Choices and Macroeconomic Performance in the Nineties." In *Financing Development in the Nineties*, edited by Anjum Nasim. Oxford University Press for the Lahore University of Management Sciences.

_____. 1992b. "Rate Policy in Pakistan: Recent Experience and Prospects." In *Financing Development in the Nineties*, edited by Anjum Nasim. Oxford University Press for the Lahore University of Management Sciences.

Hart, Oliver. 1981. "The Market Mechanism as an Incentive Scheme." University of Cambridge Economic Theory Discussion Paper.

Hasnie, R.A., and S. Iqbal. 1992. "Government of Pakistan Debt Auctioning System and the Evolution of a Tradeable Securities Market: Implications for the Financial Sector, Monetary Policy and Pakistan's Economy." October. Mimeo.

Holmstrom, Bengt. 1979. "Moral Hazard and Observability." *Bell Journal of Economics*. 10(1): pp.74–91.

Khan, Makhdoom Ali. 1992. "Land Tenure in the Punjab." A study for the "Shelter for Low Income Communities." Project being administered by the World Bank. Mimeo.

Khan, Mohsin S., Peter J. Montiel, and Nadeem Ul Haque. 1991. "Macroeconomic Models for Developing Economies." International Monetary Fund. August.

Khan, Mohsin S., and V. Sundrarajan. 1992. "Financial Sector Reforms and Monetary Policy." In *Structural Adjustment and Macroeconomic Policy Issues*, edited by V. A. Jafarey. Washington, DC: International Monetary Fund and Pakistan Administrative Staff College.

Khanna, Ashok. 1995. "South Asian Financial Sector Development." In *Financial Sector Development in Asia*, edited by Shahid N. Zahid. Hong Kong: Oxford University Press for the Asian Development Bank.

McKinnon, Ronald I. 1973. *Money and Capital in Economic Development*. Washington, DC: Brookings Institution.

Montiel, Peter, Pierre-Richard Agenor, and Nadeem Ul Haque. 1993. "Informal Finance Markets in Developing Countries." *Advances in Theoretical and Applied Economics*. Basil-Blackwell.

Nayak, P. Jayendra. 1995. "Reform of India's Financial System." In *Financial Sector Development in Asia: Country Studies*, edited by Shahid N. Zahid. Published by Asian Development Bank.

Romer, Paul. 1993. "New Goods, Old Theory, and the Welfare Costs of Trade Restrictions." Berkeley: University of California. September.

Said, Qayyim. 1993. "Finance Companies Scandals Developed as Good and Bad Firms Were Not Distinguished." *The News*. March. Islamabad.

Samad, Abdus. 1993. *Governance, Economic Policy and Reform in Pakistan*. Lahore: Vanguard Books.

Shaw, Edward. 1973. *Financial Deepening in Economic Development*. New York: Oxford University Press.

Stiglitz, Joseph, and Andrew Weiss. 1981. "Credit Rationing in Markets with Imperfect Information." *American Economic Review*. 71(2) June, pp. 393–410.

Stiglitz, Joseph. 1993. "The Role of State in Financial Markets." World Bank Annual Conference on Development Economics, May, Washington DC.

Villanueva, Delano, and Abbas Mirakhor. 1990. "Strategies for Financial Sector Reforms." IMF *Staff Papers*. 37(3) September.

Vittas, Dimitri. 1992. "Financial Regulation: Changing the Rules of the Game." Economic Development Institute, World Bank. Mimeo.

World Bank. 1990. "Financial Systems and Development." Policy and Research Series No. 15. Washington, DC: World Bank (reprint of the *World Development Report 1989*).

AUTHOR INDEX

Aiyar, C. V., 384, 385
Amsden, Alice, 39
Anant, T. C. A., 397, 425
Azariades, C., 373
Bery, S. K., 372
Bester, H., 371, 423
Bhandari, L., 381
Bhatt, V. V., 376
Blinder, A., 375
Bouman, F. J. A., 384
Calvo, Guillermo A., 461
Caprio, Gerard, Jr., 167, 214
Chen, Shang-Cheng, 106, 141
Chiu, Paul C. H., 101, 103, 119, 137
Cho, Yoon Je, 39
Cole, David C., 39, 165, 169, 170, 173, 176
Cooper, R., 371
Corbo, V., 175
Das-Gupta, A., 369, 374, 377, 378, 380, 381, 382, 383, 385
Dattels, Peter, 446
Dholokia, N., 381
Diaz-Alejandro, C., 402
Edwards, Sebastian, 172
Euh, Yoon-Dae, 39
Faruqi, Shakil, 461
Fershtman, C., 412
Friedman, J., 412
Gangopadhyaya, S., 397
Ghate, P., 377, 383
Glosten, L. R., 372, 414
Goodhart, C. A. E., 413
Goswami, O., 397
Grossman, S., 372
Haque, Nadeem Ul, 457, 464, 482
Hart, Oliver, 485
Holmstrom, Bengt, 485
Jain, A. K., 381
de Juan, A., 394
Jung, Byung-Hyu, 56

Khan, Mohsin S., 481, 482
Khanna, Ashok, 280
Khurana, R., 381
Kuo, Ping-Sing, 153
Lamberte, Mario B., 236, 237, 245
Lapar, Ma. Lucila A., 237
Larrain, M., 222
Lee, Yung-San, 106, 141
Liang, Ming-Yih, 149
Little, I. M. D., 376
Liu, Christina Y., 101, 133
Liu, Shou-Hsiang, 103
Llanto, Gilberto M., 236, 237
McKinnon, Ronald I., 108, 154, 172, 175, 373
Madhur, —, 383
Magno, Marife, 237
de Melo, J., 175
Milgrom, P. R., 372, 414
Mirakhor, Abbas, 373, 482
Montiel, Peter J., 457, 464, 482
Nam, Sang-Woo, 20
Narayana, D., 377
Nasution, Anwar, 172, 173
Nayar, C. P. S., 383
Panikar, P. G. K., 377, 378, 382
Patrick, Hugh T., 154
Phlips, L., 371
Punyashthiti, Prakid, 327
Roe, A. R., 376
Ross, T. W., 371
Rothschild, M., 371
Said, Qayyim, 452
Saldaña, Cesar G., 276, 278, 279
Samad, Abdus, 452
Sen, C., 377
Shaw, Edward S., 373
Shea, Jia-Dong, 100, 101, 117, 138, 148, 151, 153, 154
Sheng, Andrew, 222, 398, 400
Slade, Betty F., 165, 169, 170, 176

Spence, A. M., 371
Stiglitz, Joseph E., 152, 173, 214, 371, 372, 375, 394, 404
Sun, Chen, 149
Sundararajan, V., 195, 481, 482
Taylor, Lance, 374
Thingalaya, N. K., 376
Timberg, T., 384, 385
Tobin, J., 374, 428
Trairatvorakil, Prasarn, 327
Tsiang, Shoh-Chieh, 106, 108, 148

Varma, J. R., 398
Villanueva, Delano, 373, 482
Vittas, Dimitri, 461
Vora, M. N., 381
Weiss, Andrew, 371
White, Lawrence J., 170, 280
Wu, Chung-Shu, 148
Yang, Ya-Hwei, 100, 118, 145
Yang, Young-Sik, 56
Yoo, Jung-Ho, 39
Zahid, Shahid N., 170, 173

SUBJECT INDEX

accounting principles, generally accepted (GAAP), 282, 283, 285–286
Accounting Standards Council (ASC) (Philippines), 282–283
accounting systems
    accounting profession and, 97–98, 221
    accrual basis, 283–284
    bank accounting, 284–286
    and capital markets, 202
    cash basis, 284
    credibility, 96, 221
    in India, 371, 399–400, 421–422
    in Indonesia, 202, 221
    in Korea, 73–74
    in Pakistan, 462–463
    in the Philippines, 257–258, 282–287, 290, 296
    in Taipei,China, 95–98
    in Thailand, 323
adverse selection of borrowers, 13, 115, 369, 375, 377, 396, 398, 399, 407, 429, 463, 485
    and bank regulation, 373
    credit guarantees and, 423
    interest rates and, 100, 372–373, 377
agricultural cooperatives
    in Korea, 30
    in Taipei,China, 85, 86, 121
    in Thailand, 304, 306, 307, 335–336
    See also agriculture, lending to
Agricultural Development Bank of Pakistan, 467, 468, 470
agricultural, lending to
    in India, 391, 392, 406
    in Indonesia, 187
    in Korea, 4, 6, 14, 33, 34, 38
    in Pakistan, 467, 468, 470
    in the Philippines, 241, 247, 248–249, 258
    in Taipei,China, 111, 140
    in Thailand, 313, 316, 330, 335–336, 338, 360
    See also agricultural cooperatives
arbitrage, in Pakistan, 481, 485
Asia Trust Bank (Taipei,China), 104, 105, 122
Asia Trust Bank (Thailand), 326
asymmetric information
    financial infrastructure and, 173, 281–282
    and financial markets, 238
    and government intervention, 13, 460
    in India, 372–374
    in Indonesia, 216
    in Korea, 12–13
    and lending decisions, 99, 169, 216, 238, 239, 249, 294, 372–374, 382
    and nonbank financial intermediaries, 13–14
    in Pakistan, 460
    in the Philippines, 238, 239, 249, 294, 296
    in Taipei,China, 99

Bangkok Bank, 337
Bangkok International Banking Facilities (BIBF), 313, 363, 365
Bangkok Stock Exchange, 342, 353
bankers' acceptances. *See under* Financial instruments
Bankers Association of the Philippines (BAP), 279, 280
Bank for Agriculture and Agricultural Cooperatives (BAAC) (Thailand), 306–308, 336, 338
Bank Indonesia, 168, 172, 180, 183, 218–220
    interventionist policy, 175, 219
    liquidity control, 205
    monetary policy, 197–198, 228

rediscount facilities, 176, 182
refinancing loans to banks, 182, 194
banking, laws governing
    in Indonesia, 168, 181, 182, 200
    in Korea, 40
    in the Philippines, 248
    in Taipei,China, 98–99, 101, 104, 105, 106, 110, 114, 121, 134, 135
banking, offshore
    in Taipei,China, 132–133
    in Thailand. *See* Bangkok International Banking Facilities (BIBF)
banking, quasi-, 258, 260, 272
    defined, 258
banking sector
    branching policy, 29, 101, 104, 113–114, 176, 178–180, 235, 248, 293, 313, 315–316, 360–361, 392
    concentration, 253, 256, 327
    deposit insurance, 263, 423, 424
    deposits, 18, 86, 115, 192–193, 242–244, 254, 393
    efficiency, 35–40, 100, 118–119, 120, 223, 251–252
    entry and exit, 29, 101, 106, 176, 179, 235, 246–248, 253, 293–294, 367, 444
    foreign entry, 14, 29, 179, 241–242, 247, 248, 294, 304
    funding sources, 242–244
    of India, 389–396
    of Indonesia, 168, 178–198
    insolvencies, 105, 242, 326
    of Korea, 2, 5, 28–41, 68–73
    loans, 4, 18, 25, 86, 193, 194, 254, 392–393
    loans, nonperforming, 2, 5, 10, 35–37, 66, 68, 118–119, 120, 178, 194, 195–196, 216–217, 222–223, 225, 255, 371, 450–452
    management, 316–317
    ownership, 2, 28–29, 71–73, 99, 100, 111–113, 185, 191, 314, 371, 444
    of Pakistan, 444–453
    of the Philippines, 235, 236, 240–258
    privatization, 2, 10, 71, 104, 241, 246, 295, 403–404, 473
    prudential regulation, 106
    recapitalization, 223–224, 227
    restructuring, 68–71, 227, 400–407
    of Taipei,China, 85, 86, 104, 100, 106, 110–119
    of Thailand, 304, 309–311, 313, 314–317, 325–329, 336–342
Bank of Communications (Taipei,China). *See* Chiao Tung Bank (Taipei,China)
Bank of Korea, 26–28, 59, 67
    monetary policy, 27
    rediscounting policy, 27
Bank of Taiwan, 110
Bank of Thailand, 319, 329, 344–347, 357, 358, 359, 363, 365
    intervention policy, 360
    liquidity management, 350, 360
    monetary base control, 345–346
    monetary policy, 344–347, 351, 364
    supervision and examination, 321, 322, 326, 334, 362–363
Bank of Thailand Act, 326
bankruptcy laws
    in India, 399, 420–421
bankruptcy resolution. *See* commercial legal systems
BAPEPAM (Indonesian Capital Market Executive Agency), 176, 177, 200, 201, 217, 218, 221, 222
bills finance companies, 115, 124–125
Board for Financial Reconstruction (BIFR) (India), 420–421, 427, 428
Board for Financial Supervision (India), 423
Bombay Stock Exchange (BSE), 411, 413
bond markets
    in India, 411, 413–416

SUBJECT INDEX   497

in Indonesia, 202–203
in Korea, 63–64
laws governing, 125
in Pakistan, 456–457
in the Philippines, 276
in Taipei,China, 90, 92, 102, 125–126
in Thailand, 352–353
bonds, corporate. *See under* financial
    instruments
bonds, government. *See under* financial
    instruments
Bureau of Internal Revenue (Philippines),
    284
    and pension funds, 270
    and the stock exchange, 277

call loans. *See under* financial instruments
capital adequacy requirements
    in India, 399, 422
    in Indonesia, 165, 173, 177, 180,
        181, 194, 195, 209, 224
    in Korea, 40, 52–53
    in Pakistan, 464
    in the Philippines, 249–250, 294
    in Taipei,China, 106
    in Thailand, 313, 315, 318, 334, 340,
        351, 362–363, 365
capital markets
    accounting systems and, 202
    attractiveness to investors, 196–197,
        475–476
    automation, 199, 477
    badla trading, in India, 413
    capitalization, 276–277
    cotango trading, in India, 413
    fair trade practices, 70, 76, 129, 200,
        278,–279, 323, 412–413
    foreign entry, 12, 62, 130–131, 200,
        355, 458, 477, 479
    government intervention, 473
    government support, 2, 61–62, 78,
        126, 199, 355, 356

growth, 128
in India, 411–416
in Indonesia, 169, 176, 198–203,
    225–226
information regulation, 130, 278, 322
in Korea, 1, 2, 12, 15, 17, 19, 20, 61–
    66, 70, 76, 78
laws governing, 61, 129, 130, 200,
    320, 322, 353, 321–323, 365
marginal lending, 126, 129–130, 135
over-the-counter trading, 200, 202,
    323, 411
in Pakistan, 441, 443, 453, 456–458,
    473, 474–482
participants, 63, 127, 128, 129, 274
in the Philippines, 270, 274–279
privatization, 199
regulation, 200, 218, 277–278, 295,
    412, 413
support funds, 78, 354, 355, 356
in Taipei,China, 125–131, 136, 137
taxation, 277, 278
in Thailand, 309, 320–323, 353–356
trading volume, 63, 127, 128, 276,
    353, 354, 355, 356, 411, 458
See also bond markets; stock markets
Cathay Trust (Taipei,China), 104, 105,
    122
central bank certificates. *See under*
    financial instruments
Central Bank of China (CBC), 98, 107,
    108, 110, 123
    supervision and examination, 106
    redeposit policy, 123
    rediscounts, 141
Central Bank of the Philippines, 235, 236,
    241, 244, 247, 257, 260, 279, 281, 282,
    283
    laws governing, 250, 297
    rediscounting policy, 246, 252
    supervision and examination, 257,
        277, 297

498 SUBJECT INDEX

central banks
    of India. *See* Reserve Bank of India (RBI)
    of Indonesia. *See* Bank of Indonesia
    of Korea. *See* Bank of Korea
    of Pakistan. *See* State Bank of Pakistan (SBP)
    of the Philippines. *See* Central Bank of the Philippines
    of Taipei,China. *See* Central Bank of China (CBC)
    of Thailand. *See* Bank of Thailand
Central Deposit Insurance Corporation (Taipei,China), 98, 105–106
Central Trust (Taipei,China), 111, 112
certificates of deposit (CDs), negotiable. *See under* financial instruments
Chiao Tung Bank (Taipei,China), 111, 141
China Trust (Taipei,China), 104, 110, 112, 122
Citizens National Bank (Korea), 1, 30
civil and commercial codes (Thailand), 320, 321, 334, 357
collateralized lending
    asymmetric information and, 169, 216
    in India, 373, 385
    in Indonesia, 169, 216
    in Korea, 12–13
    in the Philippines, 296
    in Taipei,China, 99, 115, 116
    in Thailand, 335
    See also mortgage loans
commercial banks
    electronic banking, 242, 327
    and financial deepening, 309, 328
    and foreign exchange supply, 352
    in India, 389–390
    in Indonesia, 168, 178, 179, 184–197
    and investment banking, 69–70
    in Korea, 14, 16, 18, 28–29, 30, 31, 71–73
    laws governing, 168, 314–317, 325–326, 362
    ownership of insurance companies, 266–267
    in Pakistan, 441, 443, 444–452, 473, 483
    in the Philippines, 235, 236, 238, 241, 242–244, 245, 249, 251, 252, 254, 255, 272, 273
    in Taipei,China, 110–119
    in Thailand, 304, 306, 307, 313, 314–317, 325–329, 352, 361–362, 367
    See also universal banks
commercial bills. *See under* financial instruments
commercial legal systems
    banking tribunals, 399, 419, 465
    contract enforcement, 399, 419, 486, 487
    foreclosure, private, 399, 408, 420
    in India, 372, 399, 408–409, 419–421
    in Indonesia, 195, 196, 216, 220–221
    in Korea, 13–14
    and lending decisions, 282
    in Pakistan, 463–465, 486, 487
    in the Philippines, 282, 288–290, 296
    in Taipei,China, 98–99
commercial paper. *See under* financial instruments
Controller of Capital Issues (CCI) (Pakistan), 453
Cooperative Bank of Taipei,China, 98, 111, 121
cooperatives, savings
    in Korea, 15, 17, 18, 46, 53, 54
    in the Philippines, 260, 261, 262
    in Taipei,China, 85, 86, 121
    in Thailand, 305, 306, 307, 336
Corporate Law Authority (CLA) (Pakistan), 460–461, 462, 476, 486
credit, access to
    in India, 384–385
    in Korea, 39

SUBJECT INDEX   499

in the Philippines, 239
in Thailand, 328, 369
credit, directed
    in India, 391–392, 406
    in Indonesia, 165, 176, 177, 181–182
    in Korea, 2, 3–6, 11, 12, 27, 66, 76
    and macroeconomic imbalance, 5
    in Pakistan, 437, 454, 467, 470, 472, 473, 484
    in the Philippines, 248, 249
    in Taipei,China, 100, 141–142
    in Thailand, 313, 316, 346, 360, 361
credit allocation. *See* credit, directed
credit control system
    in Korea, 6–7, 10, 27
    in Pakistan, 467
credit foncier companies, in Thailand, 304–305, 306, 307, 334
    insolvencies, 334
    laws governing, 317, 321, 326, 331, 334, 362
credit guarantee arrangements
    and adverse selection of borrowers, 423
    in Indonesia, 219
    in Korea, 17, 33, 52
    in the Philippines, 264, 271
    in Taipei,China, 143–145
credit information arrangements
    in Korea, 52
Credit Information Bureau Inc. (CIBI) (Philippines), 290–291
Credit Information Bureau (Pakistan), 462, 482
credit market signaling, 406, 425
credit rating agencies, 282
    in Korea, 59, 75
    in Pakistan, 481–482
    in the Philippines. *See* Credit Information Bureau Inc. (CIBI) (Philippines)
    in Thailand, 313, 365
curb market. *See* finance, informal

debentures. *See under* financial instruments
debt, public
    foreclosure, in India, 397–398, 415
    in Korea, 23, 24–25
    in Taipei,China, 90, 117
debt financing
    in Indonesia, 169, 198
debt recovery. *See* commercial legal systems
Deposit Insurance and Credit Guarantee Corporation (India), 424
deposit insurance arrangements, 423
    implicit, in Korea, 13
    in India, 424
    in the Philippines, 251, 261, 264
    in Taipei,China, 98, 105–106
deposit money banks (DMBs), 16, 18, 24, 31, 32–34
deposits, demand. *See under* financial instruments
deposits, time. *See under* financial instruments
Development Bank of the Philippines (DBP), 241, 246, 252, 275
development finance institutions (DFIs)
    loan performance, 395, 452
    in India, 394–396, 407–410, 422, 423, 425–426
    in Pakistan, 441, 443, 452–453, 480
    refinance, 408, 410
    restructuring, 407–410
    supervision, 423
development institutions
    in Korea, 2, 15, 17, 18, 41–44
disclosure regulation. *See* regulation, information
distress management. *See* rehabilitation of ailing companies

Eurobonds. *See under* financial instruments
Exchange Equalization Fund (EEF) (Thailand), 348–350, 352
    "daily fixing," 349
    objectives, 350

Executive Yuan (Taipei,China), 110, 141
Export-Import Bank
    of China, 111
    of India, 394, 395, 408
    of Korea, 4, 18, 43, 77
    of Thailand, 313, 365–366

Farmers Bank of China (Taipei,China), 111
Federal Bank for Cooperatives (Pakistan), 467, 468
finance, informal
    in Korea, 1, 3, 5, 8, 15, 54–56
    in Pakistan, 437, 460
    rotating credit club, 134–135
    in Taipei,China, 92, 100, 101, 102, 103, 118, 133–138, 151, 153
    in Thailand, 338
    in India, 371, 373, 376, 376, 378–387
    vs. formal finance, 382–384
    market structure, 380–382
    nidhis, 380, 381, 383, 385
    nonbanking finance companies (NBFCs), 381–382, 386, 387
    regulation, 386–387
    risk diversification, 383–384
    rural, 379–380
    and small enterprise lending, 101, 103, 407
    sources, 380, 381
    urban, 379, 380
    uses, 379–380
    See also unregulated markets; pawnshops
finance companies, industrial and consumer
    activities, 213, 259, 329–330
    branching policy, 318
    capital adequacy requirements, 332
    in commercial banking, 365
    concentration, 215, 332
    credits and borrowing, controls on, 318

    funding sources, 214, 259, 330
    in Indonesia, 176, 212–215
    insolvencies, 319, 330
    laws governing, 259–250, 317–319, 321, 326, 331, 334, 362
    ownership, 213, 317–318
    in the Philippines, 258–260, 272
    in securities business, 342
    shares in other finance companies, 319
    in Thailand, 304, 306, 307, 317–319, 322, 329–333, 342, 362
finance ministries
    of Indonesia, 201, 210, 217
    of Pakistan, 460, 466, 482
    of Taipei,China, 98, 106, 125, 127
    of Thailand, 314, 321, 339, 340, 363
financial deepening
    nonbank financial intermediaries and, 77
financial deepening, indicators of
    in India, 387–389
    in Indonesia, 169–172
    in Korea, 1, 15–21, 44
    in Pakistan, 438–441
    in Taipei,China, 147–148
    in Thailand, 309–311, 366–367
financial infrastructure
    and asymmetric information, 173, 281–282
    in India, 399–400, 419–424
    in Indonesia, 173, 215–221, 228
    in Korea, 73–76
    in Pakistan, 460–466
    in the Philippines, 281–291
    in Taipei,China, 95–99
    See also accounting system; commercial legal systems
Financial Institutions Development Fund (Thailand). *See under* rehabilitation of ailing companies
financial instruments, 84–87, 88, 125
    bankers' acceptances, 87, 201

Subject Index

bonds, corporate, 8, 20, 63, 75, 84, 87, 125, 126
bonds, government, 25, 46, 60, 63, 84, 87, 125–126, 265, 346, 350–351, 352–353, 357, 364, 414, 441, 443, 473, 477, 478, 479
call loans, 47, 58, 133, 271, 272, 274, 350, 416, 455–456
central bank certificates, 87, 169, 176, 203, 204, 219
certificates of deposit (CDs), negotiable, 3, 23, 24, 29, 58, 60, 87, 126, 176, 203, 416
commercial bills, 61, 416
commercial paper, 24, 47, 57–58, 59, 87, 203, 260, 272, 273, 274, 291, 353, 416, 427, 457, 481
debentures, 20, 24, 25, 46, 53, 87, 91, 126
deposits, demand, 23, 24, 87, 244
deposits, time, 32, 45, 87, 191, 115, 244–245, 277
Eurobonds, 62
foreign exchange instruments, 87, 133
in Korea, 8, 10
loans, long-term, 15, 115, 116, 123, 140, 141, 241, 245, 246, 266, 274, 275, 295, 330, 392, 410, 483
loans, short-term, 245, 246, 260, 295, 483
monetary stabilization bonds (MSBs), 3, 11, 28, 58, 60–61, 63
promissory notes, 330
repurchase agreements (repos), 24, 46, 58, 59, 203, 258, 351, 456
securities, government, 23, 60, 91, 259, 272, 273, 274, 276, 411
taxation, 293, 478–480
in Thailand, 364
trade bills, 61
treasury bills, 60, 87, 245, 259, 265, 272, 273, 277, 351, 411, 416, 456, 473, 477, 479, 480

financial intermediaries, 26–57, 110–124, 178–198, 205–215, 240–271, 314–320, 325–344, 389–396, 441–455
  assets, 86, 185, 186, 191, 192, 235, 236, 258, 264, 270, 306, 307, 309, 310, 327, 329
  competition, 101, 113, 172, 179, 253, 313, 325, 328, 334, 364, 365, 444
  fund sources, 90
  government intervention and, 101, 360
  government ownership, 28–29, 99, 168, 246, 444
  in India, 389–396
  in Indonesia, 178–198, 205–215
  in Korea, 26–57
  nonperforming loans, 216, 252, 255
  in Pakistan, 441–455
  in the Philippines, 240–271
  private-sector participation, 263
  in Taipei,China, 90, 110–124
  in Thailand, 306, 307, 309, 310, 314–320, 325–344
  See also commercial banks; credit foncier companies; development finance institutions; development institutions; finance companies, industrial and consumer; financial intermediaries, nonbank; insurance companies; investment and trust companies; investment banks; leasing companies; modarabas; merchant banking corporations; mutual funds; pension funds; pawnshops; rural banks; thrift banks
financial intermediaries, nonbank
  and asymmetric information, 13
  and financial deepening, 77
  funding sources, 261–262
  in Indonesia, 205–215
  in Korea, 11–12, 14–15, 17, 18, 39, 44–57, 77

ownership, 56–57
in Pakistan, 454–455
in the Philippines, 235, 236, 258–271
private-sector participation, 263
prudential regulation, 52–54, 263
in Taipei,China, 119–124
in Thailand, 329–335
See also credit foncier companies;
  finance companies, industrial and
  consumer; insurance companies;
  investment and trust companies;
  leasing companies; modarabas;
  pension funds
financial sector
  assets, 20, 21, 24, 86, 235–238, 258
  and economic growth, 168, 312–313
  economic importance, 309, 311
  economic regulation and, 10
  efficiency, 226, 251–252, 366–367, 385–386
  entry, 10, 101, 263
  of India, 387–396
  of Indonesia, 168–169, 170–171, 225–226
  of Korea, 14–26
  of Pakistan, 437, 441–455
  of the Philippines, 235–239
  prudential regulation, 215–216
  reserves and liquidity requirements, 40, 272, 273, 361, 472
  of Taipei,China, 82–95
  of Thailand, 304–313
  See also financial intermediaries;
    capital markets; money markets
financial sector policies
  in Pakistan, 483–487
  in the Philippines, 292–297
  in Taipei,China, 106–109, 138–142
financial sector reform
  and competition, 172–173, 175, 176, 179, 184–190, 226, 253, 303
  and government intervention, 173, 175, 226, 240
  in India, 371, 397–419, 429–431

in Indonesia, 163, 164–166, 175–184, 198, 215, 226
in Korea, 10–12
in Pakistan, 472–474, 485–487
in the Philippines, 239, 240–241, 245, 246, 247–248, 249–250, 253, 272, 273, 274, 278, 279, 280–281, 292–297
in Taipei,China, 101–104
in Thailand, 303, 356–366
financial sector reform, sequencing of
  in India, 430–431
  in Indonesia, 175–178
  in Pakistan, 486–487
  in Thailand, 367–368
fiscal deficits
  financing of, 87, 93, 441, 466, 484
  in Pakistan, 441, 466, 483, 484
  in Taipei,China, 87, 93
fiscal deficits, quasi, 467–469, 471
foreign aid
  to Indonesia, 168, 172
  to Taipei,China, 106, 113
foreign borrowings
  of Indonesia, 177, 178, 183, 196, 198, 227
  of Korea, 12, 23, 25, 43
  of Pakistan, 459
  of the Philippines, 280
  of Taipei,China, 103
  of Thailand, 312, 328, 340, 341, 352
foreign exchange controls
  dismantling, 131–132, 358–359
  in Taipei,China, 108–109, 131–132
  in Thailand, 355, 358–359
foreign exchange instruments. See under financial instruments
foreign exchange markets
  deregulation, 279, 281, 295
  and hedging of business risks, 419
  hundi markets, 474
  in India, 418–419
  interbank, 279, 352
  in Korea, 60

off-floor trading, 279, 280, 295
  in Pakistan, 458–459, 473–474
  parallel trading, 279
  in the Philippines, 279–281, 295
  in Taipei,China, 131–133
  in Thailand, 352
foreign exchange rate policy
  in Indonesia, 164
  in Korea, 12
  and price stability, 108–109
  in Taipei,China, 108–109
  in Thailand, 348–350
foreign exchange retention
  in the Philippines, 279
foreign participation
  in the banking sector, 14, 29, 31, 85,
    104, 113–114, 119, 132, 133, 174,
    179, 181, 185, 186, 187, 190, 192,
    193, 195, 241–242, 247, 248, 256,
    294, 304, 325, 366, 390, 391, 422,
    444, 445, 446, 447, 448, 449–450
  in capital markets, 12, 62, 130–131,
    200, 202, 276, 355, 458, 477, 479
  foreign equity, 247
  in India, 390, 391, 422
  in Indonesia, 165, 166, 174, 179,
    181, 185, 186, 187, 190, 192, 193,
    195, 200, 202, 206, 208
  in the insurance industry, 123, 206,
    208, 264, 267, 294, 334
  in Korea, 12, 14, 29, 31, 62, 64
  in Pakistan, 444, 445, 446, 447, 448,
    449–450, 477, 479
  in the Philippines, 241–242, 247,
    248, 256, 264, 267, 276, 294
  in securities companies, 64
  in Taipei,China, 85, 104, 113–114,
    119, 123, 130–131, 132, 133
  in Thailand, 304, 325, 334, 365
  See also joint ventures
foreign trade, loans for
  in India, 391
  in Indonesia, 181
  in Korea, 4, 33, 34, 43

  in Pakistan, 468, 470
  in Taipei,China, 111, 139–140, 151
foreign trade policy
  in Indonesia, 165–166, 174
Fuh Hwa Securities Finance Company
  (Taipei,China), 126, 129, 135

Government Housing Bank (Thailand),
  306, 307, 308, 340
government intervention
  arguments for, 152
  and development of financial
    intermediaries, 360
  in Indonesia, 173
  in Korea, 2, 10–11, 13, 35, 71
  in Pakistan, 441, 473, 485, 487
  in the Philippines, 240
  and resource allocation, 240
  in Taipei,China, 99–101
  See also financial intermediaries,
    government ownership;
    regulation, economic; regulation,
    information; regulation,
    prudential; subsidy; taxation
Government Savings Bank (Thailand),
  305, 306, 307, 337–338, 351
  advantages over commercial banks,
    337
Government Service Insurance System
  (GSIS) (Philippines), 263, 264, 265,
  266, 275

Home Development Mutual Fund
  (HDMF) (Philippines), 275
Home Insurance and Guarantee
  Corporation (Philippines), 263, 264
households
  in Korea, 22–23, 24, 33
  in Pakistan, 442
  in Taipei,China, 87, 88, 89, 90, 92,
    93, 95, 103, 115, 117, 138, 150
housing finance
  in India, 382
  in Indonesia, 214, 219

in Korea, 4, 30, 34
in Pakistan, 443, 455, 482–483
in the Philippines, 245, 266
in Thailand, 330
Housing Finance Credit Guarantee Fund (Korea), 52

India
  accounting systems, 371, 421–422
  agricultural lending, 391, 392, 406
  asymmetric information, 372–374
  banking sector, 389–396
  bankruptcy law, 399
  banks, foreign, 390, 391, 422
  bond markets, 411, 413–416
  capital adequacy requirements, 399, 422
  capital markets, 411–416
  collateralized lending, 373
  commercial banks, 389–390
  commercial legal systems, 372, 399, 408–409, 419–421
  credit subsidies, 391
  deposit insurance, 424
  development finance institutions (DFIs), 394–396, 407–410, 422, 423, 425–426
  directed credit, 391–392, 406
  financial deepening, 387–389
  financial infrastructure, 399–400, 419–424
  financial sector, 387–396
  financial sector reform, 371, 397–419, 429–431
  foreign exchange markets, 418–419
  foreign trade financing, 391
  industrial lending, 392, 407
  informal finance, 371, 373, 378–387
  insurance companies, 389
  interest rates, 375–377, 390–391
  monetary policy, 398, 414
  money markets, 408, 409, 416–418
  nonperforming loans, 371
  privatization, 403–404
  reserve requirements, 391
  rehabilitation of ailing companies, 400–407, 427–428
  restructuring of banks, 400–407
  restructuring of development finance institutions, 407–410
  rural banks, 389–390
  small enterprise lending, 391, 392, 407, 410, 428–429
  stock markets, 411, 412–413

Indonesia
  accounting systems, 221
  agricultural lending, 187
  Banking Acts (1968), 168
  Banking Law Nr. 7, 181, 182
  Banking Reform (1988), 200
  banking sector, 168, 178–198
  banks, foreign, 174, 179, 181, 185, 186, 187, 190, 191, 192, 193, 195
  bond markets, 202–203
  capital adequacy requirements, 165, 173, 177, 180, 181, 195, 224
  Capital Market Law (1952), 200
  capital markets, 169, 176, 198–203, 225–226
  commercial banks, 168, 178, 179, 184–197
  commercial legal systems, 195, 196, 220–221
  debt financing, 169
  directed credit, 165, 176, 177, 181–182
  economic growth, 166, 167
  exchange rates, 164
  exchange rate subsidies, 183, 198
  finance companies, 176, 212–215
  financial deepening, 169–172
  financial infrastructure, 173, 215–221, 228
  financial intermediaries, nonbank, 205–215
  financial sector reform, 163, 164–165, 175–184, 198

financial sector structure, 168–169
fiscal policy, 164
foreign aid, 168, 172
foreign borrowing, 177, 178, 183, 196, 197, 198, 227
foreign entry, 165, 166, 206
foreign trade, 165–166, 181
government intervention, 173
insurance companies, 205–212, 225
interest rates, 176
inflation rates, 166
insurance companies, 205–212
Insurance Law (1992), 208, 209, 210
joint ventures, 166, 179, 181, 185, 187, 186, 191, 192, 193, 195, 206
monetary policy, 164–165, 197–198, 228
money markets, 203–205, 225–226
nonperforming loans, 178, 195–196, 216, 222–223, 225
Pakto, 176, 179, 180, 182, 183, 205
pension funds, 218
private sector, 166, 168
privatization, 199, 222
prudential regulation, 165, 173, 178, 180–181, 194, 209–210, 215–216
reinsurance industry, 210–211
reserve requirements, 174, 176, 180
rural banks (BPR banks), 178, 179, 180, 181, 185, 187
savings, 168, 172
small enterprise lending, 177, 182, 185, 187, 196, 226
social indicators, 166–167
Industrial Bank of Korea, 1, 29, 30
Industrial Credit and Investment Corporation of India (ICICI), 407
Industrial Development Bank of India (IDBI), 395, 407, 408, 425
Industrial Finance Corporation of India (IFCI), 407
Industrial Finance Corporation of Thailand (IFCT), 306, 307, 308, 340–341

industry, lending to
  in India, 392, 407
  in Korea, 23, 24–25, 26, 33, 37–40
  in the Philippines, 241, 247, 258, 275
  in Taipei,China, 89, 90, 92, 93, 94, 95, 96, 102, 103, 115, 116, 117, 118, 136, 137, 141–142
  in Thailand, 330, 340
inflation
  in Indonesia, 166
  and interest rates, 246
  in Korea, 9
  in Taipei,China, 106
information asymmetry. *See* asymmetric information
insolvency law
  in the Philippines, 289–290, 296
insolvencies. *See under* banking sector; credit foncier companies; finance companies, industrial and consumer
Institute of Chartered Accountants (Pakistan), 463
insurance, laws governing
  in Indonesia, 208, 209
  in the Philippines, 266, 267, 268, 269
  in Thailand, 335
Insurance Commission (Philippines), 266, 267, 268, 269, 297
insurance companies
  entry, barriers to, 208–209, 267
  foreign participation, 123, 206, 208–209, 294
  funding sources, 335
  growth, 264–266, 268, 335
  in India, 389
  in Indonesia, 205–212, 225
  in Korea, 17, 19, 24, 25, 48–49, 53, 43, 56
  life insurance, 17, 19, 24, 123, 211, 264, 265, 266
  nonlife insurance, 17, 123, 211–212, 264, 265, 266, 267
  ownership, 56, 206, 266–267

in Pakistan, 442, 443, 459
in the Philippines, 236, 263–269, 294
pre-need industry, 268
premiums, deregulation of, 210
prudential regulation, 53, 209, 295
regulatory environment, 208, 209, 210, 212, 218, 267–269
reinsurance, 210–211, 264–265
and savings mobilization, 269, 335
social insurance, 212
in Taipei,China, 85, 86, 123–124
taxation, 268, 293
in Thailand, 305, 306, 307, 334–335
interest rates
and adverse selection, 100, 372–373, 377
and cost of capital, 197
government control of, 11, 100
in India, 390–391
in Indonesia, 176
inflation and, 246
in Korea, 1, 2, 7–10, 11
in Pakistan, 466–467, 470, 472
in the Philippines, 244–245, 256
and price stability, 108
spread, 119, 253, 256, 364, 365, 369
and savings mobilization, 108, 172, 467, 472
in Taipei,China, 100, 108, 109, 119
in Thailand, 328, 359, 364, 365, 369
interest rates, deposit
in Korea, 8, 9–10, 32, 45, 46
in Pakistan, 472
in the Philippines, 244–245, 256, 274
in Taipei,China, 108
in Thailand, 328, 359
interest rates, lending
in Korea, 8, 9
laws governing, 357
in the Philippines, 245–246, 256
in Taipei,China, 136
in Thailand, 328, 359, 364

interest rates, liberalization of
in India, 375–377
in Korea, 8–10
in the Philippines, 246, 272, 292–293
in Taipei,China, 101, 119
in Thailand, 359
International Commercial Bank of China, 132
International Monetary Fund (IMF), 348, 355, 358, 415, 418
investment and trust companies
in Korea, 1, 15, 17, 19, 23, 25, 47, 52, 53, 54, 55, 56, 59, 70
laws governing, 70
in Pakistan, 443, 455, 473
in the Philippines, 236, 258, 260, 261, 262, 272
in Taipei,China, 85, 86, 115, 122
Investment Corporation of Pakistan, 443, 453

Jakarta Stock Exchange (JSE), 176, 177, 194, 199, 200, 201, 202
joint ventures
banks, 15, 48, 179, 181, 185, 186, 187, 190, 191, 195, 247, 294
development finance institutions, 452
in Indonesia, 166, 179, 181, 185, 186, 187, 190, 191, 195, 213, 214
in Korea, 15, 48, 49, 64
finance companies, 49, 213, 214
in Pakistan, 452
in the Philippines, 247, 294

Karachi Stock Exchange (KSE), 457, 458
Korea
accounting systems, 73–74
agricultural cooperatives, 30
agricultural lending, 4, 6, 14, 33, 34, 38
asymmetric information, 12–13

Subject Index

banking sector, 2, 5, 28–41
banks, foreign, 14, 29, 31
bond markets, 63–64
capital adequacy requirements, 40, 52–53
capital markets, 1, 2, 15, 17, 18, 20, 61–66, 70, 76, 78
collateralized lending, 12–13
commercial banks, 14, 16, 18, 28–29, 30, 31, 69–70, 71–73
commercial legal systems, 13–14
credit access, 39
credit control system, 6–7, 27
credit guarantee funds, 33, 52
credit information arrangements, 52
credit rating agencies, 59, 75
deposit insurance, implicit, 13
deposit money banks (DMBs), 16, 18, 31, 32–34
development institutions, 2, 15, 17, 18, 41–44, 77
directed credit, 2, 3–6, 11, 12, 27, 66, 76
financial deepening, 15–21
financial infrastructure, 73–76
financial intermediaries, nonbank, 11–12, 14–15, 17, 18, 39, 44–57
financial sector, 14–26
financial sector reform, 10–12
foreign borrowing, 12, 23, 43
foreign investments, 14, 62
foreign trade loans, 4, 33, 34, 43
General Banking Law, 40
government debt, 23, 24
government intervention, 2, 10–11, 35
industrial lending, 23, 24–25, 26, 33, 37–40
informal finance, 1, 3, 5, 8, 15, 54–56
information regulation, 74–75
insurance companies, 17, 18, 24, 25, 48–49, 53, 54, 56

interest rates, 1, 2, 7–10, 11, 32, 45, 46
investment and finance companies, 1, 15, 17, 18, 23, 25, 47, 52, 53, 54, 55, 56, 59, 70
joint ventures, 15, 48, 49
leasing companies, 17, 49, 51, 53, 54
merchant banking corporations, 15, 17, 18, 48, 53, 54, 56
monetary policy, 27
money markets, 3, 57–61, 78
mutual savings and finance companies, 1, 17, 18, 45, 53, 54
nonperforming loans, 2, 5, 10, 35–37, 66, 68
pension funds, 24
postal savings system, 17, 18, 46
privatization, 2, 10, 71
Public Corporation Inducement Act (1972), 61
rediscount policy, 27
rehabilitation of ailing companies, 13–14, 42, 66–68
reserve requirements, 27, 40, 53
savings, 20, 21, 22, 23
savings cooperatives, 15, 17, 18, 46, 53, 54
savings institutions, 17, 18, 44–46
securities companies, 17, 56, 64
securities investment trust companies, 17, 18, 64–65, 70–71
small enterprise lending, 4, 5, 6, 11, 14, 29, 30, 33, 43, 45, 52
specialized banks, 14, 16, 18, 29–32, 68
stock markets, 20, 62–63
venture capital firms, 17, 50–52
Korea Credit Guarantee Fund, 52
Korea Development Bank, 4, 18, 41–43
    funding sources, 42
    as supplier of long-term capital, 77
Korea Exchange Bank, 1, 29

Korea Housing Bank, 1, 30
Korea Insurance Development Institute, 48
Korea Long-Term Credit Bank, 18, 43–44, 77–78
Korea Nonbank Deposit Insurance Corporation, 52
Korea Securities Finance Corporation (KSFC), 15, 19, 59, 65
Korea Stock Exchange, 12, 62, 66, 75
Korea Technology Credit Guarantee Fund, 52
Krung Thai Bank, 326, 337

Land Bank of the Philippines, 241, 246, 252
Land Bank of Taipei,China, 111, 112
leasing companies
    funding sources, 49, 51
    in Korea, 17, 49, 51, 53, 54
    in Pakistan, 443, 454, 473, 480, 483
legal recourse for lenders. *See* commercial legal systems
lending, marginal. *See* under capital markets
lending investors
    in the Philippines, 236, 258, 260, 261, 262, 263, 293
loans
    to agriculture. *See* agriculture, lending to
    for foreign trade. *See* foreign trade, loans for
    to industry. *See* industry, lending to
    interest rates. *See* interest rates, lending
    loans to directors, officers, stockholders and related interests (DOSRI), 250, 257, 260, 316
    reserve requirement, 245, 272, 273, 328–329, 332
    single borrower's limit, 181, 250, 260, 318
    to the small enterprise sector. *See* small enterprises, lending to
loans, long-term. *See under* financial instruments
loans, nonperforming
    of banks, 2, 5, 10, 35–37, 66, 68, 114, 118–119, 120, 178, 195–196, 222–223, 225, 252, 255, 371, 450–452
    of development finance institutions, 371, 395, 452
loans, short-term. *See under* financial instruments

Medium Business Bank of Taiwan, 114, 141,
medium business banks, 85, 86, 114–115, 143
merchant banking corporations
    funding sources, 48
    in Korea, 15, 17, 19, 48, 53, 54, 56
    ownership, 56
modarabas, 443, 453, 454–455, 473, 480
monetary stabilization bonds (MSBs). *See under* financial instruments
money market instruments
    in India, 416
    in Indonesia, 176, 204, 219
    in the Philippines, 271–272
    in Taipei,China, 87, 124–125
    in Thailand, 350
money markets
    competition, 125
    in India, 408, 409, 416–418
    in Indonesia, 203–205, 225–226
    interest rates and, 78
    in Korea, 3, 57–61, 78
    in Pakistan, 455–456
    participants, 125, 272
    in the Philippines, 271–274
    regulatory environment, 273–274
    in Taipei,China, 87, 90, 92, 102, 103, 124–125, 136, 137

taxation, 273
   in Thailand, 309, 350–351
   trading volume, 274
money markets, interbank
   in Indonesia, 203–204
moral hazard, 13, 67, 68, 369, 373, 375, 404, 406, 415, 424, 426, 463, 485
   in pension funds, 270
   in securities trading, 415
mortgage loans
   in Pakistan, 483
   in the Philippines, 260, 288–289
   in Taipei,China, 116
   in Thailand, 334, 340, 359
multilateral lending agencies, 247, 275, 295, 312, 418
   See also International Monetary Fund (IMF); World Bank
Muslim Commercial Bank (Pakistan), 444
mutual funds
   benefits, 343
   investment limitations, 343
   investment promotion measures, 344
   ownership, 344
   in Pakistan, 441, 453, 458
   in the Philippines, 278
   in project financing, 425
   in Thailand, 342–344
mutual savings and finance companies, in Korea, 1, 17, 18, 45, 53, 54

National Bank for Agriculture and Rural Development (India), 394, 395
National Housing Bank (India), 394, 395, 408
National Housing Fund (Korea), 49
National Investment Fund (NIF) (Korea), 3, 4, 33, 34, 43, 49, 50
National Investment (Unit) Trust (NIT) (Pakistan), 443, 453
National Stock Exchange (NSE) (India), 416

Overseas Chinese Bank (Taipei,China), 105
Overseas Trust (Taipei,China), 105, 122

Pakistan
   accounting systems, 462–463
   agricultural lending, 467, 468, 470
   banking sector, 444–453
   banking tribunals, 465
   banks, foreign, 444, 445, 446, 447, 448, 449–450
   bond markets, 456–457
   capital adequacy requirements, 464
   capital markets, 441, 443, 453, 456–458, 473, 474–482
   commercial banks, 441, 443, 444–452, 473, 483
   commercial legal systems, 463–465, 486, 487
   credit information bureau, 462, 482
   credit rating system, 481–482
   development finance institutions (DFIs), 441, 443, 452–453, 480
   directed credit, 437, 454, 467, 470, 472, 473, 484
   export financing, 468, 470
   financial deepening, 438–441
   financial infrastructure, 460–466
   financial intermediaries, nonbank, 454–455
   financial sector, 437, 441–455
   financial sector policies, 483–487
   financial sector reform, 472–474, 485–487
   fiscal deficit, financing of, 441, 466, 484
   fiscal deficit, quasi-, 467–469, 471
   foreign borrowing, 459
   foreign exchange markets, 458–459, 473–474
   government intervention, 441, 473, 485, 487
   households, 442

housing finance companies, 443, 455, 482–483
informal finance, 437, 460
information regulation, 462
insurance companies, 442, 443, 459
interest rates, 466, 467, 470, 472
investment banks, 443, 455, 473
leasing companies, 443, 454, 473, 480, 483
joint ventures, 452
modarabas, 443, 453, 454–455, 473, 480
money markets, 455–456
mutual funds, 441, 453, 458
nonperforming loans, 450–452
privatization, 444, 473
prudential regulation, 451, 463, 464
savings, 438–441
scheduled banks, 442, 448
small enterprise lending, 468, 470, 484
Pakistan Banking Council, 461
pawnshops
  in India, 383
  in the Philippines, 237, 258, 260, 261, 262, 263, 293
  in Taipei,China, 135
  in Thailand, 305, 306, 307
pension funds
  in Indonesia, 218
  in Korea, 24
  in the Philippines, 269–271, 273, 293
  regulatory environment, 270–271
  taxation, 270–271
Philippine Crop Insurance Corporation (PCIC), 263, 264, 265
Philippine Dealing System, 280–281
Philippine Deposit Insurance Corporation (PDIC), 251, 261, 263, 264
Philippine National Bank, 241, 246, 247
Philippines
  accounting systems, 257–258, 282–287, 290, 296

agricultural lending, 241, 248–249, 258
banking, quasi, 258, 260
banking sector, 235, 240–248
banks, foreign, 241–242, 256, 294
bond markets, 276
capital adequacy requirements, 249–250, 294
capital markets, 274–279
Central Bank Act (1993), 250, 297
commercial banks, 235, 236, 238, 241, 242–244, 245, 249, 251, 252, 254, 255, 272, 273
commercial legal systems, 282, 288–209, 296
credit guarantee institutions, 271
credit information bureau, 290–291
directed credit, 248, 249
finance companies, 258–260, 272
financial infrastructure, 281–291
financial intermediaries, nonbank, 235, 236, 258–271
financial sector, 235–239
financial sector reform, 239, 240–241, 246, 247–248, 249–250, 253, 272, 273, 274, 278, 279, 280–281, 292–297
financial sector structure, 236–237
Financing Company Act (RA 5980), 259, 250
foreign borrowing, 280
foreign exchange markets, 279–281
foreign investments, 247, 267, 277, 278, 294
General Banking Act, 248, 259
industrial lending, 241, 247, 258, 275
insolvencies, 242
insolvency law, 289–290, 296
Insurance Code, 266, 267, 268, 269
insurance companies, 236, 263–269, 294

interest rates, 244–246
investment and trust companies, 236, 258, 260, 261, 262, 272
joint ventures, 247, 294
lending investors, 237, 258, 260, 261, 262, 263, 293
money markets, 271–273
mutual funds, 278
pawnshops, 237, 258, 260, 261, 262, 263, 293
pension funds, 269–271, 273, 293
prudential regulation, 249–250, 294–295
rural banks, 236, 238, 240, 241, 242, 243, 244, 251, 252, 254, 255
Securities Act, 273
securities companies, 237, 258, 260, 261
small enterprise lending, 240, 245, 248, 249, 261, 262, 263, 290
specialized government banks, 236, 241, 242, 243, 245, 247, 251, 252, 254, 255, 256
subsidies, to credit, 252, 292
thrift banks, 235, 236, 241, 242, 243, 244, 245, 249, 251, 252, 254, 294
thrift institutions, nonbank, 237
universal banks, 240, 241, 250, 253, 256, 294
venture capital firms, 237, 258, 260, 261, 262
Philippine Stock Exchange, 276
regulation, 277
policy-based lending. *See* credit, directed
policy loans. *See* credit, directed
postal savings system
in Korea, 17, 18, 46
in Taipei,China, 85, 87, 122–123
private sector
in the financial sector, 112, 124, 235, 263, 304, 393–394, 444
in India, 393–394

in Indonesia, 166, 168
lending to, 89–93, 94, 95, 102, 103, 115, 116, 117, 118, 136, 137
in Pakistan, 444
in the Philippines, 235, 263
in Taipei,China, 89–93, 94, 95, 102, 103, 112, 115, 116, 118, 124, 136, 137
in Thailand, 304
privatization,
in India, 403–404
in Indonesia, 199, 222
in Korea, 2, 10, 71
in Pakistan, 444, 473
in the Philippines, 241, 246, 295
in Taipei,China, 104
promissory notes. *See under* financial instruments
PT Bank Summa (Indonesia), 195, 216, 219

regulation, economic. *See* banking sector, entry and exit; banking sector, foreign entry; capital markets, foreign entry; credit, directed; financial sector, entry; insurance companies, entry, barriers to; interest rates, government control of
regulation, information
of capital markets, 74–75, 129, 322
in Korea, 70
in Pakistan, 462
in Taipei,China, 129
in Thailand, 322, 323
regulation, prudential
of banks, 106, 194
of capital markets, 295, 412, 413
of finance companies, 317–319
in Indonesia, 165, 173, 178, 180–181, 194, 209–210, 215–216
of insurance companies, 53, 209–210, 295
in Korea, 52–54

512 SUBJECT INDEX

of nonbank financial intermediaries,
  52–54
in Pakistan, 461, 463, 464
in the Philippines, 249–250, 294–
  295, 296
single borrower's limit, 294, 315, 318
in Taipei,China, 106
See also capital adequacy requirements
rehabilitation of ailing companies
  April 4 Lifeboat Scheme (Thailand),
    326, 331, 332, 333, 334
  Financial Institutions Development
    Fund (Thailand), 319–320, 326
  in India, 400–407, 427–428
  in Indonesia, 224
  in Korea, 13–14, 42, 66–68
  in Malaysia, 224
  in the Philippines, 242
  in Taipei,China, 104–105
repurchase agreements (repos). *See under*
  financial instruments
Reserve Bank of India (RBI), 382, 392,
  393, 399, 405, 413, 416, 417, 424
  monetary policy, 398
  supervision and examination, 387,
    400, 422–423
reserve requirements
  in India, 391
  in Indonesia, 174, 176, 180
  in Korea, 27, 53
  in Pakistan, 472
  in the Philippines, 245, 249
  in Thailand, 314–315, 328–329, 361
rural banks
  in India, 389–390
  in Indonesia (BPR banks), 178, 179,
    180, 181, 185, 187
  in the Philippines, 236, 238, 240, 241,
    242, 243, 244, 251, 252, 254, 255

savings
  composition, 20–21, 22, 438–441
  in Indonesia, 168, 172

interest rates and, 108, 172, 467, 472
in Korea, 20–21, 22, 23
in Pakistan, 438–441, 472
in Taipei,China, 148–151
in Thailand, 306, 345
users, 441
savings deposits
  competition, 242, 244
  in Indonesia, 193
  in Korea, 32, 46
  in the Philippines, 242, 277
  in Taipei,China, 87
schedule, banks, in Pakistan
  deposits, 448
  funds flow, 442
securities, government. *See under* financial
  instruments
securities and exchange agencies
  in India, 408, 412, 413, 425
  in Korea, 62, 65, 75
  in Pakistan, 461
  in the Philippines, 259, 260, 263, 268,
    273, 274, 277, 278, 282, 283, 297
  in Taipei,China, 125
  in Thailand, 313, 322, 323, 365
securities companies
  funding sources, 64
  in Korea, 17, 19, 56, 64
  laws governing, 317, 321, 326, 331,
    334, 362
  ownership, 56
  in the Philippines, 237, 258, 260, 261
  in Thailand, 304, 322, 342, 362
securities investment trust companies
  (SITCs)
  in Korea, 17, 19, 64–65
  conversion to securities companies,
    70–71
  funding sources, 65
securities markets. *See* capital market
Small and Medium Business Credit
  Guarantee Fund (Taipei,China), 143–
  145

SUBJECT INDEX   513

small enterprises, lending to
    in India, 391, 392, 407, 410, 428–429
    in Indonesia, 177, 182, 185, 187, 196, 226
    in Korea, 4, 5, 6, 11, 14, 29, 30, 33, 43, 45, 52
    laws governing, 249
    in Pakistan, 468, 470, 484
    in the Philippines, 240, 245, 248, 249, 261, 262, 263, 290
    in Taipei,China, 101, 103, 114, 117, 121, 138, 140, 141, 142–145
    in Thailand, 308, 316, 328, 360
Small Industries Development Bank of India (SIDBI), 391, 395, 408, 410, 429
Small Industries Finance Corporation of Thailand (SIFCT), 306, 307, 308
Social Security System (SSS) (Philippines), 263, 264, 265, 275
specialized banks
    funding sources, 29, 30, 31–32
    in Korea, 14, 16, 18, 29–32, 68
    in the Philippines, 236, 241, 242, 243, 244, 245, 247, 251, 252, 254, 255, 256
    in Taipei,China, 111, 112, 123, 140–145
    See also agriculture, lending to; industry, lending to; small enterprises, lending to
State Bank of India (SBI), 389, 404
State Bank of Pakistan (SBP), 442, 443, 450, 452, 458, 459, 460
    funds flow, 442
State Life Insurance Corporation (SLIC) (Pakistan), 459, 476
Stock Exchange of Thailand (SET), 311, 322–323, 342, 343, 353–356
stock exchanges
    of India. *See* Bombay Stock Exchange; National Stock Exchange (NSE)
    of Indonesia. *See* Jakarta Stock Exchange (JSE); Surabaya Stock Exchange (SSE)
    of Korea. *See* Korea Stock Exchange
    of Pakistan. *See* Karachi Stock Exchange (KSE)
    of the Philippines. *See* Philippine Stock Exchange
    of Taipei,China. *See* Taiwan Stock Exchange
    of Thailand. *See* Bangkok Stock Exchange; Stock Exchange of Thailand (SET)
stock markets
    in India, 411, 412–413
    in Korea, 20, 62–63
    in Pakistan, 474–477
    in the Philippines, 275–277, 295
    in Taipei,China, 89, 126–131
    in Thailand, 353–356
subsidies, credit
    in India, 391, 401
    in Indonesia, 178, 181, 196
    in Korea, 26
    in Pakistan, 467
    in the Philippines, 239, 252, 292
    in Thailand, 364
subsidies, exchange rate
    in Indonesia, 183, 198
Sumarlin shocks, 178, 183, 184, 194, 198, 204, 205
Surabaya Stock Exchange (SSE), 199, 200, 201, 202
Syndicate Bank (India), 378

Taipei,China
    accounting systems, 95–98
    agricultural lending, 111, 140
    Banking Law (1931, revised 1975, 1989), 98–99, 101, 104, 105, 106, 110, 114, 121, 134, 135
    banking sector, 85, 86, 100, 106, 110–119

banks, foreign, 85, 104, 113–114, 119, 132, 133
bills finance companies, 115, 124–125
bond markets, 90, 92, 102, 125–126
capital adequacy requirements, 106
capital markets, 125–131, 136, 137
collateralized lending, 99, 115
commercial banks, 110–119
commercial legal systems, 98–99
Company Law, 125
credit guarantees, 143–144
deposit insurance, 105–106
directed credit, 100
exchange rate policy, 108–109
financial deepening, 147–148
financial infrastructure, 95–99
financial intermediaries, nonbank, 119–124
financial sector, 82–95
financial sector policies, 106–109, 138–142
financial sector reform, 101–104
fiscal deficit, 93
foreign aid, 106, 113
foreign exchange markets, 131–133
foreign investment, 130
foreign trade financing, 111, 139–140, 151
government debt, 90, 117
government intervention, 99–101
Income Tax Law, 99
industrial lending, 89, 90, 92, 93, 94, 95, 96, 102, 103, 115, 116, 117, 118, 136, 137, 141–142
informal finance, 92, 100, 101, 102, 103, 118, 133–138, 151, 152
information regulation, 129
insolvencies, 104–105
insurance companies, 85, 86, 123–124
insurance companies, foreign, 123
interest rates, 11, 101, 108, 109, 119, 136

investment and trust companies, 85, 86, 115, 122
medium business banks, 85, 86, 114–115, 143
money markets, 87, 90, 92, 102, 103, 124–125, 136, 137
nonperforming loans, 114, 118–119, 120
offshore banking, 132–133
pawnshops, 135
postal savings system, 85, 86, 87, 122–123
private sector, lending to, 89–93, 94, 95, 102, 103, 115, 116, 117, 118, 136, 137
privatization, 104
prudential regulation, 106
rehabilitation of ailing companies, 104–105
savings, 148–151
Securities and Exchange Law (1988), 129, 130
small enterprise lending, 101, 103, 114, 117, 121, 138, 140, 141, 142–145
specialized banks, 111, 112, 123, 140–145
stock market, 89, 126–131
universal banks, 110
venture capital companies, 141
Taiwan Stock Exchange, 126
taxation
  capital gains tax, 199, 200, 277, 348, 354
  of financial instruments, 293, 478–480
  in Indonesia, 199, 200
  laws governing, 97
  in the Philippines, 249, 277, 278, 293
  in Taipei,China, 97
  in Thailand, 324–325, 348, 354
Tenth Credit Cooperative Association (Taipei,China), 104, 105
Thai Bankers' Association, 367, 369

Thailand
- accounting systems, 323
- Act on the Undertaking of Finance Business, Securities Business, and Credit Foncier Business (B. E. 2522) (1979, amended 1983, 1985, and 1992), 317, 321, 326, 331, 334, 362
- agricultural cooperatives, 305, 306, 307, 335–336
- agricultural lending, 313, 316, 330, 335–336, 338, 360
- banking sector, 304, 309–311, 314–317, 325–329, 336–342
- banks, foreign, 304, 325, 365
- bond markets, 352–353
- capital adequacy requirement, 313, 315, 318, 334, 340, 362–363, 365
- capital markets, 309, 320–323, 353–356
- Civil and Commercial Codes, 320, 321, 334, 357
- Commercial Banking Act (B. E. 2505) (1962, amended 1979, 1985, and 1992), 314–317, 325–326
- commercial banks, 304, 306, 307, 313, 314–317, 325–329, 352, 361–362, 367
- credit access, 328, 369
- credit foncier companies, 304–305, 306, 307, 334
- credit rating agencies, 313, 365
- directed credit, 313, 316, 346, 360, 361
- economic development targets, 312
- exchange rate policy, 348–350
- finance companies, 304, 306, 307, 317–319, 322, 329–333, 342, 362
- financial deepening, 309–311, 366–367
- Financial Institutions Lending Rate Act (B.E 2423) (1980), 357
- financial sector reform, 356–366, 367–368
- financial sector reform and competition, 303
- financial sector structure, 304–313
- fiscal policy, 347–348
- foreign borrowing, 312, 328, 340, 341, 352
- foreign exchange markets, 352
- foreign insurance companies, 334
- industrial lending, 330, 340
- information regulation, 322, 323
- insolvencies, 319, 326, 330
- insurance companies, 305, 306, 307, 334–335
- interest rates, 359, 364, 365, 369
- Life Insurance Act (B. E. 2510) (1967), 335
- monetary policy, 344–346
- money markets, 309, 350–351
- mortgage loans, 334, 340
- mutual funds, 342–344
- offshore banking, 313, 363, 365
- pawnshops, 305, 306, 307
- Public Company Act (B.E. 2521) (1973), 320, 321
- reserve requirements, 314–315, 328–329, 361
- savings cooperatives, 305, 306, 307, 336
- Securities and Exchange Act (1992), 321–323, 365
- Securities and Exchange of Thailand Act (B.E. 2517) (1974), 322, 353
- securities companies, 304, 322, 342, 362
- small enterprise lending, 308, 316, 328, 360
- taxation, 324–325, 348, 354

thrift banks
- in the Philippines, 235, 236, 241, 242, 243, 244, 245, 249, 251, 252, 254, 294

thrift institutions, nonbank
    in the Philippines, 237
trade bills. *See under* financial instruments
treasury bills. *See under* financial
    instruments

underwriting, securities
    in Korea, 48
    in the Philippines, 241
    in Taipei,China, 125
Unit Trust of India (UTI), 389, 407
universal banks, 110, 241
    in the Philippines, 240, 241, 250,
        253, 256, 294
    in Taipei,China, 110

unregulated markets
    in Taipei,China, 133–138
    See also finance, informal

venture capital firms
    in Korea, 17, 50–52
    in the Philippines, 236, 258, 260,
        261, 262
    in Taipei, China, 141

World Bank, 167, 172, 194, 195, 472